Statistical Inference for Engineer

A mathematically accessible and up-to-date introduction to the tools needed to address modern inference problems in engineering and data science, ideal for graduate students taking courses on statistical inference and detection and estimation, and an invaluable reference for researchers and professionals.

With a wealth of illustrations and examples to explain the key features of the theory and to connect with real-world applications, additional material to explore more advanced concepts, and numerous end-of-chapter problems to test the reader's knowledge, this textbook is the "go-to" guide for learning about the core principles of statistical inference and its application in engineering and data science.

The password-protected Solutions Manual and the Image Gallery from the book are available at www.cambridge.org/Moulin.

Pierre Moulin is a professor in the ECE Department at the University of Illinois at Urbana-Champaign. His research interests include statistical inference, machine learning, detection and estimation theory, information theory, statistical signal, image, and video processing, and information security. Moulin is a Fellow of the IEEE, and served as a Distinguished Lecturer for the IEEE Signal Processing Society. He has received two best paper awards from the IEEE Signal Processing Society and the US National Science Foundation CAREER Award. He was founding Editor-in-Chief of the *IEEE Transactions on Information Security and Forensics.*

Venugopal V. Veeravalli is the Henry Magnuski Professor in the ECE Department at the University of Illinois at Urbana-Champaign. His research interests include statistical inference and machine learning, detection and estimation theory, and information theory, with applications to data science, wireless communications, and sensor networks. Veeravalli is a Fellow of the IEEE, and served as a Distinguished Lecturer for the IEEE Signal Processing Society. Among the awards he has received are the IEEE Browder J. Thompson Best Paper Award, the National Science Foundation CAREER Award, the Presidential Early Career Award for Scientists and Engineers (PECASE), and the Wald Prize in Sequential Analysis.

Statistical Inference for Engineers and Data Scientists

PIERRE MOULIN
University of Illinois, Urbana-Champaign

VENUGOPAL V. VEERAVALLI
University of Illinois, Urbana-Champaign

CAMBRIDGE
UNIVERSITY PRESS

CAMBRIDGE
UNIVERSITY PRESS

Shaftesbury Road, Cambridge CB2 8EA, United Kingdom

One Liberty Plaza, 20th Floor, New York, NY 10006, USA

477 Williamstown Road, Port Melbourne, VIC 3207, Australia

314–321, 3rd Floor, Plot 3, Splendor Forum, Jasola District Centre, New Delhi – 110025, India

103 Penang Road, #05–06/07, Visioncrest Commercial, Singapore 238467

Cambridge University Press is part of Cambridge University Press & Assessment, a department of the University of Cambridge.

We share the University's mission to contribute to society through the pursuit of education, learning and research at the highest international levels of excellence.

www.cambridge.org
Information on this title: www.cambridge.org/9781107185920

DOI: 10.1017/9781107185920

First published 2019

A catalogue record for this publication is available from the British Library

ISBN 978-1-107-18592-0 Hardback

Additional resources for this publication at www.cambridge.org/Moulin.

We balance probabilities and choose the most likely
– Sherlock Holmes

Brief Contents

Contents

Preface

In the engineering context, statistical inference has traditionally been ubiquitous in areas as diverse as signal processing, communications, and control. Historically, one of the most celebrated applications of statistical inference theory was the development of radar systems, which was a major turning point during World War II. During the following decades, the theory has been expanded considerably and has provided solutions to an impressive variety of technical problems, including reliable detection, identification, and recovery of radio and television signals, of underwater signals, and of speech signals; reliable communication of data on point-to-point links and on information networks; and control of plants. In the last decade or so, the reach of this theory has expanded even further, finding applications in biology, security (detection of threats), and analysis of big data.

In a broad sense, statistical inference theory addresses problems of detection and estimation. The underlying theory is foundational for machine learning and data science, as it provides golden standards (fundamental performance limits), which, in some cases, can be approached asymptotically by learning algorithms. In order to develop a deep understanding of machine learning, where one does not assume a prior statistical model for the data, one first needs to thoroughly understand model-based statistical inference, which is the subject of this book.

This book is intended to provide a unifying and insightful view, and a fundamental understanding of statistical inference for engineers and data scientists. It should serve both as a textbook and as a modern reference for researchers and practitioners. The core principles of statistical inference are introduced and illustrated with numerous examples that are designed to be accessible to the broadest possible audience, without relying on domain-specific knowledge. The examples are designed to emphasize key features of the theory and the implications of the assumptions made (e.g., assuming prior distributions and cost functions) and the subtleties that arise when applying the theory.

After an introductory chapter, the book is divided into two main parts. The first part (Chapters 2–10) covers hypothesis testing, where the quantity being inferred (*state*) takes on a finite set of values. The second part (Chapters 11–15) covers estimation, where the state is not restricted to a finite set. A summary of the contents of the chapters is as follows:

- In Chapter 1, the problems of hypothesis testing and estimation are introduced through examples, and then cast in the general framework of statistical decision theory. Various approaches (e.g., Bayes, minimax) to solving statistical decision making problems are defined and compared. The notation used in the book is also defined in this chapter.

- In Chapter 2, the focus is on binary hypothesis testing, where the state takes one of two possible values. The three basic formulations of the binary hypothesis testing problem, namely, Bayesian, minimax, and Neyman–Pearson, are described along with illustrative examples.
- In Chapter 3, the methods developed in Chapter 2 are extended to the case of m-ary hypothesis testing, with $m > 2$. This chapter also includes a discussion of the problem of designing m binary tests simultaneously and obtaining performance guarantees for the collection of tests (rather than for each individual test).
- In Chapter 4, the problem of composite hypothesis testing is studied, where each hypothesis may be associated with more than one probability distribution. Uniformly most powerful (UMP), locally most powerful (LMP), generalized likelihood ratio (GLR), non-dominated, and robust tests are developed to address the composite nature of the hypotheses.
- In Chapter 5, the principles developed in the previous chapters are applied to the problem of detecting a signal, which is a finite sequence, observed in noise. Various models for the signal and noise are considered, along with a discussion of the structures of the optimal tests.
- In Chapter 6, various notions of distances between two distributions are introduced, along with their relationships. These distance metrics prove to be useful in deriving bounds on the performance of the tests for hypothesis testing problems. This chapter should also be of independent interest to researchers from other fields where such distance metrics find applications.
- In Chapter 7, analytically tractable performance bounds for hypothesis testing are derived. Of central interest are upper and lower bounds on error probabilities of optimal tests. A key tool that is used in deriving these bounds is the Chernoff bound, which is discussed in great detail in this chapter.
- In Chapter 8, large-deviations theory, whose basis is the Chernoff bound studied in Chapter 7, is used to derive performance bounds on hypothesis testing with a large number of independent and identically distributed observations under each hypothesis. The asymptotics of these methods are also studied, and tight approximations that are based on the method of exponential tilting are presented.
- In Chapter 9, the problem of hypothesis testing is studied in a sequential setting where we are allowed to choose when to stop taking observations before making a decision. The related problem of quickest change detection is also studied, where the observations undergo a change in distribution at some unknown time and the goal is to detect the change as quickly as possible, subject to false-alarm constraints.
- In Chapter 10, hypothesis testing is studied in the setting where the observations are realizations of random processes. Notions of Kullback–Leibler and Chernoff divergence rates, and Radon–Nikodym derivatives between two distributions on random processes are introduced and exploited to develop detection schemes.
- In Chapter 11, the Bayesian approach to parameter estimation is discussed, where the unknown parameter is modeled as random. The cases of scalar- and vector-valued

parameter estimation are studied separately to emphasize the similarities and differences between these two cases.

- In Chapter 12, several methods are introduced for constructing good estimators when prior probabilistic models are not available for the unknown parameter. The notions of unbiasedness and minimum-variance unbiased estimation are defined, along with notions of sufficient statistics and completeness. Exponential families are studied in detail.
- In Chapter 13, the information inequality is studied for both scalar and vector valued parameters. This fundamental inequality, when applied to unbiased estimators, results in the powerful Cramér–Rao lower bound (CRLB) on the variance.
- In Chapter 14, the focus is on the maximum likelihood (ML) approach to parameter estimation. Properties of the ML estimator are studied in the asymptotic setting where the number of observations goes to infinity. Recursive ways to compute (approximations to) ML estimators are studied, along with the practically useful expectation-maximization (EM) algorithm.
- In Chapter 15, we shift away from parameter estimation and study the problem of estimating a discrete-time random signal using noisy observations of the signal. The celebrated Kalman filter is studied in detail, along with some extensions to nonlinear filtering. The chapter ends with a discussion of estimation in finite alphabet hidden Markov models (HMMs).

The main audience for this book is graduate students and researchers that have completed a first-year graduate course in probability. The material in this book should be accessible to engineers and data scientists working in industry, assuming they have the necessary probability background.

This book is intended for a one-semester graduate-level course, as it is taught at the University of Illinois at Urbana-Champaign. The core of such a course (about two-thirds) could be formed using the material from Chapter 1, Chapter 2, Sections 3.1–3.4, Sections 4.1–4.6, Sections 6.1–6.3, Chapter 7, Sections 8.1 and 8.2, Chapter 11, Chapter 12, Sections 13.1–13.5, Sections 14.1–14.6, Section 14.11, and Sections 15.1–15.3. The remaining third of the course could cover selected topics from the remaining material at the discretion of the instructor.

Acknowledgments

This book is the result of course notes developed by the authors over a period of more than 20 years as they alternated teaching a graduate-level course on the topic in the Electrical and Computer Engineering department at the University of Illinois at Urbana-Champaign. The authors gratefully acknowledge the invaluable feedback and help that they received from the students in this course over the years.

Finally, the authors are thankful to their families for their love and support over the years.

Acronyms

a.s.	almost surely
ADD	average detection delay
AUC	area under the curve
BH	Benjamini–Hochberg
CADD	conditional average detection delay
cdf	cumulative distribution function
CFAR	constant false-alarm rate
CLT	Central Limit Theorem
cgf	cumulant-generating function
CRLB	Cramér–Rao lower bound
CuSum	cumulative sum
$\xrightarrow{\text{d.}}$	convergence in distribution
DFT	discrete fourier transform
EKF	extended Kalman filter
EM	expectation-maximization
FAR	false-alarm rate
FDR	false discovery rate
FWER	family-wise error rate
GLR	generalized likelihood ratio
GLRT	generalized likelihood ratio test
GSNR	generalized signal-to-noise ratio
HMM	hidden Markov model
i.i.d.	independent and identically distributed
i.p.	in probability
JSB	joint stochastic boundedness
KF	Kalman filter
KL	Kullback–Leibler
LFD	least favorable distribution
LLRT	log-likelihood ratio test
LMMSE	linear minimum mean squared-error
LMP	locally most powerful
LMS	least mean squares
LRT	likelihood ratio test
MAP	maximum *a posteriori*
mgf	moment-generating function

ML	maximum likelihood
MLE	maximum likelihood estimator
MMAE	minimum mean absolute-error
MMSE	minimum mean squared-error
MOM	method of moments
MPE	minimum probability of error
m.s.	mean squares
MVUE	minimum-variance unbiased estimator
MSE	mean squared-error
NLF	nonlinear filter
NP	Neyman–Pearson
pdf	probability density function
PFA	probability of false alarm
pmf	probability mass function
QCD	quickest change detection
RLS	recursive least squares
ROC	receiver operating characteristic
SAGE	space-alternating generalized EM
SND	standard noncoherent detector
SNR	signal-to-noise ratio
SPRT	sequential probability ratio test
SR	Shiryaev–Roberts
UMP	uniformly most powerful
WADD	worst-case average detection delay
w.p.	with probability

1 Introduction

In this chapter we introduce the problems of detection and estimation, and cast them in the general framework of *statistical decision theory*, using the notions of states, observations, actions, costs, and optimal decision making. We then introduce the Bayesian approach as the optimal approach in the case of random states. For nonrandom states, the minimax approach and other non-Bayesian approaches are introduced.

1.1 Background

This book is intended to be accessible to graduate students and researchers that have completed a first-year graduate course in probability and random processes. Familiarity with the notions of convergence of random sequences (in probability, almost surely, and in the mean square sense) is needed in the sections of this text that deal with asymptotics. Elementary knowledge of measure theory is helpful but not required. An excellent reference for this background material is Hajek's book [1]. Familiarity with basic tools from matrix analysis [2] and optimization [3] is also assumed.

Textbooks on detection and estimation include the classic book by Van Trees [4], and the more recent books by Poor [5], Kay [6], and Levy [7]. For a thorough treatment of the subject, the reader is referred to the classic books by Lehman [8] and Lehman and Casella [9]. While these books treat detection and estimation as distinct but loosely related topics, we stress in this text the tight connection between the underlying concepts, via Wald's statistical decision theory [10, 11, 12].

The subject matter is also related to statistical learning theory and to information theory. Statistical learning theory deals with unknown probability distributions and with the construction of decision rules whose performance approaches that of rules that know the underlying distributions. Performance analysis of the latter rules is of central interest in this book, and many of the analytic tools are also useful in learning theory [13]. Finally, the connection to information theory appears via the role of Kullback–Leibler divergence between probability distributions, Fisher information, sufficient statistics, and information geometry [14].

1.2 Notation

The following notation is used throughout this book. Random variables are denoted by uppercase letters (e.g., Y), their individual values by lowercase letters (e.g., y), and the set of values by script letters (e.g., $\mathcal{Y}$). We use *sans serif* notation for matrices (e.g., A); the identity matrix is denoted by I. The indicator function of a set $\mathcal{A}$ is denoted by $\mathbb{1}\mathcal{A}$, i.e.,

$$\mathbb{1}\mathcal{A}(x) = \begin{cases} 1 & \text{if } x \in \mathcal{A} \\ 0 & \text{otherwise.} \end{cases} \tag{1.1}$$

For simplicity of the exposition, we shall mostly be interested in discrete random variables and in continuous random variables over Euclidean spaces. In some cases where we need to distinguish between scalar and vector variables, we use boldface for vectors (e.g., $\boldsymbol{Y} = [Y_1\ Y_2]^\top$). The reader is referred to [1] for a thorough introduction to the topics in this section.

1.2.1 Probability Distributions

Consider a random variable Y taking its values in a set $\mathcal{Y}$ endowed with a σ-algebra $\mathscr{F}$. A probability measure P is a real-valued function on $\mathscr{F}$ that satisfies Kolmogorov's three axioms of probability: nonnegativity, unit measure, and σ-additivity.

If $\mathcal{Y}$ is (finitely or infinitely) countable, $\mathscr{F}$ is the collection of all subsets of $\mathcal{Y}$, and we denote by $\{p(y),\ y \in \mathcal{Y}\}$ a probability mass function (pmf) on $\mathcal{Y}$. The probability of a set $\mathcal{A} \subset \mathcal{Y}$ (hence $\mathcal{A} \in \mathscr{F}$) is $\mathrm{P}(\mathcal{A}) = \sum_{y\in\mathcal{A}} p(y)$. If $\mathcal{Y}$ is the n-dimensional Euclidean space $\mathbb{R}^n$ (with $n \geq 1$), we choose $\mathscr{F} = \mathscr{B}(\mathbb{R}^n)$, the Borel σ-algebra which contains all n-dimensional rectangles and all finite or infinite unions of such rectangles. We then assume that Y is a continuous random variable, i.e., it has a probability density function (pdf) which we denote by $\{p(y),\ y \in \mathcal{Y}\}$. The probability of a Borel set $\mathcal{A} \in \mathscr{B}(\mathbb{R}^n)$ is then $\mathrm{P}(\mathcal{A}) = \int_{\mathcal{A}} p(y)\,dy$. The cumulative distribution function (cdf) of Y is given by

$$F(y) = \mathrm{P}(\mathcal{A}_y), \tag{1.2}$$

where $\mathcal{A}_y = \{y' \in \mathcal{Y} : y'_i \leq y_i,\ 1 \leq i \leq n\}$ is an n-dimensional orthant. To summarize:

$$\mathrm{P}(\mathcal{A}) = \begin{cases} \sum_{y\in\mathcal{A}} p(y) & : \text{discrete } \mathcal{Y} \\ \int_{\mathcal{A}} p(y)\,dy & : \text{continuous } \mathcal{Y}. \end{cases} \tag{1.3}$$

1.2.2 Conditional Probability Distributions

A conditional pmf of a random variable Y is a collection of pmfs $\{p_x(y),\ y \in \mathcal{Y}\}$ indexed by a conditioning variable $x \in \mathcal{X}$. Note that x is not necessarily interpreted as a realization of random variable X, and in that sense, $p_x(y)$ is different from the more traditional conditional distribution of a random variable Y, given another random variable X. Similarly, a conditional pdf is a collection of pdfs $\{p_x(y),\ y \in \mathcal{Y}\}$ indexed by a conditioning variable $x \in \mathcal{X}$. The conditional probability of a Borel set $\mathcal{A} \in \mathscr{B}(\mathbb{R}^n)$ is $\mathrm{P}_x(\mathcal{A}) = \int_{\mathcal{A}} p_x(y)\,dy$. The conditional cdf is denoted by $F_x(y) = \mathrm{P}_x(\mathcal{A}_y)$ where $\mathcal{A}_y$ is the orthant defined in Section 1.2.1. For both the discrete and the continuous cases, we may write $p(y|x)$ instead of $p_x(y)$, especially in the case that x is a realization of a random variable X.

1.2.3 Expectations and Conditional Expectations

The expectation of a function $g(Y)$ of a random variable Y is given by

$$\mathbb{E}[g(Y)] = \begin{cases} \sum_{y \in \mathcal{Y}} p(y) g(y) & : \text{discrete } \mathcal{Y} \\ \int_{\mathcal{Y}} p(y) g(y)\, dy & : \text{continuous } \mathcal{Y}. \end{cases} \tag{1.4}$$

The expectation may be viewed as a linear function of the pmf p in the discrete case, and a linear functional of the pdf p in the continuous case.[1] The conditional expectation of $g(Y)$ given x is denoted by $\mathbb{E}_x[g(Y)]$ and takes the form

$$\mathbb{E}_x[g(Y)] = \begin{cases} \sum_{y \in \mathcal{Y}} p_x(y) g(y) & : \text{discrete } \mathcal{Y} \\ \int_{\mathcal{Y}} p_x(y) g(y)\, dy & : \text{continuous } \mathcal{Y}. \end{cases} \tag{1.5}$$

1.2.4 Unified Notation

The cases of discrete and continuous Y can be handled in a unified way, avoiding duplication of formulas. Indeed the probability of a set $\mathcal{A} \in \mathscr{F}$ may be written as the Lebesgue–Stieltjes integral

$$\mathrm{P}(\mathcal{A}) = \int_{\mathcal{A}} p(y)\, d\mu(y), \tag{1.6}$$

where μ is a finite measure on $\mathscr{F}$, equal to the Lebesgue measure in the continuous case ($d\mu(y) = dy$, or equivalently $\mu(\mathcal{A}) = \int_{\mathcal{A}} dy$ for all $\mathcal{A} \in \mathscr{F}$) and to the counting measure in the discrete case.[2] The expectation of a function $g(Y)$ is likewise given by

$$\mathbb{E}[g(Y)] = \int_{\mathcal{Y}} p(y) g(y)\, d\mu(y). \tag{1.7}$$

For conditional expectations we have

$$\mathbb{E}_x[g(Y)] = \int_{\mathcal{Y}} p_x(y) g(y)\, d\mu(y), \quad \forall x \in \mathcal{X}. \tag{1.8}$$

1.2.5 General Random Variables

The expressions defined in the previous sections can be generalized to mixed discrete-continuous random variables using the Lebesgue–Stieltjes formalism. Indeed let Y be a random variable over $\mathbb{R}^n$ with cdf F defined in (1.2). Then the probability of a Borel set $\mathcal{A}$ is

$$\mathrm{P}(\mathcal{A}) = \int_{\mathcal{A}} dF(y) \tag{1.9}$$

(the Lebesgue–Stieltjes integral) and the expectation of a function $g(Y)$ is

$$\mathbb{E}[g(Y)] = \int_{\mathcal{Y}} g(y) dF(y). \tag{1.10}$$

[1] Typical functions $g(Y)$ of interest are polynomials of Y (the expected values thereof are moments of the distribution) and indicator functions of sets (the expected values thereof are probabilities of said sets).

[2] The counting measure on a set of points $y_1, y_2, \ldots$ is defined as $\mu(\mathcal{A}) = \sum_i \mathbb{1}\{y_i \in \mathcal{A}\}$.

The random variable Y has a density with respect to a measure μ which is the sum of Lebesgue measure and a counting measure, in which case the expressions (1.6) and (1.7) hold and allow simple calculations.[3] The same concept applies to conditional probabilities and conditional expectations, with F replaced by the conditional cdf F_x. Finally, if $n \geq 2$ and the distribution P assigns positive probabilities to subsets of several dimensions, then Y is not of the mixed discrete-continuous type but (1.9) and (1.10) hold, and so do the expressions (1.6) and (1.7) using a suitable μ.

For any pair of random variables (X, Y), assuming $f(y) = \mathbb{E}[X|Y = y]$ exists for each $y \in \mathcal{Y}$, let $\mathbb{E}[X|Y] \triangleq f(Y)$ which is a function of Y. The law of iterated expectations states that $\mathbb{E}[X] = \mathbb{E}(\mathbb{E}[X|Y])$, which will be very convenient throughout this text. In particular, this law may be applied to evaluate the expectation of a function $g(X, Y)$ as $\mathbb{E}[g(X, Y)] = \mathbb{E}(\mathbb{E}[g(X, Y)|Y])$.

The (scalar) Gaussian random variable with mean μ and variance σ^2 is denoted by $\mathcal{N}(\mu, \sigma^2)$. The Gaussian random vector with mean $\boldsymbol{\mu}$ and covariance matrix C is denoted by $\mathcal{N}(\boldsymbol{\mu}, \mathsf{C})$.

1.3 Statistical Inference

The detection and estimation problems treated in this course are instances of the following general statistical inference problem. Given *observations* y taking their value in some data space $\mathcal{Y}$, infer (estimate) some unknown *state* x taking its value in a state space $\mathcal{X}$. The observations are noisy. More precisely, they are stochastically related to the state via a *conditional probability distribution* $\{\mathrm{P}_x, x \in \mathcal{X}\}$, i.e., a collection of distributions on $\mathcal{Y}$ indexed by $x \in \mathcal{X}$. Note that no assumption of randomness of the state is made at this point. The key problems are to construct a *good* estimator and to characterize its performance. This raises the following central issues:

- What performance criteria should be used?
- Can we derive an estimator that is *optimal* under some desirable performance criterion?
- Is such an estimator computationally tractable?
- If not, can the optimization be restricted to a useful class of tractable estimators?

The theory applies to a remarkable variety of statistical inference problems covering applications in diverse areas (communications, signal processing, control, life sciences, economics, etc.).

[3] For instance, let $\mathcal{Y}$ be the interval $[0, 1]$ and consider the cdf $F_0(y) = y\mathbb{1}\{0 \leq y \leq 1\}$ corresponding to the uniform distribution on $\mathcal{Y}$, the cdf $F_1(y) = \mathbb{1}\{\frac{1}{3} \leq y \leq 1\}$ corresponding to the degenerate distribution that assigns all mass at the point $y = \frac{1}{3}$, and the cdf $F_2(y) = \frac{1}{2}[F_0(y) + F_1(y)]$ corresponding to the mixture of the continuous and discrete random variables above. The means μ_0 and μ_1 of the continuous and discrete random variables are equal to $\frac{1}{2}$ and $\frac{1}{3}$, respectively. The mean of the mixture distribution is $\mu_3 = \frac{1}{2}[\mu_1 + \mu_2] = \frac{5}{12}$.

1.3.1 Statistical Model

The statistical model is a triple $(\mathcal{X}, \mathcal{Y}, \{\mathrm{P}_x, x \in \mathcal{X}\})$ consisting of a state space $\mathcal{X}$, an observation space $\mathcal{Y}$, and a family of conditional distributions P_x on $\mathcal{Y}$, indexed by $x \in \mathcal{X}$. The state space may be a continuum or a discrete set, in which cases the inference problem is conventionally called estimation or detection, respectively.

Estimation: Often the state $x \in \mathcal{X}$ is a scalar parameter, e.g., the unknown attenuation factor of a transmission system, a temperature, or some other physical quantity. More generally x could be a vector-valued parameter, e.g., x represents location in two- or three-dimensional Euclidean space, or x represents a set of individual states associated with sensors at different physical locations. Another example of vector-valued x is a discrete-time signal (speech, geophysical, etc.). Yet another version of this problem involves functions defined over some compact set, e.g., x is a continuous-time signal over a finite time window $[0, T]$. The notion of time can be extended to space, and so x could also be a two- or three-dimensional discrete-space signal, or a continuous-space signal. In all cases, the parameter space $\mathcal{X}$ is a continuum (an uncountably infinite set). In this text, we will limit our attention to the m-dimensional Euclidean space $\mathcal{X} = \mathbb{R}^m$. Many of the key insights are already apparent in the scalar case $(m - 1)$.

Detection: In other problems, the state x is a discrete quantity, e.g., a bit in a communication system ($\mathcal{X} = \{0, 1\}$), or a sequence of bits ($\mathcal{X} = \{0, 1\}^n$) where n is the length of the binary sequence. Alternatively, x could take one of m values for a signal classification problem (e.g., x represents a word in a speech recognition system, or a person in a biometric verification system). The observed signal comes from one of m possible classes or categories.

Similarly, observations can come in a variety of formats: the data space $\mathcal{Y}$ could be:

- a finite set, e.g., the observation is a single bit $y \in \mathcal{Y} = \{0, 1\}$ in a data communication system, or a sequence of bits;
- a countably infinite set, e.g., the set of all binary sequences, $\mathcal{Y} = \{0, 1\}^{\mathbb{N}}$;
- an uncountably infinite set, such as $\mathbb{R}$ for scalar observations, or $\mathbb{R}^m$ for vector-valued observations, where $m > 1$, or the space $L^2[0, T]$ of finite-energy analog signals defined over the time window $[0, T]$.

In many problems, the observations form a length-n sequence whose elements take their value in a common space $\mathcal{Y}_1$. Then $\mathcal{Y} = \mathcal{Y}_1^n$ is the n-fold product of $\mathcal{Y}_1$, and we use the boldface notation $\boldsymbol{Y} = (Y_1, Y_2, \dots, Y_n)$ to represent its elements. The sequence $\boldsymbol{Y}$ often follows the product conditional distribution $p_x(\boldsymbol{y}) = \prod_{i=1}^n p_{Y_i|X=x}(y_i)$, given the state x.

The estimation problem consists of designing a function $\hat{x} : \mathcal{Y} \to \mathcal{X}$ (*estimator*) that produces an *estimate* $\hat{x}(y) \in \mathcal{X}$, given observations $y \in \mathcal{Y}$. The detection problem admits the same formulation, the only difference being that $\mathcal{X}$ is a discrete set instead of a continuum. Constraints such as linearity can be imposed on the design of the estimator.

1.3.2 Some Generic Estimation Problems

1. Parameter estimation: estimate the mean μ and variance σ^2 of a Gaussian distribution from a sequence of observations Y_i, $1 \le i \le n$ drawn i.i.d. (independent and identically distributed) $\mathcal{N}(\mu, \sigma^2)$. Here $\mathcal{Y} = \mathbb{R}^n$ and $\mathcal{X} = \mathbb{R} \times \mathbb{R}^+$. A simple estimator of the mean and variance would be the so-called sample mean $\hat{\mu}(\boldsymbol{Y}) = \frac{1}{n}\sum_{i=1}^n Y_i$ and sample variance $\hat{\sigma}^2(\boldsymbol{Y}) = \frac{1}{n}\sum_{i=1}^n (Y_i - \hat{\mu}(\boldsymbol{Y}))^2$ which are respectively linear and quadratic functions of the data.
2. Probability mass function estimation: estimate a pmf $p = \{p(1), p(2), \dots, p(k)\}$ from a sequence of observations Y_i, $1 \le i \le n$ drawn i.i.d. p. Here $\mathcal{Y} = \{1, \dots, k\}^n$ and $\mathcal{X}$ is the probability simplex in $\mathbb{R}^k$. A simple estimator would be the empirical pmf $\hat{p}(l) = \frac{1}{n}\sum_{i=1}^n \mathbb{1}\{Y_i = l\}$ for $l = 1, 2, \dots, k$.
3. Estimation of signal in i.i.d. noise: $Y_i = s_i + Z_i$ for $i = 1, 2, \dots, n$ where Z_i are i.i.d. $\mathcal{N}(0, \sigma^2)$. Here $\mathcal{Y} = \mathcal{X} = \mathbb{R}^n$. A simple estimator would be $\hat{S}_i = cY_i$, where $c \in \mathbb{R}$ is a weight to be optimized.
4. Parametric estimation of signal in i.i.d. noise: $Y_i = s_i(\theta) + Z_i$ for $i = 1, 2, \dots, n$ where $\{s_i(\theta)\}_{i=1}^n$ is a sequence parameterized by $\theta \in \mathcal{X}$, and Z_i are i.i.d. $\mathcal{N}(0, \sigma^2)$. The unknown parameter could be estimated using the nonlinear least-squares estimator $\hat{\theta} = \operatorname{argmin}_{\theta \in \mathcal{X}} \sum_i (Y_i - s_i(\theta))^2$.
5. Signal estimation, prediction and smoothing (Figure 1.1): $Y_i = (h \star s)_i + Z_i$ for $i = 1, 2, \dots, n$, where Z_i are i.i.d. $\mathcal{N}(0, \sigma^2)$, and $h \star s$ denotes the convolution of the sequence $\boldsymbol{s}$ with the impulse response of the linear system. A candidate estimator would be $\hat{S}_i = (g \star Y)_i$, where g is the impulse response of an estimation filter. In a prediction problem, a filter is designed to estimate a future sample of the signal (e.g., estimate s_{i+1}). In a smoothing problem, a filter is designed to estimate past values of the signal.
6. Image denoising: $Y_{ij} = s_{ij} + Z_{ij}$ for $i = 1, 2, \dots, n_1$ and $j = 1, 2, \dots, n_2$ and Z_{ij} are i.i.d. $\mathcal{N}(0, \sigma^2)$.
7. Estimation of a continuous-time signal: $Y(t) = s(t) + Z(t)$ for $0 \le t \le T$ where x belongs to a separable Hilbert space such as $L^2[0, T]$, and Z is stationary Gaussian noise with mean zero and covariance function $R(t)$, $t \in \mathbb{R}$.

1.3.3 Some Generic Detection Problems

1. Binary hypothesis testing: under hypothesis H_0, the observations are a sequence Y_i, $i = 1, 2, \dots, n$ drawn i.i.d. p_0. Under the rival hypothesis H_1, the observations are drawn i.i.d. p_1.
2. Signal detection in i.i.d. Gaussian noise: $Y_i = \theta s_i + Z_i$ for $i = 1, 2, \dots, n$ where $\boldsymbol{s} \in \mathbb{R}^n$ is a known sequence, $\theta \in \{0, 1\}$, and Z_i are i.i.d. $\mathcal{N}(0, \sigma^2)$.

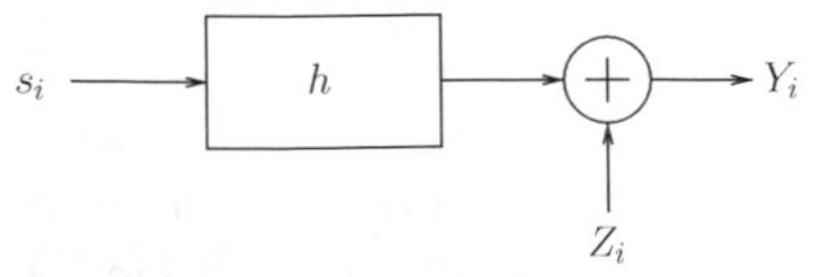

Figure 1.1 Signal prediction and smoothing

3. M-ary signal classification in i.i.d. Gaussian noise: $Y_i = s_i(\theta) + Z_i$ for $i = 1, 2, \ldots, n$ where $s(\theta) \in \mathbb{R}^n$ is a sequence parameterized by $\theta \in \{0, 1, \ldots, M-1\}$, and Z_i are i.i.d. $\mathcal{N}(0, \sigma^2)$. This generalizes Problem #2 above. A simple detector is the correlation detector, the output $\hat{\theta}$ of which maximizes the correlation score $T(\theta) = \sum_{i=1}^n Y_i s_i(\theta)$ over all $\theta \in \{0, 1, \ldots, M-1\}$.
4. Composite hypothesis testing in i.i.d. Gaussian noise: $Y_i = \alpha s_i(\theta) + Z_i$ for $i = 1, 2, \ldots, n$ where the multiplier $\alpha \in \{0, 1\}$, the sequence $s(\theta) \in \mathbb{R}^n$ is a function of some nuisance parameter $\theta \in \mathbb{R}^m$, and Z_i are i.i.d. $\mathcal{N}(0, \sigma^2)$. Here only α is of interest; it is possible (but not required) to estimate θ as a means to solve the detection problem.

1.4 Performance Analysis

For estimation problems, various metrics can be used to measure the closeness of the estimator $\hat{x}$ to the original x. For instance, let $\mathcal{X} = \mathbb{R}^m$ and consider:

- squared error $C(x, \hat{x}) = \sum_{i=1}^m (\hat{x}_i - x_i)^2$;
- absolute error $C(x, \hat{x}) = \sum_{i=1}^m |\hat{x}_i - x_i|$.

The squared-error criterion is more tractable but penalizes large errors heavily. This could be desirable in some problems but undesirable in others (e.g., in the presence of outliers in the data). In such cases, the absolute-error criterion might be preferable.

For detection problems, one may consider:

- for $\mathcal{X} = \{0, 1, \ldots, m-1\}$: zero-one loss $\mathbb{1}\{\hat{x} \neq x\}$;
- for $\mathcal{X} = \{0, 1\}^d$ (space of length-d binary sequences): Hamming distance $d_H(\hat{x}, x) = \sum_{i=1}^d \mathbb{1}\{\hat{x}_i \neq x_i\}$. The Hamming distance, normalized by d, is the bit error rate (BER) in a digital communication system.

In both cases, the performance metric can be evaluated on a specific x, on a specific y, or in some average sense. The average value of the performance metric is often used as a design criterion for the estimation (detection) system, as discussed next.

1.5 Statistical Decision Theory

Detection and estimation problems fall under the umbrella of Abraham Wald's statistical decision theory [10, 11], where the goal is to make a right (optimal) choice from a set of alternatives in a noisy environment. A general framework for statistical decision theory is depicted in Figure 1.2. The first block represents the observational model which is governed by the conditional distribution $p_x(y)$ of the observations given the state x. The second block is a decision rule δ that outputs an action $a = \delta(y)$. Each possible action a incurs a cost $C(a, x)$, to be minimized in some sense.[4]

[4] The cost is also called *loss*. Alternatively, one could take a more positive view and specify a *utility function* $U(a, x)$, to be maximized. Both problem formulations are equivalent if one specifies $U(a, x) = b - C(a, x)$ for some constant b.

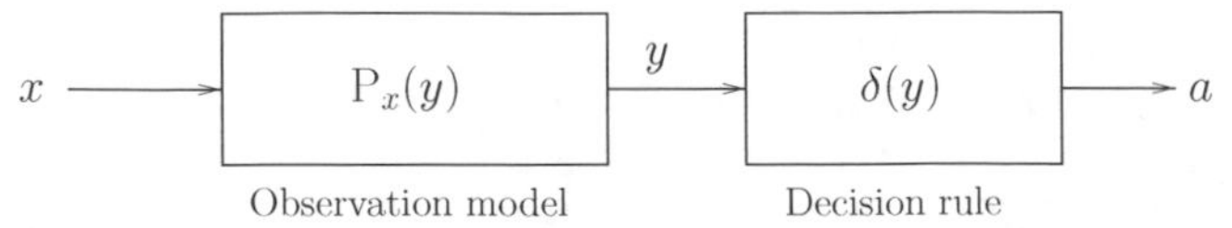

Figure 1.2 Statistical decision-making framework

Formally, there are six basic ingredients in a typical decision theory problem:

1. $\mathcal{X}$: the set of states. For detection problems, the number of states is finite, i.e., $|\mathcal{X}| = m < \infty$. For binary detection problems, we usually let $\mathcal{X} = \{0, 1\}$. We denote a typical state for detection problems by the variable $j \in \mathcal{X}$. We denote a typical state for estimation problems by the variable $x \in \mathcal{X}$.
2. $\mathcal{A}$: the set of actions, or possible decisions about the state. In most detection and estimation problems, $\mathcal{A} = \mathcal{X}$. However, in problems such as decoding with an erasure option, $\mathcal{A}$ could be larger than $\mathcal{X}$. In other problems such as composite hypothesis testing, $\mathcal{A}$ is smaller than $\mathcal{X}$. We denote a typical decision by the variable $a \in \mathcal{A}$.
3. $C(a, x)$: the cost of taking action a when the state of Nature is x. Typically $C(a, x) \geq 0$, and the cost is to be minimized in some suitable sense. Quantifying the costs incurred from decisions allows us to optimize the decision rule. An example of cost function that is relevant in many applications is the *uniform* cost function for which $|\mathcal{X}| = |\mathcal{A}| = m < \infty$ and

$$C(a, x) = \begin{cases} 0 & \text{if } a = x \\ 1 & \text{if } a \neq x, \end{cases} \qquad x \in \mathcal{X}, a \in \mathcal{A}. \tag{1.11}$$

4. $\mathcal{Y}$: the set of observations. The decision is made based on some *random* observation Y taking values in $\mathcal{Y}$. The observations could be continuous (with $\mathcal{Y} \subset \mathbb{R}^n$) or discrete.
5. $\{\mathrm{P}_x, \ x \in \mathcal{X}\}$: the observational model, a family of probability distributions on $\mathcal{Y}$ conditioned on the state of Nature x.
6. $\mathcal{D}$: the set of decision rules or tests. A decision rule is a mapping $\delta : \mathcal{Y} \mapsto \mathcal{A}$ that associates an action $a = \delta(y)$ to each possible observation $y \in \mathcal{Y}$.

Detection problems are also referred to as *hypothesis testing* problems, with the understanding that each state corresponds to a hypothesis about the nature of the observations. The hypothesis corresponding to state j is denoted by H_j.

1.5.1 Conditional Risk and Optimal Decision Rules

The cost associated with a decision rule $\delta \in \mathcal{D}$ is a random quantity (because Y is random) given by $C(\delta(Y), x)$. Therefore, to *order* decision rules according to their "merit" we use the quantity

$$R_x(\delta) \triangleq \mathbb{E}_x[C(\delta(Y), x)] = \int_{\mathcal{Y}} p_x(y) C(\delta(y), x) d\mu(y), \tag{1.12}$$

which we call the *conditional risk* associated with δ when the state is x.

The conditional risk function can be used to obtain a (partial) ordering of the decision rules in $\mathcal{D}$, in the following sense.

Table 1.1 Decision rules and conditional risks for Example 1.1

δ	$y=1$	$y=2$	$y=3$	$R_0(\delta)$	$R_1(\delta)$
δ_1	0	0	0	0	1
δ_2	0	0	1	0	0.5
δ_3	0	1	0	0	0.5
$\boxed{\delta_4}$	0	1	1	$\boxed{0}$	$\boxed{0}$
δ_5	1	0	0	1	1
δ_6	1	0	1	1	0.5
δ_7	1	1	0	1	0.5
δ_8	1	1	1	1	0

DEFINITION 1.1 A decision rule δ is *better*[5] than decision rule δ' if

$$R_x(\delta) \le R_x(\delta'), \quad \forall x \in \mathcal{X}$$

and

$$R_x(\delta) < R_x(\delta') \text{ for at least one } x \in \mathcal{X}.$$

Sometimes it may be possible to find a decision rule $\delta^\star \in \mathcal{D}$ which is better than any other $\delta \in \mathcal{D}$. In this case, the statistical decision problem is solved. Unfortunately, this usually happens only for trivial cases as in the following example.

Example 1.1 Suppose $\mathcal{X} = \mathcal{A} = \{0, 1\}$ with the uniform cost function as in (1.11). Furthermore suppose the observation Y takes values in the set $\mathcal{Y} = \{1, 2, 3\}$ and the conditional pmfs of Y are:

$$p_0(1) = 1,\ p_0(2) = p_0(3) = 0, \qquad p_1(1) = 0,\ p_1(2) = p_1(3) = 0.5.$$

Then it is easy to see that we have the conditional risks for the eight possible decision rules depicted in Table 1.1. Clearly, δ_4 is the best rule according to Definition 1.1, but this happens only because the conditional pmfs p_0 and p_1 have disjoint supports (see also Exercise 2.1).

Since conditional risks cannot be used directly in finding optimal solutions to statistical decision making problems except in trivial cases, there are two general approaches for finding optimal decision rules: *Bayesian* and *minimax*.

1.5.2 Bayesian Approach

If the state of Nature is random with known *prior* distribution $\pi(x)$, $x \in \mathcal{X}$, then one can define the *average* risk or *Bayes* risk associated with a decision rule δ, which is given by

$$r(\delta) \triangleq \mathbb{E}[R_X(\delta)] = \mathbb{E}[C(\delta(Y), X)]. \tag{1.13}$$

[5] If δ dominates δ' as in this definition, the decision rule δ' is sometimes said to be *inadmissible* [8] for the statistical inference problem.

More explicitly,

$$r(\delta) = \begin{cases} \int_{\mathcal{X}} \int_{\mathcal{Y}} \pi(x) p_x(y) C(\delta(y), x) dy dx & : \text{continuous } X, Y \\ \sum_{x\in\mathcal{X}} \sum_{y\in\mathcal{Y}} \pi(x) p_x(y) C(\delta(y), x) & : \text{discrete } X, Y, \end{cases} \tag{1.14}$$

which may also be written using the unified notation

$$r(\delta) = \int_{\mathcal{X}} \int_{\mathcal{Y}} \pi(x) p_x(y) C(\delta(y), x) d\mu(y) d\nu(x) \tag{1.15}$$

for appropriate measures μ and ν on $\mathcal{Y}$ and $\mathcal{X}$, respectively.

The optimal decision rule, in the Bayesian framework, is the one that minimizes the Bayes risk:

$$\delta_{\mathrm{B}} = \arg\min_{\delta\in\mathcal{D}} r(\delta), \tag{1.16}$$

and is known as the Bayesian decision rule. The minimizer in (1.16) might not be unique; all such minimizers are Bayes rules.

1.5.3 Minimax Approach

What if we are not given a prior distribution on the set $\mathcal{X}$? We could postulate a distribution on $\mathcal{X}$ (for example, a uniform distribution) and use the Bayesian approach. On the other hand, one may want to guarantee a certain level of performance for all choices of state. In this case, we use a minimax approach. To this end, we define the maximum (or worst-case) risk (see Figure 1.3(a)):

$$R_{\max}(\delta) \triangleq \max_{x\in\mathcal{X}} R_x(\delta). \tag{1.17}$$

The minimax decision rule minimizes the worst-case risk:

$$\begin{aligned} \delta_{\mathrm{m}} &\triangleq \arg\min_{\delta\in\mathcal{D}} R_{\max}(\delta) \\ &= \arg\min_{\delta\in\mathcal{D}} \max_{x\in\mathcal{X}} R_x(\delta). \end{aligned} \tag{1.18}$$

The minimizer in (1.18) might not be unique; all such minimizers are minimax rules.

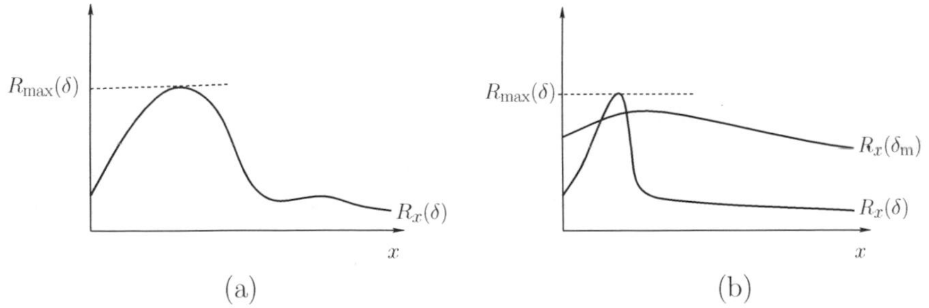

Figure 1.3 Minimax risk: (a) conditional risk and maximum risk for decision rule δ; (b) example where minimax rule is pessimistic

Table 1.2 Decision rules, and Bayes and minimax risks for Example 1.2

δ	$y=1$	$y=2$	$y=3$	$R_0(\delta)$	$R_1(\delta)$	$0.5(R_0(\delta)+R_1(\delta))$	$R_{\max}(\delta)$
δ_1	0	0	0	0	1	0.5	1
$\boxed{\delta_2}$	0	0	1	$\boxed{0}$	$\boxed{0.5}$	0.25	0.5
δ_3	0	1	0	0.5	0.5	0.5	0.5
$\boxed{\delta_4}$	0	1	1	$\boxed{0.5}$	$\boxed{0}$	0.25	0.5
δ_5	1	0	0	0.5	1	0.75	1
δ_6	1	0	1	0.5	0.5	0.5	0.5
δ_7	1	1	0	1	0.5	0.75	1
δ_8	1	1	1	1	0	0.5	1

The minimax approach is closely related to game theory [11] and more specifically a zero-sum game with payoff function $R_x(\delta)$ and state x and decision rule δ selected by Nature and by the statistician, respectively.

In game-theoretic parlance, the minimax approach is *secure* in the sense that it offers a performance guarantee even in the worst-case scenario. This property would be most useful if x were selected by an adversary. However, when x is selected by Nature, selecting δ based on the worst-case scenario could result in an overly conservative design, as illustrated in Figure 1.3(b), which compares δ_{m} with another rule δ. It appears that $R_x(\delta_{\mathrm{m}})$ is much larger than $R_x(\delta)$ for all values of x except in a small neighborhood where δ_{m} marginally outperforms δ. In this case it does not appear reasonable to select δ_{m} instead of δ.

Example 1.2 Consider the same setup as in Example 1.1 with the following conditional pmfs:

$$p_0(1) = p_0(2) = 0.5,\ p_0(3) = 0, \qquad p_1(1) = 0,\ p_1(2) = p_1(3) = 0.5.$$

We can compute the conditional risks for the eight possible decision rules as shown in Table 1.2. Clearly there is no "best" rule based on conditional risks alone in this case. However, δ_2 is better than δ_1, δ_3, δ_5, δ_6, and δ_7; and δ_4 is better than δ_3, δ_5, δ_6, δ_7, and δ_8. Now consider finding a Bayes rule for priors $\pi_0 = \pi_1 = 0.5$. It is clear from Table 1.2 that δ_2 and δ_4 are both Bayes rules. Also, δ_2, δ_3, δ_4, and δ_6 are all minimax rules with minimax risk equal to 0.5.

1.5.4 Other Non-Bayesian Rules

A third approach, which does not require the specification of costs and is often convenient for parameter estimation ($\mathcal{A} = \mathcal{X}$), is to adopt an estimation rule of the form

$$\hat{x}(y) = \max_{x \in \mathcal{X}} \Psi(x, y), \tag{1.19}$$

where the objective function Ψ should be tractable and have some good properties. For instance, $\Psi(x, y) = p_x(y)$, which, for fixed y, is called the likelihood function for x. The resulting $\hat{x}$ is called the maximum-likelihood (ML) estimator (see Chapter 14).

Often one also seeks decision rules that have some desired property such as unbiasedness (see MVUEs in Chapter 12) or invariance with respect to some natural group operation. This defines a class $\mathcal{D}$ of feasible decision rules.

1.6 Derivation of Bayes Rule

Define the *a posteriori cost* (*aka posterior cost*) of taking action $a \in \mathcal{A}$ given $y \in \mathcal{Y}$ as

$$C(a|y) \triangleq \mathbb{E}[C(a, X)|Y = y] = \int_{\mathcal{X}} C(a, x)\pi(x|y)d\nu(x), \tag{1.20}$$

where

$$\pi(x|y) = \frac{p_x(y)\pi(x)}{p(y)} = \frac{p_x(y)\pi(x)}{\int_{\mathcal{X}} p_x(y)\pi(x)d\nu(x)} \tag{1.21}$$

is the *posterior* probability of the state given the observation, and

$$p(y) = \int_{\mathcal{X}} \pi(x)p_x(y)d\nu(x)$$

is the marginal distribution on Y.

The risk $r(\delta)$ can be written as

$$r(\delta) = \mathbb{E}[C(\delta(Y), X)] = \int_{\mathcal{X}} R_x(\delta)\pi(x)d\nu(x) = \mathbb{E}[R_X(\delta)], \tag{1.22}$$

or equivalently as

$$r(\delta) = \mathbb{E}[C(\delta(Y), X)] = \int_{\mathcal{Y}} C(\delta(y)|y)p(y)d\mu(y) = \mathbb{E}[C(\delta(Y)|Y)], \tag{1.23}$$

where $C(\delta(y)|y)$ is the posterior risk, evaluated using (1.20). The problem is to minimize $r(\delta)$ over all $\delta = \{\delta(y),\ y \in \mathcal{Y}\} \in \mathcal{D}$. By (1.23), $r(\delta)$ is additive over y, and the minimizer is simply obtained by solving the minimization problem

$$\min_{a \in \mathcal{A}} C(a|y)$$

independently for each $y \in \mathcal{Y}$. The general procedure to find the solution is therefore as follows:

1. Compute the posterior distribution using Bayes rule
$$\pi(x|y) = \frac{p_x(y)\pi(x)}{p(y)}. \tag{1.24}$$
2. Compute the posterior costs $C(a|y) = \int_{\mathcal{X}} C(a, x)\pi(x|y)d\nu(x)$ for all $a \in \mathcal{A}$.
3. Find the Bayes decision rule
$$\delta_{\mathrm{B}}(y) = \arg\min_{a \in \mathcal{A}} C(a|y). \tag{1.25}$$

Hence the solution depends on the costs and on the *posterior* distribution $\pi(x \mid y)$, which is the fundamental distribution in all Bayesian inference problems.

Example 1.3 Consider the following simplified Bayesian model for the decision of buying or not buying health insurance. Say each person's health is represented by a binary variable x taking value 1 (healthy) or 0 (not healthy). Each person's belief about their own health is likewise represented by a binary random variable Y taking value 1 (believe healthy) or 0 (believe not healthy). A healthy person has no symptom and believes she is healthy, but a non-healthy person believes she is healthy with probability 1/2. Thus $p_1(y) = \mathbb{1}\{y = 1\}$ and $p_0(y) = \frac{1}{2}$ for $y = 0, 1$. The prior belief that a person is healthy is $\pi(1) = 0.8$. Each person must decide whether or not to buy health insurance covering the cost of medical treatment for themselves (as well as for uninsured persons). The cost of the insurance premium is 1 and no further cost arises whether or not the person needs medical treatment. If a person does not buy health insurance, the cost is 0 if the person is healthy. However, the cost is 10 if the person is not healthy, reflecting the cost of future medical treatments.

The question we would like to answer is: should people purchase insurance?

Solution The state space is $\mathcal{X} = \{0, 1\}$ and the action space is $\mathcal{A} = \{\text{buy}, \text{not}\}$. The cost function is given by

$$C(a, x) = \begin{cases} 1, & a = \text{buy}, \forall x \\ 10, & a = \text{not}, \ x = 0 \\ 0, & a = \text{not}, x = 1. \end{cases}$$

The marginal pmf of Y is given by

$$p(y) = \begin{cases} p_0(0)\pi(0) + p_1(0)\pi(1) = 0.2/2 + 0 = 0.1 & \text{if } y = 0 \\ 0.9 & \text{if } y = 1. \end{cases}$$

Note that $Y = 0$ implies that $X = 0$. The posterior distribution of X is given by

$$\pi(x|0) = \begin{cases} 1 & \text{if } x = 0 \\ 0 & \text{if } x = 1, \end{cases}$$

and

$$\pi(x|1) = \begin{cases} \frac{p_0(1)\pi(0)}{p(1)} = \frac{1}{9} & \text{if } x = 0 \\ \frac{8}{9} & \text{if } x = 1. \end{cases}$$

The posterior cost of taking action a given y is $C(a|y) = \sum_{x=0,1} C(a, x)\pi(x|y)$. The posterior costs for $Y = 0$ (which implies $X = 0$) are

$$C(\text{buy}|0) = 1,$$
$$C(\text{not}|0) = 10 > 1,$$

hence the Bayes decision for a person who believes she is not healthy is $\delta_B(0) = \text{buy}$. The posterior costs for $Y = 1$ are

$$\begin{aligned} C(\text{buy}|1) &= 1, \\ C(\text{not}|1) &= \sum_{x=0,1} \pi(x|1)C(\text{not}, x) = \frac{1}{9}10 + \frac{8}{9}0 = \frac{10}{9} > 1, \end{aligned}$$

hence the Bayes decision for a person who believes she is healthy is $\delta_B(1) = \text{buy}$.

1.7 Link Between Minimax and Bayesian Decision Theory

The Bayesian and minimax approaches are seemingly very different but there exists a fundamental relation between them. When the state is random with prior $\pi(x)$, $x \in \mathcal{X}$, the Bayes risk of decision rule $\delta \in \mathcal{D}$ is given by

$$\begin{aligned} r(\delta, \pi) &= \int_{\mathcal{X}} R_x(\delta)\pi(x)d\nu(x) \\ &\stackrel{(a)}{\leq} \left(\max_{x\in\mathcal{X}} R_x(\delta)\right) \int_{\mathcal{X}} \pi(x)d\nu(x) \\ &= \max_{x\in\mathcal{X}} R_x(\delta), \end{aligned} \tag{1.26}$$

which is upper bounded by the maximum conditional risk. This inequality holds for all π. Moreover, equality in (a) is achieved by any prior π_δ that puts all its mass on the set of maximizer(s) of $R_x(\delta)$. Hence we obtain the fundamental equality

$$r(\delta, \pi_\delta) = \max_{\pi} r(\delta, \pi) = \max_{x\in\mathcal{X}} R_x(\delta), \quad \forall\delta, \tag{1.27}$$

which shows that the worst-case conditional risk for rule δ is the same as its Bayes risk under prior π_δ.

Minimizing both sides of (1.27) over δ, we obtain the minimax risk, which is achieved by the minimax rule δ^*:

$$\begin{aligned} \min_{\delta} \max_{\pi} r(\delta, \pi) &= \min_{\delta} \max_{x\in\mathcal{X}} R_x(\delta) \\ &= R_{\max}(\delta^*). \end{aligned} \tag{1.28}$$

1.7.1 Dual Concept

The Bayes rule achieves the minimal risk, $\min_\delta r(\delta, \pi)$, over all decision rules. Denote by δ_π a Bayes rule under prior π. The worst Bayes risk (over all priors) is given by

$$\max_{\pi} r(\delta_\pi, \pi) = \max_{\pi} \min_{\delta} r(\delta, \pi). \tag{1.29}$$

DEFINITION 1.2 A prior π^* is said to be *least favorable* if it achieves the maximum risk in (1.29).

1.7.2 Game Theory

The decision models in this section are closely related to game theory. Assume a game is played between the decision maker and Nature, with risk $r(\delta, \pi)$ as the payoff function, minimizing variable δ, and maximizing variable π. The expressions min max and max min obtained in (1.28) and (1.29) reflect the different knowledge available to the two players. The player with more information (who knows the variable selected by the opponent) has a potential advantage, as reflected in the following fundamental result, which states that minimax risk (among all $\delta \in \mathcal{D}$) is at least as large as worst-case Bayes risk.

PROPOSITION 1.1

$$\max_{\pi} \min_{\delta} r(\delta, \pi) \leq \min_{\delta} \max_{\pi} r(\delta, \pi). \tag{1.30}$$

Proof For all δ and π, we have

$$\max_{\pi'} r(\delta, \pi') \geq r(\delta, \pi).$$

In particular, this inequality holds for δ^* that achieves $\min_\delta \max_{\pi'} r(\delta, \pi')$. Thus, for all π, we have

$$\min_{\delta} \max_{\pi'} r(\delta, \pi') = \max_{\pi'} r(\delta^*, \pi') \geq r(\delta^*, \pi) \geq \min_{\delta} r(\delta, \pi). \tag{1.31}$$

In particular, for $\pi = \pi^*$ that achieves $\max_\pi \min_\delta r(\delta, \pi)$, (1.31) becomes

$$\min_{\delta} \max_{\pi} r(\delta, \pi) \geq \min_{\delta} r(\delta, \pi^*) = \max_{\pi} \min_{\delta} r(\delta, \pi).$$

This concludes the proof. □

1.7.3 Saddlepoint

In some cases, the minmax and maxmin values are equal, and a saddlepoint of the game is obtained, as illustrated in Figure 1.4. The following result is fundamental and its proof follows directly from the definitions [12, Ch. 5].

THEOREM 1.1 Saddlepoint Theorem for Deterministic Decision Rules. *If $r(\delta, \pi)$ admits a saddlepoint (δ^*, π^*), i.e.,*

$$r(\delta^*, \pi) \overset{(a)}{\leq} r(\delta^*, \pi^*) \overset{(b)}{\leq} r(\delta, \pi^*), \quad \forall \delta, \pi, \tag{1.32}$$

then δ^* is a minimax decision rule, π^* is a least-favorable prior, and

$$\max_{\pi} \min_{\delta} r(\delta, \pi) = \min_{\delta} \max_{\pi} r(\delta, \pi) = r(\delta^*, \pi^*).$$

A useful notion for identifying potential saddlepoints is introduced next.

DEFINITION 1.3 An *equalizer rule* δ is a decision rule such that $R_x(\delta)$ is independent of $x \in \mathcal{X}$.

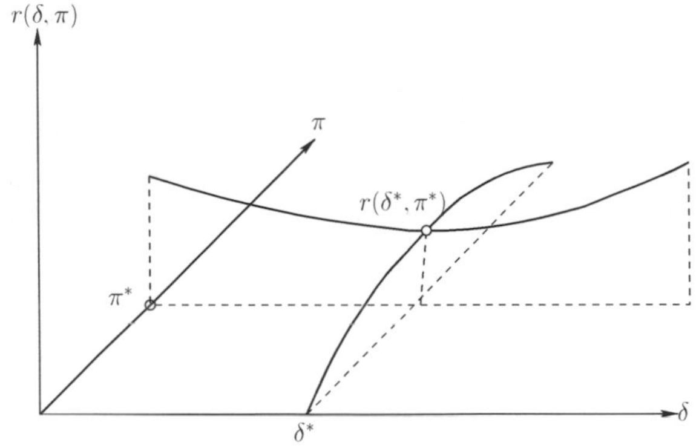

Figure 1.4 Saddlepoint of risk function $r(\delta, \pi)$

PROPOSITION 1.2 *If there exists an equalizer rule* δ^* *that is also a Bayes rule for some* π^**, then* (δ^*, π^*) *is a saddlepoint of the risk function* $r(\delta, \pi)$.

Proof To prove the proposition, we need to show that (1.32) holds. Since δ^* is by assumption a Bayes rule for π^*, the second inequality (b) holds. Since δ^* is also an equalizer rule, we have

$$r(\delta^*, \pi) = \int_{\mathcal{X}} R_x(\delta)\pi(x)d\nu(x) = R_x(\delta) \int_{\mathcal{X}} \pi(x)d\nu(x) = R_x(\delta) \quad \forall \pi, x, \tag{1.33}$$

hence the first inequality (a) holds with equality. □

1.7.4 Randomized Decision Rules

As already discussed, the player who has more knowledge generally has an advantage, which is reflected by the fact that strict inequality might hold in (1.30). In such cases one can remedy the situation and construct a better decision rule by randomly choosing within a set of deterministic decision rules.

DEFINITION 1.4 A *randomized* decision rule $\tilde{\delta}$ chooses randomly between a finite set of deterministic decision rules:[6]

$$\tilde{\delta} = \delta_\ell \text{ with probability } \gamma_\ell, \quad \ell = 1, \dots, L,$$

for some L and some $\{\gamma_\ell, \delta_\ell \in \mathcal{D},\ 1 \le \ell \le L\}$ such that $\sum_{\ell=1}^{L} \gamma_\ell$.

This definition implies that $\tilde{\delta}(y)$ is a discrete random variable taking values in $\mathcal{A}$ for each $y \in \mathcal{Y}$ with probabilities $\gamma_1, \dots, \gamma_L$. Therefore, given an observation $y \in \mathcal{Y}$ and a state $x \in \mathcal{X}$, the cost is now a random variable that can take values $C(\delta_\ell(y), x)$ with probabilities γ_ℓ for $\ell = 1, \dots, L$. Performance of a decision rule is evaluated by taking the expected value of that random variable, as in the following definition.

[6] We can define the randomized decision rule alternatively as a mapping from the observation space $\mathcal{Y}$ to a distribution on the action space $\mathcal{A}$ [11], where the distribution assigns probability mass γ_ℓ to the action $\delta_\ell(y)$, $\ell = 1, \dots, L$.

DEFINITION 1.5 The *expected cost* for randomized decision rule $\tilde{\delta}$ is

$$C(\tilde{\delta}(y), x) \triangleq \sum_{\ell=1}^{L} \gamma_l \, C(\delta_\ell(y), x),$$

for each $x \in \mathcal{X}$ and $y \in \mathcal{Y}$.

One can define conditional risk and Bayes risk in exactly the same way as before, if cost is replaced with expected cost. In particular, the conditional risk for randomized decision rule $\tilde{\delta}$ is $R_x(\tilde{\delta}) \triangleq \mathbb{E}_x[C(\tilde{\delta}(Y), x)]$ for each $x \in \mathcal{X}$, and the Bayes risk for $\tilde{\delta}$ under prior π is $r(\tilde{\delta}, \pi) = \int_{\mathcal{X}} R_x(\tilde{\delta})\pi(x)d\nu(x)$.

The following proposition follows directly from Definition 1.5.

PROPOSITION 1.3 *The conditional risk for randomized decision rule $\tilde{\delta}$ is*

$$R_x(\tilde{\delta}) = \sum_{\ell=1}^{L} \gamma_\ell R_x(\delta_\ell), \quad x \in \mathcal{X}.$$

The Bayes risk for $\tilde{\delta}$ under prior π is

$$r(\tilde{\delta}, \pi) = \sum_{\ell=1}^{L} \gamma_\ell \, r(\delta_\ell, \pi). \tag{1.34}$$

Note that a deterministic decision rule is a degenerate randomized decision rule with $L = 1$. Hence the set $\tilde{\mathcal{D}}$ of randomized decision rules contains the set $\mathcal{D}$ of deterministic decision rules, and optimizing over $\tilde{\mathcal{D}}$ will necessarily result in at least as good a decision rule as that obtained by optimizing over $\mathcal{D}$.

THEOREM 1.2 *Randomization does not improve Bayes rules:*

$$\min_{\delta \in \mathcal{D}} r(\delta, \pi) = \min_{\tilde{\delta} \in \tilde{\mathcal{D}}} r(\tilde{\delta}, \pi). \tag{1.35}$$

Proof See Exercise 1.8. □

Theorem 1.2 shows that randomization is not useful in a Bayesian setting. However, we can see in Example 1.2 that randomizing between δ_2 and δ_4 with equal probability results in a rule with minimax risk equal to 0.25. Thus, we see that randomization can improve minimax rules. We will explore this further in the next chapter. We will also see that randomization can yield better Neyman–Pearson rules for binary detection problems. More importantly finding an optimal rule within the class of randomized decision rules $\tilde{\mathcal{D}}$ can often be more tractable than finding an optimal rule within the class of deterministic decision rules $\mathcal{D}$. In particular, within the class of randomized decision rules, the minimax risk equals the worst-case Bayes risk and the following result holds.

THEOREM 1.3 Existence of Saddlepoint. *Suppose $\mathcal{Y}$ and $\mathcal{A}$ are finite sets. Then, the function $r(\tilde{\delta}, \pi)$ admits a saddlepoint $(\tilde{\delta}^*, \pi^*)$, i.e.,*

$$r(\tilde{\delta}^*, \pi) \le r(\tilde{\delta}^*, \pi^*) \le r(\tilde{\delta}, \pi^*), \quad \forall \tilde{\delta}, \pi, \tag{1.36}$$

and $\tilde{\delta}^*$ is a minimax (randomized) decision rule, π^* is a least-favorable prior, and

$$\max_{\pi} \min_{\delta \in \mathcal{D}} r(\delta, \pi) = \max_{\pi} \min_{\tilde{\delta} \in \tilde{\mathcal{D}}} r(\tilde{\delta}, \pi) = \min_{\tilde{\delta}} \max_{\pi} r(\tilde{\delta}, \pi) = r(\tilde{\delta}^*, \pi^*). \tag{1.37}$$

Proof Let $L = |\mathcal{A}|^{|\mathcal{Y}|}$ and let δ_ℓ, $1 \le \ell \le L$ range over all possible deterministic decision rules. Then the randomized decision rule $\tilde{\delta}$ is parameterized by the probability vector $\gamma \triangleq \{\gamma_\ell, 1 \le \ell \le L\}$. By Proposition 1.3, the risk function $r(\tilde{\delta}, \pi)$ of (1.34) is linear both in $\tilde{\delta}$ (via γ) and in π. Moreover the feasible sets for γ and π are probability simplices and are thus convex and compact. By the fundamental theorem of game theory, a saddlepoint exists [15]. The first equality in (1.37) follows from Theorem 1.1. □

The notion of equalizer rules (Definition 1.3) also applies to randomized decision rules. The following proposition is proved the same way as Proposition 1.2 (see Exercise 1.9).

PROPOSITION 1.4 *If there exists an equalizer rule $\tilde{\delta}^*$ that is also a Bayes rule for some π^*, then $(\tilde{\delta}^*, \pi^*)$ is a saddlepoint of the risk function $r(\tilde{\delta}, \pi)$.*

Example 1.4 Assume the following simplified model for penalty kicks in football. The penalty shooter aims either left or right. The goalie dives left or right at the very moment the ball is kicked. If the goalie dives in the same direction as the ball, the probability of scoring is 0.55. Otherwise the probability of scoring is 0.95. We denote by x the direction selected by the penalty shooter (left or right) and by a the direction selected by the goalie (also left or right). We define the cost $C(a, x) = 0.55$ if $a = x$, and $C(a, x) = 0.95$ otherwise. In this problem setup, no observations are available. A deterministic decision rule for the goalkeeper reduces to an action, hence there exist only two possible deterministic rules: δ_1 (dive left) and δ_2 (dive right).

The worst-case risk for both δ_1 and δ_2 is 0.95. (A goalie who systematically chooses one side will be systematically punished by his opponent, who would simply shoot in the opposite direction.) However, the randomized decision rule $\delta^\star$ that picks δ_1 and δ_2 with equal probability is an equalizer rule, since the conditional risk is $\frac{1}{2}(0.95 + 0.55) = 0.75$ for both $x =$ left and $x =$ right. Moreover, if x is random with uniform probability, we have $r(\delta_1) = r(\delta_2) = 0.75$ (both deterministic rules have the same risk) and therefore $r(\delta^\star) = 0.75$ as well. Since $\delta^\star$ is an equalizer rule as well as a Bayes rule, we conclude that $\delta^\star$ achieves the minimax risk and that the uniform distribution on x is the least-favorable prior.

Exercises

1.1 *Binary Decision Making.* Consider the following decision problem with $\mathcal{X} = \mathcal{A} = \{0, 1\}$. Suppose the cost function is given by

$$C(a, x) = \begin{cases} 0 & \text{if } a = x \\ 1 & \text{if } x = 0, a = 1 \\ 10 & \text{if } x = 1, a = 0. \end{cases}$$

The observation Y takes values in the set $\mathcal{Y} = \{1, 2, 3\}$ and the conditional pmfs of Y are:

$$p_0(1) = 0.4,\ p_0(2) = 0.6,\ p_0(3) = 0, \qquad p_1(1) = p_1(2) = 0.25,\ p_1(3) = 0.5.$$

(a) Is there a best decision rule based on conditional risks?

(b) Find Bayes (for equal priors) and minimax rules within the set of deterministic decision rules.

(c) Now consider the set of randomized decision rules. Find a Bayes rule (for equal priors). Also find a randomized rule whose maximum risk is smaller than that of the minimax rule of part (b).

1.2 *Health Insurance.* Find the minimax decision rule for the health insurance problem given in Example 1.3.

1.3 *Binary Communication with Erasures.* Let $\mathcal{X} = \{0, 1\}$, and $\mathcal{A} = \{0, 1, \mathrm{e}\}$. This example models communication of a bit $x \in \mathcal{X}$ with an erasure option at the receiver. Now suppose

$$p_j(y) = \frac{1}{\sqrt{2\pi\sigma^2}} \exp\left[-\frac{(y - (-1)^{j+1})^2}{2\sigma^2}\right], \quad j \in \mathcal{X},\ y \in \mathbb{R}.$$

That is, Y has distribution $\mathcal{N}(-1, \sigma^2)$ when the state is 0, and Y has distribution $\mathcal{N}(1, \sigma^2)$ when the state is 1. Assume a cost structure

$$C(a, j) = \begin{cases} 0 & \text{if } a = 0,\ j = 0 \text{ or } a = 1,\ j = 1 \\ 1 & \text{if } a = 1,\ j = 0 \text{ or } a = 0,\ j = 1 \\ c & \text{if } a = \mathrm{e}. \end{cases}$$

Furthermore, assume that the two states are equally likely.

(a) First assume that $c < 0.5$. Show that the Bayes rule for this problem has the form

$$\delta_{\mathrm{B}}(y) = \begin{cases} 0 & y \le -t \\ \mathrm{e} & -t < y < t \\ 1 & y \ge t. \end{cases}$$

Also give an expression for t in terms of the parameters of the problem.

(b) Now find $\delta_{\mathrm{B}}(y)$ when $c \ge 0.5$.

1.4 *Medical Diagnosis.* Let X be the health state of a patient, who is healthy ($x = 0$) with probability 0.8 and sick ($x = 1$) with probability 0.2. A medical test produces a measurement Y following a Poisson law with parameter $x + 1$. The physician may take one of three actions: declare the patient healthy ($a = 0$), sick ($a = 1$), or order a new medical test ($a = \mathrm{t}$). The cost structure is as follows:

$$C(a, x) = \begin{cases} 0 & \text{if } a = x \\ 1 & \text{if } a = 1, x = 0 \\ e & \text{if } a = 0, x = 1 \\ e/4 & \text{if } a = \mathrm{t}. \end{cases}$$

Derive the Bayes decision rule for the physician.

1.5 *Optimal Routing.* A driver needs to decide between taking the highway (h) or side streets (s) based on prior information and input from a map application on her phone about congestion on the highway. If the highway is congested, it would be worse than taking the side streets, but if the highway has normal traffic it would better than taking the side streets. Let X denote the state of the highway, with value 0 denoting normal traffic, and value 1 denoting congestion. The prior information that the driver has tells her that at the time she is planning to drive

$$\pi(0) = \mathrm{P}\{X = 0\} = 0.3, \quad \pi(1) = \mathrm{P}\{X = 1\} = 0.7.$$

The map application displays a recommendation (using colors such as blue and red), which is modeled as a binary observation $Y \in \{0, 1\}$, with

$$p_0(0) = 0.8, \;\; p_0(1) = 0.2, \quad p_1(0) = 0.4, \;\; p_1(1) = 0.6.$$

The cost function that penalizes the driver's actions is given by:

$$C(a, x) = \begin{cases} 10 & \text{if } a = h, x = 1 \\ 1 & \text{if } a = h, x = 0 \\ 5 & \text{if } a = s, x = 0 \\ 2 & \text{if } a = s, x = 1. \end{cases}$$

Derive the Bayes decision rule for the driver.

1.6 *Job Application.* Don, Marco, and Ted are CEO candidates for their company. Each candidate incurs a cost if another candidate is selected for the job. If Don is selected, the cost to Marco and Ted is 10. If Marco or Ted is selected, the cost to the other two candidates is 2. The cost to a candidate is 0 if he is selected.

(a) The selection will be made by a committee; it is believed that the committee will select Don with probability 0.8 and Marco with probability 0.1. Compute the expected cost for each candidate.
(b) While Marco and Ted are underdogs, if either one of them drops out of contention, Don's probability of being selected will drop to 0.6. Marco and Ted discuss that possibility and have three possible actions (Marco drops out, Ted drops out, or both stay in the race). What decision minimizes their expected cost?

1.7 *Wise Investor.* Ann has received \$1000 and must decide whether to invest it in the stock market or to safely put it in the bank where it collects no interest. Assume the economy is either in recession or in expansion, and will remain in the same state for the next year. During that period, the stock market will go up 10 percent if the economy expands, and down 5 percent otherwise. The economy has a 40 percent prior probability to be in expansion. An investment guru claims to know the state of the economy and reveals it to Ann. Whatever the state is, there is a probability $\lambda \in [0, 1]$ that the guru is incorrect. Ann will act on the binary information provided by the guru but is well aware of the underlying probabilistic model. Give the Bayesian and minimax decision rules and Ann's expected financial gains in terms of λ. Perform a sanity check to verify your answer is correct in the extreme cases $\lambda = 0$ and $\lambda = 1$.

1.8 *Randomized Bayes Rules.* Prove Theorem 1.2.

1.9 *Equalizer Rule.* Prove Proposition 1.4.

References

[1] B. Hajek, *Random Processes for Engineers*, Cambridge University Press, 2015.

[2] R. A. Horn and C. R. Johnson, *Matrix Analysis*, Cambridge University Press, 1990.

[3] S. Boyd and L. Vandenberghe, *Convex Optimization*, Cambridge University Press, 2004.

[4] H. L. Van Trees, *Detection, Estimation and Modulation Theory, Part I*, Wiley, 1968.

[5] H. V. Poor, *An Introduction to Signal Detection and Estimation*, 2nd edition, Springer-Verlag, 1994.

[6] S. M. Kay, *Fundamentals of Statistical Signal Processing: Detection Theory*, Prentice-Hall, 1998.

[7] B. C. Levy, *Principles of Signal Detection and Parameter Estimation*, Springer-Verlag, 2008.

[8] E. L. Lehman, *Testing Statistical Hypotheses*, 2nd edition, Springer-Verlag, 1986.

[9] E. L. Lehman and G. Casella, *Theory of Point Estimation*, 2nd edition, Springer-Verlag, 1998.

[10] A. Wald, *Statistical Decision Functions*, Wiley, New York, 1950.

[11] T. S. Ferguson, *Mathematical Statistics: A Decision Theoretic Approach*, Academic Press, 1967.

[12] J. O. Berger, *Statistical Decision Theory and Bayesian Analysis*, Springer-Verlag, 1985.

[13] L. Devroye, L. Györfig, and G. Lugosi, *A Probabilistic Theory of Pattern Recognition*, Springer, 1996.

[14] T. M. Cover and J. A. Thomas, *Elements of Information Theory*, 2nd edition, 2006.

[15] J. von Neumann, "Sur la théorie des jeux," *Comptes Rendus Acad. Sci.*, pp. 1689–1691, 1928.

Part I

Hypothesis Testing

2 Binary Hypothesis Testing

In this chapter we cover the foundations of binary hypothesis testing or binary detection. The three basic formulations of the binary hypothesis testing problem, namely, Bayesian, minimax, and Neyman–Pearson, are described along with illustrative examples.

2.1 General Framework

We first specialize the general decision-theoretic framework of Section 1.5 to the binary hypothesis testing problem. In particular, the six ingredients of the framework specialize as follows:

1. The state space, $\mathcal{X} = \{0, 1\}$, and its elements will be generically denoted by $j \in \mathcal{X}$. We refer equivalently to "state j" or to "hypothesis H_j."
2. The action space $\mathcal{A}$ is the same as the state space $\mathcal{X}$, and its elements will be generically denoted by $i \in \mathcal{A}$.
3. The cost function $C : \mathcal{A} \times \mathcal{X} \to \mathbb{R}$ can be viewed as a square matrix whose element C_{ij} is the cost of declaring hypothesis H_i is true when H_j is true. The case of uniform costs, $C_{ij} = \mathbb{1}\{i \neq j\}$, will be particularly insightful.
4. The decision is made based on observation Y taking values in $\mathcal{Y}$. The observations could be continuous (with $\mathcal{Y} \subset \mathbb{R}^n$) or discrete.
5. The conditional pdf (or pmf) for the observations given state j (hypothesis H_j) is denoted by $p_j(y)$, $y \in \mathcal{Y}$.
6. We will consider both deterministic decision rules $\delta \in \mathcal{D}$ and randomized decision rules $\tilde{\delta} \in \tilde{\mathcal{D}}$. Note there are $2^{|\mathcal{Y}|}$ possible deterministic decision rules when $\mathcal{Y}$ is finite. Any deterministic decision rule δ partitions the observation space into disjoint sets $\mathcal{Y}_0$ and $\mathcal{Y}_1$, corresponding to decisions $\delta(y) = 0$ and $\delta(y) = 1$, respectively.

The hypotheses H_0 and H_1 are sometimes referred to as the *null hypothesis* and the *alternative hypothesis*, respectively. The regions $\mathcal{Y}_0$ and $\mathcal{Y}_1$ are then referred to as the *acceptance region* and the *rejection region* for H_0, respectively.

The conditional risk associated with deterministic decision rule δ when the state is j is

$$
\begin{aligned}
R_j(\delta) = \mathbb{E}_j\left[C(\delta(Y), j)\right] &= \int_{\mathcal{Y}} C(\delta(y), j) p_j(y) d\mu(y) \\
&= C_{0j}\, \mathrm{P}_j(\mathcal{Y}_0) + C_{1j}\, \mathrm{P}_j(\mathcal{Y}_1), \quad j = 0, 1.
\end{aligned} \tag{2.1}
$$

We define the probability of false alarm $\mathrm{P_F}(\delta)$ and the probability of miss $\mathrm{P_M}(\delta)$ as

$$
\mathrm{P_F}(\delta) \triangleq \mathrm{P}_0(\mathcal{Y}_1), \quad \mathrm{P_M}(\delta) \triangleq \mathrm{P}_1(\mathcal{Y}_0). \tag{2.2}
$$

ASSUMPTION 2.1 *The cost of a correct decision about the state is strictly smaller than that of a wrong decision:*

$$
C_{00} < C_{10}, \qquad C_{11} < C_{01}.
$$

Under this assumption, the following inequalities hold for any deterministic decision rule δ:[1]

$$
R_0(\delta) = C_{00}\, \mathrm{P}_0(\mathcal{Y}_0) + C_{10}\, \mathrm{P}_0(\mathcal{Y}_1) \ge C_{00}\, \mathrm{P}_0(\mathcal{Y}_0) + C_{00}\, \mathrm{P}_0(\mathcal{Y}_1) = C_{00}, \tag{2.3}
$$

$$
R_1(\delta) = C_{01}\, \mathrm{P}_1(\mathcal{Y}_0) + C_{11}\, \mathrm{P}_1(\mathcal{Y}_1) \ge C_{11}\, \mathrm{P}_1(\mathcal{Y}_0) + C_{11}\, \mathrm{P}_1(\mathcal{Y}_1) = C_{11}. \tag{2.4}
$$

Consider the rules $\delta_0(y) = 0$ for all y, and $\delta_1(y) = 1$ for all y. It is clear that δ_0 achieves the lower bound in (2.3) and δ_1 achieves the lower bound in (2.4). However, $R_1(\delta_0)$ and $R_0(\delta_1)$ are typically unacceptably large because the rules δ_0 and δ_1 ignore the observations.

Now, we explore under what conditions one can find a *best* rule based on Definition 1.1. For a rule $\delta^\star$ to be best, it must be *better* than both δ_0 and δ_1, and therefore must satisfy $R_0(\delta^\star) = C_{00}$ and $R_1(\delta^\star) = C_{11}$. It is easy to show that such a $\delta^\star$ exists if and only if p_0 and p_1 have disjoint supports as in Example 1.1 (see Exercise 2.1).

2.2 Bayesian Binary Hypothesis Testing

Assuming priors π_0, π_1 for the states, the Bayes risk for a decision rule δ is given by

$$
\begin{aligned}
r(\delta) &= \pi_0 R_0(\delta) + \pi_1 R_1(\delta) \\
&= \pi_0\, (C_{10} - C_{00})\, \mathrm{P}_0(\mathcal{Y}_1) + \pi_1\, (C_{01} - C_{11})\, \mathrm{P}_1(\mathcal{Y}_0) \\
&\qquad + \pi_0\, C_{00} + \pi_1\, C_{11}.
\end{aligned} \tag{2.5}
$$

The posterior probability $\pi(j|y)$ of state j, given observation y, is given by

$$
\pi(j|y) = \frac{p_j(y)\pi_j}{p(y)}. \tag{2.6}
$$

The posterior cost of decision $i \in \mathcal{X}$, given observation y, is given by

$$
C(i|y) \triangleq \sum_{j \in \mathcal{X}} \pi(j|y) C_{ij}. \tag{2.7}
$$

[1] These lower bounds can be shown to hold for randomized rules as well.

Using (1.25), we can find a Bayes decision rule for binary detection as:

$$\delta_{\mathrm{B}}(y) = \arg\min_{i\in\{0,1\}} C(i|y) = \begin{cases} 1 & < \\ \text{any} \quad : C(1|y) & = \quad C(0|y), \\ 0 & > \end{cases} \tag{2.8}$$

where "any" can be arbitrarily chosen to be 0 or 1, without any effect on Bayes risk. Hence the Bayes solution need not be unique. By (2.7) we have

$$C(0|y) - C(1|y) = \pi(1|y)[C_{01} - C_{11}] - \pi(0|y)[C_{10} - C_{00}]$$

and thus (2.8) yields

$$\delta_{\mathrm{B}}(y) = \begin{cases} 1 & > \\ \text{any} \quad : \pi(1|y)[C_{01} - C_{11}] & = \quad \pi(0|y)[C_{10} - C_{00}]. \\ 0 & < \end{cases} \tag{2.9}$$

Under Assumption 2.1, it follows from (2.6) and (2.9) that

$$\delta_{\mathrm{B}}(y) = \begin{cases} 1 & > \\ \text{any} \quad : \dfrac{p_1(y)}{p_0(y)} & = \quad \dfrac{\pi_0}{\pi_1}\dfrac{C_{10} - C_{00}}{C_{01} - C_{11}}. \\ 0 & < \end{cases} \tag{2.10}$$

2.2.1 Likelihood Ratio Test

DEFINITION 2.1 The *likelihood ratio* is given by

$$L(y) = \frac{p_1(y)}{p_0(y)}, \quad y \in \mathcal{Y}, \tag{2.11}$$

with the notational convention that $\frac{0}{0} = 0$, and $\frac{x}{0} = \infty$, for $x > 0$.

If we further define the *threshold* η by:

$$\eta = \frac{\pi_0}{\pi_1}\frac{C_{10} - C_{00}}{C_{01} - C_{11}}, \tag{2.12}$$

then we can write (2.10) as

$$\delta_{\mathrm{B}}(y) = \begin{cases} 1 & > \\ \text{any} \quad : L(y) & = \quad \eta. \\ 0 & < \end{cases} \tag{2.13}$$

Hence the Bayes rule is a *likelihood ratio test* (LRT) with threshold η.

The likelihood ratio $L(y)$ depends only on the observations, not the cost matrix or the prior. The LRT threshold η depends only on the cost matrix and the prior – not the observations. It is remarkable that $L(y)$ (a single number) contains all the information about Y that is needed to make optimal decisions.

2.2.2 Uniform Costs

For uniform costs, $C_{00} = C_{11} = 0$, and $C_{01} = C_{10} = 1$. The posterior risks (2.7) are $C(1|y) = \pi(0|y)$ and $C(0|y) = \pi_1(y)$. Therefore, (2.12) simplifies to $\eta = \frac{\pi_0}{\pi_1}$, and

$$\delta_B(y) = \begin{cases} 1 & & > & \\ \text{any} & : \pi(1|y) & = & \pi(0|y). \\ 0 & & < & \end{cases} \tag{2.14}$$

Thus, for uniform costs, the Bayes rule is a *maximum a posteriori* (MAP) rule.

Specializing (2.5) to the case of uniform costs, we obtain the Bayes risk of any decision rule δ as

$$r(\delta) = \pi_0 \mathrm{P}_0(\mathcal{Y}_1) + \pi_1 \mathrm{P}_1(\mathcal{Y}_0),$$

which is average probability of error $\mathrm{P_e}(\delta)$. Thus, for uniform costs, the Bayes rule is also a *minimum probability of error* (MPE) rule.

Finally, if we have uniform costs and equal priors ($\pi_0 = \pi_1 = \frac{1}{2}$) then $\eta = 1$, and it follows from (2.12), (2.11), and (2.13) that the Bayes rule is a *maximum likelihood* (ML) decision rule.

2.2.3 Examples

Example 2.1 *Signal Detection in Gaussian Noise.* This detection problem arises in a number of engineering applications, including radar and digital communications, and can be described by the hypothesis test

$$\begin{cases} H_0 : Y = Z \\ H_1 : Y = \mu + Z, \end{cases} \tag{2.15}$$

where μ is a real-valued constant, and Z is a zero-mean Gaussian random variable with variance σ^2, denoted by $Z \sim \mathcal{N}(0, \sigma^2)$. Without loss of generality, we may assume that $\mu > 0$.

The conditional pdfs for the observation Y are given by

$$p_0(y) = \frac{1}{\sqrt{2\pi\sigma^2}} \exp\left\{-\frac{y^2}{2\sigma^2}\right\},$$
$$p_1(y) = \frac{1}{\sqrt{2\pi\sigma^2}} \exp\left\{-\frac{(y-\mu)^2}{2\sigma^2}\right\}, \quad y \in \mathbb{R}.$$

Hence the likelihood ratio is given by

$$L(y) = \frac{p_1(y)}{p_0(y)} = \exp\left\{-\frac{(y-\mu)^2 - y^2}{2\sigma^2}\right\} = \exp\left\{\frac{\mu}{\sigma^2}\left(y - \frac{\mu}{2}\right)\right\}.$$

Taking the logarithm or applying any other monotonically increasing function to both sides of the LRT (2.13) does not change the decision. Hence (2.13) is equivalent to a threshold test on the observation y:

$$\delta_B(y) = \begin{cases} 1 : & y \geq \tau \\ 0 : & y < \tau, \end{cases} \tag{2.16}$$

with threshold

$$\tau = \frac{\sigma^2}{\mu} \ln \eta + \frac{\mu}{2}.$$

Note we have ignored the "any" case which has zero probability under both H_0 and H_1 here. The acceptance and rejection regions for the test are the half-lines $\mathcal{Y}_1 = [\tau, \infty)$ and $\mathcal{Y}_0 = (-\infty, \tau)$, respectively, as illustrated in Figure 2.1.

Performance Evaluation Denote by

$$\Phi(t) = \int_{-\infty}^{t} \frac{1}{\sqrt{2\pi}} e^{-x^2/2} \, dx, \quad t \in \mathbb{R}, \tag{2.17}$$

the cdf of the $\mathcal{N}(0, 1)$ random variable, and by Q its complement, i.e., $Q(t) = 1 - \Phi(t) = \Phi(-t)$, for $t \in \mathbb{R}$. The Q function is plotted in Figure 2.2. Then the probability of false alarm and the probability of miss are respectively given by

$$\mathrm{P_F}(\delta_B) = \mathrm{P_0}(\mathcal{Y}_1) = \mathrm{P_0}\{Y \geq \tau\} = Q\left(\frac{\tau}{\sigma}\right), \tag{2.18}$$

$$\mathrm{P_M}(\delta_B) = \mathrm{P_1}(\mathcal{Y}_0) = \mathrm{P_1}\{Y < \tau\} = Q\left(\frac{\mu - \tau}{\sigma}\right). \tag{2.19}$$

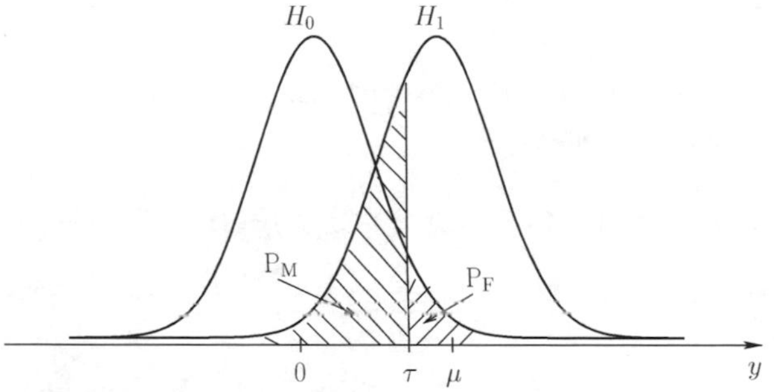

Figure 2.1 Gaussian binary hypothesis testing

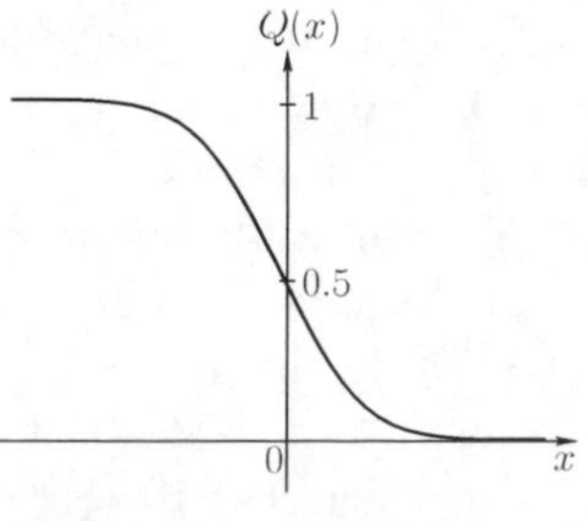

Figure 2.2 The Q function

Table 2.1 All possible decision rules for Example 2.2, with the corresponding rejection and acceptance regions, conditional risks under uniform costs, Bayes risk, and worst-case risk

δ	$\mathcal{Y}_0$	$\mathcal{Y}_1$	$R_0(\delta)$	$R_1(\delta)$	$\pi_0 R_0(\delta) + \pi_1 R_1(\delta)$	$\max\{R_0(\delta), R_1(\delta)\}$
δ_1	$\{1, 2, 3\}$	$\emptyset$	0	1	π_1	1
$\boxed{\delta_2}$	$\{1, 2\}$	$\{3\}$	0	0.5	$\boxed{\pi_1/2}$	$\boxed{0.5}$
δ_3	$\{1, 3\}$	$\{2\}$	0.5	0.5	$1/2$	0.5
$\boxed{\delta_4}$	$\{1\}$	$\{2, 3\}$	0.5	0	$\boxed{\pi_0/2}$	$\boxed{0.5}$
δ_5	$\{2, 3\}$	$\{1\}$	0.5	1	$\pi_1 + \pi_0/2$	1
δ_6	$\{2\}$	$\{1, 3\}$	0.5	0.5	$1/2$	0.5
δ_7	$\{3\}$	$\{1, 2\}$	1	0.5	$\pi_0 + \pi_1/2$	1
δ_8	$\emptyset$	$\{1, 2, 3\}$	1	0	π_0	1

Hence, the average error probability is given by

$$\begin{aligned} \mathrm{P_e}(\delta_\mathrm{B}) &= \pi_0 \mathrm{P_F}(\delta_\mathrm{B}) + \pi_1 \mathrm{P_M}(\delta_\mathrm{B}) \\ &= \pi_0 Q\left(\frac{\mu}{2\sigma} + \frac{\sigma}{\mu} \ln \eta\right) + \pi_1 Q\left(\frac{\mu}{2\sigma} - \frac{\sigma}{\mu} \ln \eta\right). \end{aligned} \tag{2.20}$$

The error probability depends on μ and σ only via their ratio, $d \triangleq \mu/\sigma$. If equal priors are assumed ($\pi_0 = \pi_1 = \frac{1}{2}$), then $\eta = 1$, $\tau = \frac{\mu}{2}$, and $\mathrm{P_e}(\delta_\mathrm{B}) = Q(d/2)$, which is a decreasing function of d. The worst value of d is $d = 0$, in which case the two hypotheses are indistinguishable and $\mathrm{P_e}(\delta_\mathrm{B}) = \frac{1}{2}$. For large d, we have $Q(d/2) \sim \frac{\exp\{-d^2/8\}}{d\sqrt{\pi/2}}$, hence $\mathrm{P_e}(\delta_\mathrm{B})$ vanishes exponentially with d.

Example 2.2 *Discrete Observations.* We revisit Example 1.2 with $\mathcal{X} = \mathcal{A} = \{0, 1\}$, $\mathcal{Y} = \{1, 2, 3\}$, and conditional pmfs:

$$\begin{aligned} p_0(1) &= \frac{1}{2}\ p_0(2) = \tfrac{1}{2}\ p_0(3) = 0, \\ p_1(1) &= 0\ p_1(2) = \tfrac{1}{2}\ p_1(3) = \frac{1}{2}. \end{aligned}$$

As we saw before, there are $2^3 = 8$ possible decision rules. We rewrite these decision rules using the decision regions $\mathcal{Y}_0$ and $\mathcal{Y}_1$ in Table 1.2, and provide their conditional risks for uniform costs and Bayes risks for arbitrary priors.

By inspection of Table 2.1, rules $\delta_1, \delta_3, \delta_5, \delta_6, \delta_7$ cannot be better than δ_2 for any value of π_0, and are strictly worse for any $\pi_0 \notin \{0, 1\}$. Similarly, rules $\delta_3, \delta_5, \delta_6, \delta_7, \delta_8$ cannot be better than δ_4 for any value of π_0, and are strictly worse for any $\pi_0 \notin \{0, 1\}$. This leaves only $\delta_2 = \mathbb{1}\{3\}$ and $\delta_4 = \mathbb{1}\{2, 3\}$ as admissible rules for $\pi_0 \notin \{0, 1\}$. Observe that for both rules, one conditional risk is equal to 0 and the other one is 0.5. The rules have respective Bayes risks $\mathrm{P_e}(\delta_2) = \frac{1-\pi_0}{2}$ and $\mathrm{P_e}(\delta_4) = \frac{\pi_0}{2}$. Hence δ_4 is a Bayes rule for $0 \le \pi_0 \le \frac{1}{2}$, and is unique for $0 < \pi_0 < \frac{1}{2}$; and δ_2 is a Bayes rule for $\frac{1}{2} \le \pi_0 \le 1$, and is unique for $\frac{1}{2} < \pi_0 < 1$. The Bayes risk is equal to $\mathrm{P_e}(\delta_\mathrm{B}) = \frac{1}{2}\min\{\pi_0, 1 - \pi_0\}$.

While generally inefficient, exhaustive listing and comparison of all decision rules is easy here because $\mathcal{Y}$ is small. A more principled approach is to derive the LRT. Here the likelihood ratio is given by

$$L(y) = \frac{p_1(y)}{p_0(y)} = \begin{cases} 0 & : \ y = 1 \\ 1 & : \ y = 2 \\ \infty & : \ y = 3. \end{cases} \tag{2.21}$$

Under uniform costs, the LRT threshold is $\eta = \frac{\pi_0}{1-\pi_0}$. Thus we see that δ_2 is a Bayes rule for $0 \le \pi_0 \le \frac{1}{2}$, and δ_4 is a Bayes rule for $\frac{1}{2} \le \pi_0 \le$. For $\pi_0 = \frac{1}{2}$, we have $\eta = 1$. By (2.13), the decision in the case $y = 2$ is immaterial, and again we see that both δ_2 and δ_4 are Bayes rules.

Example 2.3 *Binary Communication Channel.* Consider the binary communication channel with input $j \in \mathcal{X} = \{0, 1\}$ and output $y \in \mathcal{Y} = \{0, 1\}$ viewed as transmitted and received bits, respectively. The channel has crossover error probabilities λ_0 and λ_1 shown in Figure 2.3. There $p_0(1) = \lambda_0$ and $p_1(0) = \lambda_1$.

There are $2^2 = 4$ possible decision rules. Their conditional risks for uniform costs are shown in Table 2.2. There $\lambda = \frac{1}{2}(\lambda_0 + \lambda_1)$, $\overline{\lambda} = \max(\lambda_0, \lambda_1)$, and $\underline{\lambda} = \min(\lambda_0, \lambda_1)$. Clearly $\underline{\lambda} \le \lambda \le \overline{\lambda}$.

The likelihood ratio is given by

$$L(y) = \frac{p_1(y)}{p_0(y)} = \begin{cases} \frac{\lambda_1}{1-\lambda_0} & : \ y = 0 \\ \frac{1-\lambda_1}{\lambda_0} & : \ y = 1. \end{cases}$$

The LRT is given by (2.13). Assume now uniform costs and equal priors, in which case $\eta = 1$ again.

Case I: $\lambda_0 + \lambda_1 < 1$. Hence $\frac{\lambda_1}{1-\lambda_0} < 1 < \frac{1-\lambda_1}{\lambda_0}$, and the Bayes rule is given by $\delta_B(y) = y$ (output the received bit). This is rule δ_2 in Table 2.2. Then

$$P_F(\delta_B) = p_0(1) = \lambda_0, \quad P_M(\delta_B) = p_1(0) = \lambda_1, \quad P_e(\delta_B) = \frac{\lambda_0 + \lambda_1}{2}.$$

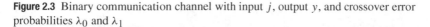

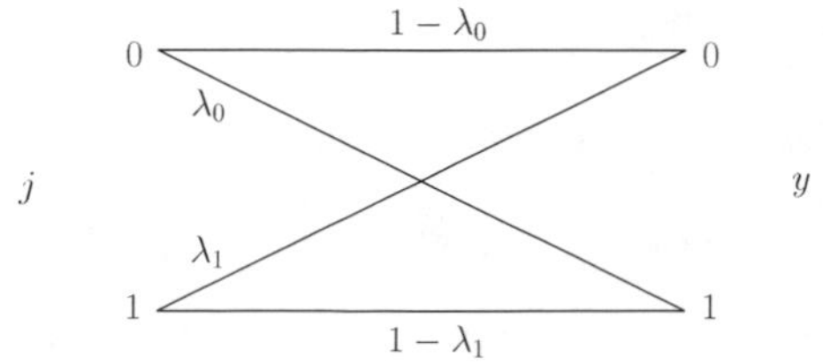

Figure 2.3 Binary communication channel with input j, output y, and crossover error probabilities λ_0 and λ_1

Table 2.2 All possible decision rules for Example 2.3, with the corresponding rejection and acceptance regions, conditional risks under uniform costs, Bayes risk under the uniform prior, and worst-case risk

δ	$\mathcal{Y}_0$	$\mathcal{Y}_1$	$R_0(\delta)$	$R_1(\delta)$	$0.5(R_0(\delta) + R_1(\delta))$	$\max(R_0(\delta), R_1(\delta))$
δ_1	$\{0, 1\}$	$\emptyset$	0	1	0.5	1
δ_2	$\{0\}$	$\{1\}$	λ_0	λ_1	λ	$\overline{\lambda}$
δ_3	$\{1\}$	$\{0\}$	$1 - \lambda_0$	$1 - \lambda_1$	$1 - \lambda$	$1 - \underline{\lambda}$
δ_4	$\emptyset$	$\{0, 1\}$	1	0	0.5	1

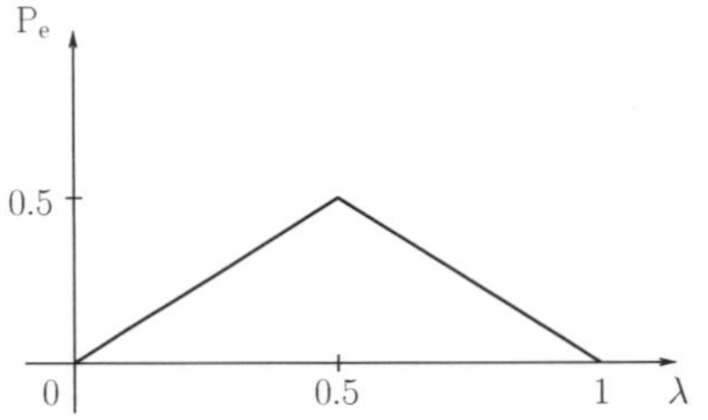

Figure 2.4 Risk (error probability) as a function of λ for the binary communication channel

Case II: $\lambda_0 + \lambda_1 > 1$. The same derivation yields $\delta_B(y) = 1 - y$ (flip the received bit, i.e., rule δ_3 in Table 2.2),

$$\mathrm{P_F}(\delta_\mathrm{B}) = p_0(0) = 1 - \lambda_0, \quad \mathrm{P_M}(\delta_\mathrm{B}) = p_1(1) = 1 - \lambda_1, \quad \mathrm{P_e}(\delta_\mathrm{B}) = 1 - \frac{\lambda_0 + \lambda_1}{2}.$$

Case III: $\lambda_0 + \lambda_1 = 1$. Here all possible decisions have the same risk and are Bayes rules, and $\mathrm{P_e}(\delta_\mathrm{B}) = \frac{1}{2}$. The channel is so noisy that the observation does not convey any information about the transmitted bit.

Combining all three cases, we express the error probability as

$$\mathrm{P_e}(\delta_\mathrm{B}) = \min\{\lambda, 1 - \lambda\},$$

where $\lambda = (\lambda_0 + \lambda_1)/2$. As shown in Figure 2.4, the error probability is a continuous function of λ, but the decision rule changes abruptly at $\lambda = 0.5$, due to the transition from Case I to Case II.

2.3 Binary Minimax Hypothesis Testing

The deterministic minimax rule is obtained from (1.18) as

$$\delta_\mathrm{m} = \arg \min_{\delta \in \mathcal{D}} R_{\max}(\delta), \tag{2.22}$$

where $R_{\max}(\delta) = \max\{R_0(\delta), R_1(\delta)\}$. Several approaches can be considered to find minimax rules. The most naive one is to evaluate every possible decision rule, as done in Tables 1.2 and 2.2. A second approach is to seek an equalizer rule that is also a Bayes rule; by Proposition 1.2 such a rule is a minimax rule. The third approach, which is generally an improvement over (2.22), consists of seeking a randomized decision rule:

$$\tilde{\delta}_{\mathrm{m}} = \arg\min_{\tilde{\delta} \in \tilde{\mathcal{D}}} R_{\max}(\tilde{\delta}), \tag{2.23}$$

where $R_{\max}(\tilde{\delta}) = \max\{R_0(\tilde{\delta}), R_1(\tilde{\delta})\}$.

2.3.1 Equalizer Rules

To develop intuition, we first solve the minimax problem for the three problems of Section 2.2.3.

We have established in Example 2.1 that, for Gaussian hypothesis testing, Bayes rules are threshold rules of the form

$$y \underset{H_0}{\overset{H_1}{\gtrless}} \tau = \frac{\mu}{2} + \frac{\sigma^2}{\mu} \ln \frac{\pi_0}{\pi_1}. \tag{2.24}$$

The conditional risks are given by

$$R_0(\delta) = Q\left(\frac{\tau}{\sigma}\right) \quad \text{and} \quad R_1(\delta) = Q\left(\frac{\mu - \tau}{\sigma}\right).$$

To be not only a Bayes rule but also an equalizer rule, δ must be of the form (2.24) and satisfy $R_0(\delta) = R_1(\delta)$, i.e., τ satisfies the equality $\frac{\tau}{\sigma} = \frac{\mu-\tau}{\sigma}$. Therefore $\tau = \frac{\mu}{2}$, which corresponds to the least favorable prior $\pi_0^* = \frac{1}{2}$. The corresponding Bayes rule δ^* is given by (2.24) with threshold $\tau = \frac{\mu}{2}$. Since δ^* is both an equalizer rule and a Bayes rule for prior π^*, it follows from Proposition 1.2 that (δ^*, π^*) is a saddlepoint of the risk function and therefore δ^* is a minimax rule. The minimax risk is $R_{\max}(\delta^*) = Q\left(\frac{\mu}{2\sigma}\right)$, which is the same as Bayes risk under the least-favorable prior.

For the binary communication problem (Example 2.3), in the symmetric case when $\lambda_0 = \lambda_1$, we have $\underline{\lambda} = \lambda = \overline{\lambda}$. For $\lambda \leq \frac{1}{2}$, decision rule $\delta^*(y) = y$ is an equalizer rule and is a Bayes rule under the uniform prior π^*. It follows again from Proposition 1.2 that δ^* is a minimax rule and π^* a least-favorable prior. Likewise, for $\lambda \geq \frac{1}{2}$, the deterministic decision rule $\delta^*(y) = 1 - y$ is minimax. Hence the minimax risk is $\min(\lambda, 1 - \lambda)$ for all $\lambda \in [0, 1]$. The minimax rule is unique for $\lambda \neq \frac{1}{2}$, and there are two deterministic minimax rules for $\lambda = \frac{1}{2}$.

For the other problem with discrete observations (Example 2.2) it is seen from Table 1.2 that no deterministic equalizer rule exists, and that the minimax risk over $\mathcal{D}$ is $\frac{1}{2}$, which is *twice* the Bayes risk under the least-favorable prior ($\pi_0^* = \frac{1}{2}$). However, consider the randomized detection rule $\tilde{\delta}^*$ that picks δ_2 or δ_4 with equal probability ($\gamma_2 = \gamma_4 = \frac{1}{2}$). In this case we have $R_j(\tilde{\delta}^*) = \frac{1}{2}[R_j(\delta_2) + R_j(\delta_4)] = \frac{1}{4}$ for both $j = 0$ and $j = 1$, hence $\tilde{\delta}^*$ is a (randomized) equalizer rule. Moreover $\tilde{\delta}^*$ is also a Bayes rule for $\pi_0^* = \frac{1}{2}$. It follows from Proposition 1.4 that $(\tilde{\delta}^*, \pi_0^*)$ is a saddlepoint of the risk function and therefore a minimax rule. The minimax risk is $R_{\max}(\tilde{\delta}^*) = \frac{1}{4}$, which is the same as Bayes risk under the least-favorable prior. Hence randomized decision rules outperform deterministic rules in the minimax sense here.

In the next subsections we present a systematic procedure for finding minimax rules, again by relating the minimax problem to a Bayesian problem.

2.3.2 Bayes Risk Line and Minimum Risk Curve

Denote by δ_{π_0} the Bayes rule under prior π_0 on H_0. We begin with the following definitions.

DEFINITION 2.2 For any $\delta \in \mathcal{D}$, the *Bayes risk line* is

$$r(\pi_0; \delta) \triangleq \pi_0 \, R_0(\delta) + (1 - \pi_0) \, R_1(\delta), \quad \pi_0 \in [0, 1]. \tag{2.25}$$

DEFINITION 2.3 The *Bayes minimum-risk curve* is

$$V(\pi_0) \triangleq \min_{\delta \in \mathcal{D}} r(\pi_0; \delta) = r(\pi_0; \delta_{\pi_0}), \quad \pi_0 \in [0, 1]. \tag{2.26}$$

Bayes risk lines and the minimum risk curve are illustrated in Figure 2.5. The end points of the Bayes risk line are $r(0; \delta) = R_1(\delta)$ and $r(1; \delta) = R_0(\delta)$. The following lemma states some useful properties of $V(\pi_0)$.

LEMMA 2.1

(i) V is a concave (continuous) function on $[0, 1]$.
(ii) If $\mathcal{D}$ is finite, then V is piecewise linear.
(iii) $V(0) = C_{11}$ and $V(1) = C_{00}$.

Proof The minimum of a collection of linear functions is concave; therefore, the concavity of V follows from the fact that each of the risk lines $r(\pi_0; \delta)$ is linear in π_0. The functions in the collection are indexed by $\delta \in \mathcal{D}$, hence V is piecewise linear if $\mathcal{D}$ is finite. As for the end point properties,

$$V(0) = \min_{\delta \in \mathcal{D}} R_1(\delta) = \min_{\delta \in \mathcal{D}} [C_{01} \mathrm{P}_1(\mathcal{Y}_0) + C_{11} \mathrm{P}_1(\mathcal{Y}_1)] = C_{11},$$

where the minimizing rule is $\delta^*(y) = 1$, for all $y \in \mathcal{Y}$. Similarly $V(1) = C_{00}$. □

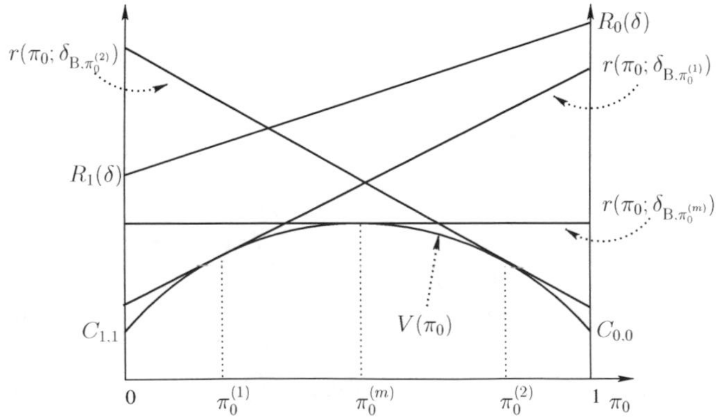

Figure 2.5 Bayes risk lines and a differentiable minimum risk curve $V(\pi_0)$. The deterministic equalizer rule $\delta_{\pi_0^*}$ is the minimax rule

We can write $V(\pi_0)$ in terms of the likelihood ratio $L(y)$ and threshold η as:

$$V(\pi_0) = \pi_0[C_{10}\mathrm{P}_0\{L(Y) \geq \eta\} + C_{00}\mathrm{P}_0\{L(Y) < \eta\}] \\ + (1 - \pi_0)[C_{11}\mathrm{P}_1\{L(Y) \geq \eta\} + C_{01}\mathrm{P}_1\{L(Y) < \eta\}].$$

If $L(y)$ has no point masses under P_0 or P_1, then V is differentiable in π_0 (since η is differentiable in π_0).[2]

2.3.3 Differentiable $V(\pi_0)$

Let us first consider the case where V is differentiable for all $\pi_0 \in [0, 1]$. Then $V(\pi_0)$ achieves its maximum value at either the end point $\pi_0 = 0$ or $\pi_0 = 1$ or within the interior $\pi_0 \in (0, 1)$. Usually $C_{00} = C_{11}$ in which case $V(0) = V(1)$ and the maximum is achieved at an interior point π_0^*, as illustrated in Figure 2.5. In all cases, the Bayes risk line $r(\pi_0; \delta)$ associated with any decision rule δ must lie entirely above the curve $V(\pi_0)$. Otherwise there would exist some π_0 such that $r(\pi_0; \delta) < V(\pi_0) = \min_\delta r(\pi_0; \delta)$, contradicting the definition of the minimum. If the Bayes risk line $r(\pi_0; \delta)$ is tangent to the curve $V(\pi_0)$ at some point π_0', then $r(\pi_0'; \delta) = V(\pi_0')$, hence δ is a Bayes rule for π_0'. Moreover, any δ whose Bayes risk line lies strictly above the curve $V(\pi_0)$ can be improved by a Bayes rule. The statement of Theorem 2.1 below follows directly.

THEOREM 2.1 *Assume V is differentiable on* $[0, 1]$.

(i) *If V has an interior maximum, then the minimax rule δ_{m} is the deterministic equalizer rule*

$$\delta_{\mathrm{m}} = \delta_{\pi_0^*}, \tag{2.27}$$

where $\pi_0^* = \arg\max_{\pi_0 \in [0,1]} V(\pi_0)$ is obtained by solving $dV(\pi_0)/d\pi_0 = 0$, i.e., δ_{m} is a Bayes rule for the worst-case prior.

(ii) *If the maximum of $V(\pi_0)$ occurs at $\pi_0^* = 0$ or 1, then (2.27) holds but δ_{m} is generally not an equalizer rule.*

(iii) *The minimax risk is equal to $V(\pi_0^*)$.*

(iv) *Randomization cannot improve the minimax rule.*

2.3.4 Nondifferentiable $V(\pi_0)$

If V is *not* differentiable for all π_0, the conclusions of Theorem 2.1 still hold if V is differentiable at its maximum. If V is not differentiable at its (interior) maximum π_0^*, we have the scenario depicted in Figure 2.6. It is important to note that, in this case, finding the minimax rule within the class of deterministic decision rules $\mathcal{D}$ may be quite challenging except in simple examples like the one considered in Example 1.2. In particular, the minimax rule within $\mathcal{D}$ is *not* necessarily a deterministic Bayes rule (LRT), as the following example shows.

[2] This condition typically holds for continuous observations when $p_0(y)$ and $p_1(y)$ are pdfs with the same support, but not necessarily even in this case.

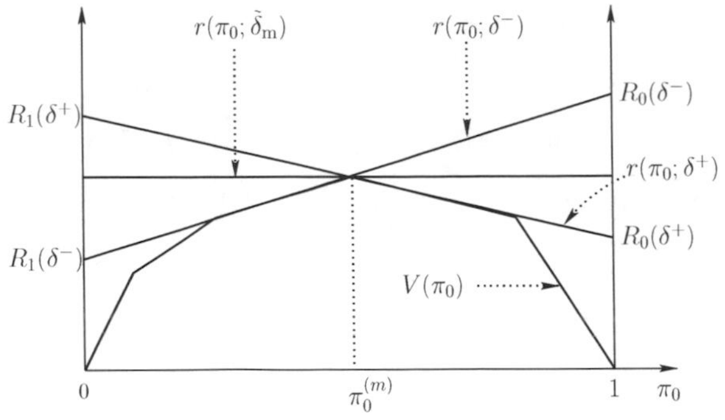

Figure 2.6 Minimax rule when V is *not* differentiable at π_0^*

Example 2.4 Consider $\mathcal{Y} = \{1, 2, 3\}$ with the following conditional pmfs:

$$p_0(1) = 0.04,\ p_0(2) = 0.95,\ p_0(3) = 0.01,$$
$$p_1(1) = 0.001,\ p_1(2) = 0.05,\ p_1(3) = 0.949.$$

Assuming uniform costs we can compute the conditional risks for the eight possible deterministic decision rules as shown in the table below.

δ	$y = 1$	$y = 2$	$y = 3$	$R_0(\delta)$	$R_1(\delta)$	$R_{\max}(\delta)$
δ_1	0	0	0	0	1	1
δ_2	0	0	1	0.01	0.051	0.051
δ_3	0	1	0	0.95	0.95	0.95
δ_4	0	1	1	0.96	0.001	0.96
δ_5	1	0	0	0.04	.999	0.999
δ_6	1	0	1	0.05	0.05	0.05
δ_7	1	1	0	0.99	0.949	0.99
δ_8	1	1	1	1	0	1

Clearly δ_6 is minimax within the class of deterministic rules for this problem. The likelihood ratio takes the values $L(1) = \frac{1}{40}$, $L(2) = \frac{1}{19}$, and $L(3) = 94.9$. Since the likelihood ratio strictly increases from 1 to 3, the only deterministic Bayes rules (likelihood ratio tests) are given by $\{\delta_1, \delta_2, \delta_4, \delta_8\}$, and this set does not include the deterministic minimax rule δ_6.

We therefore extend our search for minimax rules to the class of randomized decision rules $\tilde{\mathcal{D}}$.

Denote by δ^- and δ^+ the (deterministic) Bayes rules corresponding to priors approaching π_0^* from below and above, respectively. Their risk lines are shown in

Figure 2.6 and have slopes $R_0(\delta^-) - R_1(\delta^-) > 0$ and $R_0(\delta^+) - R_1(\delta^+) < 0$, respectively. Both rules have the same Bayes risk $V(\pi_0^*)$, but unequal conditional risks: $R_0(\delta^-) > R_0(\delta^+)$ while $R_1(\delta^-) < R_1(\delta^+)$. Consider now the randomized decision rule

$$\tilde{\delta} = \begin{cases} \delta^- & \text{with probability } \gamma \\ \delta^+ & \text{with probability } (1-\gamma). \end{cases} \tag{2.28}$$

The conditional risks of $\tilde{\delta}$ are given by

$$\begin{aligned} R_0(\tilde{\delta}) &= \gamma R_0(\delta^-) + (1-\gamma)R_0(\delta^+), \\ R_1(\tilde{\delta}) &= \gamma R_1(\delta^-) + (1-\gamma)R_1(\delta^+). \end{aligned} \tag{2.29}$$

In particular, $\tilde{\delta}$ reduces to the deterministic rules δ^- and δ^+ if $\gamma = 1$ and $\gamma = 0$, respectively. For any $\gamma \in [0, 1]$, the risk line of $\tilde{\delta}$ is a tangent to the V curve. All these risk lines go through the point $(\pi_0^*, V(\pi_0^*))$. The risk line can be made horizontal (i.e, the conditional risks can be make equal) by choosing

$$\gamma = \frac{R_1(\delta^+) - R_0(\delta^+)}{(R_1(\delta^+) - R_0(\delta^+)) + (R_0(\delta^-) - R_1(\delta^-))} \triangleq \gamma^*. \tag{2.30}$$

The resulting rule $\tilde{\delta}^*$ is an equalizer rule and is a Bayes rule for prior π_0^*, hence is a minimax rule.

2.3.5 Randomized LRTs

Since Bayes rules are LRTs, the randomized minimax rule of (2.28) is a randomized LRT which we now derive explicity. Both δ^- and δ^+ are deterministic LRTs with threshold η determined by π_0^*:

$$\delta^-(y) = \begin{cases} 1 & : \; L(y) > \eta \\ 1 & : \; L(y) = \eta \\ 0 & : \; L(y) < \eta \end{cases}, \qquad \delta^+(y) = \begin{cases} 1 & : \; L(y) > \eta \\ 0 & : \; L(y) = \eta \\ 0 & : \; L(y) < \eta. \end{cases} \tag{2.31}$$

The tests δ^- and δ^+ differ only in the decision made in the event $L(Y) = \eta$, respectively selecting H_1 and H_0 in that case. For δ^- and δ^+ to be different, $L(Y)$ must have a point mass at η, i.e., $\mathrm{P}_j\{L(Y) = \eta\} \neq 0$, for $j = 0, 1$. This also implies that V is *not* differentiable at π_0^* and that neither δ^- nor δ^+ is an equalizer rule.

A randomized LRT takes the form

$$\tilde{\delta}(y) = \begin{cases} 1 & : \quad > \\ 1 \text{ w.p. } \gamma & : \; L(y) = \eta. \\ 0 & : \quad < \end{cases} \tag{2.32}$$

THEOREM 2.2 *Assume that V attains its maximum over* $[0, 1]$ *at an interior point π_0^* and is not differentiable at that point. Then*

(i) *The minimax solution within $\tilde{\mathcal{D}}$ is given by the randomized equalizer rule*

$$\tilde{\delta}_m(y) = \begin{cases} 1 & : \quad > \\ 1 \text{ w.p. } \gamma^* & : \quad L(y) = \eta^*, \\ 0 & : \quad < \end{cases} \tag{2.33}$$

where the threshold η^* is related to π_0^* via (2.12), and the randomization parameter γ^* is given by (2.30).

(ii) *The minimax risk is equal to $V(\pi_0^*)$.*

2.3.6 Examples

Example 2.5 *Signal Detection in Gaussian Noise (continued).* In this example we study the minimax solution to the detection problem described in Example 2.1 using the systematic procedure we have developed. We assume uniform costs. The minimum Bayes risk curve is given by

$$\begin{aligned} V(\pi_0) &= \pi_0 \mathrm{P}_0\{Y \geq \tau\} + (1 - \pi_0)\mathrm{P}_1\{Y < \tau\} \\ &= \pi_0 Q\left(\frac{\tau}{\sigma}\right) + (1 - \pi_0)\Phi\left(\frac{\tau - \mu}{\sigma}\right) \end{aligned}$$

with

$$\tau = \frac{\sigma^2}{\mu} \ln\left(\frac{\pi_0}{1 - \pi_0}\right) + \frac{\mu}{2}.$$

Clearly V is a differentiable function in π_0, and therefore the deterministic equalizer rule is minimax. We can solve for the equalizer rule without explicitly maximizing V. In particular, if we denote the LRT with threshold τ by δ_τ, then

$$R_0(\delta_\tau) = Q\left(\frac{\tau}{\sigma}\right), \quad R_1(\delta_\tau) = \Phi\left(\frac{\tau - \mu}{\sigma}\right) = Q\left(\frac{\mu - \tau}{\sigma}\right).$$

Setting $R_0(\delta_\tau) = R_1(\delta_\tau)$ yields

$$\tau^* = \frac{\mu}{2},$$

from which we can conclude that $\eta^* = 1$ and $\pi_0^* = 0.5$.

Thus the minimax decision rule is given by

$$\delta_{\mathrm{m}}(y) = \delta_{0.5}(y) = \begin{cases} 1 & : \; y \geq \frac{\mu}{2} \\ 0 & : \quad < \end{cases}$$

and the minimax risk is given by

$$R_{\max}(\delta_{\mathrm{m}}) = V(0.5) = Q\left(\frac{\mu}{2\sigma}\right).$$

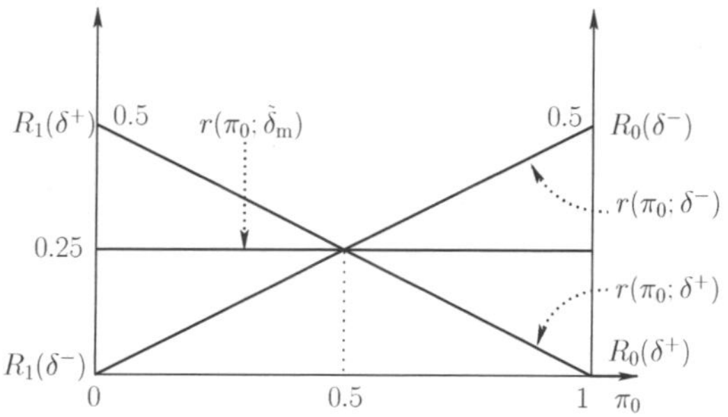

Figure 2.7 Minimax rule for Example 2.6

Example 2.6 *Discrete Observations (continued).* In this example, we study the (randomized) minimax solution to the detection problem described in Example 2.2. Recall from (2.21) that $L(1) = 0$, $L(2) = 1$, and $L(3) = \infty$. Assuming uniform costs, randomized LRTs are given by (2.32) with $\eta = \pi_0/(1 - \pi_0)$ and $\gamma \in [0, 1]$.

For $\pi_0 \in (0, 0.5)$, $\eta \in (0, 1)$, and thus all the Bayes rules in (2.13) collapse to the single deterministic rule:

$$\delta^-(y) = \delta_4(y) = \begin{cases} 1 & : \; y = 2, 3 \\ 0 & : \; y = 1. \end{cases}$$

Similarly, for $\pi_0 \in (0.5, 1)$, $\eta \in (1, \infty)$, and thus all the Bayes rules in (2.13) collapse to the single deterministic rule:

$$\delta^+(y) = \delta_2(y) = \begin{cases} 1 & : \; y = 3 \\ 0 & : \; y = 1, 2. \end{cases}$$

For $\pi_0 = 0.5$, the randomized LRTs with threshold $\eta = 1$ are all Bayes rules. and are obtained by randomizing between δ^+ and δ^-. The minimum Bayes risk curve $V(\pi_0) = \frac{1}{2}\min\{\pi_0, 1-\pi_0\}$ is shown in Figure 2.7 and attains its maximum at $\pi_0^* = 0.5$ where it is not differentiable. Furthermore, $R_1(\delta^-) = R_0(\delta^+) = 0$, and $R_0(\delta^-) = R_1(\delta^+) = 0.5$. Therefore (2.30) yields $\gamma^* = 0.5$, and the minimax decision rule $\tilde{\delta}_m$ is given by (2.33) with $\eta^* = 1$ and $\gamma^* = 0.5$. The minimax risk is equal to the worst-case Bayes risk: $r(\tilde{\delta}_m) = V(0.5) = 0.25$.

Example 2.7 *Z Channel.* Consider the binary communication problem of Example 2.3 with $\lambda_0 = 0$ and $\lambda_1 = \lambda$. This is called the Z channel: $\mathcal{Y} = \{0, 1\}$ with $p_0(0) = 1$ and $p_1(0) = \lambda$. By inspection of the conditional risks in Table 2.2, we see that rule δ_1 is never better than δ_2, and δ_3 is never better than δ_4. This leaves only two rules to consider: $\delta_2(y) = y$ (reproduce observed bit) and $\delta_4(y) = 1$ (ignore observation and output 1). The Bayes risk lines for δ_2 and δ_4 are given by $r(\pi_0; \delta_2) = (1 - \pi_0)\lambda$ and π_0, respectively. Hence $V(\pi_0) = \min\{\pi_0, (1 - \pi_0)\lambda\}$ where δ_4 is the Bayes rule for $0 \le \pi_0 \le \pi_0^* = \frac{\lambda}{1+\lambda}$ (least-favorable prior) and δ_2 is the Bayes rule for $\pi_0^* \le \pi_0 \le 1$.

Hence we can identify δ_4 with δ^- and δ_2 with δ^+. Picking the randomized rule $\tilde{\delta}$ of (2.28), we obtain the minimax choice by selecting

$$\gamma^* = \frac{R_1(\delta^+) - R_0(\delta^+)}{(R_1(\delta^+) - R_0(\delta^+)) + (R_0(\delta^-) - R_1(\delta^-))} = \frac{\lambda - 0}{(\lambda - 0) + (1 - 0)} = \frac{\lambda}{\lambda + 1}.$$

The minimax risk is $r(\tilde{\delta}_{\rm m}) = V(\pi_0^*) = \frac{\lambda}{1+\lambda}$.

2.4 Neyman–Pearson Hypothesis Testing

For binary detection problems without a prior on the state, a commonly used alternative to the minimax formulation is the Neyman–Pearson (NP) formulation, which trades off the probability of false alarm and the probability of miss. The tradeoff was already apparent in Figure 2.1 where $\mathrm{P_F}$ decreases and $\mathrm{P_M}$ increases as the test threshold τ increases. For this family of threshold decision rules (which encompasses all Bayes rules), the tradeoff between $\mathrm{P_F}$ and $\mathrm{P_M}$ is parameterized by $\tau \in \mathbb{R}$.

For deterministic decision rule δ, we defined in (2.2) $\mathrm{P_F}(\delta) = \mathrm{P}_0(\mathcal{Y}_1)$ and $\mathrm{P_M}(\delta) = \mathrm{P}_1(\mathcal{Y}_0)$. For a randomized decision rule as defined in Definition 1.4, we can similarly define the *probability of false alarm*

$$\mathrm{P_F}(\tilde{\delta}) \triangleq \mathrm{P}_0\{\text{say } H_1\} = \sum_{\ell=1}^{L} \gamma_\ell \mathrm{P_F}(\delta_\ell) \tag{2.34}$$

and the *probability of miss*

$$\mathrm{P_M}(\tilde{\delta}) \triangleq \mathrm{P}_1\{\text{say } H_0\} = \sum_{\ell=1}^{L} \gamma_\ell \mathrm{P_M}(\delta_\ell). \tag{2.35}$$

Instead of $\mathrm{P_M}$, an equivalent measure of performance that is commonly used in radar and surveillance applications is the *probability of detection*:

$$\mathrm{P_D}(\tilde{\delta}) \triangleq 1 - \mathrm{P_M}(\tilde{\delta}), \tag{2.36}$$

which is also called the *power* of the decision rule $\tilde{\delta}$. The Neyman–Pearson (NP) problem is generally stated in terms of $\mathrm{P_D}$ and $\mathrm{P_F}$ as

$$\tilde{\delta}_{\rm NP} = \arg \max_{\tilde{\delta} \in \tilde{\mathcal{D}}:\, \mathrm{P_F}(\tilde{\delta}) \le \alpha} \mathrm{P_D}(\tilde{\delta}), \tag{2.37}$$

where $\alpha \in (0, 1)$ is called the *significance level* of the test. We wish to find a rule $\tilde{\delta}_{\rm NP}$ with a probability of false alarm not exceeding α that maximizes the probability of correct detection. The choice of α is application-dependent and implicitly depends on the "cost" of a false alarm. For instance, in a public surveillance system, a false alarm may be costly if it causes serious disruptions and antagonizes individuals who are incorrectly deemed suspect. In the medical field, a false alarm may be relatively benign as it causes additional tests to be ordered for the patient, whereas the cost of a miss could have life-threatening consequences.

Note that unlike the Bayesian and minimax optimization problems, which are formulated in terms of conditional risks, the NP optimization problem is stated in terms of conditional error probabilities. In particular, the problem implicitly assumes uniform costs, which means $P_F(\tilde{\delta}) = R_0(\tilde{\delta})$ and $P_D(\tilde{\delta}) = 1 - R_1(\tilde{\delta})$, and so the NP optimization problem is to minimize $R_1(\tilde{\delta})$ subject to $R_0(\tilde{\delta}) \le \alpha$.

2.4.1 Solution to the NP Optimization Problem

To solve the NP optimization problem, we once again resort to Bayesian risk lines and the minimum risk curve $V(\pi_0)$ with uniform costs. As depicted in Figures 2.8 and 2.9, we can infer two properties of $\tilde{\delta}_{NP}$ that hold whether $V(\pi_0)$ is differentiable or not. We assume that $\alpha \le V'(0) = \frac{dV}{d\pi_0}(0)$.[3]

1. The risk line $r(\pi_0; \tilde{\delta})$ must be a *tangent* to the curve $V(\pi_0)$, otherwise $\tilde{\delta}_{NP}$ could be improved by the Bayes rule with the same P_F. (For instance, δ improves over $\tilde{\delta}'$ in Figure 2.8.) Hence $\tilde{\delta}_{NP}$ must be a Bayes rule for some prior π_0^*.
2. $\tilde{\delta}_{NP}$ must satisfy the significance level constraint with equality. Otherwise one could simultaneously increase $P_F(\tilde{\delta}_{NP})$ – while still satisfying the constraint $P_F(\tilde{\delta}_{NP}) \le \alpha$ – and *increase* the power of the test (reduce $P_M(\tilde{\delta}_{NP})$).

THEOREM 2.3 *The solution to the NP problem* (2.37) *is a randomized LRT:*

$$\tilde{\delta}_{NP}(y) = \tilde{\delta}_{\eta,\gamma}(y) = \begin{cases} 1 & > \\ 1 \text{ w.p. } \gamma^* & : \; L(y) = \eta^*, \\ 0 & < \end{cases} \tag{2.38}$$

where the threshold η^* *and randomization parameter* γ^* *are chosen so that*

$$P_F(\tilde{\delta}_{NP}) = P_0\{L(Y) > \eta^*\} + \gamma^* P_0\{L(Y) = \eta^*\} = \alpha. \tag{2.39}$$

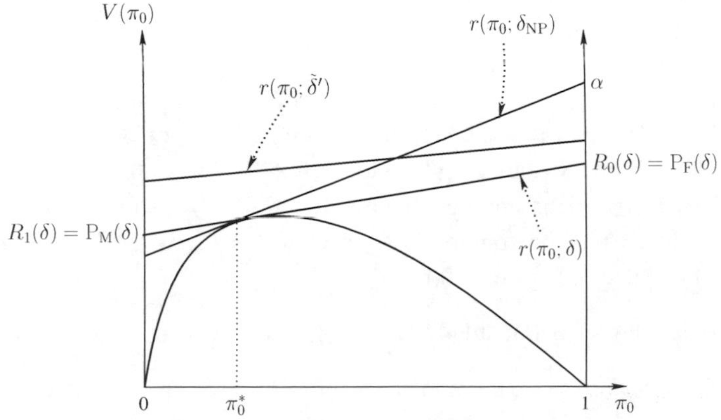

Figure 2.8 Differentiable minimum-risk curve $V(\pi_0)$

[3] If $\alpha > V'(0)$ then $\tilde{\delta}_{NP}$ need not satisfy $P_F = \alpha$ and need not be a Bayes rule. See Exercise 2.11.

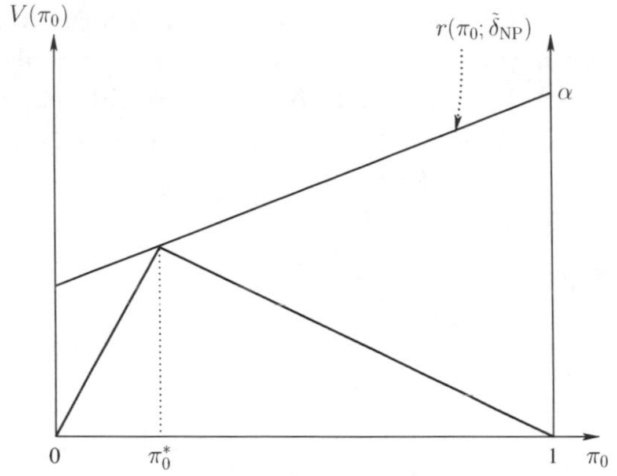

Figure 2.9 The minimum-risk curve $V(\pi_0)$ is not everywhere differentiable

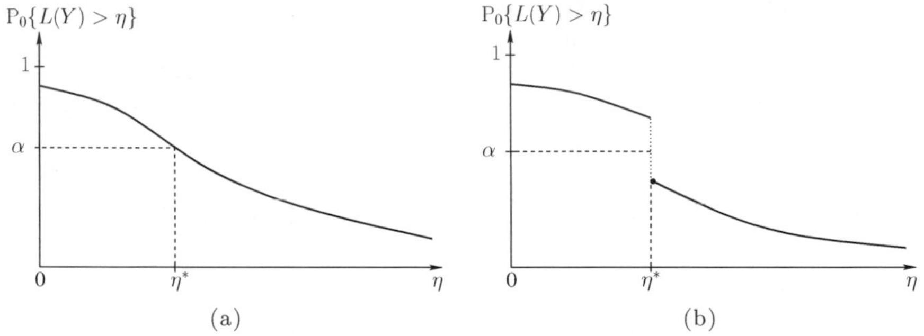

Figure 2.10 Complementary cdf of $L(Y)$ under H_0: (a) $P_0\{L(Y) > \eta\}$ is continuous; (b) $P_0\{L(Y) > \eta\}$ is discontinuous

2.4.2 NP Rule

The procedure for finding the parameters η^* and γ^* of the NP solution is illustrated in Figure 2.10, which displays the complementary cdf $P_0\{L(Y) > \eta\}$ as a function of η. This is a right-continuous function.

Case I: $P_0\{L(Y) > \eta\}$ is a continuous function of η. Referring to Figure 2.10(a), the equation $P_0\{L(Y) > \eta^*\} = \alpha$ admits a solution for η^*. The randomization parameter γ can be chosen arbitrarily (including 0) since $P_0\{L(Y) = \eta\} = 0$.

Case II: $P_0\{L(Y) > \eta\}$ is not a continuous function of η. If there exists η^* such that $P_0\{L(Y) > \eta^*\} = \alpha$, the solution is as in Case I. Otherwise, we use the following procedure illustrated in Figure 2.10(b):

1. Find the smallest η such that $P_0\{L(Y) > \eta\} \le \alpha$:

$$\eta^* = \min\{\eta : P_0\{L(Y) > \eta\} \le \alpha\}. \tag{2.40}$$

2. Then

$$\gamma^* = \frac{\alpha - P_0\{L(Y) > \eta\}}{P_0\{L(Y) = \eta\}}. \tag{2.41}$$

The probability of detection (power) of $\tilde{\delta}_{NP}$ at significance level α can be computed as

$$P_D(\tilde{\delta}_{NP}) = P_1\{L(y) > \eta^*\} + \gamma^* P_1\{L(y) = \eta^*\}, \tag{2.42}$$

while $P_F(\tilde{\delta}_{NP})$ is given by (2.39).

2.4.3 Receiver Operating Characteristic

A plot of $P_D(\tilde{\delta})$ versus $P_F(\tilde{\delta}) = \alpha$ is called the receiver operating characteristic (ROC) of a decision rule $\tilde{\delta} \in \tilde{\mathcal{D}}$. The "best" ROC is achieved by $\tilde{\delta}_{NP}$ as illustrated in Figure 2.11 for a differentiable ROC. Specific examples of ROC are given in the next section. Each point on the ROC is indexed by α or equivalently, by (η^*, γ^*).

A "good" ROC plot is close to a reverse L, as one can simultaneously achieve high P_D and low P_F. Conversely, in the case of indistinguishable hypotheses ($p_0 = p_1$, hence $L(Y) = 1$ with probability 1), the ROC of $\tilde{\delta}_{NP}$ is the straight-line segment $P_D = P_F$. Note that the error probability for the minimax rule can be obtained from the ROC curve of $\tilde{\delta}_{NP}$ as the intersection of the curve with the straight line $P_D = 1 - P_F$.

THEOREM 2.4 *The ROC of $\tilde{\delta}_{NP}$ has the following properties:*

(i) *The ROC is a concave, nondecreasing function.*
(ii) *The ROC lies above the 45° line, i.e., $P_D(\tilde{\delta}_{NP}) \geq P_F(\tilde{\delta}_{NP})$.*
(iii) *The slope of the ROC at a point of differentiability is equal to the value of the threshold η required to achieve the P_D and P_F at that point, i.e.,*

$$\frac{dP_D}{dP_F} = \eta.$$

(iv) *The right derivative of the ROC at a point of nondifferentiability is equal to the value of the threshold η required to achieve the P_D and P_F at that point.*

Proof See Exercise 2.24. □

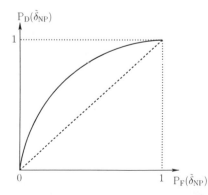

Figure 2.11 Receiver Operating Characteristic (ROC) of $\tilde{\delta}_{NP}$

2.4.4 Examples

Example 2.8 *Signal Detection in Gaussian Noise.* In this example we study the NP solution to the detection problem described in Example 2.1. As noted before, LRTs are equivalent to threshold tests on Y. Moreover, $L(Y)$ does not have point masses under either H_0 or H_1, so Case I of Section 2.4.2 applies, and no randomization is needed. Thus

$$\tilde{\delta}_{\mathrm{NP}}(y) = \delta_\tau(y) = \begin{cases} 1 & : \ y \geq \tau \\ 0 & : \ \ < \end{cases}$$

with $\tau = \frac{\sigma^2}{\mu} \ln \eta + \frac{\mu}{2}$. Now

$$\mathrm{P_F}(\delta_\tau) = \mathrm{P}_0\{Y \geq \tau\} = Q\left(\frac{\tau}{\sigma}\right).$$

Therefore, the $\mathrm{P_F}$ constraint of α is met by setting

$$\tau = \sigma Q^{-1}(\alpha),$$

where $Q^{-1} : (0, 1) \mapsto \mathbb{R}$ is the inverse of thc Q function. The power of δ_τ is given by

$$\mathrm{P_D}(\delta_\tau) = \mathrm{P}_1\{Y \geq \tau\} = Q\left(\frac{\tau - \mu}{\sigma}\right) = Q(Q^{-1}(\alpha) - d),$$

where $d = \mu/\sigma$. The ROC is plotted in Figure 2.12. As d increases, $\mathrm{P_D}$ increases for a given level of $\mathrm{P_F}$, and the ROC approaches a reverse L.

Example 2.9 *Discrete Observations.* In this example we study the NP solution to the detection problem described in Example 2.2. Due to the fact that $L(1) = 0$, $L(2) = 1$, and $L(3) = \infty$, we have

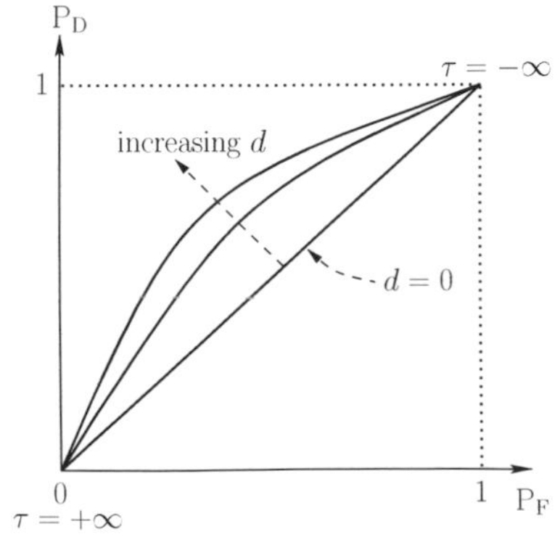

Figure 2.12 ROC for Gaussian hypothesis testing

$$\mathrm{P}_0\{L(Y > \eta\} = \begin{cases} \frac{1}{2} & : \ \eta \in [0, 1) \\ 0 & : \ \eta \in [1, \infty). \end{cases}$$

Thus, for $\alpha \in [0, \frac{1}{2})$, we have $\eta = 1$ and $\gamma = \frac{\alpha - 0}{1/2} = 2\alpha$, which yields

$$\tilde{\delta}_{\mathrm{NP}}(y) = \begin{cases} 1 & : \ y = 3 \\ 1 \text{ w.p. } 2\alpha & : \ y = 2 \\ 0 & : \ y = 1 \end{cases}$$

and

$$\mathrm{P}_{\mathrm{D}}(\tilde{\delta}_{\mathrm{NP}}) = p_1(3) + 2\alpha p_1(2) = \frac{1}{2} + \alpha.$$

For $\alpha \in [\frac{1}{2}, 1]$, $\eta = 0$ and $\gamma = \frac{\alpha - \frac{1}{2}}{1/2} = 2\alpha - 1$, which yields

$$\tilde{\delta}_{\mathrm{NP}}(y) = \begin{cases} 1 & : \ y = 2, 3 \\ 1 \text{ w.p. } 2\alpha - 1 & : \ y = 1 \end{cases}$$

and $\mathrm{P}_{\mathrm{D}}(\tilde{\delta}_{\mathrm{NP}}) = 1$. The ROC is plotted in Figure 2.13 and has a point of nondifferentiability at $\alpha = \frac{1}{2}$.

Example 2.10 *Z Channel.* We study the NP solution for the Z channel problem of Example 2.7. The likelihood ratio is given by

$$L(y) = \frac{p_1(y)}{p_0(y)} = \begin{cases} \lambda & : \ y = 0 \\ \infty & : \ y = 1 \end{cases}$$

and $\mathrm{P}_0\{Y = 0\} = 1$. Hence $\mathrm{P}_0\{L(y) > \eta\} = \mathbb{1}\{\eta < \lambda\}$. We seek the smallest η such that $\alpha \geq \mathrm{P}_0\{L(y) > \eta\}$. First assume $\alpha < 1$. From Figure 2.14 it can be seen that $\eta = \lambda$. Next, to find γ, we use (2.41) and obtain $\gamma = \alpha$. Hence the Neyman–Pearson rule at significance level α is:

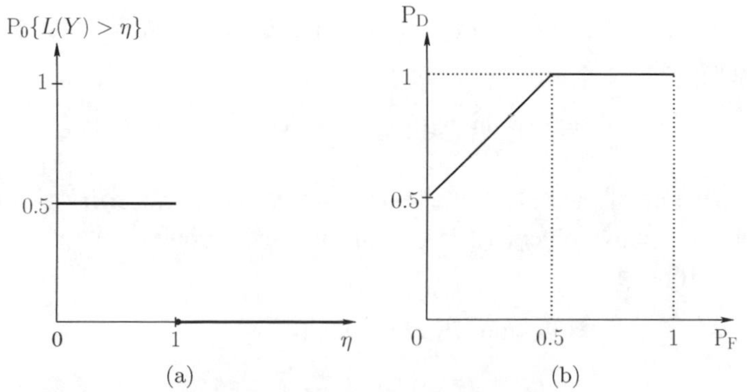

Figure 2.13 Example 2.9: (a) $\mathrm{P}_0\{L(Y) > \eta\}$; (b) ROC

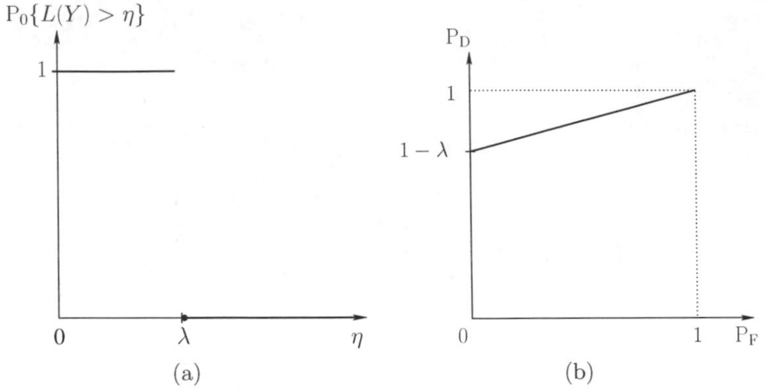

Figure 2.14 Z channel: (a) $P_0\{L(Y) > \eta\}$; (b) ROC

$$\tilde{\delta}_{NP}(y) = \begin{cases} 1 & : L(y) > \lambda \quad (y = 1) \\ 1 \text{ w.p. } \alpha & : L(y) = \lambda \quad (y = 0) \end{cases} \tag{2.43}$$

(note the event $L(Y) < \lambda$ has probability zero). For $\alpha = 1$, we obtain $\eta = 0$ and $\gamma = 1$, corresponding to the deterministic rule $\delta(y) \equiv 1$ (always say H_1). Note that this is also the limit of (2.43) as $\alpha \uparrow 1$.

Therefore, for all $\alpha \in [0, 1]$, the power of the NP test is

$$\begin{aligned} P_D(\tilde{\delta}_{NP}) &= P_1\{L(Y) > \lambda\} + \alpha P_1\{L(Y) = \lambda\} \\ &= P_1\{Y = 1\} + \alpha P_1\{Y = 0\} \\ &= 1 - \lambda + \alpha\lambda. \end{aligned}$$

The ROC for the Z channel is given by $P_D = 1-\lambda+P_F\lambda$ and is plotted in Figure 2.14(b). The straight line segment with slope λ joins the two points with coordinates (P_F, P_D) = $(0, 1-\lambda)$ and $(1, 0)$, which are respectively achieved by the deterministic rules $\delta(y) = y$ and $\delta(y) = 1$.

2.4.5 Convex Optimization

The NP problem (2.37) can also be solved using convex optimization. Indeed (2.37) may be written as

$$\tilde{\delta}_{NP} = \arg\min_{\tilde{\delta}} P_M(\tilde{\delta}) \qquad \text{subject to } P_F(\tilde{\delta}) \le \alpha. \tag{2.44}$$

The expressions (2.34) and (2.36) for $P_F(\tilde{\delta})$ and $P_D(\tilde{\delta})$ are linear in $\tilde{\delta}$, hence the constrained optimization problem (2.37) is linear and is equivalent to the Lagrangian optimization problem

$$\begin{aligned} \tilde{\delta}_\lambda &= \arg\min_{\tilde{\delta}} [P_M(\tilde{\delta}) + \lambda P_F(\tilde{\delta})] \\ &= \arg\min_{\tilde{\delta}} \left[\frac{1}{1+\lambda} P_M(\tilde{\delta}) + \frac{\lambda}{1+\lambda} P_F(\tilde{\delta}) \right], \end{aligned}$$

where the Lagrange multiplier $\lambda \geq 0$ is chosen so as to satisfy the condition $P_F(\tilde{\delta}_\lambda) = \alpha$. Equivalently, the NP rule is given by

$$\tilde{\delta}_\lambda = \arg\min_{\tilde{\delta}} r(\pi_0; \tilde{\delta}),$$

where we have reparameterized the Lagrange multiplier:

$$\lambda = \frac{\pi_0}{1-\pi_0} \geq 0 \quad \Leftrightarrow \quad \pi_0 = \frac{\lambda}{1+\lambda} \in [0, 1].$$

If the equality $P_F(\tilde{\delta}_\lambda) = \alpha$ cannot be achieved (because $\alpha > V'(0)$) then the P_F constraint is inactive, $\lambda = 0$, and $\tilde{\delta}_{NP}$ is the Bayes rule corresponding to $\pi_0 = 0$.

Exercises

2.1 *Best Decision Rules.* For the binary hypothesis testing problem, with $C_{00} < C_{10}$ and $C_{11} < C_{01}$, show there is no "best" rule based on conditional risks, except in the trivial case where $p_0(y)$ and $p_1(y)$ have disjoint supports.

2.2 *Bayesian Test with Nonuniform Costs.* Consider the binary detection problem where p_0 and p_1 are Rayleigh densities with scale parameters 1 and $\sigma > 1$, respectively:

$$p_0(y) = y\, e^{-\frac{y^2}{2}}\, \mathbb{1}\{y \geq 0\} \quad \text{and} \quad p_1(y) = \frac{y}{\sigma^2} e^{-\frac{y^2}{2\sigma^2}}\, \mathbb{1}\{y \geq 0\}.$$

(a) Find the Bayes rule for equal priors and a cost structure of the form $C_{00} = C_{11} = 0$, $C_{10} = 1$, and $C_{01} = 4$.
(b) Find the Bayes risk for the Bayes rule of part (a).

2.3 *Minimum Bayes Risk Curve and Minimax Rule.* For Exercise 1.1, find the minimum Bayes risk function $V(\pi_0)$, and then find a minimax rule in the set of randomized decision rules using $V(\pi_0)$.

2.4 *Minimum Bayes Risk Curve.* Suppose

$$V(\pi_0) = \begin{cases} 2\pi_0 & \text{if } \pi_0 \in [0, \frac{1}{4}) \\ \frac{1}{2} & \text{if } \pi_0 \in [\frac{1}{4}, \frac{3}{4}) \\ 2(1-\pi_0) & \text{if } \pi_0 \in [\frac{3}{4}, 1]. \end{cases}$$

Is $V(\pi_0)$ a valid minimum Bayes risk function for *uniform costs*? Explain your reasoning.

2.5 *Nonuniform Costs.* Consider the binary hypothesis testing problem with cost function

$$C_{ij} = \begin{cases} 0 & \text{if } i = j \\ 1 & \text{if } j = 0, i = 1 \\ 10 & \text{if } j = 1, i = 0. \end{cases}$$

The observation Y takes values in the set $\mathcal{Y} = \{a, b, c\}$ and the conditional pmfs of Y are:

$$p_0(a) = p_0(b) = 0.5 \quad \text{and} \quad p_1(a) = p_1(b) = 0.25, \;\; p_1(c) = 0.5.$$

(a) Is there a best decision rule based on conditional risks?
(b) Find Bayes (for equal priors) and minimax rules within the set of deterministic decision rules.
(c) Now consider the set of randomized decision rules. Find a Bayes rule (for equal priors). Also construct a randomized rule whose maximum risk is smaller than that of the minimax rule of part (b).

2.6 *Bayes and Minimax Decision Rules.* Consider the binary detection problem with

$$p_0(y) = \frac{y+3}{18} \mathbb{1}\{y \in [-3, 3]\}, \text{ and } p_1(y) = \frac{1}{4} \mathbb{1}\{y \in [0, 4]\}.$$

(a) Find a Bayes rule for uniform costs and equal priors and the corresponding minimum Bayes risk.
(b) Find a minimax rule for uniform costs, and the corresponding minimax risk.

2.7 *Binary Detection.* Consider the binary detection problem with

$$p_0(y) = \frac{1}{2}\mathbb{1}\{y \in [-1, 1]\}, \text{ and } p_1(y) = \frac{3y^2}{2}\mathbb{1}\{y \in [-1, 1]\}.$$

(a) Find a Bayes rule for equal priors and uniform costs, and the corresponding Bayes risk.
(b) Find a minimax rule for uniform costs, and the corresponding minimax risk. (You may leave your answer in terms of a solution to a polynomial equation.)
(c) Find a NP rule for level $\alpha \in (0, 1)$, and plot its ROC.

2.8 *Multiple Predictions.* Consider a binary hypothesis problem where the decision maker is allowed to make n predictions of the hypothesis, and the cost is the number of prediction errors (hence is an integer between 0 and n). Assume $\mathcal{Y} = \{a, b, c\}$ and the conditional pmfs of Y satisfy $p_0(a) = p_0(b) = p_1(b) = p_1(c) = 0.5$. Derive the deterministic minimax risk and the minimax risk. Explain why these risks are equal for even n.

2.9 *Penalty Kicks.* Assume the following simplified model for penalty kicks in football. The penalty shooter aims either left or right. The goalie dives left or right at the very moment the ball is kicked. If the goalie dives in the same direction as the ball, the probability of scoring is 0.55. Otherwise the probability of scoring is 0.95. We denote by x the direction selected by the penalty shooter (left or right) and by a the direction selected by the goalie (also left or right). We define the cost $C(a, x) = 0.55$ if $a = x$, and $C(a, x) = 0.95$ otherwise. In this problem setup, no observations are available. A deterministic decision rule for the goalkeeper reduces to an action, hence there exist only two possible deterministic rules: δ_1 (dive left) and δ_2 (dive right).

(a) Determine the minimax decision rule (and the minimax risk) for the goalie.
(b) Now assume the shooter is more accurate and powerful when aiming at the left side. If the goalie dives in the same direction (left), the probability of scoring is 0.6. If the shooter aims at the right side and the goalie dives in the same direction, the probability of scoring is 0.5. In all other cases, the probability of scoring is 0.95. Determine the minimax decision rule for the goalie, the optimal strategy for the shooter, and the minimax risk.

2.10 *Nonuniform Costs.* The minimum Bayes risk for a binary hypothesis testing problem with costs $C_{00} = 1$, $C_{11} = 2$, $C_{10} = 2$, $C_{01} = 4$ is given by

$$V(\pi_0) = 2 - \pi_0^2,$$

where π_0 is the prior probability of hypothesis H_0.

(a) Find the minimax risk and the least favorable prior.
(b) What is the conditional probability of error given H_0 for the minimax rule in (a)?
(c) Find the α-level NP rule, and compute its ROC.

2.11 *Choice of Significance Level for NP Test.*

(a) Show that the test $\delta(y) = \mathbb{1}\{y \in \text{supp}(P_1)\}$ is a Bayes test for $\pi_0 = 0$.
(b) Derive the risk line for δ.
(c) Show that the slope of this risk line at the origin is given by $V'(0) = P_0\{\text{supp}(P_1)\}$.
(d) Assume there exists a subset $\mathcal{S}$ of $\mathcal{Y}$ such that $P_0(\mathcal{S}) = 0$ and $P_1(\mathcal{S}) > 0$. Explain why there cannot be any advantage in choosing a significance level $\alpha > V'(0)$ for the NP test.
(e) If $V'(0) < \alpha < 1$, find a deterministic Bayes rule that is an NP rule, and show there exist NP rules that are not Bayes rules.

2.12 *Laplacian Distributions.* Consider the binary detection problem with

$$p_0(y) = \frac{1}{2}e^{-|y|}, \text{ and } p_1(y) = e^{-2|y|}, \quad y \in \mathbb{R}.$$

(a) Find the Bayes rule and Bayes risk for equal priors and a cost structure of the form $C_{00} = C_{11} = 0$, $C_{10} = 1$, and $C_{01} = 2$.
(b) Find a NP rule at significance level $\alpha = 1/4$.
(c) Find the probability of detection for the rule of part (b).

2.13 *Bayes and Neyman–Pearson Decision Rules.* Consider the binary hypothesis test with

$$p_0(y) = e^{-y}\,\mathbb{1}\{y \geq 0\}, \text{ and } p_1(y) = \frac{1}{2}\,\mathbb{1}\{0 \leq y \leq 1\} + \frac{1}{2}\,e^{-(y-1)}\,\mathbb{1}\{y > 1\}.$$

(a) Assume uniform costs, and derive the Bayes rules and risk curve $V(\pi_0)$, $0 \leq \pi_0 \leq 1$.
(b) Give the NP rule at significance level 0.01.

2.14 *Discrete Observations.* Consider a binary hypothesis test between p_0 and p_1, where the observation space $\mathcal{Y} = \{a, b, c, d\}$ and the probability vectors p_0 and p_1 are equal to $(\frac{1}{2}, \frac{1}{4}, \frac{1}{4}, 0)$ and $(0, \frac{1}{4}, \frac{1}{4}, \frac{1}{2})$, respectively.

(a) Assume the cost matrix $C_{00} = C_{11} = 0$, $C_{10} = 1$, $C_{01} = 2$. Give the Bayes minimum-risk curve $V(\pi_0)$, $0 \le \pi_0 \le 1$, and identify the least-favorable prior.
(b) Prove or disprove the following claim: "The decision rule $\delta(y) = \mathbb{1}\{y \in \{c, d\}\}$ is a NP rule."

2.15 *Receiver Operating Characteristic.* Derive the ROC curve for the hypothesis test

$$\begin{cases} H_0: & Y \sim p_0(y) = \text{Uniform } [0, 1] \\ H_1: & Y \sim p_1(y) = e^{-y}, \;\; y \ge 0. \end{cases}$$

2.16 *ROCs and Bayes Risk.* Consider a binary hypothesis testing problem in which the ROC for the NP rule is given by $\mathrm{P_D} = \mathrm{P_F}^{1/4}$. For the same hypotheses, consider a Bayesian problem with cost structure $C_{10} = 3$, $C_{01} = 2$, $C_{00} = 2$, $C_{11} = 0$.

(a) Show that the decision rule $\delta(y) \equiv 1$ is Bayes for $\pi_0 \le 1/3$.
(b) Find the minimum Bayes risk for $\pi_0 = 1/2$.

2.17 *Radar.* A radar sends out n short pulses. To decide whether a target is present at a given hypothesized distance, it correlates its received signal against delayed versions of these pulses. The correlator output corresponding to the ith pulse is denoted by $Y_i = B_i\theta + N_i$, where $\theta > 0$ under hypothesis H_1 (target present), and $\theta = 0$ under H_0 (target absent), B_i takes values ± 1 with equal probability (a simple model of a random phase shift in the ith pulse), and $\{N_i\}_{i=1}^n$ are i.i.d. $\mathcal{N}(0, \sigma^2)$.

Show that the NP rule can be written as

$$\delta(y) = \begin{cases} 1 & \sum_{i=1}^n g(\frac{\theta |Y_i|}{\sigma^2}) > \tau \\ 0 & \text{else,} \end{cases}$$

where $g(x) = \ln \cosh(x)$.

2.18 *Faulty Sensors.* A number n of sensors are deployed in a field in order to detect an acoustic event. Each sensor transmits a bit whose value is 1 if the event has occurred and the sensor is operational, and is 0 otherwise. However, each sensor is defective with 50 percent probability, independently of the other sensors. The observation $Y \in \{0, 1, \ldots, n\}$ is the sum of bit values transmitted by the active sensors. Derive the ROC for this problem.

2.19 *Photon Detection.* The emission of photons follows a Poisson law. Consider the detection problem

$$\begin{cases} H_0: Y \sim \text{Poi}(1) \\ H_1: Y \sim \text{Poi}(\lambda), \end{cases}$$

where $Y \in \{0, 1, 2, \ldots\}$ is the photon count, and Poi(λ) is the Poisson law with parameter λ. Assume $\lambda > 1$. The null hypothesis models background noise, and the alternative models some phenomenon of interest. This model arises in various modalities, including optical communications and medical imaging.

(a) Show that the LRT can be reduced to the comparison of Y with a threshold.
(b) Give an expression for the maximum achievable probability of correct detection at significance level $\alpha = 0.05$.
Hint: A randomized test is needed.
(c) Determine the value of λ needed so that the probability of correct detection (at significance level 0.05) is 0.95.

2.20 *Geometric Distributions.* Consider the detection problem for which $\mathcal{Y} = \{0, 1, 2, \ldots\}$ and the pmfs of the observations under the two hypotheses are:

$$p_0(y) = (1 - \beta_0)\beta_0^y, \quad y = 0, 1, 2, \ldots$$

and

$$p_1(y) = (1 - \beta_1)\beta_1^y, \quad y = 0, 1, 2, \ldots.$$

Assume that $0 < \beta_0 < \beta_1 < 1$.

(a) Find the Bayes rule for uniform costs and equal priors.
(b) Find the NP rule with false-alarm probability $\alpha \in (0, 1)$. Also find the corresponding probability of detection as a function of α.

2.21 *Discrete Observations.* Consider the binary detection problem with $\mathcal{Y} = \{0, 1, 2, 3\}$ and

$$p_1(y) = \frac{y}{6} \quad \text{and} \quad p_0(y) = \frac{1}{4}, \quad y \in \mathcal{Y}.$$

(a) Find a Bayes rule for equal priors and uniform costs.
(b) Find the Bayes risk for the rule in part (a).
(c) Assuming uniform costs, plot the Bayes risk lines for the following two tests:

$$\delta^{(1)}(y) = \begin{cases} 1 & \text{if } y = 2, 3 \\ 0 & \text{if } y = 0, 1 \end{cases} \quad \text{and} \quad \delta^{(2)}(y) = \begin{cases} 1 & \text{if } y = 3 \\ 0 & \text{if } y = 0, 1, 2. \end{cases}$$

(d) Find a minimax rule for uniform costs.
(e) Find a NP rule for level $\alpha = \frac{1}{3}$.
(f) Find P_D for the rule in part (e).

2.22 *Exotic Optimality Criterion.* Consider a binary detection problem, where the goal is to minimize the following risk measure

$$\rho(\tilde{\delta}) = [P_F(\tilde{\delta})]^2 + P_M(\tilde{\delta}).$$

(a) Show that the solution is a (possibly randomized) LRT.

(b) Find the solution for the observation model

$$p_0(y) = 1, \quad p_1(y) = 2y, \quad y \in [0, 1].$$

2.23 *Random Priors.* Consider the variation of the Bayesian detection problem where the prior probability π_0 (and hence π_1) is a random variable that is independent of the observation Y and the hypothesis. Further assume that π_0 has mean $\overline{\pi}_0 \in (0, 1)$. We consider the problem of minimizing the average risk

$$\overline{r}(\tilde{\delta}) = \mathbb{E}[\pi_0 R_0(\tilde{\delta}) + (1 - \pi_0) R_1(\tilde{\delta})],$$

where the expectation is over the distribution of π_0. Find a decision rule that minimizes $\overline{r}$. Comment on your solution.

2.24 *Properties of ROC.* Prove Theorem 2.4.
Hints: One way to solve Part (ii) of this problem is to consider the cases $\eta \leq 1$ and $\eta > 1$ separately and use the following change of measure argument: For any event $\mathcal{A}$,

$$\mathrm{P}_1(\mathcal{A}) = \int_{\mathcal{A}} p_1(y) d\mu(y) = \int_{\mathcal{A}} L(y) p_0(y) d\mu(y).$$

To solve Part (iii), start with

$$\frac{d\mathrm{P}_\mathrm{D}}{d\mathrm{P}_\mathrm{F}} = \frac{d\mathrm{P}_\mathrm{D}}{d\eta} \Big/ \frac{d\mathrm{P}_\mathrm{F}}{d\eta},$$

and then express P_D and P_F in terms of the pdf of $L(Y)$ under H_0 using a change of measure argument similar to that in Part (ii).

2.25 *AUC.* Since "best" tests exist only in trivial cases, the performance of different tests is often compared using the *area under the curve* (AUC) where the curve is the ROC. Derive the ROC and compute the AUC for the detection problems in Exercises 2.12 and 2.13.

2.26 *Worst Tests.*

(a) Find a test $\tilde{\delta}$ that *minimizes* $\mathrm{P}_\mathrm{D}(\tilde{\delta})$ subject to the constraint $\mathrm{P}_\mathrm{F}(\tilde{\delta}) \leq \alpha$.
Hint: Relate this problem to the NP problem at significance level $1 - \alpha$.
(b) Derive the set of all achievable probability pairs $(\mathrm{P}_\mathrm{F}(\tilde{\delta}), \mathrm{P}_\mathrm{D}(\tilde{\delta}))$.
Hint: This set can be expressed in terms of the ROC for the NP family of tests.

2.27 *Improving Suboptimal Tests.* Consider the pdfs

$$p_0(y) = \frac{1}{2} e^{-|y|} \quad \text{and} \quad p_1(y) = e^{-2|y|}, \quad y \in \mathbb{R}.$$

Let $\delta_0(y) = 0$ and $\delta_1(y) = 1$ for all y, and consider a randomized test $\tilde{\delta}_3$ that returns 1 with probability 0.5 if the observation $y < 0$, and returns $\mathbb{1}\{0 \leq y \leq \tau\}$ if $y \geq 0$; the threshold $\tau \geq 0$. All three tests are suboptimal.

(a) Derive and sketch the ROC for $\tilde{\delta}_3$.

(b) Show that suitable randomization between δ_0, δ_1, and $\tilde{\delta}_3$ improves the ROC for $P_F \in [0, \alpha_0] \cup [\alpha_1, 1]$ where $\alpha_0 = \frac{\sqrt{3}}{4}$ and $\alpha_1 = 1 - \alpha_0$. Show that the resulting ROC is a straight line segment for $P_F \in [0, \alpha_0]$, coincides with the ROC of $\tilde{\delta}_3$ for $P_F \in [\alpha_0, \alpha_1]$, and is another straight line segment for $P_F \in [\alpha_1, 1]$.

2.28 *Neyman–Pearson Hypothesis Testing With Laplacian Observations.* Consider the binary detection problem with

$$p_0(y) = \frac{1}{2}e^{-|y|}, \quad \text{and} \quad p_1(y) = \frac{1}{2}e^{-|y-2|}, \quad y \in \mathbb{R}.$$

This is a case where the likelihood ratio has point masses even though p_0 and p_1 have no point masses.

(a) Show that the Neyman–Pearson solution, which is a randomized LRT, is equivalent to a deterministic *threshold* test on the observation y.
(b) Find the ROC for the test in part (a).

3 Multiple Hypothesis Testing

The methods developed for binary hypothesis testing are extended to more than two hypotheses in this chapter. We first consider the problem of distinguishing between $m > 2$ hypotheses using a single test, which is called the m-ary detection problem. We study the Bayesian, minimax, and generalized Neyman–Pearson versions of the m-ary detection problem, and provide a geometric interpretation of the solutions. We then consider the problem of designing m binary tests and obtaining performance guarantees for the collection of tests (rather than for each individual test).

3.1 General Framework

In the basic multiple-hypothesis testing problem, both the state space and the action space have m elements each, say $\mathcal{X} = \mathcal{A} = \{0, 1, \dots, m-1\}$. The problem is called m-ary detection, or *classification*. It is assumed that under hypothesis H_j, the observation is drawn from a distribution P_j with associated density $p_j(y),\ y \in \mathcal{Y}$. The problem is to test between the m hypotheses

$$H_j \,:\, Y \sim p_j, \quad j = 0, 1, \dots, m-1 \tag{3.1}$$

given a cost matrix $\{C_{ij}\}_{0 \le i,j \le m-1}$.

The problem (3.1) arises in many types of applications ranging from handwritten digit recognition ($m = 10$, Y = image) to optical character recognition, speech recognition, biological and medical image analysis, sonar, radar, decoding, system identification, etc.

To any deterministic decision rule $\delta : \mathcal{Y} \to \{0, 1, \dots, m-1\}$, we may associate a partition $\mathcal{Y}_i,\ i = 0, 1, \dots, m-1$ of the observation space $\mathcal{Y}$ such that

$$\delta(y) = i \quad \Leftrightarrow \quad y \in \mathcal{Y}_i,$$

as illustrated in Figure 3.1. Hence we may write δ in the compact form

$$\delta(y) = \sum_{i=0}^{m-1} i\, \mathbb{1}\{y \in \mathcal{Y}_i\} \tag{3.2}$$

and refer to $\{\mathcal{Y}_i\}_{i=1}^{m}$ as the *decision regions*.

The conditional risks for δ are given by

$$R_j(\delta) = \sum_{i=0}^{m-1} C_{ij} \mathrm{P}_j(\mathcal{Y}_i), \quad j = 0, 1, \dots, m-1, \tag{3.3}$$

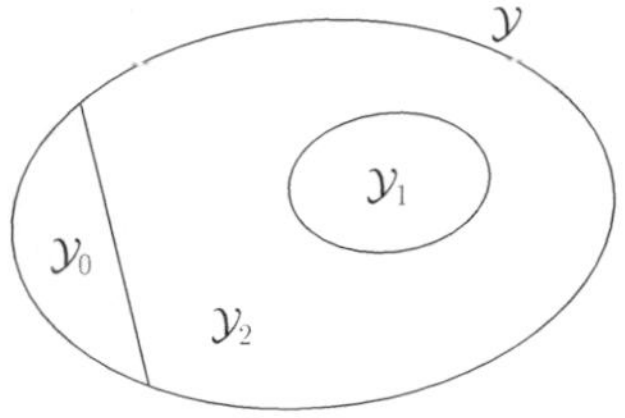

Figure 3.1 Ternary hypothesis testing: decision regions $\mathcal{Y}_0, \mathcal{Y}_1, \mathcal{Y}_2$ associated with the decision rule δ

where $\mathrm{P}_j(\mathcal{Y}_i) = \mathrm{P}\{\text{Say } H_i | H_j \text{ is true}\}$ is the probability of confusing hypothesis i with the actual hypothesis j when $i \neq j$. In the case of uniform costs, $C_{ij} = \mathbb{1}\{i \neq j\}$, the conditional risks are conditional error probabilities:

$$R_j(\delta) = 1 - \mathrm{P}_j(\mathcal{Y}_j), \quad j = 0, 1, \ldots, m-1, \tag{3.4}$$

where $\mathrm{P}_j(\mathcal{Y}_j)$ is the probability that observation Y is correctly classified when H_j is true.

3.2 Bayesian Hypothesis Testing

In the Bayesian setting, each hypothesis H_j is assumed to have prior probability π_j. The posterior probability distribution

$$\pi(j|y) = \frac{\pi_j p_j(y)}{p(y)}, \quad j = 0, 1, \ldots, m-1$$

plays a central role in the Bayesian problem. As seen in (2.7), the posterior cost of decision i is

$$C(i|y) = \sum_{j=0}^{m-1} C_{ij}\pi(j|y) = \sum_{j=0}^{m-1} C_{ij}\frac{\pi_j p_j(y)}{p(y)} \tag{3.5}$$

and the Bayes decision rule is

$$\delta_{\mathrm{B}}(y) = \arg\min_{0 \le i \le m-1} C(i|y) = \arg\min_{0 \le i \le m-1} \sum_{j=0}^{m-1} C_{ij}\frac{\pi_j p_j(y)}{p(y)}. \tag{3.6}$$

The Bayes risk for any decision rule δ is given by

$$\begin{aligned} r(\delta) &= \sum_{j=0}^{m-1} \pi_j R_j(\delta) \\ &= \sum_{i,j=0}^{m-1} C_{ij}\pi_j \mathrm{P}_j(\mathcal{Y}_i). \end{aligned} \tag{3.7}$$

Uniform costs For uniform costs, (3.4) and (3.7) yield

$$r(\delta) = 1 - \sum_{j=0}^{m-1} \pi_j \mathrm{P}_j(\mathcal{Y}_j). \tag{3.8}$$

The posterior cost of (3.5) is given by

$$C(i|y) = 1 - \pi(i|y),$$

hence

$$\delta_{\mathrm{B}}(y) = \arg\min_{0 \le i \le m-1} C(i|y) = \arg\max_{0 \le i \le m-1} \pi(i|y), \tag{3.9}$$

which is the MAP decision rule. Its risk is the decision error probability,

$$\mathrm{P_e}(\delta_{\mathrm{B}}) = r(\delta_{\mathrm{B}}) = \mathbb{E}[\min_i C(i|Y)] = 1 - \mathbb{E}[\max_i \pi(i|Y)]. \tag{3.10}$$

3.2.1 Optimal Decision Regions

Observe that $p(y)$ in the denominator of the right hand side of (3.6) does not depend on i and therefore has no effect on the minimization. It will sometimes be convenient to use $p_0(y)$ as a normalization factor instead of $p(y)$ and write

$$\begin{aligned} \delta_{\mathrm{B}}(y) &= \arg\min_{0 \le i \le m-1} \sum_{j=0}^{m-1} C_{ij} \frac{\pi_j p_j(y)}{p_0(y)} \\ &= \arg\min_{0 \le i \le m-1} \sum_{j=0}^{m-1} C_{ij} \pi_j L_j(y), \end{aligned}$$

where we have introduced the likelihood ratios

$$L_j(y) \triangleq \frac{p_j(y)}{p_0(y)}, \quad i = 0, 1, \ldots, m-1, \tag{3.11}$$

and $L_0(y) \equiv 1$.

In the sequel, we consider the case $m = 3$ (ternary hypothesis testing) for which the concepts can be easily visualized. Then

$$\delta_{\mathrm{B}}(y) = \arg\min_{i=0,1,2} \underbrace{[C_{i0}\pi_0 + C_{i1}\pi_1 L_1(y) + C_{i2}\pi_2 L_2(y)]}_{\triangleq f_i(y)}, \tag{3.12}$$

where we have introduced the *discriminant functions*

$$f_i(y) = C_{i0}\pi_0 + C_{i1}\pi_1 L_1(y) + C_{i2}\pi_2 L_2(y), \quad i = 0, 1, 2,$$

which are linear in the likelihood ratios $L_1(y)$ and $L_2(y)$.

The two-dimensional (2-D) random vector $T \triangleq [L_1(Y), L_2(Y)]$ contains all the information needed to make an optimal decision. We may express the minimization problem of (3.12) in the 2-D decision space with coordinates $L_1(y)$ and $L_2(y)$, as shown in Figure 3.2. The decision regions for H_0, H_1, and H_2 are separated by lines. For instance,

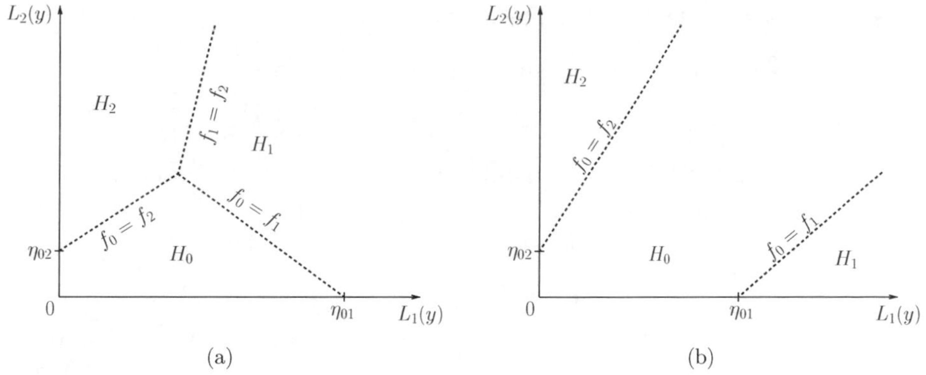

Figure 3.2 Typical decision regions for ternary hypothesis testing

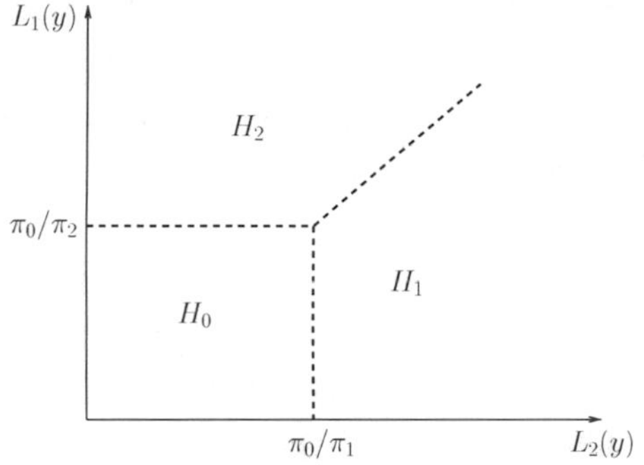

Figure 3.3 Decision regions for ternary hypothesis testing with uniform costs

the decision boundary for H_0 versus H_1 satisfies the equality $f_0(y) = f_1(y)$, which is the linear equation

$$(C_{00} - C_{10})\pi_0 + (C_{01} - C_{11})\pi_1 L_1(y) + (C_{02} - C_{12})\pi_2 L_2(y) = 0. \tag{3.13}$$

In the decision space of Figure 3.2, this is a line with slope equal to $\frac{C_{01}-C_{11}}{C_{12}-C_{02}}\frac{\pi_1}{\pi_2}$ and x-intercept equal to $\eta_{01} = \frac{C_{10}-C_{00}}{C_{01}-C_{11}}\frac{\pi_0}{\pi_1}$.

In the special case of uniform costs, $C_{ij} = \mathbb{1}\{i \neq j\}$, we have

$$f_i(y) = \frac{p(y)}{p_0(y)} - \pi_i L_i(y), \quad i = 0, 1, 2,$$

hence

$$\delta_{\mathrm{B}}(y) = \arg\max_{i=0,1,2} \{\pi_0, \pi_1 L_1(y), \pi_2 L_2(y)\}, \tag{3.14}$$

which is the MAP classification rule. The decision space is shown in Figure 3.3.

For $m > 3$, the same approach can be used, with a $m-1$ dimensional vector of likelihood ratios $T = [L_1(Y), \ldots, L_{m-1}(Y)]$, m discriminant functions $f_i(y), i = 0, 1, \ldots, m-1$, and $\frac{m(m-1)}{2}$ hyperplanes separating each pair of hypotheses (comparing pairs of discriminant functions).

3.2.2 Gaussian Ternary Hypothesis Testing

Assume uniform costs and consider the test

$$H_j : Y \sim \mathcal{N}(\mu_j, \mathsf{K}_j), \quad j = 0, 1, 2,$$

where $Y \in \mathbb{R}^d$ follows an n-dimensional Gaussian distribution with mean $\mu_j = \mathbb{E}_j[Y]$ and covariance matrix $\mathsf{K}_j = \mathrm{Cov}_j[Y] \triangleq \mathbb{E}_j[(Y-\mu_j)(Y-\mu_j)^\top]$ under H_j. Assume each $\pi_j > 0$ and each K_j is full rank. The likelihood ratios of (3.11) are given by

$$L_i(y) = \frac{p_i(y)}{p_0(y)} = \frac{(2\pi)^{-d/2}|\mathsf{K}_i|^{-1/2}\exp\{-\frac{1}{2}(y-\mu_i)^\top \mathsf{K}_i^{-1}(y-\mu_i)\}}{(2\pi)^{-d/2}|\mathsf{K}_0|^{-1/2}\exp\{-\frac{1}{2}(y-\mu_0)^\top \mathsf{K}_0^{-1}(y-\mu_0)\}}, \quad i = 0, 1, 2.$$

It is more convenient to work with the log-likelihood ratios, which are quadratic functions of y for each $i = 0, 1, 2$:

$$\ln L_i(y) = -\frac{1}{2}\ln|\mathsf{K}_i| + \frac{1}{2}\ln|\mathsf{K}_0| - \frac{1}{2}(y-\mu_i)^\top \mathsf{K}_i^{-1}(y-\mu_i) + \frac{1}{2}(y-\mu_0)^\top \mathsf{K}_0^{-1}(y-\mu_0).$$

Since each $\pi_i > 0$ and the logarithm function is monotonically increasing, the Bayes rule of (3.14) can be expressed in terms of the test statistic $T = (\ln L_1(Y), \ln L_2(Y))$:

$$\delta_{\mathrm{B}}(y) = \arg\min_{i=0,1,2} \{\ln \pi_0, \ln \pi_1 + \ln L_1(y), \ln \pi_2 + \ln L_2(y)\}.$$

Alternatively, for $d = 2$, we may view the decision boundaries directly in the observation space $\mathbb{R}^2$, as shown in Figure 3.4. The decision boundaries are quadratic functions of y. If all three covariance matrices are identical ($\mathsf{K}_0 = \mathsf{K}_1 = \mathsf{K}_2$), the decision boundaries are linear in y.

3.3 Minimax Hypothesis Testing

We turn our attention to minimax hypothesis testing. As in the binary case, it will generally be useful to consider randomized decision rules. Recall from Definition 1.4 that a randomized decision rule $\tilde{\delta}$ picks a rule at random from a set of deterministic decision rules $\{\delta_l\}_{l=1}^L$ with respective probabilities $\{\gamma_l\}_{l=1}^L$. The worst-case risk for a randomized decision rule $\tilde{\delta}$ is

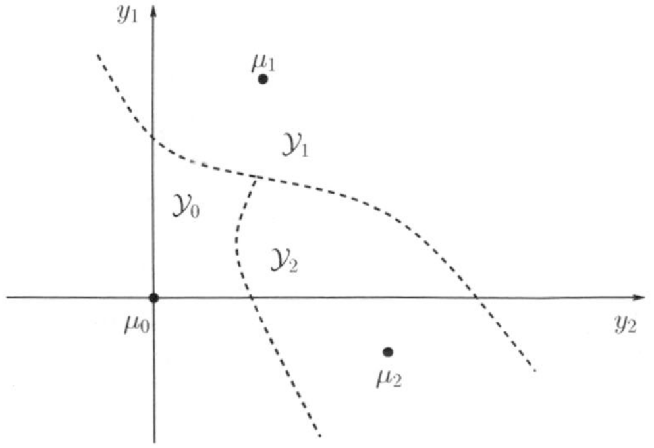

Figure 3.4 Quadratic decision boundaries for Gaussian hypothesis testing ($Y \in \mathbb{R}^2$)

$$R_{\max}(\tilde{\delta}) = \max_{0 \le j \le m-1} R_j(\tilde{\delta}). \tag{3.15}$$

The sets of deterministic and randomized decision rules are denoted by $\mathcal{D}$ and $\widetilde{\mathcal{D}}$, respectively.

Similarly to the notion of risk lines and risk curves in Section 2.3.2, we can gain intuition by considering a geometric representation of the problem.

DEFINITION 3.1 The *risk vector* of a decision rule $\tilde{\delta} \in \widetilde{\mathcal{D}}$ is

$$R(\tilde{\delta}) = [R_0(\tilde{\delta}), R_1(\tilde{\delta}), \dots, R_{m-1}(\tilde{\delta})] \in \mathbb{R}^m.$$

DEFINITION 3.2 The *risk set* for $\mathcal{D}$ is $R(\mathcal{D}) \triangleq \{R(\delta),\ \delta \in \mathcal{D}\} \quad \subset \mathbb{R}^m$.

DEFINITION 3.3 The *risk set* for $\widetilde{\mathcal{D}}$ is $R(\widetilde{\mathcal{D}}) \triangleq \{R(\tilde{\delta}),\ \tilde{\delta} \in \widetilde{\mathcal{D}}\} \quad \subset \mathbb{R}^m$.

By Proposition 1.3, we have $R(\tilde{\delta}) = \sum_{\ell=1}^{L} \gamma_\ell R(\delta_\ell)$ for a randomized rule $\tilde{\delta}$ that picks deterministic rule δ_ℓ with probability γ_ℓ. Hence the following proposition:

PROPOSITION 3.1 *The risk set $R(\widetilde{\mathcal{D}})$ is the convex hull of $R(\mathcal{D})$.*

The risk set $R(\mathcal{D})$ for a finite $\mathcal{D}$ is collection of at most $|\mathcal{D}|$ points, and its convex hull $R(\widetilde{\mathcal{D}})$ is a convex m-polytope, as illustrated in Figure 3.5 for the binary communication Example 2.3 ($m = 2$). The polytope has at most $|\mathcal{D}|$ vertices. If $\mathcal{D}$ is a continuum, the risk set $R(\widetilde{\mathcal{D}})$ is convex but does not have vertices.[1]

In the Bayesian setting, the probabilities assigned to the m hypotheses can be represented by a probability vector π on the probability simplex in $\mathbb{R}^m$. The risk of a rule $\tilde{\delta}$ under prior π is then given by

$$r(\pi, \tilde{\delta}) = \sum_j \pi_j R_j(\tilde{\delta}) = \pi^\top R(\tilde{\delta}). \tag{3.16}$$

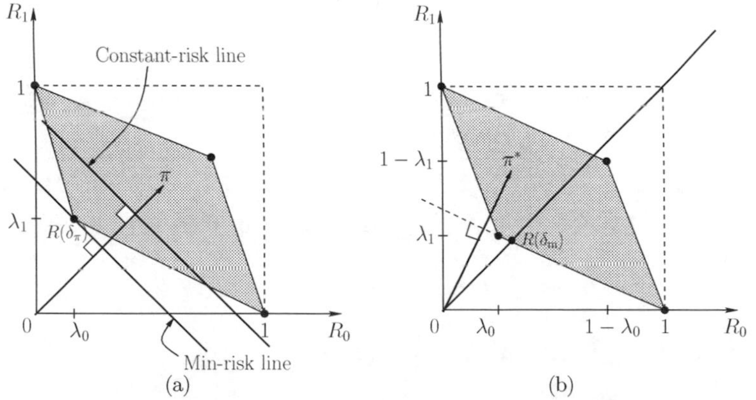

Figure 3.5 Risk set $R(\widetilde{\mathcal{D}})$ for binary hypothesis testing Example 2.3. The four possible deterministic rules are the vertices of the (shaded) risk set $R(\widetilde{\mathcal{D}})$. Geometric view of (a) probability vector π and risk vector $R(\delta_\pi)$ for Bayes rule δ_π under prior π; and (b) probability vector π^* and risk vector $R(\delta_m)$ for minimax rule δ_m and least-favorable prior π^*

[1] Recall that a set $\mathcal{R}$ is said to be convex if for all $u, u' \in \mathcal{R}$, the element $\lambda u + (1-\lambda)u' \in \mathcal{R}$ for all $\lambda \in [0, 1]$.

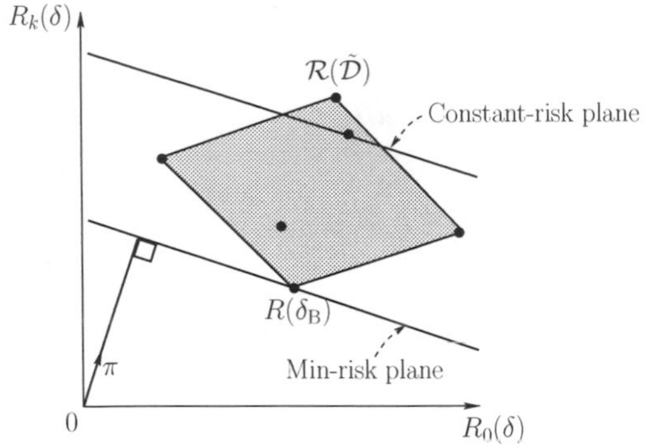

Figure 3.6 Two-dimensional section of risk set $R(\widetilde{\mathcal{D}})$: geometric view of Bayes rule δ_B

As illustrated in Figure 3.6, all the risk vectors that lie in a plane orthogonal to the probability vector π have the same Bayes risk which is proportional to the distance from the plane to the origin.[2] The risk vector for the Bayes rule δ_B is therefore a point of tangency of such a plane to (from below) the risk set $R(\widetilde{\mathcal{D}})$. The minimum risk is always achieved at an extremal point of $R(\widetilde{\mathcal{D}})$, hence no randomization is needed.

In the minimax setting, all the risk vectors that lie in the closed set

$$\mathcal{O}(r) \triangleq \{R \in \mathbb{R}^m \ : \ R_j \le r \ \forall j, \text{ with equality for at least one } j\}$$

(boundary of an m-dimensional orthant) have the same worst-case risk (3.15). The vector $(r, r, \dots, r)$ is the "corner point" of the orthant, as illustrated in Figure 3.7. The minimax solution δ_m is therefore a point of tangency of the risk set $R(\mathcal{D})$ with $\mathcal{O}(r)$ that has the lowest possible value of r, and that value is the minimax risk. In some problems, the minimax solution is an equalizer rule, represented by a "corner point" of the set $R(r)$. This situation occurs in Figures 3.5(b) and 3.7(a) but not in Figure 3.7(b). The least favorable prior π^* can also be found using geometric reasoning. If $V(\pi) \triangleq r(\pi; \delta_B)$ is the minimum Bayes risk under prior π, then the point $(V(\pi), \dots, V(\pi))$ with equal coordinates is in the tangent plane to $R(\widetilde{\mathcal{D}})$ at $R(\delta_B)$. Hence the least favorable prior is the one that maximizes $V(\pi)$. Moreover, if the point $(V(\pi), \dots, V(\pi)) \in R(\widetilde{\mathcal{D}})$, that point is a Bayes rule as well as an equalizer rule.

Example 3.1 Let $\mathcal{X} = \mathcal{Y} = \{0, 1, 2\}$ and consider the following three distributions:

$$p_0(0) = \frac{1}{2} \ p_0(1) = \tfrac{1}{2} \ p_0(2) = 0,$$
$$p_1(0) = \frac{1}{3} \ p_1(1) = \tfrac{1}{3} \ p_1(2) = \frac{1}{3},$$
$$p_2(0) = 0 \ p_2(1) = \tfrac{1}{2} \ p_2(2) = \frac{1}{2}.$$

Assume uniform costs. Find the minimax decision rule and the least-favorable prior.

[2] The proportionality constant is $1/\|\pi\|$.

Table 3.1 All admissible decision rules for Example 3.1, with the corresponding decision regions, conditional error probabilities, worst-case risk, and Bayes risk under least favorable prior $\pi^* = [2/7, 3/7, 2/7]^\top$

δ	$\mathcal{Y}_0$	$\mathcal{Y}_1$	$\mathcal{Y}_2$	$R_0(\delta)$	$R_1(\delta)$	$R_2(\delta)$	$r_{\max}(\delta)$	$r(\pi^*, \delta)$
δ_1	$\{0, 1\}$	$\{2\}$	$\emptyset$	0	2/3	1	1	4/7
δ_2	$\{0, 1\}$	$\emptyset$	$\{2\}$	0	1	1/2	1	4/7
δ_3	$\{0\}$	$\{1, 2\}$	$\emptyset$	1/2	1/3	1	1	4/7
δ_4	$\{0\}$	$\{1\}$	$\{2\}$	1/2	2/3	1/2	2/3	4/7
δ_5	$\{0\}$	$\{2\}$	$\{1\}$	1/2	2/3	1/2	2/3	4/7
δ_6	$\{0\}$	$\emptyset$	$\{1, 2\}$	1/2	1	0	1	4/7
δ_7	$\{1\}$	$\{0, 2\}$	$\emptyset$	1/2	1/3	1	1	4/7
δ_8	$\{1\}$	$\{0\}$	$\{2\}$	1/2	2/3	1/2	2/3	4/7
δ_9	$\emptyset$	$\{0, 1, 2\}$	$\emptyset$	1	0	1	1	4/7
δ_{10}	$\emptyset$	$\{0, 1\}$	$\{2\}$	1	1/3	1/2	1	4/7
δ_{11}	$\emptyset$	$\{0, 2\}$	$\{1\}$	1	1/3	1/2	1	4/7
δ_{12}	$\emptyset$	$\{0\}$	$\{1, 2\}$	1	2/3	0	1	4/7

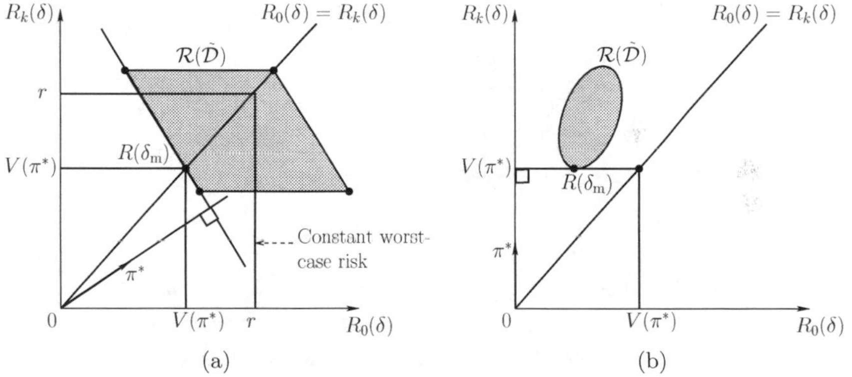

Figure 3.7 Two-dimensional section of risk set $R(\tilde{\mathcal{D}})$: geometric view of minimax rule δ_{m} in the case δ_{m} is (a) an equalizer rule; (b) not an equalizer rule

While there are $|\mathcal{D}| = 3^3 = 27$ possible decision rules, we may exclude any rule such that $\delta(0) = 2$ and $\delta(2) = 0$. Hence there are two possible actions to consider when $Y = 0$ or $Y = 2$, and three actions when $Y = 1$, for a total of 12 decision rules. Their conditional risks (conditional error probabilities $R_j(\delta) = 1 - \mathrm{P}_j(\mathcal{Y}_j)$ for $j = 0, 1, 2$) are listed in Table 3.1.

The best deterministic rules are δ_4, δ_5, and δ_8, with a worst-case risk of $\frac{2}{3}$; moreover, all three rules have the same risk vector $R = [1/2, 2/3, 1/2]^\top$. However, the worst-case risk can be reduced using randomization. Consider $\delta_4(y) = y$ and $\delta_9(y) = 1$; the latter has risk vector $R(\delta_9) = [1, 0, 1]^\top$. If we pick δ_4 with probability γ and δ_9 with probability $1 - \gamma$, we obtain a randomized rule $\tilde{\delta}$ with risk vector

$$R(\tilde{\delta}) = \gamma R(\delta_4) + (1-\gamma)R(\delta_9) = \left[1 - \frac{\gamma}{2}, \frac{2\gamma}{3}, 1 - \frac{\gamma}{2}\right]^\top .$$

An equalizer rule is obtained by choosing $\gamma = \frac{6}{7}$. All three conditional risks are then equal to $\frac{4}{7}$, hence $R_{\max}(\tilde{\delta}) = \frac{4}{7}$ is lower than $\frac{2}{3}$ achieved by the best deterministic rules of Table 3.1.

While $\tilde{\delta}$ constructed above is an equalizer rule, the question is whether it is also a Bayes rule. The Bayes risks of rules δ_4 and δ_9 are respectively

$$r(\pi; \delta_4) = \frac{1}{2}\pi_0 + \frac{2}{3}\pi_1 + \frac{1}{2}\pi_2 = \frac{3+\pi_1}{6},$$
$$r(\pi; \delta_9) = \pi_0 + \pi_2 = 1 - \pi_1.$$

If $\tilde{\delta}$ is a Bayes rule, we must have $r(\pi; \delta_4) = r(\pi; \delta_9)$, hence $\pi_1 = \frac{3}{7}$. Then $r(\pi; \delta_4) = r(\pi; \delta_9) = \frac{4}{7} = R_{\max}(\tilde{\delta})$. By symmetry, we may guess that the least favorable prior satisfies $\pi_0^* = \pi_2^*$, hence $\pi^* = [2/7, 3/7, 2/7]^\top$. Computing the risks of all decision rules under π^* in the last column of Table 3.1, we see that none of them is lower than $\frac{4}{7}$.[3] We conclude that $\tilde{\delta}$ is a minimax rule and that π^* is a least-favorable prior.

In this problem the minimax rule $\tilde{\delta}$ is not unique: one could have randomized between δ_5 and δ_9, or between δ_8 and δ_9. One could even have randomized among any subset of $\{\delta_4, \delta_5, \delta_8, \delta_9\}$, provided δ_9 is included in the mix and receives a probability weight of $\frac{1}{7}$. However, there is no advantage in randomizing between more than two rules in this problem.

3.4 Generalized Neyman–Pearson Detection

The Neyman–Pearson optimality criterion can be extended to m-ary detection in several ways. For instance, when $m \ge 3$ one may want to minimize the conditional risk $R_{m-1}(\tilde{\delta})$ subject to the inequality constraints $R_k(\tilde{\delta}) \le \alpha_k$ for $k = 0, 1, \ldots, m-2$. Since the constraints are linear in $\tilde{\delta}$, the problem can be formulated as a linear optimization problem as in Section 2.4.5. The solution admits a geometric interpretation: each constraint $R_k(\tilde{\delta}) \le \alpha_k$ defines a half-plane. The feasible set, which is the intersection of the (convex) risk set $R(\widetilde{\mathcal{D}})$ with those $m-1$ half spaces, is convex. The minimum of a linear function over a convex set is achieved at an extremal point on the feasible set. The solution is a randomized Bayes rule. Since there are $m-1$ constraints, the Bayes rule randomizes between m deterministic rules at most.

3.5 Multiple Binary Tests

We now consider the problem of designing m binary tests

$$H_{0i} \quad \text{versus} \quad H_{1i}, \qquad i = 1, 2, \ldots, m. \tag{3.17}$$

[3] In fact $r(\pi^*, \delta) = \frac{4}{7}$ for all 12 rules listed in Table 3.1. Note, however, that some of the other 15 rules are worse, for instance $\delta(y) = 0$ has risk vector $R = [0, 1, 1]$ and therefore $r(\pi^*, \delta) = \frac{5}{7}$.

Hence the state space and the action space are $\mathcal{X} = \mathcal{A} = \{0, 1\}^m$. This inference model occurs in a number of important applications, including data mining, medical testing, signal analysis, and genomics. Typically each test corresponds to a question about the observed dataset $\mathcal{Y}$, e.g., does $\mathcal{Y}$ satisfy a certain property or not? The number of tests could be very large, each of them leading to a potential *discovery* about $\mathcal{Y}$.

As we have seen in Chapter 2, a central problem in binary hypothesis testing is to control the tradeoff between false alarms and missed detections (missed discoveries). The problem is clearly more complex for multiple binary tests. The classical analogue to the probability of false alarm is the *family-wise error rate*

$$\text{FWER} \triangleq \text{P}_0\left\{\cup_{i=1}^m \{\text{say } H_{1i}\}\right\}, \tag{3.18}$$

where P_0 is the distribution of the observations when *all* null hypotheses are true. The missed detections can be quantified by the average power of the tests, defined as

$$\overline{\beta}_m \triangleq \frac{1}{m}\sum_{i=1}^m \text{P}_{1i}\{\text{say } H_{1i}\}, \tag{3.19}$$

where P_{1i} is the distribution of the test statistic for the ith test under H_{1i}. As we shall see, a simple and classical solution is to pick a significance level $\alpha \in (0, 1)$ and apply the *Bonferroni correction* [1], which guarantees that $\text{FWER} \le \alpha$. A more recent and powerful approach is the *Benjamini–Hochberg procedure* [2] which is adaptive and controls the *false discovery rate*, i.e., the expected fraction of false discoveries, to be defined precisely in Section 3.5.2.

First we introduce the notion of p-value of a binary test. Assume a single binary test H_0 versus H_1 taking the form $\delta(y) = \mathbb{1}\{t(y) > \eta\}$ where $t(y)$ is the test statistic, and η is the threshold of the test. For instance, $t(y)$ could be the likelihood ratio $p_1(y)/p_0(y)$, but other choices are possible.

DEFINITION 3.4 The p-value of the test statistic is $p \triangleq \text{P}_0\{T \ge t(y)\}$, where $T \sim \text{P}_0$.

In other words, the p-value is the right tail probability of the distribution P_0, evaluated at $t(y)$. If $t(Y)$ follows the distribution P_0, the p-value is a random variable, and is uniformly distributed on $[0, 1]$. The smaller the p-value, the more likely it is that H_1 is true. To guarantee a probability of false alarm equal to α (as in a NP test), one can select the test threshold η such that $\text{P}_0\{t(Y) > \eta\} = \alpha$. (For simplicity we assume that P_0 has no mass at $\{t(Y) = \eta\}$, hence no randomization is needed.) Equivalently, the test can be expressed in terms of the p-value as $\delta(p) = \mathbb{1}\{p < \alpha\}$.

Now consider m tests $\delta_i : [0, 1] \to \{0, 1\}$, $i = 1, 2, \dots, m$, each of which uses a p-value Q_i as its test statistic.[4] Denote by P_0 the joint distribution of $\{Q_i\}_{i=1}^m$ when all null hypotheses are true.

3.5.1 Bonferroni Correction

If one selects α as the significance level of each individual test, then FWER equals α if $m = 1$ but increases with m. More precisely,

[4] These random p-values are not denoted by $\{P_i\}$ so as not to cause confusion with probability measures.

Table 3.2 Notation for multiple binary hypothesis testing

	Declared true (accepted)	Declared false (rejected)	Total
True null hypotheses	A_c	R_e	m_0
False null hypotheses	A_e	R_c	$m - m_0$
Total	$A = m - R$	R	m

$$\mathrm{FWER} \le \sum_{i=1}^{m} \mathrm{P}_0\{\text{say } H_{1i}\} = m\alpha,$$

where the inequality follows from the union bound (*aka* the Bonferroni inequality), and holds whether the test statistics $\{Q_i\}_{i=1}^m$ are independent or not. In the special case of independent p-values $\{Q_i\}_{i=1}^m$, we have $\mathrm{FWER} = 1 - (1-\alpha)^m \approx m\alpha$ for small $m\alpha$.

To reduce FWER, one could set the significance level of each test to α/m. This is the *Bonferroni correction*. Unfortunately, doing so reduces the average power of the tests.

3.5.2 False Discovery Rate

Table 3.2 introduces some notation. Denote by $\boldsymbol{\delta} = \{\delta_i\}_{i=1}^m$ a collection of m binary tests and by m_0 the number of true null hypotheses. Of these, a number A_c are correctly accepted, and R_e are erroneously rejected. There are also $m - m_0$ false null hypotheses. Of these, a number A_e are erroneously accepted, and R_c are correctly rejected. Denote by $R = R_\mathrm{c} + R_\mathrm{e}$ the total number of rejected null hypotheses, and by $A = m - R$ the total number of accepted null hypotheses. The random variables $A_\mathrm{c}, A_\mathrm{e}, R_\mathrm{e}, R_\mathrm{c}$ are unobserved, but A and R are.

With this notation, we have $\mathrm{FWER}(\boldsymbol{\delta}) = \mathrm{P}_0\{R_\mathrm{e} \ge 1\}$. The fraction of errors occurring by erroneously rejecting null hypotheses is the random variable

$$F \triangleq \begin{cases} \frac{R_\mathrm{e}}{R_\mathrm{e}+R_\mathrm{c}} = \frac{R_\mathrm{e}}{R} & : R > 0 \\ 0 & : R = 0. \end{cases} \tag{3.20}$$

The false discovery rate (FDR) is defined as the expected value of F:

$$\mathrm{FDR}(\boldsymbol{\delta}) \triangleq \mathbb{E}[F],$$

where the expectation is with respect to the joint distribution P of $\{Q_i\}_{i=1}^m$ – which depends on the underlying combination of true and false null hypotheses. Note that if $m_0 = m$ (all null hypotheses are true) then $R_\mathrm{e} = R$, implying that in this case $\mathrm{FDR}(\boldsymbol{\delta}) = \mathrm{P}_0\{R_\mathrm{e} \ge 1\} = \mathrm{FWER}(\boldsymbol{\delta})$.

The Benjamini–Hochberg procedure selects a target $\alpha \in (0, 1)$ and makes decisions on each hypothesis in a way that guarantees that $\mathrm{FDR} \le \alpha$ for all P. Hence one might have to contend with several false alarms (in average), but this allows an increase in the average power of the tests (enabling more discoveries). This procedure is particularly attractive for problems involving large m, and has been found relevant in several types of problems:

- overall decision making based on multiple inferences, e.g., target detection in a network [6];
- multiple individual decisions (an overall decision is not required), e.g., voxel classification in brain imaging [7].

3.5.3 Benjamini–Hochberg Procedure

Fix $\alpha \in (0, 1)$. Given m p-values $Q_1, \ldots, Q_m$, reorder them in nondecreasing order as $Q_{(1)} \leq Q_{(2)} \leq \cdots \leq Q_{(m)}$, and denote the corresponding null hypotheses by $H_{(1)}, H_{(2)}, \ldots, H_{(m)}$. Let k be the largest i such that $Q_{(i)} \leq \frac{i}{m}\alpha$. Then reject the null hypotheses $H_{(1)}, H_{(2)}, \ldots, H_{(k)}$ (report k discoveries). Denote by $\boldsymbol{\delta}_{\mathrm{BH}}$ the resulting collection of tests.

THEOREM 3.1 *For independent* $Q_1, \ldots, Q_m$ *and any configuration of false null hypotheses,* $\mathrm{FDR}(\boldsymbol{\delta}_{\mathrm{BH}}) \leq \alpha$.

The theorem has been extended to the case of positively correlated test statistics [3].

Example 3.2 The following example is taken from [2, Sec. 4]. Let m be a multiple of 4 and assume that

$$\mathrm{P}_{0i} = \mathcal{N}(0, 1),\ 1 \leq i \leq m, \quad \mathrm{P}_{1i} = \mathcal{N}(j\mu, 1),\ j = 1, 2, 3, 4,\ \frac{(j-1)m}{4} < i \leq \frac{jm}{4}.$$

Let $\alpha = 0.05$ and $\mu = 1.25$. The center column of Figure 3.8 shows the average power $\overline{\beta}_m$ of the Benjamini–Hochberg (BH) procedure for $m = 4, 8, 16, 32, 64$. The simulations show that for large m, the average power is substantially larger than that obtained using the Bonferroni correction method.

Applications of the BH procedure reported in the literature include sparse signal analysis [4], detection of the number of incoming signals in array signal processing [5], target detection in networks [6], neuro-imaging [7], and genomics [8].

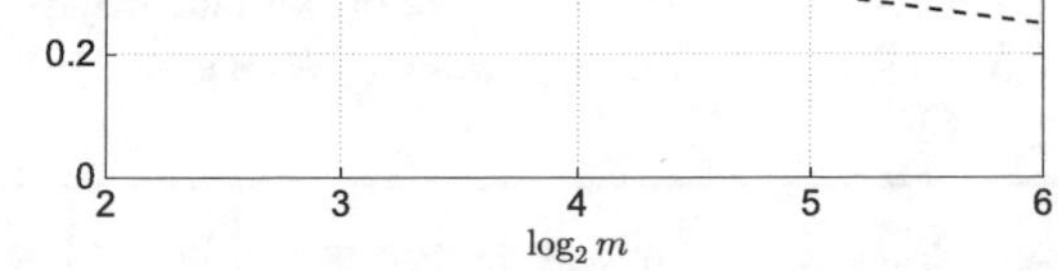

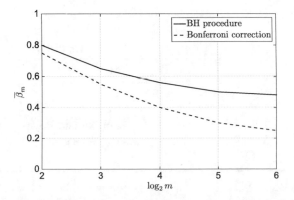

Figure 3.8 Application of BH procedure to Example 3.2 with 50 percent true null hypotheses: average power $\overline{\beta}_m$ versus $\log_2 m \in \{2, 3, 4, 5, 6\}$.

3.5.4 Connection to Bayesian Decision Theory

While the BH procedure may seem *ad hoc*, it admits an interesting connection to Bayes theory. To see this, consider the simplified problem where the distributions $P_{0i} = P_0$ and $P_{1i} = P_1$ are the same for all i, and each null hypothesis H_{0i} is true with probability π_0. Denote by F_0 and F_1 the cdfs of the test statistic T under distributions P_0 and P_1, respectively, and by $F(t) = \pi_0 F_0(t) + \pi_1 F_1(t)$ the cdf of the mixture distribution. The conditional probability of the null hypothesis given $T \ge t$ is given by

$$\pi(0|T \ge t) = \pi_0 \frac{P_0\{T \ge t\}}{P\{T \ge t\}} = \pi_0 \frac{1 - F_0(t)}{1 - F(t)}, \tag{3.21}$$

where $P = \pi_0 P_0 + \pi_1 P_1$ is the mixture distribution. Now for $T \sim P_0$, the p-values defined by $p = 1 - F_0(t)$ are uniformly distributed, hence their cdf is given by $G_0(\gamma) = \gamma$ for $\gamma \in [0, 1]$. For $T \sim P$, the p-values follow the mixture distribution with cdf equal to $G(\gamma) = F(F_0^{-1}(1 - \gamma))$.

Given the p-values $\{Q_i\}_{i=1}^m$, we may construct the following empirical cdf:

$$\widehat{G}(\gamma) = \frac{1}{m} \sum_{i=1}^m \mathbb{1}\{Q_i \le \gamma\}, \quad \gamma \in [0, 1],$$

which converges uniformly to $G(\gamma)$ as $m \to \infty$ by the Kolmogorov–Smirnov limit theorem. Observe that $\widehat{G}(Q_{(i)}) = i/m$, hence the BH procedure may be written as: Let k be the largest i such that $Q_{(i)} \le \widehat{G}(Q_{(i)})\alpha$. For large m, the ratio k/m converges to the solution of the nonlinear equation

$$\gamma = G(\gamma)\,\alpha. \tag{3.22}$$

Moreover, for large m, the conditional probability of (3.21) may be approximated as

$$\pi(0|P \le \gamma) \approx \pi_0 \frac{\gamma}{G(\gamma)} = \pi_0 \alpha.$$

Since this conditional probability can be interpreted as the false discovery rate of the tests $\{Q_i \le \gamma\}$, we denote it by $\mathrm{Fdr}(\gamma)$. Moreover,

$$\begin{aligned}
\mathrm{FDR}(\boldsymbol{\delta}_{\mathrm{BH}}) = \mathbb{E}[F] &\approx \mathbb{E} \frac{\frac{1}{m}\sum_{i=1}^m \mathbb{1}\{Q_i \le \gamma,\ H_{0i} \text{ true}\}}{\frac{1}{m}\sum_{i=1}^m \mathbb{1}\{Q_i \le \gamma\}} \\
&\approx \frac{\mathbb{E}[\frac{1}{m}\sum_{i=1}^m \mathbb{1}\{Q_i \le \gamma,\ H_{0i} \text{ true}\}]}{\mathbb{E}[\frac{1}{m}\sum_{i=1}^m \mathbb{1}\{Q_i \le \gamma\}]} \\
&= \frac{\pi_0 \gamma}{G(\gamma)} = \mathrm{Fdr}(\gamma) = \pi_0 \alpha.
\end{aligned}$$

Hence one may interpret the BH procedure as an approximate Bayesian procedure that assumes very few discoveries ($\pi_0 \approx 1$) and guarantees a false discovery rate of α.

Example 3.3 Let $\sigma > 1$ and assume the Bayesian setup where $P_{0i} = P_0 = \mathrm{Exp}(1)$ and $P_{1i} = P_1 = \mathrm{Exp}(\sigma)$ are the same for all i. Hence

$$p_0(t) = e^{-t}, \qquad p_1(t) = \frac{1}{\sigma} e^{-t/\sigma},$$

$$1 - F_0(t) = e^{-t}, \qquad 1 - F_1(t) = e^{-t/\sigma}, \qquad t \geq 0.$$

The p-value associated with t is $\gamma = 1 - F_0(t) = e^{-t}$. Moreover,

$$G(\gamma) = \pi_0\gamma + \pi_1 \left[1 - F_1(t)\right]|_{t=\ln(1/\gamma)} = \pi_0\gamma + \pi_1\gamma^{1/\sigma}.$$

Solving (3.22), we obtain

$$\gamma = \left(\frac{(1-\pi_0)\alpha}{1-\pi_0\alpha}\right)^{\sigma/\sigma-1)}.$$

For large m, the average power of the BH procedure is therefore

$$\overline{\beta}_m \approx 1 - F_1(t)|_{t=\ln(1/\gamma)} = \gamma^{1/\sigma} = \left(\frac{(1-\pi_0)\alpha}{1-\pi_0\alpha}\right)^{1/\sigma-1)}.$$

Note that this expression is independent of m and is larger (closer to 1) for $\sigma \gg 1$, as expected since the rival distributions are more dissimilar in this case. In contrast, the average power of the m NP tests at significance level α/m (guaranteeing that FWER approaches α as $m \to \infty$) is $(\alpha/m)^{1/\sigma}$, which vanishes polynomially with m (see Exercise 3.9).

Exercises

3.1 *Ternary Poisson Hypothesis Testing.* Derive the MPE rule for ternary hypothesis testing assuming uniform prior and

$$H_0 : Y \sim \text{Poi}(\lambda), \qquad H_1 : Y \sim \text{Poi}(2\lambda), \qquad H_2 : Y \sim \text{Poi}(3\lambda),$$

where $\lambda > 0$. Infer a condition on λ under which the MPE rule never returns H_1.

3.2 *Ternary Hypothesis Testing.* Consider the ternary hypothesis testing problem where p_0, p_1, and p_2 are uniform pdfs over the intervals $[0, 1]$, $[0, \frac{1}{2})$, and $[\frac{1}{2}, 1]$, respectively. Assume uniform costs.

(a) Find the Bayes classification rule and the minimum Bayes classification error probability assuming a prior $\pi = \left(\pi_0, \frac{1-\pi_0}{2}, \frac{1-\pi_0}{2}\right)$, for all values of $\pi_0 \in [0, 1]$.
(b) Find the minimax classification rule, the least-favorable prior, and the minimax classification error.
(c) Find the classification rule that minimizes the probability of classification error under H_0 subject to the constraint that the error probability under H_1 does not exceed $\frac{1}{5}$, and the error probability under H_2 does not exceed $\frac{2}{5}$.

3.3 *5-ary Classification.* A communication system uses multi-amplitude signaling to send one of five signals. After correlation of the received signal with the carrier, the correlation statistic Y follows one of the following hypotheses:

$$H_j : Y = (j-2)\mu + Z, \qquad j = 0, 1, 2, 3, 4,$$

where $\mu > 0$ and $Z \sim \mathcal{N}(0, \sigma^2)$. Assume uniform costs.

(a) Assume the hypotheses are equally likely, and find the decision rule with minimum probability of error. Also find the corresponding minimum Bayes risk.
Hint: Find the probability of correct decision making first.

(b) Find the minimax classification rule for $d = \frac{\mu}{\sigma} = 3$. To do this, seek a Bayes rule which is an equalizer.
Hint: Guess what the acceptance regions for each hypothesis are, up to two parameters which can be determined by numerical optimization.

3.4 *Decision Regions for Ternary Hypothesis Testing.* Show the decision regions for a ternary Bayesian hypothesis test with uniform priors and costs $C_{ij} = |i - j|$.

3.5 *Ternary Gaussian Hypothesis Testing.* Consider the following ternary hypothesis testing problem with two-dimensional observation $\boldsymbol{Y} = [Y_1\ Y_2]^\top$:

$$H_0 : \ \boldsymbol{Y} = \boldsymbol{N}, \qquad H_1 : \boldsymbol{Y} = \boldsymbol{s} + \boldsymbol{N}, \qquad H_2 : \boldsymbol{Y} = -\boldsymbol{s} + \boldsymbol{N},$$

where $\boldsymbol{s} = \frac{1}{\sqrt{2}}[1\,1]^\top$, and the noise vector $\boldsymbol{N}$ is Gaussian $\mathcal{N}(\mathbf{0}, \mathsf{K})$ with covariance matrix

$$\mathsf{K} = \begin{pmatrix} 1 & \frac{1}{4} \\ \frac{1}{4} & 1 \end{pmatrix}.$$

(a) Assuming that all hypotheses are equally probable, show that the minimum error probability rule can be written as

$$\delta_{\mathrm{B}}(y) = \begin{cases} 1 & \boldsymbol{s}^\top \boldsymbol{y} \geq \eta \\ 0 & -\eta < \boldsymbol{s}^\top \boldsymbol{y} < \eta \\ 2 & \boldsymbol{s}^\top \boldsymbol{y} \leq -\eta. \end{cases}$$

(b) Specify the value of η that minimizes the error probability and find the minimum error probability.

3.6 *MPE Rule.* Derive the MPE rule for the following ternary hypothesis testing problem with uniform prior and two-dimensional observation $\boldsymbol{Y} = (Y_1, Y_2)^\top$ where Y_1 and Y_2 are independent under each hypothesis and

$$H_j : \ Y_1 \sim \mathrm{Poi}(2\,e^j), \ Y_2 \sim \mathcal{N}(2j, 1), \quad j = 0, 1, 2.$$

Show the acceptance regions of the test in the Y plane.

3.7 *LED Display.* The following figure depicts a seven-segment LED display. Each segment can be turned on or off to display digits. For instance, digit 0 is displayed by turning on segments a through f and leaving segment g turned off.

Some segments may die (can no longer be turned on), impairing the ability of the device to properly display digits. We assume that each of the seven segments may die independently, with probability θ. Hence there is a possibility that digits are confused: for instance a displayed 0 might also be an 8 with dead segment g. In this problem, we analyze various error probabilities using Bayesian inference, uniforms costs, and uniform prior.

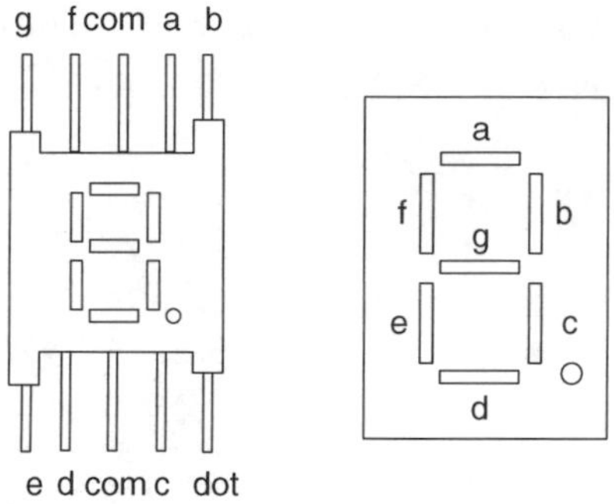

Seven-segment display

Denote by $y \in \{0, 1\}^7$ the displayed configuration, i.e., the sequence of on/off states for a particular digit. To each digit $j \in \{0, 1, \dots, 9\}$ we may associate a probability distribution $p_j(y)$. For instance, if $j = 8$ and $y = (1, 1, 1, 1, 1, 1, 1)$, we have $p_j(y) = (1 - \theta)^7$.

(a) Derive a Bayes test and the Bayes error probability for testing H_0 versus H_8.
(b) Repeat with H_4 versus H_8.
(c) Repeat with H_0 versus H_4.
(d) Give a lower and an upper bound on Bayes error probability for testing H_0 versus H_4 versus H_8.

3.8 *False Discovery Rate.* Show that $\mathrm{FDR}(\boldsymbol{\delta}) \le \mathrm{FWER}(\boldsymbol{\delta})$ for any $\boldsymbol{\delta}$.

3.9 *Bonferroni Correction.* Show that the average power of the m NP tests at significance level α/m in Example 3.3 is equal to $(\alpha/m)^{1/\sigma}$.

3.10 *Bayesian Version of BH Procedure.* Derive the asymptotic FDR for the BH procedure at level α when the m test statistics are independent and $m \to \infty$. For each test there are 3 equiprobable hypotheses, with the following pdfs on [0,1]:

$$p_{0i}(t) = 1, \quad p_{1i}(t) = 2t, \quad p_{2i}(t) = 3t^2, \quad 1 \le i \le m.$$

3.11 *Computer Simulation of BH Procedure.* Consider m binary tests. For each test i, the observation is the absolute value of a random variable Y_i drawn from either $\mathrm{P}_0 = \mathcal{N}(0, 1)$ or $\mathrm{P}_1 = \frac{1}{2}[\mathcal{N}(-3, 1) + \mathcal{N}(3, 1)]$. The random variables $\{Y_i\}_{i=1}^m$ are mutually independent.

(a) Give the asymptotic FWER (as $m \to \infty$) that results when each test is a NP test at significance level α/m (i.e., Bonferroni correction is used).
Hint. This value is close to, but not exactly equal to α.
(b) Determine the significance level ensuring that FWER $= \alpha$ (exactly) for each m.
(c) Fix $\alpha = 0.05$ and implement the Benjamini–Hochberg procedure with FDR $= \alpha$ and the procedure of Part (b) with FWER $= \alpha$. Let the first $m/2$ null hypotheses be true. To estimate the average power $\overline{\beta}_m$ of (3.19), perform 10 000 independent

simulations and average the number of true discoveries over the 10 000 m tests. Plot the estimated average power $\overline{\beta}_m$ against m ranging between 1 and 512.

References

[1] O. J. Dunn, "Multiple Comparisons Among Means," *J. Am. Stat. Assoc.*, Vol. 56, No. 293, pp. 52–64, 1961.

[2] Y. Benjamini and Y. Hochberg, "Controlling the False Discovery Rate: A Practical and Powerful Approach to Multiple Testing," *J. Roy. Stat. Soc.* Series B, Vol. 57, No. 1, pp. 289–300, 1995.

[3] Y. Benjamini and D. Yekutieli, "The Control of the False Discovery Rate in Multiple Testing under Dependency," *Annal. Stat.*, Vol. 29, No. 4, pp. 1165–1188, 2001.

[4] F. Abramovich, Y. Benjamini, D. L. Donoho, and I. M. Johnstone, "Adapting to Unknown Sparsity by Controlling the False Discovery Rate," *Ann. Stat.*, Vol. 34, No. 2, pp. 584–653, 2006.

[5] P. J. Chung, J. F. Böhme, M. F. Mecklenbräucker, and A. O. Hero, III, "Detection of the Number of Signals Using the Benjamini–Hochberg Procedure," *IEEE Trans. Signal Process.*, 2006.

[6] P. Ray and P. K. Varshney, "False Discovery Rate Based Sensor Decision Rules for the Network-Wide Distributed Detection Problem," *IEEE Trans. Aerosp. Electron. Syst.*, Vol. 47, No. 3, pp. 1785–1799, 2011.

[7] C. R. Genovese, N. A. Lazar, and T. Nichols, "Thresholding of Statistical Maps in Functional Neuroimaging Using the False Discovery Rate," *Neuroimage*, Vol. 15, No. 4, pp. 870–878, 2002.

[8] J. D. Storey and R. Tibshirani, "Statistical Significance for Genomewide Studies," *Proc. Nat. Acad. Sci.*, Vol. 100, No. 6, pp. 9440–9445, 2003.

4 Composite Hypothesis Testing

Composite hypothesis testing is studied in this chapter, where each hypothesis may be associated with more than one probability distribution. We begin with the case of two hypotheses. The Bayesian approach to this problem is first considered. Then the notions of uniformly most powerful test, locally most powerful test, generalized likelihood ratio test, and non-dominated test are introduced and their advantages and limitations are identified. The approach is then extended to the m-ary case. The robust hypothesis testing problem is also introduced, along with approaches to solving this problem.

4.1 Introduction

For the binary detection problem of Chapter 2 we have assumed that conditional densities p_0 and p_1 are specified completely. Under this assumption, we saw that all three classical formulations of the binary detection problem (Bayes, minimax, Neyman–Pearson) led to the same solution structure, the likelihood ratio test (LRT), which is a comparison of the likelihood ratio $L(y)$ to an appropriately chosen threshold. We now study the situation where p_0 and p_1 are not specified explicitly, but come from a parameterized family of densities $\{p_\theta, \theta \in \mathcal{X}\}$, with $\mathcal{X}$ being either a discrete set or a subset of a Euclidean space. The hypothesis H_j corresponds to $\theta \in \mathcal{X}_j$, $j = 0, 1$, where the sets $\mathcal{X}_0$ and $\mathcal{X}_1$ form a partition of the state space $\mathcal{X}$, namely they are disjoint, and their union is $\mathcal{X}$.

Composite binary detection (hypothesis testing) may therefore be viewed as a statistical decision theory problem where the set of states $\mathcal{X}$ is nonbinary, but the set of actions $\mathcal{A} = \{0, 1\}$ is still binary. The cost function relating the actions and states will be denoted by $C(i, \theta)$, $i \in \{0, 1\}$, $\theta \in \mathcal{X}$. A deterministic decision rule $\delta : \mathcal{Y} \to \{0, 1\}$ is a function of the observations hence *may not* depend on θ. The same holds for a randomized decision rule. We formulate binary composite hypothesis testing as follows:

$$\begin{cases} H_0 : Y \sim p_\theta, & \theta \in \mathcal{X}_0 \\ H_1 : Y \sim p_\theta, & \theta \in \mathcal{X}_1. \end{cases} \tag{4.1}$$

The mapping from the parameter space $\mathcal{X}$ to the space $\mathcal{P}[\mathcal{Y}]$ of probability distributions on Y is depicted in Figure 4.1. Note that it is possible to have $p_\theta = p_{\theta'}$ for some $\theta \neq \theta'$.

We illustrate the setting of composite hypothesis testing with an example before studying the general problem:

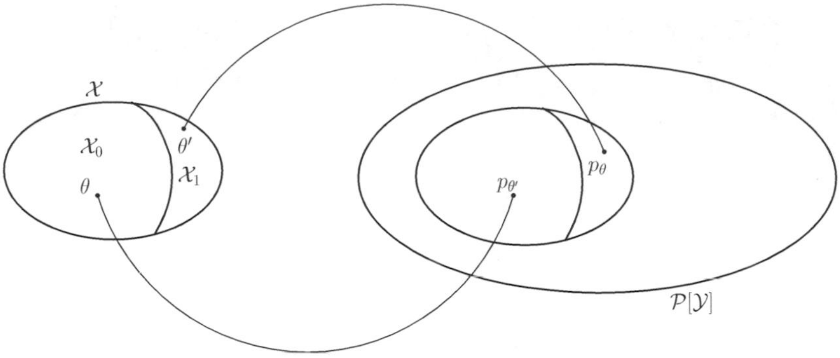

Figure 4.1 Mapping from parameter set $\mathcal{X}$ to set of probability distributions $\mathcal{P}[\mathcal{Y}]$

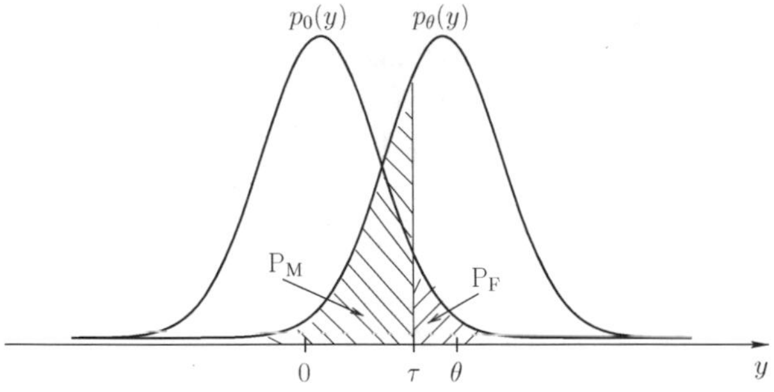

Figure 4.2 Gaussian hypothesis testing with $\mathcal{X}_0 = \{0\}$ and $\mathcal{X}_1 = (0, \infty)$

$$\begin{cases} H_0 : Y \sim \mathcal{N}(\theta, \sigma^2), & \theta \in \mathcal{X}_0 = \{0\} \\ H_1 : Y \sim \mathcal{N}(\theta, \sigma^2), & \theta \in \mathcal{X}_1 = (0, \infty). \end{cases} \tag{4.2}$$

This setup has classically been used to model a radar detection problem with Gaussian receiver noise. We try to determine whether a target is present, i.e., whether H_1 is true. The state θ represents the strength of the target signature and depends on the distance to the target, the target type, and other physical factors. In (4.2), H_0 is the so-called null hypothesis which consists of a single distribution for the observations, so H_0 is a simple hypothesis. Since H_1 has more than one (here infinitely many) possible distributions, H_1 is a composite hypothesis. As an example of a possible decision rule, consider a threshold rule for Y, in which case the threshold τ is to be determined, but is not allowed to depend on the unknown θ. As illustrated in Figure 4.2, the probability of false alarm is $P_F = Q(\frac{\tau}{\sigma})$ and the probability of miss is $P_M = Q(\frac{\theta-\tau}{\sigma})$.

Different notions of optimality are considered in the following sections to select the decision rule.

4.2 Random Parameter Θ

Assume the state θ is a realization of a random variable Θ with known prior $\pi(\theta)$, $\theta \in \mathcal{X}$. From (1.25), we immediately have that the Bayes rule for composite hypothesis testing is given by

$$\delta_{\rm B}(y) = \arg\min_{i \in \{0,1\}} C(i|y), \tag{4.3}$$

where, using the notation introduced in (1.20),

$$C(i|y) = \int_{\mathcal{X}} C(i,\theta)\pi(\theta|y)d\nu(\theta),$$

with

$$\pi(\theta|y) = \frac{p_\theta(y)\,\pi(\theta)}{p(y)}.$$

4.2.1 Uniform Costs Over Each Hypothesis

In the special case where costs are uniform over each hypothesis, i.e.,

$$C(i,\theta) = C_{ij}, \quad \forall \theta \in \mathcal{X}_j \tag{4.4}$$

(but it is not necessary that $C_{ij} = \mathbb{1}_{\{i \neq j\}}$), the posterior cost has the same form as the Bayes rule for simple hypothesis testing. Indeed we can expand $C(i|y)$ as

$$C(i|y) = C_{i0}\int_{\mathcal{X}_0} \pi(\theta|y)d\nu(\theta) \;+\; C_{i1}\int_{\theta \in \mathcal{X}_1} \pi(\theta|y)d\nu(\theta),$$

from which we can easily conclude that

$$C(1|y) \leq C(0|y) \quad \Longleftrightarrow \quad \frac{\int_{\mathcal{X}_1} p_\theta(y)\pi(\theta)d\nu(\theta)}{\int_{\mathcal{X}_0} p_\theta(y)\pi(\theta)d\nu(\theta)} \geq \frac{C_{10}-C_{00}}{C_{01}-C_{11}}. \tag{4.5}$$

Now, we define the priors on the hypotheses as

$$\pi_j \triangleq \int_{\mathcal{X}_j} \pi(\theta)d\nu(\theta), \quad j = 0, 1, \tag{4.6}$$

the conditional density of Θ given H_j as

$$\pi_j(\theta) = \frac{\pi(\theta)}{\pi_j}, \quad \theta \in \mathcal{X}_j, \quad j = 0, 1, \tag{4.7}$$

and the conditional densities for the hypotheses as

$$p(y|H_j) \triangleq \int_{\mathcal{X}_j} p_\theta(y \leq C(0|y) \quad \Longleftrightarrow \quad L(y) \geq \eta \tag{4.8}$$

with $\eta = \frac{C_{10}-C_{00}}{C_{01}-C_{11}}\frac{\pi_0}{\pi_1}$ as defined in (2.12) and $L(y) = p(y|H_1)/p(y|H_0)$. This means that

$$\delta_{\rm B}(y) = \begin{cases} 1 & L(y) \geq \eta. \\ 0 & \quad < \end{cases} \tag{4.9}$$

Therefore, we conclude that the Bayes rule for composite detection is nothing but a LRT for the (simple) binary detection problem:

$$\begin{cases} H_0 : Y \sim p(y|H_0) \\ H_1 : Y \sim p(y|H_1) \end{cases}$$

with priors π_0 and π_1 as defined in (4.6).

Example 4.1 *Uniform Costs with only H_1 Composite.* Assume $C(i,\theta) = C_{ij} = \mathbb{1}\{i \neq j\}$, equal priors: $\pi_0 = \pi_1 = \frac{1}{2}$, fixed $\theta = \theta_0$ under H_0, and conditional prior $\pi_1(\theta) = \beta e^{-\beta(\theta-1)}\mathbb{1}\{\theta > 1\}$ under H_1. Consider the following hypotheses:

$$\begin{cases} H_0 : p_\theta(y) = \theta e^{-\theta y}\,\mathbb{1}\{y \geq 0\}, & \theta \in \mathcal{X}_0 = \{1\} \\ H_1 : p_\theta(y) = \theta e^{-\theta y}\,\mathbb{1}\{y \geq 0\}, & \theta \in \mathcal{X}_1 = (1,\infty). \end{cases}$$

The exponential distribution p_θ and the prior π_1 are shown in Figure 4.3.

Evaluating the conditional marginals (4.8), we have, for $y \geq 0$,

$$\begin{aligned} p(y|H_0) &= e^{-y}, \\ p(y|H_1) &= \int_{\mathcal{X}_1} p_\theta(y)\pi_1(\theta)d\theta = \int_1^\infty \theta e^{-\theta y}\beta e^{-\beta(\theta-1)}\,d\theta \\ &= \beta e^{-y}\int_1^\infty \theta e^{-(\theta-1)(y+\beta)}\,d\theta \\ &= \beta e^{-y}\left[\int_1^\infty (\theta-1)e^{-(\theta-1)(y+\beta)}d\theta + \int_1^\infty e^{-(\theta-1)(y+\beta)}d\theta\right] \\ &= \beta e^{-y}\left[\frac{1}{(y+\beta)^2} + \frac{1}{y+\beta}\right]. \end{aligned}$$

The likelihood ratio is given by

$$L(y) = \frac{p(y|H_1)}{p(y|H_0)} = \beta\left[\frac{1}{(y+\beta)^2} + \frac{1}{y+\beta}\right], \quad y \geq 0.$$

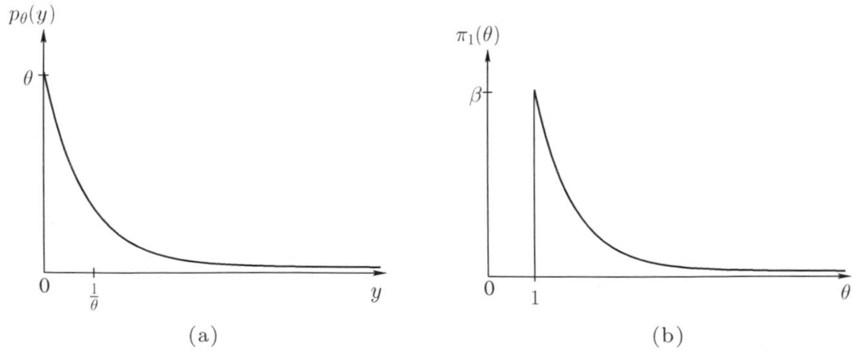

Figure 4.3 Example 4.1: (a) $p_\theta(y)$; and (b) $\pi_1(\theta)$

Observe that $L(y) \to 1$ as $\beta \to \infty$, and thus H_0 and H_1 become indistinguishable in the limit as $\beta \to \infty$. This should be expected because $\pi_1(\theta)$ concentrates around $\theta = 1$ in that case. The decision rule is

$$\delta_{\mathrm{B}}(y) = \begin{cases} 1 & \text{if } L(y) = \beta\left[\frac{1}{(y+\beta)^2} + \frac{1}{y+\beta}\right] \geq \eta = 1 \\ 0 & \text{otherwise.} \end{cases}$$

Since the function $L(y)$ is monotonically decreasing, we obtain, equivalently,

$$\delta_{\mathrm{B}}(y) = \begin{cases} 1 & \text{if } y \leq \tau \\ 0 & \text{if } y > \tau, \end{cases}$$

where

$$\tau = \left(-\frac{1}{2} + \sqrt{\frac{1}{4} + \beta^{-1}}\right)^{-1} - \beta.$$

Example 4.2 *Uniform Costs with Both Hypotheses Composite.* Consider the composite detection problem in which $\mathcal{X} = [0, \infty)$, $\mathcal{X}_0 = [0, 1)$, and $\mathcal{X}_1 = [1, \infty)$, with $C(i, \theta) = C_{ij} = \mathbb{1}\{i \neq j\}$, and

$$p_\theta(y) = \theta e^{-\theta y}\, \mathbb{1}_{\{y \geq 0\}}, \quad \pi(\theta) = e^{-\theta}\, \mathbb{1}_{\{\theta \geq 0\}}.$$

To compute the Bayes rule for this problem, we first compute

$$\int_{\mathcal{X}_1} p_\theta(y)\pi(\theta)d\theta = \int_1^\infty \theta e^{-\theta(y+1)}d\theta = \frac{(y+2)e^{-(y+1)}}{(y+1)^2}$$

and

$$\int_{\mathcal{X}_0} p_\theta(y)\pi(\theta)d\theta = \int_0^1 \theta e^{-\theta(y+1)}d\theta = \frac{1-(y+2)e^{-(y+1)}}{(y+1)^2}.$$

Then, from (4.5), we get that

$$\delta_{\mathrm{B}}(y) = \begin{cases} 1: & (y+2) \geq 0.5e^{(y+1)}, \\ 0: & \quad\quad < \end{cases}$$

which can be simplified to

$$\delta_{\mathrm{B}}(y) = \begin{cases} 1: & 0 \leq y \leq \tau, \\ 0: & y > \tau \end{cases}$$

where τ is a solution to the transcendental equation $(y+2) = 0.5e^{(y+1)}$, which yields $\tau \approx 0.68$.

We may also compute the Bayes risk for δ_{B} as:

$$\begin{aligned} r(\delta_{\mathrm{B}}) &= \pi_0 R_0(\delta_{\mathrm{B}}) + \pi_1 R_1(\delta_{\mathrm{B}}) \\ &= \pi_0 \mathrm{P}(\mathcal{Y}_1|H_0) + \pi_1 \mathrm{P}(\mathcal{Y}_0|H_1) \end{aligned}$$

$$= \pi_0 \int_0^\tau \int_0^1 \frac{1}{\pi_0}\theta e^{-\theta(y+1)} d\theta dy + \int_\tau^\infty \int_1^\infty \frac{1}{\pi_1}\theta e^{-\theta(y+1)} d\theta dy$$

$$= \int_0^\tau \frac{1-(y+2)e^{-(y+1)}}{(y+1)^2} dy + \int_\tau^\infty \frac{(y+2)e^{-(y+1)}}{(y+1)^2} dy.$$

4.2.2 Nonuniform Costs Over Hypotheses

We now give an example of the case where the cost function depends on θ.

Example 4.3 Assume the same setting as Example 4.1 except that the cost function is now

$$C(i,\theta) = \begin{cases} \frac{1}{\theta} & \text{if } \ i=0, \quad \theta > 1 \qquad (H_1) \\ 1 & \text{if } \ i=1, \quad \theta = 1 \qquad (H_0) \\ 0 & \text{else.} \end{cases} \tag{4.10}$$

The cost for saying H_0 when H_1 is true varies with θ and is lower than that in Example 4.2.

The posterior risks are given by

$$C(1|y) = \int_{\mathcal{X}} \pi(\theta|y) C(1,\theta) d\theta = \int_{\mathcal{X}_0} \pi(\theta|y) d\theta$$

$$= \mathrm{P}(H_0|Y=y) = \frac{\mathrm{P}(H_0)\, p(y|H_0)}{p(y)} = \frac{\frac{1}{2}e^{-y}}{p(y)}$$

and

$$C(0|y) = \int_{\mathcal{X}} \pi(\theta|y) C(0,\theta) d\theta = \int_{\mathcal{X}_1} \pi(\theta|y)\frac{1}{\theta} d\theta = \frac{1}{2}\int_{\mathcal{X}_1} \pi_1(\theta|y)\frac{1}{\theta} d\theta$$

$$= \frac{1}{2}\int_1^\infty \frac{p_\theta(y)\pi_1(\theta)}{p(y)}\frac{1}{\theta} d\theta = \frac{\frac{1}{2}}{p(y)}\int_1^\infty \theta e^{-\theta y}\beta e^{-\beta(\theta-1)}\theta^{-1} d\theta$$

$$= \frac{\frac{1}{2}e^{-y}}{p(y)}\beta\int_1^\infty e^{-(\theta-1)(y+\beta)} d\theta = \frac{\frac{1}{2}e^{-y}}{p(y)}\frac{\beta}{y+\beta} \le C(1|y) \quad \forall y \ge 0. \tag{4.11}$$

Therefore, we have that $\delta_{\mathrm{B}}(y) = 0$, for all $y \ge 0$, i.e., the Bayes rule always picks H_0.

Interestingly, if we change θ^{-1} into $k\theta^{-1}$ with $k > 1$ in the definition of the cost (4.10), the optimal decision rule becomes

$$\delta_{\mathrm{B}}(y) = \begin{cases} 1: & y \le (k-1)\beta. \\ 0: & > \end{cases}$$

This illustrates the potential sensitivity of the Bayes rule to the cost assignments.

4.3 Uniformly Most Powerful Test

When a prior distribution for the parameter θ is not available, we may attempt to extend the Neyman–Pearson approach to select the decision rule. The probabilities of false alarm and correct detection depend on both the decision rule $\tilde{\delta}$ and the state θ and may be written as

$$\begin{aligned} P_F(\tilde{\delta}, \theta_0) &= P_{\theta_0}\{\text{say } H_1\} = P_{\theta_0}\{\tilde{\delta}(Y) = 1\}, \quad \theta_0 \in \mathcal{X}_0, \\ P_D(\tilde{\delta}, \theta_1) &= P_{\theta_1}\{\text{say } H_1\} = P_{\theta_1}\{\tilde{\delta}(Y) = 1\}, \quad \theta_1 \in \mathcal{X}_1. \end{aligned}$$

Note that this is actually the same function which is given a different name under H_0 and under H_1.

Apparently a natural extension of the NP approach would be

$$\begin{aligned} &\text{Maximize } P_D(\tilde{\delta}, \theta_1), \ \forall \theta_1 \in \mathcal{X}_1 \\ &\text{subject to } P_F(\tilde{\delta}, \theta_0) \leq \alpha, \ \forall \theta_0 \in \mathcal{X}_0. \end{aligned} \tag{4.12}$$

Unfortunately, in general no such $\tilde{\delta}$ exists because the same $\tilde{\delta}$ cannot always simultaneously maximize different functions. If a solution does exist, it is said to be a universal most powerful (UMP) test at level α. Such a test, denoted by $\tilde{\delta}_{\text{UMP}}$, satisfies the following two properties. First,

$$\sup_{\theta_0 \in \mathcal{X}_0} P_F(\tilde{\delta}, \theta_0) \leq \alpha \tag{4.13}$$

for $\tilde{\delta} = \tilde{\delta}_{\text{UMP}}$. Second, for any $\tilde{\delta} \in \tilde{\mathcal{D}}$ that satisfies (4.13), we must have

$$P_D(\tilde{\delta}, \theta_1) \leq P_D(\tilde{\delta}_{\text{UMP}}, \theta_1), \quad \forall \theta_1 \in \mathcal{X}_1. \tag{4.14}$$

Consider the special case where H_0 is a simple hypothesis, i.e., $\mathcal{X}_0 = \{\theta_0\}$. Then we have only one constraint $P_F(\tilde{\delta}, \theta_0) \leq \alpha$ in (4.12). Denote by

$$L_{\theta_1}(y) = \frac{p_{\theta_1}(y)}{p_{\theta_0}(y)}, \quad \theta_1 \in \mathcal{X}_1$$

the likelihood ratio for p_{θ_1} versus p_{θ_0}. The corresponding NP rule is given by

$$\tilde{\delta}_{\text{NP}}^{\theta_1}(y) = \begin{cases} 1 & & > & \\ 1 \text{ w.p. } \gamma & : L_{\theta_1}(y) & = & \eta \\ 0 & & < & \end{cases}, \quad \theta_1 \in \mathcal{X}_1, \tag{4.15}$$

where the test parameters η and γ (which could depend on θ_1) are chosen to satisfy the constraint $P_F(\tilde{\delta}, \theta_0) \leq \alpha$ with equality, as derived in Section 2.4. If a UMP test exists, then $\tilde{\delta}_{\text{UMP}}$ must coincide with the NP test $\tilde{\delta}_{\text{NP}}^{\theta_1}$ for each $\theta_1 \in \mathcal{X}_1$.

4.3.1 Examples

Example 4.4 *Detection of One-Sided Composite Signal in Gaussian Noise.* We revisit the Gaussian example of (4.2). We have $\theta_0 = 0$ and consider three scenarios for $\mathcal{X}_1$.

Case I: $\mathcal{X}_1 = (0, \infty)$, as illustrated in Figure 4.2. Then all NP rules $\tilde{\delta}_{\text{NP}}^{\theta_1}$ (where $\theta_1 > 0$) are of the form

$$\tilde{\delta}_{\text{NP}}^{\theta_1}(y) = \begin{cases} 1 : & y \geq \tau, \\ 0 : & < \end{cases}$$

where τ is positive and satisfies $\text{P}_\text{F} = Q(\frac{\tau}{\sigma}) = \alpha$. Hence $\tau = \sigma Q^{-1}(\alpha)$ does not depend on θ_1, and $\tilde{\delta}_{\text{NP}}^{\theta_1}$ is UMP:

$$\delta^+(y) = \begin{cases} 1 : & y \geq \sigma Q^{-1}(\alpha). \\ 0 : & < \end{cases} \tag{4.16}$$

Notice, however, that its power strongly depends on θ_1:

$$\text{P}_\text{D}(\delta_{\text{UMP}}, \theta_1) = \text{P}_{\theta_1}\{Y \geq \sigma Q^{-1}(\alpha)\} = Q(Q^{-1}(\alpha) - \theta_1/\sigma).$$

Case II: $\mathcal{X}_1 = (-\infty, 0)$, as shown in Figure 4.4. All NP rules $\tilde{\delta}_{\text{NP}}^{\theta_1}$ (where $\theta_1 < 0$) are of the form

$$\tilde{\delta}_{\text{NP}}^{\theta_1}(y) = \begin{cases} 1 : & y \leq \tau, \\ 0 : & > \end{cases}$$

where τ is now negative and satisfies $\text{P}_\text{F} = Q(\frac{-\tau}{\sigma}) = \alpha$. Hence $\tau = -\sigma Q^{-1}(\alpha)$ does not depend on θ_1, and the decision rule is UMP:

$$\delta^-(y) = \begin{cases} 1 : & y \leq -\sigma Q^{-1}(\alpha). \\ 0 : & < \end{cases} \tag{4.17}$$

Its power is $\text{P}_\text{D}(\delta_{\text{UMP}}, \theta_1) = Q(Q^{-1}(\alpha) + \theta_1/\sigma)$.
Case III: $\mathcal{X}_1 = \mathbb{R}\backslash\{0\}$. Now the NP test depends on the sign of θ_1. For example, the α-level tests corresponding to $\theta_1 = 1$ and $\theta_1 = -1$ are different, as in Cases I and II, respectively. Hence no UMP test exists.

Example 4.5 *Detection of One-Sided Composite Signal in Cauchy Noise.* From the Gaussian example (Example 4.4), we may be tempted to conclude that for problems involving signal detection in noise, UMP tests exist as long as H_1 is one-sided. To see

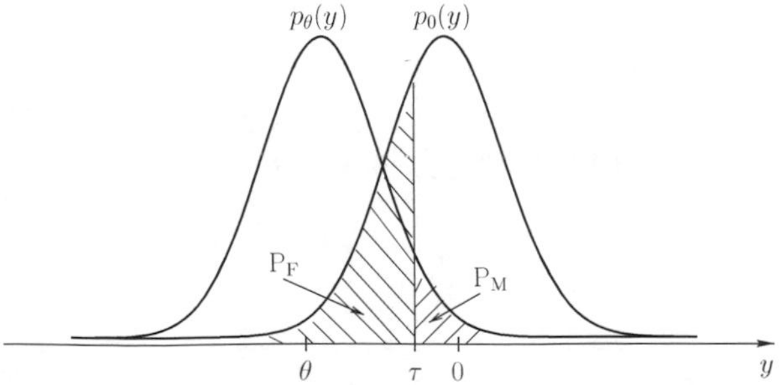

Figure 4.4 Gaussian hypothesis testing with $\mathcal{X}_0 = \{0\}$ and $\mathcal{X}_1 = (-\infty, 0)$

that this is not true in general, we consider the example where the noise has a Cauchy distribution centered at θ, i.e.,

$$p_\theta(y) = \frac{1}{\pi[1 + (y - \theta)^2]},$$

where $\mathcal{X}_0 = \{0\}$ and $\mathcal{X}_1 = (0, \infty)$. Then

$$L_{\theta_1}(y) = \frac{1 + y^2}{1 + (y - \theta_1)^2}.$$

It can be verified that the α-level NP tests for $\theta_1 = 1$ and $\theta_1 = 2$ are different, and hence there is no UMP solution.

To be UMP, the rule $\tilde{\delta}_{\text{NP}}^{\theta_1}$ must be independent of the unknown θ_1. This may seem paradoxical because η and γ in (4.15) generally depend on θ_1; however, so does $L_{\theta_1}(y)$. The apparent paradox is elucidated in Section 4.3.2 and illustrated in Figure 4.5.

4.3.2 Monotone Likelihood Ratio Theorem

THEOREM 4.1 *If there exists a function $g : \mathcal{Y} \to \mathbb{R}$ and a collection of functions $F_{\theta_1} : \mathbb{R} \to \mathbb{R}^+$ that are strictly increasing $\forall \theta_1 \in \mathcal{X}_1$ such that $L_{\theta_1}(y) = F_{\theta_1}(g(y))$, the UMP test at significance level α exists for testing the simple hypothesis $H_0 : \theta = \theta_0$ versus the composite hypothesis $H_1 : \theta \in \mathcal{X}_1$, and is of the form*

$$\tilde{\delta}_{\text{UMP}}(y) = \begin{cases} 1 & > \\ 1 \text{ w.p. } \gamma \; : \; g(y) & = \tau, \\ 0 & < \end{cases} \tag{4.18}$$

where τ and γ are such that $\mathbb{E}_{\theta_0}[\tilde{\delta}(Y)] = \mathrm{P}_{\theta_0}\{g(Y) > \tau\} + \gamma \mathrm{P}_{\theta_0}\{g(Y) = \tau\} = \alpha$.

Proof Since each function F_{θ_1} is strictly increasing,

$$L_{\theta_1}(y) > \eta \iff g(y) > \tau,$$
$$L_{\theta_1}(y) = \eta \iff g(y) = \tau, \quad \forall \theta_1 \in \mathcal{X}_1,$$

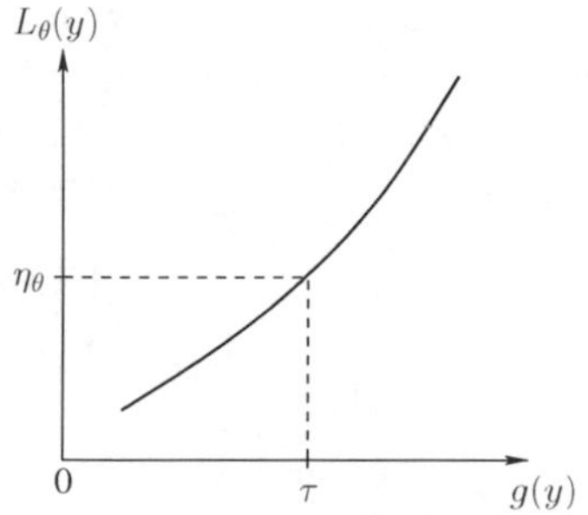

Figure 4.5 Monotone likelihood ratio: for each θ, the likelihood ratio is an increasing function of $g(y)$

for some τ. Hence the NP rule (4.15) can be written in the form given in (4.18). Now, to achieve a significance level of α, we set

$$\alpha = \mathrm{P}_{\theta_0}\{g(Y) > \tau\} + \gamma \mathrm{P}_{\theta_0}\{g(Y) = \tau\}.$$

The solution to this equation yields τ and γ that are independent of θ_1, which implies that the rule in (4.18) is UMP. □

Example 4.6 *Example 4.4, revisited.* The likelihood ratio is given by

$$L_{\theta_1}(y) = \frac{p_{\theta_1}(y)}{p_0(y)} = \frac{\frac{1}{\sqrt{2\pi\sigma^2}} \exp\{-\frac{(y-\theta_1)^2}{2\sigma^2}\}}{\frac{1}{\sqrt{2\pi\sigma^2}} \exp\{-\frac{y^2}{2\sigma^2}\}} = \exp\left\{-\frac{\theta_1^2}{2\sigma^2}\right\} \exp\left\{\frac{\theta_1 y}{\sigma^2}\right\}.$$

Case I: $\theta_1 > 0$, choose $g(y) = y$, and thus

$$F_{\theta_1}(g) = \exp\left\{-\frac{\theta_1^2}{2\sigma^2}\right\} \exp\left\{\frac{\theta_1 g}{\sigma^2}\right\},$$

which is strictly increasing in g for all $\theta_1 > 0$. Alternatively one could have taken $g(y) = e^y$, amongst other choices.

Case II: $\theta_1 < 0$, choose $g(y) = -y$, thus $F_{\theta_1}(g)$ is the same as Case I.

Case III: $\theta_1 \neq 0$ we cannot find an appropriate g, and no UMP test exists as we showed in Example 4.4.

4.3.3 Both Composite Hypotheses

We now consider the more general case where both the hypotheses are composite. Finding a UMP solution in this case, i.e., a test that satisfies (4.13) and (4.14), is considerably more complicated.

The following example illustrates a case where a UMP solution can be found, and a procedure to find it. Also see Exercises 4.14 and 4.15.

Example 4.7 *Testing Between Two One-Sided Composite Signals in Gaussian Noise.* This is an extension of the Gaussian example (Example 4.4) where $\mathcal{X}_0 = (-\infty, 0]$ and $\mathcal{X}_1 = (0, \infty)$. For fixed $\theta_0 \in \mathcal{X}_0$ and $\theta_1 \in \mathcal{X}_1$, $L_{\theta_0,\theta_1}(y)$ has no point masses under P_{θ_0} or P_{θ_1}. Therefore $\tilde{\delta}_{\mathrm{NP}}^{\theta_0,\theta_1}$ is a deterministic LRT, which can be reduced to a threshold test on Y:

$$\delta_{\mathrm{NP}}^{\theta_0,\theta_1} = \begin{cases} 1 : & L_{\theta_0,\theta_1}(y) \geq \eta \\ 0 : & \qquad\qquad < \end{cases} = \begin{cases} 1 : & y \geq \tau, \\ 0 : & \quad < \end{cases}$$

where the threshold τ_{θ_0,θ_1} is given by

$$\tau = \frac{\sigma^2 \log \eta}{\theta_1 - \theta_0} + \frac{\theta_0 + \theta_1}{2}.$$

Now in order to set the threshold τ to meet the constraint on $\mathrm{P_F}$ given in (4.13), we first compute

$$\mathrm{P_F}(\delta_{\mathrm{NP}}^{\theta_0,\theta_1}, \theta_0) = \mathrm{P}_{\theta_0}\{Y \geq \tau\} = Q\left(\frac{\tau - \theta_0}{\sigma}\right),$$

which is an increasing function of θ_0. Therefore

$$\sup_{\theta_0 \leq 0} \mathrm{P_F}(\delta_{\mathrm{NP}}^{\theta_0,\theta_1}, \theta_0) = Q\left(\frac{\tau}{\sigma}\right)$$

and we can meet the $\mathrm{P_F}$ constraint with equality by setting τ such that:

$$Q\left(\frac{\tau}{\sigma}\right) = \alpha \quad \Longrightarrow \quad \tau = \sigma Q^{-1}(\alpha).$$

Note that τ is independent of θ_0 and θ_1. Define the test

$$\delta_\tau(y) = \begin{cases} 1 : y & \geq \tau. \\ 0 : & < \end{cases}$$

We will now establish that δ_τ is a UMP test, by showing that conditions (4.13) and (4.14) hold. By construction,

$$\sup_{\theta_0 \leq 0} \mathrm{P_F}(\delta_\tau, \theta_0) = \mathrm{P_F}(\delta_\tau, 0) = \alpha,$$

and so (4.13) holds. Also, δ_τ is an α-level UMP test between the simple hypothesis $H_0 : \theta = 0$ and composite hypothesis $H_1 : \theta > \theta_0$ (see Case I in Example 4.4). Now, consider any test $\tilde{\delta} \in \tilde{\mathcal{D}}$ that satisfies $\sup_{\theta \leq 0} \mathrm{P_F}(\tilde{\delta}, \theta) \leq \alpha$. Then clearly it is also true that $\mathrm{P_F}(\tilde{\delta}, 0) \leq \alpha$. This means that $\tilde{\delta}$ is an α-level test for testing between the simple hypothesis $H_0 : \theta = 0$ and composite hypothesis $H_1 : \theta > \theta_0$, and it cannot be more powerful than δ_τ, i.e.,

$$\mathrm{P_D}(\tilde{\delta}, \theta_1) \leq \mathrm{P_D}(\delta_\tau, \theta_1) \quad \forall \theta_1 > 0.$$

Therefore (4.14) holds and we have

$$\delta_{\mathrm{UMP}}(y) = \delta_\tau(y) = \begin{cases} 1 : y & \geq \sigma Q^{-1}(\alpha). \\ 0 : & < \end{cases}$$

Again, while the test δ_{UMP} is independent of the θ_1, the performance of the test in terms of the $\mathrm{P_D}$ depends on θ_1. In particular,

$$\mathrm{P_D}(\delta_{\mathrm{UMP}}, \theta_1) = \mathrm{P}_{\theta_1}\{Y \geq \sigma Q^{-1}(\alpha)\} = Q\left(Q^{-1}(\alpha) - \frac{\theta_1}{\sigma}\right).$$

What if no UMP test exists? We examine two classical approaches: locally most powerful (LMP) test, and generalized LRT.

4.4 Locally Most Powerful Test

With the LMP test we want to maximize the probability of correct detection for values of θ in some critical neighborhood. For example, consider the following test, in which θ is a scalar parameter and $\mathcal{X}_0 = \{\theta_0\}$ (simple hypothesis H_0) and $\mathcal{X}_1 = [\theta_0, \infty)$:

$$\begin{cases} H_0 : Y \sim p_{\theta_0} \\ H_1 : Y \sim p_\theta, \quad \theta > \theta_0. \end{cases} \tag{4.19}$$

Here we want to maximize the probability of correct detection in the neighborhood of θ_0 because the two hypotheses are least distinguishable in that region. Stated explicitly:

$$\begin{aligned} &\max_{\tilde{\delta}} \mathrm{P_D}(\tilde{\delta}, \theta) \quad \text{as } \theta \downarrow \theta_0 \\ &\text{subject to} \quad \mathrm{P_F}(\tilde{\delta}, \theta_0) \le \alpha. \end{aligned} \tag{4.20}$$

Assuming the first two derivatives of $\mathrm{P_D}(\tilde{\delta}, \cdot)$ exist and are finite in the interval $[\theta_0, \theta]$, we write the first-order Taylor series expansion around θ_0 as

$$\mathrm{P_D}(\tilde{\delta}, \theta) = \mathrm{P_D}(\tilde{\delta}, \theta_0) + (\theta - \theta_0) P_D'(\tilde{\delta}, \theta_0) + O((\theta - \theta_0)^2),$$

where $\mathrm{P_D}(\tilde{\delta}, \theta_0) = \mathrm{P_F}(\tilde{\delta}, \theta_0) = \alpha$. Thus we have the approximation

$$\mathrm{P_D}(\tilde{\delta}, \theta) \cong \alpha + (\theta - \theta_0) P_D'(\tilde{\delta}, \theta_0). \tag{4.21}$$

Since $\theta - \theta_0 \ge 0$, the optimization in (4.20) is replaced by

$$\begin{aligned} &\max_{\tilde{\delta}} P_D'(\tilde{\delta}, \theta_0) \\ &\text{subject to} \quad \mathrm{P_F}(\tilde{\delta}, \theta_0) \le \alpha. \end{aligned} \tag{4.22}$$

For a randomized decision rule as in Definition 1.4,

$$\mathrm{P_D}(\tilde{\delta}, \theta) = \sum_{\ell=1}^{L} \gamma_\ell \mathrm{P_D}(\delta_\ell, \theta) = \int_{\mathcal{Y}} \sum_{\ell=1}^{L} \gamma_\ell \mathbb{1}\{\delta_\ell(y) = 1\} p_\theta(y)\, dy$$

and

$$\mathrm{P_D'}(\tilde{\delta}, \theta_0) = \int_{\mathcal{Y}} \sum_{\ell=1}^{L} \gamma_\ell \mathbb{1}\{\delta_\ell(y) = 1\} \left.\frac{\partial p_\theta(y)}{\partial \theta}\right|_{\theta=\theta_0} dy.$$

The solution to the locally optimal detection problem of (4.22) can be shown to be equivalent to NP testing between $p_{\theta_0}(y)$ and

$$\left.\frac{\partial}{\partial \theta} p_\theta(y)\right|_{\theta=\theta_0}.$$

Even though the latter quantity is not necessarily a pdf (or pmf), the steps that we followed in deriving the NP solution in Section 2.4 can be repeated to show that the solution to (4.22) has the form

$$\tilde{\delta}_{\text{LMP}}(y) = \begin{cases} 1 & > \\ 1 \text{ w.p. } \gamma : \quad L_\ell(y) & = \tau, \\ 0 & < \end{cases}$$

where

$$L_\ell(y) = \frac{\left.\frac{\partial}{\partial\theta} p_\theta(y)\right|_{\theta=\theta_0}}{p_{\theta_0}(y)}.$$

Note that if in the problem formulation (4.19) we change $\mathcal{X}_1 = (\theta_0, \infty)$ into $\mathcal{X}_1 = (-\infty, \theta_0)$, the maximization of (4.22) should be changed to a minimization.

Example 4.8 *Detection of One-Sided Composite Signal in Cauchy Noise (continued).* This problem was introduced in Example 4.5. We are testing $H_0 : \theta = 0$ against the one-sided composite hypothesis $H_1 : \theta > 0$. As indicated in Example 4.5, there is no UMP solution to this problem. We therefore examine the LMP solution:

$$p_\theta(y) = \frac{1}{\pi[1 + (y - \theta)^2]} \quad \Longrightarrow \quad \left.\frac{\partial}{\partial\theta} p_\theta(y)\right|_{\theta=0} = \frac{2y}{\pi(1 + y^2)^2}.$$

Thus

$$L_\ell(y) = \frac{2y}{1 + y^2}$$

and

$$\tilde{\delta}_{\text{LMP}}(y) = \begin{cases} 1 & : \; L_\ell(y) \geq \eta. \\ 0 & \quad\quad < \end{cases}$$

Randomization is not needed since $L_\ell(y)$ does not have point masses under P_0. To find the threshold η to meet a false-alarm constraint of $\alpha \in (0, 1)$, we first note that $L_\ell(y) \in [-1, 1]$. Therefore $\text{P}_0\{L_\ell(Y) > \eta\} = 1$ for $\eta \leq -1$, and $\text{P}_0\{L_\ell(Y) > \eta\} = 0$ for $\eta \geq 1$. For $\eta \in [0, 1)$, it is easy to show that

$$\text{P}_0\{L_\ell(Y) > \eta\} = \frac{1}{\pi} \int_a^b \frac{1}{1 + y^2} dy = \frac{1}{\pi} \left[\tan^{-1}(b) - \tan^{-1}(a)\right],$$

where

$$a = \frac{1}{\eta} - \sqrt{\frac{1}{\eta^2} - 1}, \quad b = \frac{1}{\eta} + \sqrt{\frac{1}{\eta^2} - 1}.$$

Similarly for $\eta \in (-1, 0)$, it is easy to show that

$$\text{P}_0\{L_\ell(Y) > \eta\} = 1 + \frac{1}{\pi} \left[\tan^{-1}(a) - \tan^{-1}(b)\right].$$

Using these expressions for $\text{P}_0\{L_\ell(Y) > \eta\}$, we can find η to meet a false-alarm constraint of $\alpha \in (0, 1)$.

4.5 Generalized Likelihood Ratio Test

Since UMP tests often do not exist, and LMP tests are unsuitable for two-sided composite hypothesis testing, another technique is often needed. One approach is based on the *generalized likelihood ratio* (GLR), which is defined by

$$L_G(y) \triangleq \frac{\max_{\theta\in\mathcal{X}_1} p_\theta(y)}{\max_{\theta\in\mathcal{X}_0} p_\theta(y)} = \frac{p_{\hat{\theta}_1}(y)}{p_{\hat{\theta}_0}(y)}, \tag{4.23}$$

where $\hat{\theta}_1(y)$ and $\hat{\theta}_0(y)$ are the maximum-likelihood estimates of θ given y under H_1 and H_0, respectively. Note that evaluation of $L_G(y)$ involves a maximization over θ_0 and θ_1, and so this test statistic can be considerably more complex than the LRT. Also the result of the maximization generally does not produce a pdf (or pmf) in the numerator and denominator. We can use the statistic $L_G(y)$ to produce a test, which is called the *generalized likelihood ratio test* (GLRT):

$$\tilde{\delta}_{\mathrm{GLRT}}(y) = \begin{cases} 1 & > \\ 1 \text{ w.p. } \gamma : \quad L_G(y) & = \eta. \\ 0 & < \end{cases} \tag{4.24}$$

This may be thought of as a joint estimation (of θ) and detection (of $i = 0, 1$). It should also be mentioned that the maximization problems as stated in (4.23) might not have a solution if either $\mathcal{X}_0$ or $\mathcal{X}_1$ is an open set; the most general definition of the GLR is therefore

$$L_G(y) \triangleq \frac{\sup_{\theta\in\mathcal{X}_1} p_\theta(y)}{\sup_{\theta\in\mathcal{X}_0} p_\theta(y)}. \tag{4.25}$$

4.5.1 GLRT for Gaussian Hypothesis Testing

The problem is posed as

$$\begin{aligned} H_0 &: Y \sim \mathcal{N}(\theta, \sigma^2), \quad \theta = 0, \\ H_1 &: Y \sim \mathcal{N}(\theta, \sigma^2), \quad \theta \neq 0, \end{aligned}$$

(recall no UMP test exists here) and the solution is as follows. The GLR is

$$L_G(y) = \frac{\sup_{\theta\neq 0} p_\theta(y)}{p_0(y)} = \frac{\frac{1}{\sqrt{2\pi\sigma^2}}}{\frac{1}{\sqrt{2\pi\sigma^2}}\exp\{-\frac{y^2}{2\sigma^2}\}} = \exp\left\{\frac{y^2}{2\sigma^2}\right\},$$

and so the GLRT takes the form

$$\delta_{\mathrm{GLRT}}(y) = \begin{cases} 1 : & \exp\left\{\frac{y^2}{2\sigma^2}\right\} \geq \eta \\ 0 : & < \end{cases} = \begin{cases} 1 : & |y| > \sigma\sqrt{2\ln\eta} \triangleq \tau. \\ 0 : & < \end{cases} \tag{4.26}$$

(Randomization is not needed here since L_G does not have point masses under either hypothesis.)

Performance Analysis Figure 4.6 illustrates the GLRT (4.26). Its performance is as follows:

$$P_F = \alpha = P_0\{|Y| > \tau\} = 2Q\left(\frac{\tau}{\sigma}\right). \tag{4.27}$$

Hence the threshold

$$\tau = \sigma Q^{-1}\left(\frac{\alpha}{2}\right).$$

The power of δ_{GLRT} is given by

$$\begin{aligned} P_D(\delta_{\text{GLRT}}, \theta) &= P_\theta\{|Y| > \tau\} \\ &= Q\left(\frac{\tau+\theta}{\sigma}\right) + Q\left(\frac{\tau-\theta}{\sigma}\right) \\ &= Q\left(Q^{-1}\left(\frac{\alpha}{2}\right) + \frac{\theta}{\sigma}\right) + Q\left(Q^{-1}\left(\frac{\alpha}{2}\right) - \frac{\theta}{\sigma}\right). \end{aligned}$$

Figure 4.7 compares the power of the UMP rules δ^+ of (4.16), δ^- of (4.17), and the GLRT rule, and Figure 4.8 compares their ROCs. As should be expected, the GLRT performs somewhat worse than the UMP test. However, it should be recalled that, unlike the GLRT, the UMP test requires knowledge of the sign of θ. The GLRT has found a great variety of applications. In general, the GLRT does perform well when the estimates $\hat{\theta}_0$ and $\hat{\theta}_1$ are reliable.

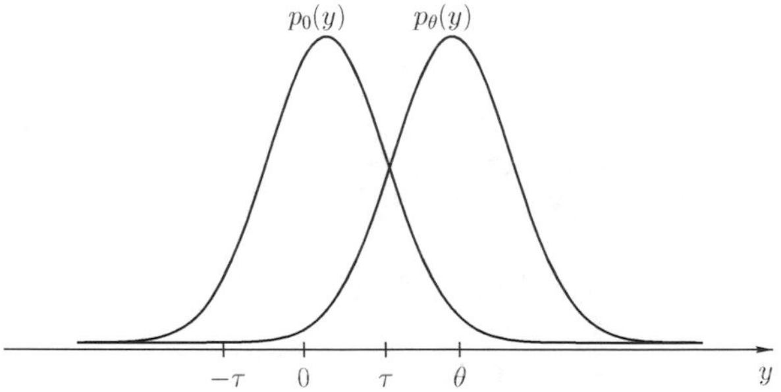

Figure 4.6 Thresholds for GLRT on Gaussian distributions

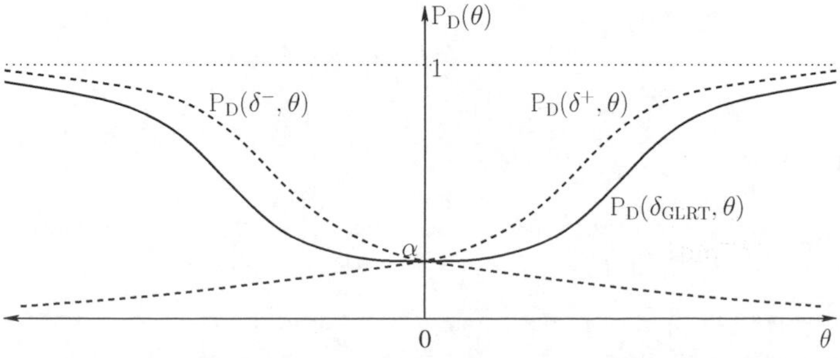

Figure 4.7 P_D versus θ for UMP and GLRT

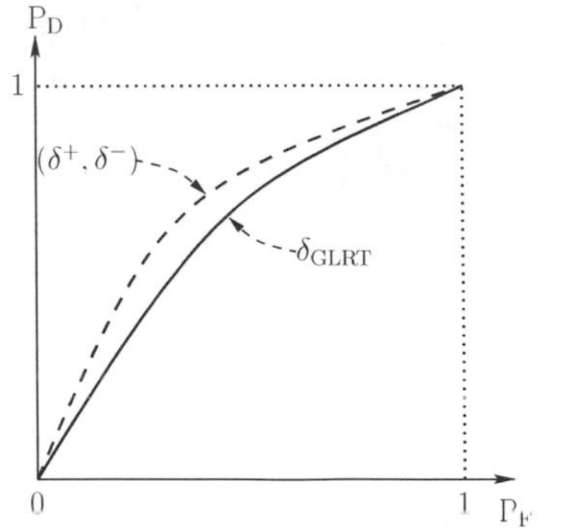

Figure 4.8 ROC for (δ^+, δ^-) and δ_{GLRT} for fixed θ

4.5.2 GLRT for Cauchy Hypothesis Testing

Example 4.9 *Detection of One-Sided Composite Signal in Cauchy Noise (continued).* This problem was introduced in Example 4.5, and we saw that there is no UMP solution. We studied the LMP solution in Example 4.8, and now we examine the GLRT solution. The GLR statistic is given by

$$L_G(y) = \frac{\sup_{\theta>0} p_\theta(y)}{p_0(y)}$$

with

$$\sup_{\theta>0} p_\theta(y) = \sup_{\theta>0} \frac{1}{\pi[1+(y-\theta)^2]} = \begin{cases} \frac{1}{\pi} & : y \geq 0 \\ \frac{1}{\pi(1+y^2)} & : y < 0. \end{cases}$$

Thus

$$L_G(y) = \begin{cases} 1+y^2 & : y \geq 0 \\ 1 & : y < 0. \end{cases}$$

To find an α-level test we need to evaluate $\text{P}_0\{L_G(Y) > \eta\}$. Clearly

$$\text{P}_0\{L_G(Y) > \eta\} = 1 \quad \text{for } 0 \leq \eta < 1.$$

For $\eta \geq 1$

$$\text{P}_0\{L_G(Y) > \eta\} = \int_{\sqrt{\eta-1}}^{\infty} \frac{1}{\pi} \frac{1}{1+y^2} dy = \frac{1}{2} - \frac{\tan^{-1}(\sqrt{\eta-1})}{\pi}.$$

There is a point of discontinuity in $\text{P}_0\{L_G(Y) > \eta\}$ at $\eta = 1$ as the value drops from 1 to the left to $\frac{1}{2}$ to the right. For $\alpha \in (\frac{1}{2}, 1]$, we would need to randomize to meet the P_F constraint with equality. For $\alpha \in (0, \frac{1}{2}]$, which would be more relevant in practice, the GLRT is a deterministic test:

$$\delta_{\mathrm{GLRT}}(y) = \begin{cases} 1 & : L_G(y) \ge \eta, \\ 0 & : \quad\quad\ < \end{cases}$$

where

$$\eta = \left[\tan(\pi(\frac{1}{2} - \alpha)) \right]^2 + 1.$$

This reduces to the following threshold test on Y:

$$\delta_{\mathrm{GLRT}}(y) = \begin{cases} 1 & : y \ge \tau, \\ 0 & : \quad < \end{cases}$$

where

$$\tau = \tan(\pi(\frac{1}{2} - \alpha)).$$

4.6 Random versus Nonrandom θ

In Sections 4.3–4.5, we assumed θ was deterministic but unknown. We now revisit the setup in Section 4.2 where θ is a realization of a random variable Θ with conditional pdf $\pi_i(\theta)$ under H_i, for $i = 0, 1$. The likelihood ratio is of the form:

$$L(y) = \frac{p(y|H_1)}{p(y|H_0)} = \frac{\int_{\mathcal{X}_1} p_\theta(y)\pi_1(\theta)\, d\nu(\theta)}{\int_{\mathcal{X}_0} p_\theta(y)\pi_0(\theta)\, d\nu(\theta)}.$$

Assume that $\mathcal{X}_0$ and $\mathcal{X}_1$ are compact sets and that for each $y \in \mathcal{Y}$, $p_\theta(y)$ is a continuous function of θ. Then by the mean value theorem, there exist $\tilde{\theta}_1(y)$ and $\tilde{\theta}_0(y)$ such that

$$p_{\tilde{\theta}_1}(y) = \int_{\mathcal{X}_1} p_\theta(y)\pi_1(\theta)\, d\nu(\theta) \quad \text{and} \quad p_{\tilde{\theta}_0}(y) = \int_{\mathcal{X}_0} p_\theta(y)\pi_0(\theta)\, d\nu(\theta).$$

We may thus write

$$L(y) = \frac{p_{\tilde{\theta}_1}(y)}{p_{\tilde{\theta}_0}(y)},$$

which is reminiscent of the GLR of (4.23), using different estimators of the parameter. The connection is examined in the following example.

Example 4.10 *Gaussian Hypothesis Testing with Gaussian Prior on θ.* Assume that $Y = \Theta + N$ where $N \sim \mathcal{N}(0, \sigma^2)$. Moreover, $\theta = 0$ under H_0, and $\Theta \sim \mathcal{N}(0, a^2)$ is independent of N under H_1. The hypothesis test becomes

$$\begin{cases} H_0 : Y \sim \mathcal{N}(0, \sigma^2) \\ H_1 : Y \sim \mathcal{N}(0, \sigma^2 + a^2). \end{cases}$$

The likelihood ratio is

$$L(y) = \frac{p(y|H_1)}{p(y|H_0)} = \frac{\frac{1}{\sqrt{2\pi(\sigma^2+a^2)}}\exp\{-\frac{y^2}{2(\sigma^2+a^2)}\}}{\frac{1}{\sqrt{2\pi\sigma^2}}\exp\{-\frac{y^2}{2\sigma^2}\}}$$
$$= \sqrt{\frac{\sigma^2}{\sigma^2+a^2}}\exp\left\{-\frac{y^2}{2\sigma^2}\left(\frac{\sigma^2}{\sigma^2+a^2}-1\right)\right\}$$
$$= \sqrt{\beta}\exp\left\{\frac{y^2}{2\sigma^2}(1-\beta)\right\},$$

where $\beta = \sqrt{\frac{\sigma^2}{\sigma^2+a^2}}$. The LRT is given by

$$\delta_{\text{LRT}}(y) = \begin{cases} 1 & : \sqrt{\beta}\exp\left\{\frac{y^2}{2\sigma^2}(1-\beta)\right\} \ge \eta \\ 0 & : \qquad\qquad < \end{cases} = \begin{cases} 1 & : |y| \ge \tau, \\ 0 & : \quad < \end{cases}$$

where

$$\tau = \sigma\sqrt{\frac{2}{1-\beta}\left(\ln\eta - \frac{1}{2}\ln\beta\right)}.$$

Notice that this optimal LRT is exactly the same as the GLRT in (4.26) despite the fact that these tests were derived under different assumptions.

4.7 Non-Dominated Tests

If θ is not modeled as random, the choice of the decision rule implies some tradeoff between the various conditional risks (indexed by $\theta \in \mathcal{X}$). Recall the notion of an admissible rule in Definition 1.1: if δ is admissible, there exists no rule δ' that is better in the sense that $R_\theta(\delta') \le R_\theta(\delta)$ for all $\theta \in \mathcal{X}$, with strict inequality for at least one value of θ. This definition extends straightforwardly to randomized decision rules. An admissible rule is also called *non-dominated.* Following Lehmann [1], we say that a class $\mathcal{C}$ of decision rules is *complete* if for any $\delta \notin \mathcal{C}$ there exists $\delta' \in \mathcal{C}$ dominating it. A class $\mathcal{C}$ is *minimal complete* if it contains only non-dominated rules.

We now characterize the minimal complete class of composite hypothesis tests. For simplicity of the exposition, we assume that $\mathcal{X}$ is a finite set. Define the risk vector $R(\tilde{\delta}) = (R_\theta(\tilde{\delta}))_{\theta\in\mathcal{X}}$ for a randomized rule $\tilde{\delta} \in \widetilde{\mathcal{D}}$, and the risk set $R(\widetilde{\mathcal{D}}) = \{R(\tilde{\delta}),\ \tilde{\delta} \in \widetilde{\mathcal{D}}\}$, which is convex. To any randomized test $\tilde{\delta}$, one can associate a conditional pmf $q(i|y) = \text{P}\{\text{Say } H_i | Y = y\}$. One may then write the conditional risk under P_θ as

$$R_\theta(\tilde{\delta}) = \mathbb{E}_\theta\left[\sum_i C(i,\theta)q(i|Y)\right], \quad \theta \in \mathcal{X}. \tag{4.28}$$

For uniform costs, this simplifies to

$$R_\theta(\tilde{\delta}) = \mathbb{E}_\theta[1 - q(j|Y)] = 1 - \mathbb{E}_\theta[q(j|Y)].$$

THEOREM 4.2 *Any non-dominated test is a Bayes test for some prior probability vector* $w = (w_\theta)_{\theta \in \mathcal{X}}$.

COROLLARY 4.1 *Under uniform costs, any non-dominated test is a MAP test for some prior probability vector* w.

Proof It follows from a well-known lemma in convex geometry [2, Sec. 2.6.3, p. 55] that

$$\tilde{\delta}^* \text{ is non-dominated} \quad \Leftrightarrow \quad \exists w \geq \mathbf{0} : w^\top R(\tilde{\delta}^*) \leq w^\top R(\tilde{\delta}), \quad \forall \tilde{\delta} \in \widetilde{\mathcal{D}}. \tag{4.29}$$

Geometrically, this implies that the vector $R(\tilde{\delta}^*)$ lies on the boundary of the risk set, and that w is a vector orthogonal to the hyperplane tangent to the risk set at $\tilde{\delta}^*$. Moreover, without loss of optimality, we can normalize this vector so that $\sum_\theta w_\theta = 1$, in which case w is a probability vector.

Denote by $q^*(i|y)$ the conditional pmf associated with $\tilde{\delta}^*$. Substituting (4.28) into (4.29), we obtain the following necessary condition for $\tilde{\delta}^*$ to be non-dominated:

$$\begin{aligned} &\int_{\mathcal{Y}} \sum_i q^*(i|y) \sum_\theta w_\theta C(i,\theta) p_\theta(y)\, d\mu(y) \\ &\leq \int_{\mathcal{Y}} \sum_i q(i|y) \sum_\theta w_\theta C(i,\theta) p_\theta(y)\, d\mu(y), \quad \forall q. \end{aligned}$$

Hence for each y and each $q(\cdot|y)$,

$$\sum_i q^*(i|y) \sum_\theta w_\theta C(i,\theta) p_\theta(y) \leq \sum_i q(i|y) \sum_\theta w_\theta C(i,\theta) p_\theta(y).$$

Write $C(i|y) = \sum_\theta w_\theta C(i,\theta) p_\theta(y)$, which can be interpreted as a posterior cost (up to a normalization factor dependent on y) since w is a probability vector. Then the previous inequality takes the form

$$\sum_i q^*(i|y) C(i|y) \leq \sum_i q(i|y) C(i|y), \quad \forall q.$$

If $\arg\min_i C(i|y)$ is unique, $q^*(\cdot|y)$ must put all mass at the minimizing i. Else $q^*(\cdot|y)$ is any distribution that puts all mass on the set of minimizers of $C(i|y)$. We conclude that $\tilde{\delta}^*$ is a Bayes test with respect to the prior probability vector w. The corollary follows. □

Example 4.11 *Composite Binary Gaussian Hypothesis Testing.* Let $P_\theta = \mathcal{N}(\theta, 1)$ and test $H_0 : \theta = 0$ versus $H_1 : \theta \in \mathcal{X}_1 = \{\theta_1, \theta_2\}$, where $\theta_2 < 0 < \theta_1$. Assume uniform costs. Note there exists no UMP test here because $\theta_2 < 0 < \theta_1$. We claim that all non-dominated tests are of the form

$$\delta(y) = \mathbb{1}\{y < \tau_2 \text{ or } y > \tau_1\} \tag{4.30}$$

where $\tau_2 \leq \tau_1$. Hence the minimal complete class of tests is two-dimensional, and its elements are parameterized by (τ_1, τ_2).

To show (4.30), by Theorem 4.2, it suffices to derive the Bayes tests where the three possible values of θ have prior probabilities given by $\pi(0)$, $\pi(\theta_1)$, and $\pi(\theta_2)$, respectively. The Bayes test compares the likelihood ratio $L(y)$ with the threshold $\eta = \frac{\pi(0)}{1-\pi(0)}$ where

$$\begin{aligned} L(y) &= \frac{p(y|H_1)}{p(y|H_0)} \\ &= \frac{\pi_1(\theta_1)\exp\{-\frac{1}{2}(y-\theta_1)^2\} + \pi_1(\theta_2)\exp\{-\frac{1}{2}(y-\theta_2)^2\}}{\exp\{-\frac{1}{2}y^2\}} \\ &= \pi_1(\theta_1)\exp\left\{\theta_1 y - \frac{1}{2}\theta_1^2\right\} + \pi_1(\theta_2)\exp\left\{\theta_2 y - \frac{1}{2}\theta_2^2\right\}. \end{aligned}$$

Note that $L(y)$ is a strictly convex function which tends to infinity as $|y| \to \infty$. Solving the equation $L'(y) = 0$, one obtains the minimizing y as

$$y^* = \frac{\theta_1 + \theta_2}{2} + \frac{\ln \frac{|\theta_2|\pi(\theta_2)}{\theta_1 \pi(\theta_1)}}{\theta_1 - \theta_2}.$$

If $L(y^*) \geq \eta$, then $\delta_B(y) \equiv 1$ is a Bayes rule and satisfies (4.30) with any $\tau_1 = \tau_2$. Else there exist τ_1 and τ_2 satisfying $L(\tau_1) = L(\tau_2) = \eta$ and $\tau_2 < y^* < \tau_1$, and the Bayes rule is given by (4.30).

Exercise 4.9 shows that the GLRT is of the form (4.30), with τ_1 and τ_2 being functions of the test threshold η.

4.8 Composite m-ary Hypothesis Testing

We now extend the composite hypothesis testing problem to the case of $m > 2$ hypotheses. The test is formulated as

$$H_j : Y \sim p_\theta, \quad \theta \in \mathcal{X}_j, \quad j = 0, 1, \ldots, m-1, \tag{4.31}$$

where the sets $\{\mathcal{X}_j\}_{j=0}^{m-1}$ form a partition of the state space $\mathcal{X}$.

4.8.1 Random Parameter Θ

Assume a prior π on $\mathcal{X}$ and a cost function $C(i, \theta)$, $0 \leq i \leq m-1$, $\theta \in \mathcal{X}$. Then $\pi_j = \int_{\mathcal{X}_j} \pi(\theta) d\nu(\theta)$ is the prior probability that hypothesis H_j is true. The Bayes rule takes the form

$$\delta_B(y) = \arg\min_{0 \leq i \leq m-1} C(i|y),$$

where

$$C(i|y) = \int_{\mathcal{X}} C(i, \theta)\pi(\theta|y) d\nu(\theta)$$

is the posterior cost of deciding in favor of H_i.

If the costs are uniform over each hypothesis, i.e.,

$$C(i,\theta) = C_{ij} \quad \forall i, j,\ \forall \theta \in \mathcal{X}_j,$$

the classification problem reduces to simple m-ary hypothesis testing with conditional distributions

$$p(y|H_j) = \int_{\mathcal{X}_j} \pi_j(\theta|y) d\nu(\theta), \quad 0 \le j \le m-1,$$

and $\pi_j(\theta|y) = \pi(\theta|y)/\pi_j$. For uniform costs, $C_{ij} = \mathbb{1}_{\{i \neq j\}}$, this reduces to MAP detection:

$$\delta_{\mathrm{B}}(y) = \arg\max_{0 \le i \le m-1} [\pi_i\, p(y|H_i)]. \tag{4.32}$$

All this is a straightforward extension of the binary case ($m = 2$) in Section 4.2.

4.8.2 Non-Dominated Tests

If θ is not modeled as random, the choice of the decision rule implies some tradeoff between the various types of error. The exposition of non-dominated tests in Section 4.7 extends directly to the m-ary case: neither Theorem 4.2 nor its corollary required that m be equal to 2.

Example 4.12 *Composite Ternary Hypothesis Testing.* Let $m = 3$ and $P_\theta = \mathcal{N}(\theta, \sigma^2)$. Assume that $\mathcal{X}_0 = \{0\}$, $\mathcal{X}_1$ is a finite subset of $(0, \infty)$, $\mathcal{X}_2$ is a finite subset of $(-\infty, 0)$, and the costs are uniform. We claim that all non-dominated tests are of the form

$$\delta(y) = \begin{cases} 0 & : \tau_2 \le y \le \tau_1 \\ 1 & : y > \tau_1 \\ 2 & : y < \tau_2, \end{cases} \tag{4.33}$$

where $\tau_2 < \tau_1$. This defines a two-dimensional family of tests parameterized by (τ_1, τ_2).

The conditional risks for these tests are given by

$$R_0(\delta) = Q\left(\frac{\tau_1}{\sigma}\right) + Q\left(\frac{-\tau_2}{\sigma}\right),$$

$$R_\theta(\delta) = Q\left(\frac{\theta - \tau_1}{\sigma}\right), \quad \theta > 0,$$

$$R_\theta(\delta) = Q\left(\frac{\tau_2 - \theta}{\sigma}\right), \quad \theta < 0.$$

To show (4.33), we apply Theorem 4.2 and derive the MAP tests for any prior π on $\mathcal{X}$. We have $\delta_{\mathrm{B}}(y) = \arg\min_{i=0,1,2} f_i(y)$ where

$$f_i(y) = \pi_i \frac{p(y|H_i)}{p(y|H_0)} = \begin{cases} \pi(0) & : i = 0 \\ \sum_{\theta \in \mathcal{X}_i} \pi(\theta) \exp\{-y\theta - \theta^2/2\} & : i = 1, 2. \end{cases}$$

Observe that f_0 is constant, f_1 is monotonically increasing, and f_2 is monotonically decreasing. Hence the decision regions $\mathcal{Y}_1$ and $\mathcal{Y}_2$ are semi-infinite intervals extending to $+\infty$ and $-\infty$, respectively. If $\pi(0)$ is small enough, the test never returns 0 and satisfies (4.33) with $\tau_1 = \tau_2$ satisfying $f_1(\tau_1) = f_2(\tau_1)$. Else the test satisfies (4.33) with $\tau_1 > \tau_2$ satisfying $f_1(\tau_1) = \pi(0) = f_2(\tau_2)$.

4.8.3 m-GLRT

There is no commonly used m-ary version of the GLRT. The following prescription seems natural enough in view of Corollary 4.1. Choose a probability vector π on $\{0, 1, \ldots, m-1\}$ and define

$$\delta_{\text{GLRT}}(y) = \arg\min_{0 \le i \le m-1} [\pi_i \max_{\theta \in \mathcal{X}_i} p_\theta(y)]. \tag{4.34}$$

Note this reduces to the GLRT (4.24) in the case $m = 2$, if one chooses $\pi_0 = \frac{\eta}{1+\eta}$. As with the GLRT, good performance can be expected when the parameter estimates $\hat{\theta}_i = \max_{\theta \in \mathcal{X}_i} p_\theta(y)$, $0 \le i \le m-1$, are reliable.

4.9 Robust Hypothesis Testing

Thus far we have studied the case where the distributions of the observations under the two hypotheses are not specified explicitly but come from a parametric family of densities $\{p_\theta, \theta \in \mathcal{X}\}$. We now generalize this notion of uncertainty to the case where the probability distributions under the two hypotheses belong to classes (sets) of distributions, often termed as *uncertainty classes*, which are not necessarily parametric. We denote the uncertainty class for hypothesis H_j by $\mathcal{P}_j$, $j = 0, 1$, and assume that the sets $\mathcal{P}_0$ and $\mathcal{P}_1$ are disjoint (have no overlap). The hypotheses can then be described by

$$\begin{cases} H_0 : \mathrm{P}_0 \in \mathcal{P}_0 \\ H_1 : \mathrm{P}_1 \in \mathcal{P}_1. \end{cases} \tag{4.35}$$

We could look for a UMP solution to (4.35) by trying to find a test $\tilde{\delta}_{\text{UMP}}$ that satisfies the following two properties. First,

$$\sup_{\mathrm{P}_0 \in \mathcal{P}_0} \mathrm{P}_\mathrm{F}(\tilde{\delta}, \mathrm{P}_0) \le \alpha \tag{4.36}$$

for $\tilde{\delta} = \tilde{\delta}_{\text{UMP}}$. Second, for any $\tilde{\delta} \in \tilde{\mathcal{D}}$ that satisfies (4.36),

$$\mathrm{P}_\mathrm{D}(\tilde{\delta}, \mathrm{P}_1) \le \mathrm{P}_\mathrm{D}(\tilde{\delta}_{\text{UMP}}, \mathrm{P}_1), \quad \forall \mathrm{P}_1 \in \mathcal{P}_1, \tag{4.37}$$

where $\mathrm{P}_\mathrm{F}(\tilde{\delta}, \mathrm{P}_0) \triangleq \mathrm{P}_0\{\tilde{\delta}(Y) = 1\}$ and $\mathrm{P}_\mathrm{D}(\tilde{\delta}, \mathrm{P}_1) \triangleq \mathrm{P}_0\{\tilde{\delta}(Y) = 1\}$. However, as we saw in Section 4.3, such UMP solutions are rare even when the distributions under the two hypotheses belong to parametric families, and so proceeding along these lines is generally futile.

An alternative approach to dealing with model uncertainty, first proposed by Huber [3], is the *minimax* approach, where the goal is to minimize the worst-case detection performance over the uncertainty classes. The decision rules thus obtained are said to be *robust* to the uncertainties in the probability distributions. The minimax robust version of the Neyman–Pearson formulation is given by[1]

$$\underset{\tilde{\delta}\in\tilde{\mathcal{D}}}{\text{maximize}} \inf_{\mathrm{P}_1\in\mathcal{P}_1} \mathrm{P_D}(\tilde{\delta},\mathrm{P}_1) \text{ subject to } \sup_{\mathrm{P}_0\in\mathcal{P}_0} \mathrm{P_F}(\tilde{\delta},\mathrm{P}_0) \le \alpha. \tag{4.38}$$

One approach to finding a solution to robust NP problem in (4.38) is based on the notion of stochastic ordering, described in the following definitions.

DEFINITION 4.1 A random variable U is said to be *stochastically larger* than a random variable V under probability distribution P if

$$\mathrm{P}\{U > t\} \ge \mathrm{P}\{V > t\}, \quad \text{for all } t \in \mathbb{R}.$$

DEFINITION 4.2 Given two candidate probability distributions P and Q for a random variable X, X is stochastically larger under distribution Q than under distribution P if

$$\mathrm{Q}\{X > t\} \ge \mathrm{P}\{X > t\}, \quad \text{for all } t \in \mathbb{R}.$$

These definitions lead to the following definition of *joint stochastic boundedness* (JSB) of a pair of classes of probability distributions.

DEFINITION 4.3 *Joint Stochastic Boundedness* [4]. A pair $(\mathcal{P}_0, \mathcal{P}_1)$ of classes of probability distributions defined on a measurable observation space $\mathcal{Y}$ is said to be jointly stochastically bounded by $(\mathrm{Q}_0, \mathrm{Q}_1)$, if there exist distributions $\mathrm{Q}_0 \in \mathcal{P}_0$ and $\mathrm{Q}_1 \in \mathcal{P}_1$ such that for any $(\mathrm{P}_0, \mathrm{P}_1) \in \mathcal{P}_0 \times \mathcal{P}_1$, $Y \in \mathcal{Y}$, and all $t \in \mathbb{R}$,

$$\mathrm{P}_0\{\ln L_q(Y) > t\} \le \mathrm{Q}_0\{\ln L_q(Y) > t\}$$

and

$$\mathrm{P}_1\{\ln L_q(Y) > t\} \ge Q_1\{\ln L_q(Y) > t\},$$

where $L_q(y) = \frac{q_1(y)}{q_0(y)}$ is the likelihood ratio between Q_1 and Q_0.

The pair $(\mathrm{Q}_0, \mathrm{Q}_1)$ is called the pair of *least favorable distributions* (LFDs) for the pair of uncertainty classes $\{\mathcal{P}_0, \mathcal{P}_1\}$. (See Figure 4.9.)

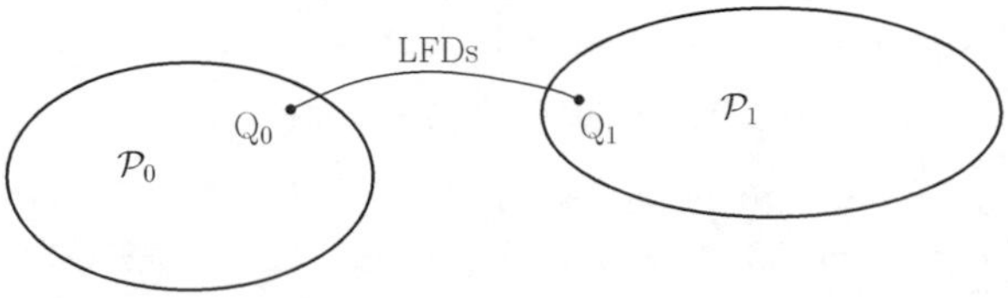

Figure 4.9 Least favorable distributions

[1] Minimax robust versions of the Bayesian and minimax hypothesis testing problems can similarly be defined.

REMARK 4.1 Note that it is not difficult to see that Definitions 4.1–4.3 remain unchanged if we replace "$> t$" with "$\geq t$" throughout; this will be relevant to the case where the random variables involved in the definitions have point masses at some points on the real line. For example, by taking convex combinations of the definitions with "$> t$" and "$\geq t$," we can see that Definition 4.2 is equivalent to the following condition:

$$\mathrm{Q}\{X > t\} + \gamma \mathrm{Q}\{X = t\} \geq \mathrm{P}\{X > t\} + \gamma \mathrm{P}\{X = t\}, \ \forall t \in \mathbb{R}, \forall \gamma \in [0, 1].$$

Definition 4.3 says that the log-likelihood ratio $\ln L_q(Y)$ is stochastically the smallest under Q_1 within class $\mathcal{P}_1$, and stochastically the largest under Q_0 within class $\mathcal{P}_0$. The following example, which considers the special case where $\mathcal{P}_0$ and $\mathcal{P}_1$ are parametric classes illustrates the notion of joint stochastic boundedness.

Example 4.13 *Least Favorable Distributions.* Let G_θ denote the Gaussian probability distribution with mean θ and variance 1, and consider the parametric classes of distributions

$$\mathcal{P}_0 = \{\mathrm{G}_\theta : \theta \in [-1, 0]\}, \quad \mathcal{P}_1 = \{\mathrm{G}_\theta : \theta \in [1, 2]\}.$$

We will establish that the pair $\mathrm{Q}_0 = \mathrm{G}_0$ and $\mathrm{Q}_1 = \mathrm{G}_1$ is a least favorable pair of distributions for the pair of uncertainty classes $(\mathrm{P}_0, \mathrm{P}_1)$. To see this, we first compute the log-likelihood ratio:

$$\ln L_q(Y) = \ln \frac{q_1(Y)}{q_0(Y)} = Y.$$

Now consider $\mathrm{P}_0 \in \mathcal{P}_0$, say $\mathrm{P}_0 = \mathrm{G}_{\theta_0}$, $\theta_0 \in [-1, 0]$. Then for all $t \in \mathbb{R}$,

$$\begin{aligned}\mathrm{P}_0\{\ln L_q(Y) > t\} &= \mathrm{P}_0\{Y > t\} = 1 - \Phi(t - \theta_0) \\ &\leq 1 - \Phi(t) = \mathrm{Q}_0\{Y > t\} = \mathrm{Q}_0\{\ln L_q(Y) > t\},\end{aligned}$$

where Φ is the cdf of the $\mathcal{N}(0, 1)$ distribution. Similarly it is easy to see that for any $\mathrm{P}_1 \in \mathcal{P}_1$, and all $t \in \mathbb{R}$

$$\mathrm{P}_1\{\ln L_q(Y) > t\} \geq \mathrm{Q}_1\{\ln L_q(Y) > t\}.$$

A *saddlepoint* solution to the minimax robust NP testing problem of (4.38) can easily be constructed if the pair $\{\mathcal{P}_0, \mathcal{P}_1\}$ is jointly stochastically bounded, as the following result shows. Similar solutions can also be obtained for the minimax robust versions of the Bayesian and minimax binary hypothesis testing problems.

THEOREM 4.3 *Suppose the pair of uncertainty classes* $\{\mathcal{P}_0, \mathcal{P}_1\}$ *is jointly stochastically bounded, with least favorable distributions given by* $(\mathrm{Q}_0, \mathrm{Q}_1)$. *Then the solution to the minimax robust NP testing problem of* (4.38) *is given by the solution to the simple hypothesis testing problem*

$$\underset{\tilde{\delta} \in \tilde{\mathcal{D}}}{\text{maximize}}\ \mathrm{P_D}(\tilde{\delta}, \mathrm{Q}_1) \text{ subject to } \mathrm{P_F}(\tilde{\delta}, \mathrm{Q}_0) \leq \alpha,$$

which is the randomized LRT

$$\tilde{\delta}^*(y) = \begin{cases} 1 & > \\ 1 \text{ w.p. } \gamma^* : \ \ln L_q(y) & = \eta^*, \\ 0 & < \end{cases}$$

where the threshold η^* and randomization parameter γ^* are chosen so that

$$\mathrm{P_F}(\tilde{\delta}^*, \mathrm{Q}_0) = \mathrm{Q}_0\{\ln L_q(Y) > \eta^*\} + \gamma^* \mathrm{Q}_0\{\ln L_q(Y) = \eta^*\} = \alpha.$$

Proof By the JSB property (also see Remark 4.1), for all $\mathrm{P}_0 \in \mathcal{P}_0$,

$$\begin{aligned} \mathrm{P_F}(\tilde{\delta}^*, \mathrm{P}_0) &= \mathrm{P}_0\{\ln L_q(Y) > \eta^*\} + \gamma^* \mathrm{P}_0\{\ln L_q(Y) = \eta^*\} \\ &\le \mathrm{Q}_0\{\ln L_q(Y) > \eta^*\} + \gamma^* \mathrm{Q}_0\{\ln L_q(Y) = \eta^*\} \\ &= \mathrm{P_F}(\tilde{\delta}^*, \mathrm{Q}_0) \\ &\le \alpha. \end{aligned}$$

Therefore $\tilde{\delta}^*$ satisfies the constraint for the minimax robust NP testing problem of (4.38), i.e.,

$$\sup_{\mathrm{P}_0 \in \mathcal{P}_0} \mathrm{P_F}(\tilde{\delta}^*, \mathrm{P}_0) \le \alpha.$$

Furthermore, for all $\mathrm{P}_1 \in \mathcal{P}_1$, and for all $\tilde{\delta} \in \tilde{\mathcal{D}}$ such that $\sup_{\mathrm{P}_0 \in \mathcal{P}_0} \mathrm{P_F}(\tilde{\delta}, \mathrm{P}_0) \le \alpha$,

$$\begin{aligned} \mathrm{P_D}(\tilde{\delta}^*, \mathrm{P}_1) &= \mathrm{P}_1\{\ln L_q(Y) > \eta^*\} + \gamma^* \mathrm{P}_1\{\ln L_q(Y) = \eta^*\} \\ &\overset{(a)}{\ge} \mathrm{Q}_1\{\ln L_q(Y) > \eta^*\} + \gamma^* \mathrm{Q}_1\{\ln L_q(Y) = \eta^*\} \\ &= \mathrm{P_D}(\tilde{\delta}^*, \mathrm{Q}_1) \\ &\overset{(b)}{\ge} \mathrm{P_D}(\tilde{\delta}, \mathrm{Q}_1) \\ &\ge \inf_{\mathrm{P}_1 \subset \mathcal{P}_1} \mathrm{P_D}(\tilde{\delta}, \mathrm{P}_1), \end{aligned} \tag{4.39}$$

where (a) follows from the JSB property, and (b) follows from the optimality of $\tilde{\delta}^*$ for N-P testing between Q_0 and Q_1 at level α. Taking the infimum over $\mathrm{P}_1 \in \mathcal{P}_1$ on the left hand side of (4.39) establishes that $\tilde{\delta}^*$ is a solution to (4.38). □

The uncertainty classes considered in Huber's original work [3] were neighborhood classes containing, under each hypothesis, a nominal distribution and distributions in its vicinity, in particular the ϵ-contamination and the total variation classes. In a later paper, Huber and Strassen [5] generalized the notion of neighborhood classes to those that can be described in terms of *alternating capacities of order* 2. When the observation space is *compact*, several uncertainty models such as ϵ-contaminated neighborhoods, total variation neighborhoods, band-classes, and p-point classes are special cases of this model with different choices of capacity. All of these uncertainty classes can be shown to possess the JSB property of Definition 4.3, and hence the corresponding minimax robust hypothesis testing problems can be shown to have saddlepoint solutions constructed from the LFDs in a unified way using Theorem 4.3 (or its Bayesian or minimax counterparts).

4.9.1 Robust Detection with Conditionally Independent Observations

Now consider the case where the observation $\boldsymbol{Y}$ is a vector (block) of observations $(Y_1, \ldots, Y_n)$, where each observation Y_k takes values in a measurable space $\mathcal{Y}_k$ and has a distribution which belongs to the class $\mathcal{P}_j^{(k)}$ when the hypothesis is H_j, $j = 0, 1$. Let the set

$$\mathcal{P}_j \triangleq \mathcal{P}_j^{(1)} \times \cdots \times \mathcal{P}_j^{(n)}$$

represent a class of distributions on $\mathcal{Y} = \mathcal{Y}_1 \times \cdots \times \mathcal{Y}_n$, which are products of distributions in $\mathcal{P}_j^{(k)}$, $k = 1, \ldots, n$. To further clarify this point, let $\mathrm{P}_j^{(k)}$ denote a typical element of $\mathcal{P}_j^{(k)}$. Then $\mathrm{P}_j \in \mathcal{P}_j$ represents the product distribution $\mathrm{P}_j^{(1)} \times \cdots \times \mathrm{P}_j^{(n)}$, i.e., the observations $(Y_1, \ldots, Y_n)$ are independent under P_j. We have the following result, which says that if the uncertainty classes corresponding to each of the observations satisfy the JSB property, and if the observations are conditionally independent, given the hypothesis, then the uncertainty classes corresponding to the entire block of observations also satisfy the JSB property.

LEMMA 4.1 *For each k, $k = 1, \ldots, n$, let the pair $(\mathcal{P}_0^{(k)}, \mathcal{P}_1^{(k)})$ be jointly stochastically bounded by $(\mathrm{Q}_0^{(k)}, \mathrm{Q}_1^{(k)})$. Then the pair $(\mathcal{P}_0, \mathcal{P}_1)$ is jointly stochastically bounded by $(\mathrm{Q}_0, \mathrm{Q}_1)$, where $\mathrm{Q}_j = \mathrm{Q}_j^{(1)} \times \cdots \times \mathrm{Q}_j^{(n)}$, $j = 0, 1$.*

The proof of Lemma 4.1 follows quite easily from the following result, whose proof is left to the reader as Exercise 4.16.

LEMMA 4.2 *Let $Z_1, Z_2, \ldots, Z_n$ be independent random variables. Suppose that Z_k is stochastically larger under distribution $\mathrm{Q}^{(k)}$ than under $\mathrm{P}^{(k)}$, i.e.,*

$$\mathrm{Q}^{(k)}\{Z_k > t_k\} \geq \mathrm{P}^{(k)}\{Z_k > t_k\}, \quad \text{for all } t_k \in \mathbb{R}.$$

Let $\mathrm{P} = \mathrm{P}^{(1)} \times \cdots \times \mathrm{P}^{(n)}$ and $\mathrm{Q} = \mathrm{Q}^{(1)} \times \cdots \times \mathrm{Q}^{(n)}$. Then $\sum_{k=1}^{n} Z_k$ is stochastically larger under Q than under P, i.e.,

$$\mathrm{Q}\left(\sum_{k=1}^{n} Z_k > t\right) \geq \mathrm{P}\left(\sum_{k=1}^{n} Z_k > t\right) \quad \text{for all } t \in \mathbb{R}.$$

Proof of Lemma 4.1 Let P_0 be any (product) distribution in the set $\mathcal{P}_0$, and P_1 be any (product) distribution in the set $\mathcal{P}_1$. Let L_q denote the likelihood ratio between Q_1 and Q_0, and let $L_q^{(k)}$ denote the likelihood ratio between $\mathrm{Q}_1^{(k)}$ and $\mathrm{Q}_0^{(k)}$. Then

$$\ln L_q(\boldsymbol{Y}) = \sum_{k=1}^{n} \ln L_q^{(k)}(Y_k).$$

By the joint stochastic boundedness property of $(\mathcal{P}_0^{(k)}, \mathcal{P}_1^{(k)})$, $\ln L_q^{(k)}(Y_k)$ is stochastically larger under $\mathrm{Q}_0^{(k)}$ than under $\mathrm{P}_0^{(k)}$. Hence, by Lemma 4.2, $\ln L_q(\boldsymbol{Y})$ is stochastically larger under Q_0 than under P_0. This proves the first condition required for the joint stochastic boundedness of $(\mathcal{P}_0, \mathcal{P}_1)$ by $(\mathrm{Q}_0, \mathrm{Q}_1)$. The other condition is proved similarly. □

4.9.2 Epsilon-Contamination Class

As we saw in Theorem 4.3, the solution to the minimax robust detection problem relied on the JSB property of the uncertainty classes. Establishing that a pair of uncertainty classes satisfies the JSB property, and finding the corresponding LFDs, is a nontrivial task. In this section, we focus on the most commonly studied pair of uncertainty classes, which was introduced in Huber's seminal work [3], the ϵ-contamination classes:

$$\mathcal{P}_j = \{\mathrm{P}_j = (1-\epsilon)\bar{\mathrm{P}}_j + \epsilon \mathrm{M}_j\}, \; j = 0, 1, \tag{4.40}$$

where ϵ is small enough so that $\mathcal{P}_0$ and $\mathcal{P}_1$ do not overlap.[2] Here $\bar{\mathrm{P}}_0$ and $\bar{\mathrm{P}}_1$ can be considered to be known *nominal* distributions for the observations under the hypotheses, but these nominal distributions are "contaminated" by unknown and arbitrary distributions M_0 and M_1, respectively. Such an uncertainty model might arise, for example, in communications systems where an extraneous unknown source of interference (e.g., lightning strike) may occur with probability ϵ every time one tries to communicate information on the channel.

THEOREM 4.4 *The uncertainty classes $(\mathcal{P}_0, \mathcal{P}_1)$ given by the ϵ-contamination model of (4.40) satisfy the JSB property of Definition 4.3, with LFDs that are given by*

$$q_0(y) = \begin{cases} (1-\epsilon)\bar{p}_0(y) & \text{if } \bar{L}(y) < b \\ \frac{(1-\epsilon)}{b}\bar{p}_1(y) & \text{if } \bar{L}(y) \geq b \end{cases} \tag{4.41}$$

and

$$q_1(y) = \begin{cases} (1-\epsilon)\bar{p}_1(y) & \text{if } \bar{L}(y) > a \\ a(1-\epsilon)\bar{p}_0(y) & \text{if } \bar{L}(y) \leq a \end{cases} \tag{4.42}$$

if $0 < a < 1 < b$ can be chosen such that q_0 and q_1 are valid densities, i.e., $\int q_j(y)d\mu(y) = 1$, $j = 0, 1$.

Proof We begin by noting that the log-likelihood ratio between q_1 and q_0 is given by

$$L_q(y) = \frac{q_1(y)}{q_0(y)} = \begin{cases} b & \text{if } \bar{L}(y) \geq b \\ \bar{L}(y) & \text{if } a < \bar{L}(y) < b \\ a & \text{if } \bar{L}(y) \leq a, \end{cases}$$

i.e., L_q is a "clipped" or "censored" version of the nominal likelihood ratio $\bar{L}$.

[2] For example, if $\bar{\mathrm{P}}_0 = \mathcal{N}(0, 1)$, $\bar{\mathrm{P}}_1 = \mathcal{N}(1, 1)$, and $\epsilon = 0.5$, then it is easily seen that $\mathcal{P}_0$ and $\mathcal{P}_1$ have the distribution $0.5\mathrm{P}_0 + 0.5\mathrm{P}_1$ in common.

To establish the JSB property, we need to show that for any $(\mathrm{P}_0, \mathrm{P}_1) \in \mathcal{P}_0 \times \mathcal{P}_1$, $Y \in \mathcal{Y}$, and all $t \geq 0$,

$$\mathrm{P}_0\{L_q(Y) > t\} \leq \mathrm{Q}_0\{L_q(Y) > t\} \tag{4.43}$$

and

$$\mathrm{P}_1\{L_q(Y) > t\} \geq Q_1\{L_q(Y) > t\}. \tag{4.44}$$

Now if $t \leq a$ or $t \geq b$, then both sides of these inequalities are equal to 0, or both sides are equal to 1, and the inequalities are met trivially. If $a < t < b$, then

$$\begin{aligned}
\mathrm{P}_0\{L_q(Y) \leq t\} &= (1-\epsilon)\bar{\mathrm{P}}_0\{L_q(Y) \leq t\} + \epsilon \mathrm{M}_0\{L_q(Y) \leq t\} \\
&\geq (1-\epsilon)\bar{\mathrm{P}}_0\{L_q(Y) \leq t\} \\
&\overset{(a)}{=} \mathrm{Q}_0\{L_q(Y) \leq t\},
\end{aligned}$$

where (a) follows because

$$L_q(Y) \leq t \implies \bar{L}(Y) < b \implies q_0(y) = (1-\epsilon)\bar{p}_0(y).$$

This establishes (4.43). A similar proof can be given for (4.44). □

REMARK 4.2 It is easy to verify that for q_0 and q_1 as defined in (4.41) and (4.42), $\mathrm{Q}_0 \in \mathcal{P}_0$ and $\mathrm{Q}_1 \in \mathcal{P}_1$, assuming of course that a and b can be chosen such that q_0 and q_1 are valid densities. Indeed we can write

$$q_0(y) = (1-\epsilon)\bar{p}_0(y) + \epsilon m_0(y),$$

where

$$m_0(y) = \frac{1-\epsilon}{\epsilon}\left[\frac{\bar{p}_1(y)}{b} - \bar{p}_0(y)\right] \mathbb{1}\{\bar{L}(y) \geq b\}.$$

A similar argument can be given for q_1.

REMARK 4.3 Note that the densities q_0 and q_1 can be rewritten as

$$\begin{aligned}
q_0(y) &= (1-\epsilon)\max\left\{\frac{\bar{p}_1(y)}{b}, \bar{p}_0(y)\right\}, \\
q_1(y) &= (1-\epsilon)\max\left\{\bar{p}_1(y), a\bar{p}_0(y)\right\}.
\end{aligned}$$

Therefore in order for q_0 and q_1 to be valid densities, i.e., $\int q_j(y)d\mu(y) = 1$, $j = 0, 1$, we must have

$$\int \max\left\{\frac{\bar{p}_1(y)}{b}, \bar{p}_0(y)\right\} d\mu(y) = \int \max\left\{\bar{p}_1(y), a\bar{p}_0(y)\right\} d\mu(y) = \frac{1}{1-\epsilon},$$

which implies that $a = \frac{1}{b}$. Also, it is clear the solution for a increases with ϵ. Therefore, for ϵ large enough, the solution a could be greater than 1. In this case, we can see that the corresponding $\mathrm{Q}_1 \in \mathcal{P}_0$, i.e., $\mathcal{P}_1$ and $\mathcal{P}_0$ have an overlap and there is no robust solution to the hypothesis testing problem. In particular, if ϵ is such that $a = b = 1$, then it is clear that $q_0 = q_1$. See Exercise 4.17 for an example.

REMARK 4.4 The robust test for the ϵ-contamination model is a LRT based on clipping the nominal likelihood ratio:

$$L_q(y) = \begin{cases} b & \text{if } \bar{L}(y) \geq b \\ \bar{L}(y) & \text{if } a < \bar{L}(y) < b \\ a & \text{if } \bar{L}(y) \leq a. \end{cases}$$

Some intuition for why such a test might be effective comes looking at the hypothesis testing problem with i.i.d. observations under each hypothesis:

$$\begin{cases} H_0 : \{Y_k\}_{k=1}^n \sim \text{i.i.d. } \mathrm{P}_0 \\ H_1 : \{Y_k\}_{k=1}^n \sim \text{i.i.d. } \mathrm{P}_1, \end{cases}$$

where $\mathrm{P}_0 \in \mathcal{P}_0$ and $\mathrm{P}_1 \in \mathcal{P}_1$ as in (4.40). Since the contaminating distributions M_0 and M_1 are not known to the decision-maker, a naive approach to constructing a test might be based on simply using the nominal likelihood ratio:

$$\bar{L}^{(n)}(\boldsymbol{y}) = \prod_{k=1}^{n} \bar{L}(y_k),$$

with the corresponding LRT given by:

$$\tilde{\delta}_{\mathrm{LRT}}(\boldsymbol{y}) = \begin{cases} 1 & > \\ 1 \text{ w.p. } \gamma : \ \bar{L}^{(n)}(\boldsymbol{y}) & = \eta. \\ 0 & < \end{cases}$$

Note that $\bar{L}^{(n)}(\boldsymbol{y})$ is sensitive to observations for which $\bar{L}(y_k) \gg 1$ or $\bar{L}(y_k) \ll 1$. The worst-case performance for the test based on $\bar{L}^{(n)}(\boldsymbol{Y})$ will occur when M_0 has all of its mass for values of y where $\bar{L}(y) \gg 1$ and M_1 has all of its mass for values of y where $\bar{L}(y) \ll 1$. With this choice:

$$\sup_{\mathrm{P}_0 \in \mathcal{P}_0} \mathrm{P_F}(\tilde{\delta}_{\mathrm{LRT}}, \mathrm{P}_0) \approx 1 - (1-\epsilon)^n, \text{ and } \sup_{\mathrm{P}_1 \in \mathcal{P}_1} \mathrm{P_M}(\tilde{\delta}_{\mathrm{LRT}}, \mathrm{P}_1) \approx 1 - (1-\epsilon)^n,$$

i.e., the worst-case performance of the LRT based on $\bar{L}^{(n)}(\boldsymbol{y})$ can be arbitrarily bad for large n.

Using $L_q(y)$ in place of $\bar{L}(y)$ limits the damage done by the contamination, and the worst-case performance is then simply the performance of the LRT between LFDs Q_1 and Q_0 for testing between the simple hypotheses where the observations follow the LFD under each hypothesis.

Exercises

4.1 *Bayesian Composite Hypothesis Testing with only H_1 Composite.* Consider the composite binary hypothesis testing problem in which

$$p_\theta(y) = \theta e^{-\theta y} \mathbb{1}\{y \geq 0\},$$

$\pi_0 = \pi_1 = \frac{1}{2}$, and the conditional prior under H_1 is

$$\pi_1(\theta) = 2\,\mathbb{1}\{0.5 \le \theta \le 1\}.$$

Find a Bayes rule and corresponding minimum Bayes risk for the hypotheses

$$\begin{cases} H_0 & : \mathcal{X}_0 = \{0.5\} \\ H_1 & : \mathcal{X}_1 = (0.5, 1]. \end{cases}$$

Assume $C(i, \theta) = C_{ij} = \mathbb{1}\{i \ne j\}$.

4.2 *Bayesian Composite Hypothesis Testing with Both Hypotheses Composite.* Consider the composite binary hypothesis testing problem in which

$$p_\theta(y) = \theta e^{-\theta y}\mathbb{1}\{y \ge 0\},$$
$$\pi(\theta) = \mathbb{1}\{0 \le \theta \le 1\},$$

and $C(i, \theta) = C_{ij} = \mathbb{1}\{i \ne j\}$.

Find a Bayes rule and corresponding minimum Bayes risk for the hypotheses

$$\begin{cases} H_0 & : \mathcal{X}_0 = [0, 0.5] \\ H_1 & : \mathcal{X}_1 = (0.5, 1]. \end{cases}$$

4.3 *Composite Hypothesis Testing with Nonuniform Costs over Hypotheses.* Rework Exercise 4.1 under the assumption that the cost function is:

$$C(i, \theta) = \begin{cases} \frac{1}{\theta} & \text{if} \quad i = 0, \quad \theta \in (0.5, 1] \qquad (H_1) \\ 1 & \text{if} \quad i = 1, \quad \theta = 0.5 \qquad (H_0) \\ 0 & \text{else.} \end{cases}$$

4.4 *GLRT and UMP for Uniform Distributions.* Consider the composite binary hypothesis test

$$\begin{cases} H_0 & : Y \sim \text{Uniform}[0, 1] \\ H_1 & : Y \sim \text{Uniform}[0, \theta], \quad \theta \in (0, 1). \end{cases} \tag{4.45}$$

(a) Derive the GLRT.
(b) Is there a UMP test for this problem? If so, give it. If not, explain why not.

4.5 *UMP for Poisson Distributions.* Let Y be Poisson(θ), i.e.,

$$p_\theta(y) = \frac{\theta^y}{y!}e^{-\theta}, \text{ for } y = 0, 1, 2, \ldots.$$

Consider the composite hypothesis testing problem

$$\begin{cases} H_0 : & Y \sim \text{Poisson}(1) \\ H_1 : & Y \sim \text{Poisson}(\theta), \quad \theta > 1. \end{cases}$$

Does a UMP rule exist? Explain.

4.6 *GLRT/UMP for Exponential Distributions.* Consider the composite hypothesis test

$$\begin{cases} H_0 : & Y \sim p_0(y) = e^{-y}\,\mathbb{1}\{y \ge 0\} \\ H_1 : & Y \sim p_\theta(y) = \theta\, e^{-\theta y}\,\mathbb{1}\{y \ge 0\}, \quad \theta > 1. \end{cases}$$

Is there a UMP test here? If so, what form does it take?

4.7 *GLRT/UMP for One-Sided Gaussian Distributions.* Consider the hypothesis test

$$\begin{cases} H_0: & Y = Z \\ H_1: & Y = s + Z, \end{cases}$$

where $s > 0$, and Z is a random variable distributed as a "one-sided Gaussian":

$$p(z) = \begin{cases} \frac{2}{\sqrt{2\pi\sigma^2}} \exp\{-\frac{z^2}{2\sigma^2}\} & : z \geq 0 \\ 0 & : \text{else.} \end{cases}$$

(a) Give the Bayes rule for deciding between H_0 and H_1 assuming known s, equal priors, and uniform costs.
(b) Now assume s is unknown, and derive the GLRT for this problem.
(c) Is the GLRT a UMP test here?

4.8 *GLRT/UMP for Gaussian Hypothesis testing.* Consider the hypothesis test

$$\begin{cases} H_0 & : \boldsymbol{Y} = \boldsymbol{Z} \\ H_1 & : \boldsymbol{Y} = \boldsymbol{s} + \boldsymbol{Z}, \end{cases}$$

where $\boldsymbol{s}$ is on the unit circle in $\mathbb{R}^2$ (you do not know where), and the vector $\boldsymbol{Z}$ is normal.

(a) Is there a UMP test for this problem? Explain why.
(b) Derive the GLRT for this problem, and sketch the boundary of the decision region.

4.9 *Composite Gaussian Hypothesis with Two Alternative Distributions.* Let $\theta_2 < 0 < \theta_1$. Derive the GLRT for the composite detection problem with $Y \sim P_\theta = \mathcal{N}(\theta, 1)$ and

$$\begin{cases} H_0 : \theta = 0 \\ H_1 : \theta \in \{\theta_1, \theta_2\}. \end{cases}$$

4.10 *Composite Gaussian Hypothesis Testing.* Consider the hypothesis testing problem:

$$\begin{cases} H_0 : \theta = 0 \\ H_1 : \theta > 0, \end{cases}$$

where

$$p_\theta(\boldsymbol{y}) = \frac{1}{2\pi} \exp\left[-\frac{(y_1 - \theta)^2 + (y_2 - \theta^2)^2}{2}\right].$$

(a) Is there a UMP test between H_0 and H_1? If so, find it for a level of $\alpha \in (0, 1)$. If not, explain clearly why not.
(b) Find an α-level LMP test for this problem, for $\alpha \in (0, 1)$.
(c) Find $\mathrm{P_D}$ as a function of α and θ for the LMP test.

4.11 *Composite Hypothesis Testing.* Consider the hypothesis testing problem:

$$\begin{cases} H_0 : \boldsymbol{Y} = \boldsymbol{Z} \\ H_1 : \boldsymbol{Y} = \boldsymbol{s}(\theta) + \boldsymbol{Z}, \end{cases}$$

where θ is a deterministic but unknown parameter that takes the value 1 or 2, $\mathbf{Z} \sim \mathcal{N}(\mathbf{0}, I)$, and

$$s(1) = \begin{bmatrix} 1 \\ 0 \\ 2 \\ 0 \end{bmatrix}, \quad s(2) = \begin{bmatrix} 0 \\ 1 \\ 0 \\ 2 \end{bmatrix}.$$

(a) Is there a UMP test between H_0 and H_1? If so, find it (for a level of α). If not, explain why not.
(b) Find an α-level GLRT for testing between H_0 and H_1.
(c) Argue that the probability of detection for the GLRT of part (b) is independent of θ, and then find the probability of detection as a function of α.

4.12 *UMP for Two-Sided Gaussian Hypothesis testing.* Here we switch the roles of H_0 and H_1 in Section 4.5.1. Consider the composite detection problem with $Y \sim P_\theta = \mathcal{N}(\theta, \sigma^2)$ and

$$\begin{cases} H_0 : \theta \neq 0 \\ H_1 : \theta = 0. \end{cases}$$

Does a UMP test exist for this problem? If so, find it for significance level α.

4.13 *UMP/LMP testing with Laplacian Observations.* Consider the composite binary detection problem in which

$$p_\theta(y) = \frac{1}{2} e^{-|y-\theta|}, \quad y \in \mathbb{R},$$

and

$$\begin{cases} H_0 : \theta = 0 \\ H_1 : \theta > 0. \end{cases}$$

(a) Show that a UMP test exists for this problem, even though the Monotone Likelihood Ratio Theorem does not apply. Find an α-level UMP test, and find P_D as a function of α for this test.
Hint: Even though the likelihood ratio has point masses under H_0 and H_1, it can be shown that randomization can be avoided in the LRT by rewriting the test in terms of Y. This step is crucial to establishing the existence of a UMP solution.
(b) Find a locally most powerful α-level test, and find P_D as a function of α for this test.
(c) Plot the ROCs for the tests in parts (a) and (b) on the same figure and compare them.

4.14 *Monotone Likelihood Ratio Theorem with Two Composite Hypotheses.* Consider the composite detection problem:

$$\begin{cases} H_0 : \theta \in \mathcal{X}_0 \\ H_1 : \theta \in \mathcal{X}_1. \end{cases}$$

Now suppose that for each fixed $\theta_0 \in \mathcal{X}_0$ and each fixed $\theta_1 \in \mathcal{X}_1$, we have

$$\frac{p_{\theta_1}(y)}{p_{\theta_0}(y)} = F_{\theta_0,\theta_1}(g(y)),$$

where the (real-valued) function g does not depend on θ_1 or θ_0, and the function F_{θ_0,θ_1} is *strictly increasing* in its argument.

Assume that for tests of the form

$$\tilde{\delta}(y) = \begin{cases} 1 & > \\ 1 \text{ w.p. } \gamma \ : \ g(y) & = \tau \\ 0 & < \end{cases} \quad \text{for some } \tau \in \mathbb{R}, \gamma \in [0, 1],$$

the supremum

$$\sup_{\theta_0 \in \mathcal{X}_0} \mathrm{P_F}(\tilde{\delta}; \theta_0) = \sup_{\theta_0 \in \mathcal{X}_0} \left(\mathrm{P}_{\theta_0}\{g(Y) > \tau\} + \gamma \mathrm{P}_{\theta_0}\{g(Y) = \tau\} \right)$$

is achieved at a fixed $\theta_0^* \in \mathcal{X}_0$, irrespective of the values of τ and γ.

Show that for any level $\alpha \in (0, 1)$, a UMP test between H_1 and H_0 exists, and specify it.

4.15 *UMP versus GLRT.* Consider the composite binary detection problem in which

$$p_\theta(y) = \theta e^{-\theta y} \, \mathbb{1}\{y \geq 0\}.$$

(a) For $\alpha \in (0, 1)$, show that a UMP test of level α exists for testing the hypotheses

$$\begin{cases} H_0 : \mathcal{X}_0 = [1, 2] \\ H_1 : \mathcal{X}_1 = (2, \infty). \end{cases}$$

Find this UMP test as a function of α.

(b) Find an α-level GLRT for testing between H_0 and H_1.

4.16 *Stochastic Ordering.* Prove Lemma 4.2.

4.17 *Robust Hypothesis Testing with ϵ-Contamination Model.* Consider the ϵ-contamination model, where the nominal distributions are $\bar{\mathrm{P}}_0 = \mathcal{N}(0, 1)$ and $\bar{\mathrm{P}}_1 = \mathcal{N}(\mu, 1)$.

(a) Find an equation for the critical value of ϵ (call it ϵ_{crit}) as function of ρ for which q_0 and q_1 as defined in (4.41) and (4.42) are equal.

(b) Compute numerically the ϵ_{crit} for $\rho = 1$.

(c) For $\epsilon = 0.5\epsilon_{\text{crit}}$, find the least favorable distributions q_0 and q_1.

References

[1] E. L. Lehmann, *Testing Statistical Hypotheses*, 2nd edition, Springer-Verlag, 1986.

[2] S. Boyd and L. Vandenberghe, *Convex Optimization*, Cambridge University Press, 2004.

[3] P. J. Huber, "A Robust Version of the Probability Ratio Test," *Annal. Math. Stat.*, Vol. 36, No. 6, pp. 1753–1758, 1965.
[4] V. V. Veeravalli, T. Başar, and H. V. Poor, "Minimax Robust Decentralized Detection," *IEEE Trans. Inf. Theory*, Vol. 40, No. 1, pp. 35–40, 1994.
[5] P. J. Huber and V. Strassen, "Minimax Tests and the Neyman–Pearson Lemma for Capacities," *Annal. Stat.*, Vol. 1, No. 2, pp. 251–263, 1973.

5 Signal Detection

In this chapter, we apply the basic principles developed so far to the problem of detecting a signal, consisting of a sequence of n values, observed with noise. The questions of interest are to determine the structure of the optimal detector, and to analyze the error probabilities. We first consider known signals corrupted by i.i.d. noise, and derive the matched filter for Gaussian noise and nonlinear detectors for non-Gaussian noise. We then introduce noise whitening methods for signal detection problems involving correlated Gaussian noise, and present the problem of signal selection. We also analyze the problem of detecting (possibly correlated) Gaussian signals in Gaussian noise. We study m-ary signal detection (classification) and its relation to optimal binary hypothesis tests via the union-of-events bound. We introduce techniques for analyzing the performance of the GLRT and Bayes tests for problems of composite hypothesis testing involving a signal known up to some parameters. Finally, we consider an approach to detecting non-Gaussian signals in Gaussian noise via the performance metric of deflection.

5.1 Introduction

In the detection problems we have studied so far, we did not make any explicit assumptions about the observation space, although most examples were restricted to scalar observations. The theory that we have developed applies equally to scalar and vector observations. In particular, consider the binary detection problem:

$$\begin{cases} H_0 : \boldsymbol{Y} \sim p_0, \\ H_1 : \boldsymbol{Y} \sim p_1, \end{cases}$$

where $\boldsymbol{Y} = (Y_1\ Y_2 \dots Y_n) \in \mathbb{R}^n$ and $\boldsymbol{y} = (y_1\ y_2 \dots y_n)$. The optimum detector for this problem, no matter which criterion (Bayes, Neyman–Pearson, or minimax) we choose, is of the form

$$\tilde{\delta}_{\mathrm{OPT}}(y) = \begin{cases} 1 & > \\ 1 \text{ w.p. } \gamma \ : \ \ln L(\boldsymbol{y}) & = \ln \eta, \\ 0 & < \end{cases} \tag{5.1}$$

where $L(\boldsymbol{y}) = p_1(\boldsymbol{y})/p_0(\boldsymbol{y})$ is the likelihood ratio. The threshold η and randomization parameter γ are chosen based on the criterion used for detection. In the Bayesian setting, η is given in (2.12), and $\gamma = 1$.

In the special case where the observations are (conditionally) independent under each hypothesis,

$$p_j(\boldsymbol{y}) = \prod_{k=1}^{n} p_{jk}(y_k), \quad j = 0, 1,$$

the log-likelihood ratio (LLR) in (5.1) can be written as

$$\log L(\boldsymbol{y}) = \sum_{k=1}^{n} \log L_k(y_k),$$

where $L_k(y_k) = p_{1,k}(y_k)/p_{0,k}(y_k)$.

5.2 Problem Formulation

The basic problem is to distinguish between two discrete-time signals $\boldsymbol{S}_0 = (S_{0,1}, \dots, S_{0,n}) \in \mathbb{R}^n$ and $\boldsymbol{S}_1 = (S_{0,1}, \dots, S_{0,n}) \in \mathbb{R}^n$ degraded by noise $\boldsymbol{Z} = (Z_1, \dots, Z_n) \in \mathbb{R}^n$ with pdf $p_{\boldsymbol{Z}}$, given the observation sequence $\boldsymbol{Y} = (Y_1, \dots, Y_n) \in \mathbb{R}^n$. The hypotheses are formulated as follows:

$$\begin{cases} H_0 : \boldsymbol{Y} = \boldsymbol{S}_0 + \boldsymbol{Z} \\ H_1 : \boldsymbol{Y} = \boldsymbol{S}_1 + \boldsymbol{Z}. \end{cases} \tag{5.2}$$

If $\boldsymbol{S}_j$ is deterministic and known, i.e., $\boldsymbol{S}_j = \boldsymbol{s}_j$ with probability 1, $j = 0, 1$, the likelihood ratio for (5.2) is given by

$$L(\boldsymbol{y}) = \frac{p_1(\boldsymbol{y})}{p_0(\boldsymbol{y})} = \frac{p_{\boldsymbol{Z}}(\boldsymbol{y} - \boldsymbol{s}_1)}{p_{\boldsymbol{Z}}(\boldsymbol{y} - \boldsymbol{s}_0)}. \tag{5.3}$$

If $\boldsymbol{S}$ is random with pdf $p_{\boldsymbol{S}}$ and is independent of $\boldsymbol{Z}$, the likelihood ratio for (5.2) is given by

$$L(\boldsymbol{y}) = \frac{p_1(\boldsymbol{y})}{p_0(\boldsymbol{y})} = \frac{\int_{\mathbb{R}^n} p_{\boldsymbol{S}_1}(\boldsymbol{s}) p_{\boldsymbol{Z}}(\boldsymbol{y} - \boldsymbol{s}) d\boldsymbol{s}}{\int_{\mathbb{R}^n} p_{\boldsymbol{S}_0}(\boldsymbol{s}) p_{\boldsymbol{Z}}(\boldsymbol{y} - \boldsymbol{s}) d\boldsymbol{s}}. \tag{5.4}$$

If $\boldsymbol{S}_j = \boldsymbol{s}_j(\theta)$, where θ is an unknown parameter taking values in a set $\mathcal{X}$, then we have a composite hypothesis testing problem (see Chapter 4).

In general, (5.3) and (5.4) can be complicated to compute and analyze if no further assumptions about the distributions of $\boldsymbol{S}$ and $\boldsymbol{Z}$ can be made. Moreover the threshold for the LRT can be difficult to determine. For example, the Neyman–Pearson rule at significance level α takes the form

$$\tilde{\delta}(\boldsymbol{y}) = \begin{cases} 1 & > \\ 1 \text{ w.p. } \gamma \; : \; L(\boldsymbol{y}) & = \eta, \\ 0 & < \end{cases}$$

where η and γ are derived from α using the procedure from Section 2.4.2:

$$\alpha = \int_{\{L(\boldsymbol{y})>\eta\}} p_0(\boldsymbol{y}) d\mu(\boldsymbol{y}) + \gamma \int_{\{L(\boldsymbol{y})=\eta\}} p_0(\boldsymbol{y}) d\mu(\boldsymbol{y}) . \tag{5.5}$$

However, the integrals are high-dimensional for large n which makes it difficult to solve (5.5) for η.

5.3 Detection of Known Signal in Independent Noise

If the signal $\boldsymbol{s}_j$, $j = 0, 1$, is deterministic and known to the detector, the problem is often called *coherent detection*. In this case, the hypothesis test of (5.2) is written as

$$\begin{cases} H_0 : \boldsymbol{Y} = \boldsymbol{s}_0 + \boldsymbol{Z} \\ H_1 : \boldsymbol{Y} = \boldsymbol{s}_1 + \boldsymbol{Z}. \end{cases} \tag{5.6}$$

Note that since $\boldsymbol{s}_0$ is known at the detector, we can subtract it out of the observations $\boldsymbol{Y}$ to get the following equivalent hypothesis test:

$$\begin{cases} H_0 \ : \boldsymbol{Y} = \boldsymbol{Z} \\ H_1 \ : \boldsymbol{Y} = \boldsymbol{s} + \boldsymbol{Z}, \end{cases} \tag{5.7}$$

with $\boldsymbol{s} = \boldsymbol{s}_1 - \boldsymbol{s}_0$.

Considerable simplifications arise when the noise samples $Z_1, \ldots, Z_n$ are statistically independent. Then the noise pdf factors as $p_{\boldsymbol{Z}}(\boldsymbol{z}) = \prod_{k=1}^{n} p_{Z_k}(z_k)$, hence the likelihood ratio of (5.3) also factors,

$$L(\boldsymbol{y}) = \prod_{k=1}^{n} \frac{p_{Z_k}(y_k - s_k)}{p_{Z_k}(y_k)}, \tag{5.8}$$

and the log-likelihood ratio test (LLRT) takes the additive form

$$\ln L(\boldsymbol{y}) = \sum_{k=1}^{n} \underbrace{\ln \frac{p_{Z_k}(y_k - s_k)}{p_{Z_k}(y_k)}}_{L_k(y_k)} \underset{H_0}{\overset{H_1}{\gtrless}} \ln \eta. \tag{5.9}$$

The general structure of the detector is shown in Figure 5.1. In the next section, we further specialize this structure to the case of typical noise distributions.

5.3.1 Signal in i.i.d. Gaussian Noise

If Z_k's are i.i.d. Gaussian random variables, i.e., $Z_k \sim \mathcal{N}(0, \sigma^2)$ for $1 \leq k \leq n$, then

$$\begin{aligned} \ln L_k(y_k) &= \ln \frac{\frac{1}{\sqrt{2\pi\sigma^2}} \exp\{-\frac{(y_k - s_k)^2}{2\sigma^2}\}}{\frac{1}{\sqrt{2\pi\sigma^2}} \exp\{-\frac{y_k^2}{2\sigma^2}\}} \\ &= \frac{1}{\sigma^2} s_k \left(y_k - \frac{s_k}{2} \right) \\ &= \frac{1}{\sigma^2} s_k \tilde{y}_k, \end{aligned} \tag{5.10}$$

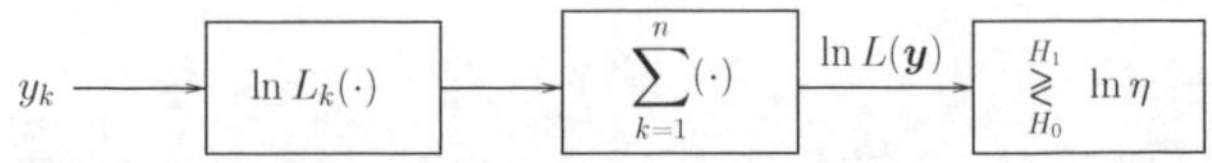

Figure 5.1 Optimal detector structure for known signal in independent noise

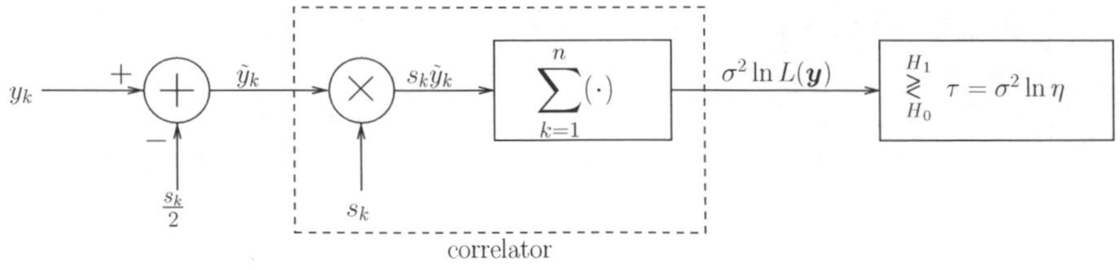

Figure 5.2 Correlator detector, optimal for i.i.d. Gaussian noise

where $\tilde{\boldsymbol{y}} = \boldsymbol{y} - \frac{1}{2}\boldsymbol{s}$ are the *centered* observations. Combining (5.9) and (5.10) yields the optimal test

$$\sum_{k=1}^{n} s_k \tilde{y}_k \overset{H_1}{\underset{H_0}{\gtrless}} \tau = \sigma^2 \ln \eta \, .$$

The detector consists of a correlator that computes the correlation between the known signal $\boldsymbol{s}$ and the centered observations $\tilde{\boldsymbol{y}}$. The resulting detector is known as a *correlator detector* and is shown in Figure 5.2.

Alternatively, the observation $\boldsymbol{y}$ can be directly correlated with $\boldsymbol{s}$, and the decision rule in this case can be written as

$$\sum_{k=1}^{n} s_k y_k \overset{H_1}{\underset{H_0}{\gtrless}} \ln \eta + \frac{1}{2} \sum_{k=1}^{n} s_k^2. \tag{5.11}$$

The correlation statistic may also be viewed as the output at time n of a linear, time-invariant filter whose impulse response is the time-reversed signal

$$h_k = s_{n-k}, \quad 0 \le k \le n-1.$$

This filter is the celebrated *matched filter*. The correlation statistic can be written as

$$\begin{aligned} \sum_{k=1}^{n} s_k \tilde{y}_k &= \sum_{k=1}^{n} h_{n-k}\, \tilde{y}_k \\ &= (\boldsymbol{h} \star \tilde{\boldsymbol{y}})_n, \end{aligned} \tag{5.12}$$

where $\star$ denotes the convolution of two sequences.

5.3.2 Signal in i.i.d. Laplacian Noise

The Laplacian pdf has heavier tails than the Gaussian pdf and is often used to model noise in applications including underwater communication and biomedical signal processing. If the Z_k's are i.i.d. Laplacian random variables, the noise samples are distributed as

$$p_Z(z_k) = \frac{a}{2} e^{-a|z_k|} \, . \tag{5.13}$$

In this case, the LLR for each observed sample y_k is

$$\ln L_k(y_k) = \ln \frac{\frac{a}{2}\, e^{-a|y_k - s_k|}}{\frac{a}{2}\, e^{-a|y_k|}} \\ = a(|y_k| - |y_k - s_k|)\ , \tag{5.14}$$

which is diagrammed in Figure 5.3. Equivalently,

$$\ln L_k(y_k) = a f_k\left(y_k - \frac{s_k}{2}\right) \operatorname{sgn}(s_k), \tag{5.15}$$

where we have introduced the *soft limiter*

$$f_k(x) = \begin{cases} |s_k| & : x > |s_k| \\ x & : |x| \leq |s_k| \\ -|s_k| & : x < -|s_k|, \end{cases} \tag{5.16}$$

which is diagrammed in Figure 5.4.

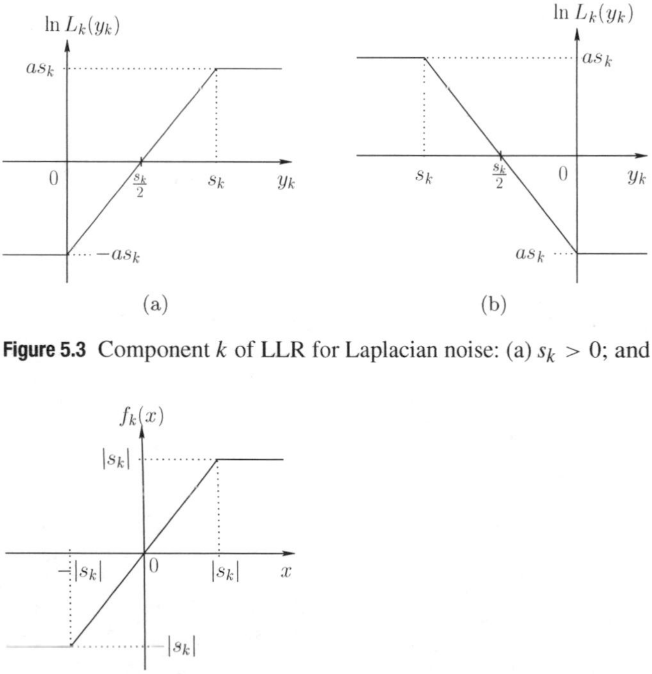

Figure 5.3 Component k of LLR for Laplacian noise: (a) $s_k > 0$; and (b) $s_k < 0$

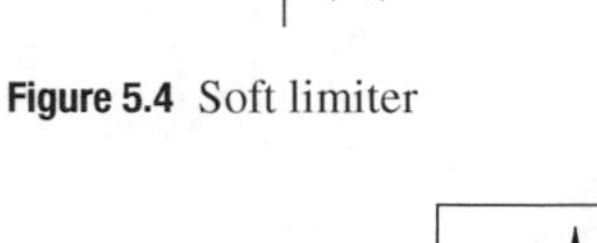

Figure 5.4 Soft limiter

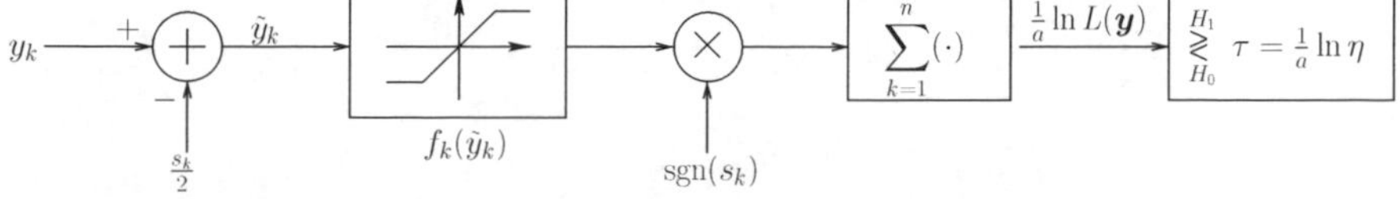

Figure 5.5 Detector structure for i.i.d. Laplacian noise

The structure of the optimal detector is shown in Figure 5.5. Note that large inputs are clipped by the soft limiter. Then the soft limiter output is correlated with the sequence of signs of s_k (instead of the sequence s_k itself). Clipping reduces the influence of large samples, which are less trustworthy than they were in the Gaussian case. Clipping makes the detector more robust against large noise samples (see also Remark 4.4).

5.3.3 Signal in i.i.d. Cauchy Noise

The Cauchy pdf has extremely heavy tails; its variance is infinite. Cauchy noise models have been used, e.g., in the area of software testing. If the noise is i.i.d. and follows the standard Cauchy distribution, we have

$$p_{Z_k}(z_k) = \frac{1}{\pi(1+z_k^2)} \quad \forall k. \tag{5.17}$$

In this case, the LLR of each observed sample is

$$\begin{aligned} \ln L_k(y_k) &= \ln \frac{\pi(1+y_k^2)}{\pi(1+(y_k - s_k)^2)} \\ &= \ln \frac{1+(\tilde{y}_k + \frac{1}{2}s_k)^2}{1+(\tilde{y}_k - \frac{1}{2}s_k)^2} \\ &\triangleq f_k(\tilde{y}_k), \end{aligned} \tag{5.18}$$

with $\tilde{y}_k = y_k - \frac{s_k}{2}$. Figure 5.6 shows a plot of f_k.

The detector structure in this case is shown in Figure 5.7. Note that large observed samples are heavily suppressed. This is because the Cauchy distribution has extremely heavy tails, and large observed values cannot be trusted.

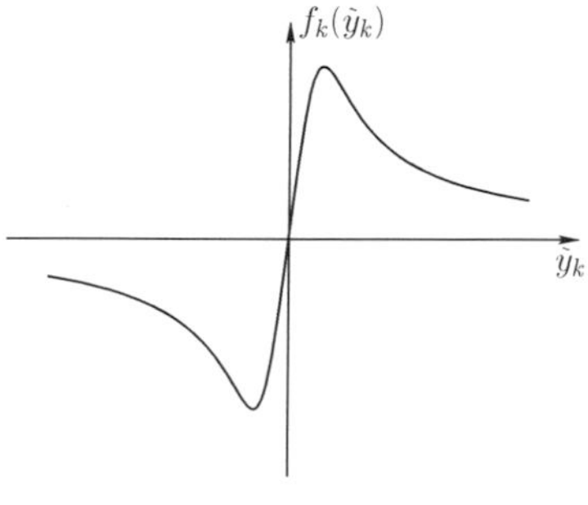

Figure 5.6 Component k of LLR for Cauchy noise

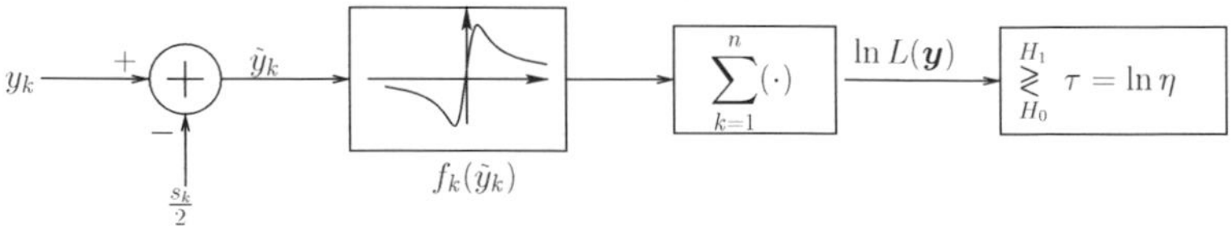

Figure 5.7 Detector structure for Cauchy noise

5.3.4 Approximate NP Test

The probabilities of false alarm and correct detection are respectively given by

$$\mathrm{P_F}(\delta_\eta) = \mathrm{P}_0\left\{\sum_{k=1}^{n} \ln L_k(Y_k) > \eta\right\},$$

$$\mathrm{P_D}(\delta_\eta) = \mathrm{P}_1\left\{\sum_{k=1}^{n} \ln L_k(Y_k) \geq \eta\right\},$$

where $\{\ln L_k(Y_k)\}_{k=1}^{n}$ are independent random variables under each hypothesis.

We defer the evaluation of these probabilities for arbitrary η to Section 5.4.2 (for the Gaussian case) and to Section 5.7.4 (for the general case). We focus our analysis here on a setting where τ is chosen to approximately obtain a target probability of false alarm.

For i.i.d. Gaussian noise, (5.10) shows that $\ln L_k(Y_k)$ is a linear function of Y_k, hence is a Gaussian random variable under both H_0 and H_1. Under H_0, the mean and variance of $\ln L(\boldsymbol{Y})$ are respectively given by

$$\mu_0 \triangleq -\frac{\|\boldsymbol{s}\|^2}{2\sigma^2} \quad \text{and} \quad v_0 \triangleq \|\boldsymbol{s}\|^2.$$

Hence the normalized log-likelihood ratio

$$U \triangleq \frac{\sum_{k=1}^{n} \ln L_k(Y_k) - \mu_0}{\sqrt{v_0}}$$

is distributed as $\mathcal{N}(0,1)$. Consider setting the threshold η so as to achieve a false-alarm probability of α:

$$\mathrm{P_F}(\delta_\eta) = \mathrm{P}_0\left\{U > \frac{\eta-\mu_0}{\sqrt{v_0}}\right\} = Q\left(\frac{\eta-\mu_0}{\sqrt{v_0}}\right) = \alpha,$$

hence

$$\eta = \mu_0 + \sqrt{v_0}\, Q^{-1}(\alpha). \tag{5.19}$$

If the noise distribution is non-Gaussian, the mean and variance of $\ln L(\boldsymbol{Y})$ under H_0 are respectively given by

$$\mu_0 \triangleq \sum_{k=1}^{n} \mathbb{E}_0[\ln L_k(Y_k)] \quad \text{and} \quad v_0 \triangleq \sum_{k=1}^{n} \mathrm{Var}_0[\ln L_k(Y_k)]. \tag{5.20}$$

If n is large and certain mild conditions on the signal are satisfied, then by application of the Central Limit Theorem [1, Sec. VIII.4], the normalized log-likelihood ratio U is approximately $\mathcal{N}(0,1)$. Under these conditions, the choice (5.19) for η yields $\mathrm{P_F}(\delta_\eta) \approx \alpha$. However, in Chapter 8 we shall see that such "Gaussian approximations" to $\mathrm{P_F}(\delta_\eta)$ are generally not valid when $\eta - \mu_0 \gg \sqrt{v_0}$.

5.4 Detection of Known Signal in Correlated Gaussian Noise

Now consider the hypothesis test (5.7) where the signal $\boldsymbol{s} \in \mathbb{R}^n$ is known, the noise $\boldsymbol{Z}$ is $\mathcal{N}(0, \mathsf{C}_Z)$, and the noise covariance matrix C_Z has full rank. The likelihood ratio of (5.3) takes the form

$$\begin{aligned} L(\boldsymbol{y}) &= \frac{p_Z(\boldsymbol{y}-\boldsymbol{s})}{p_Z(\boldsymbol{y})} \\ &= \frac{(2\pi)^{-n/2}|\mathsf{C}_Z|^{-1/2}\exp\{-\frac{1}{2}(\boldsymbol{y}-\boldsymbol{s})^\top \mathsf{C}_Z^{-1}(\boldsymbol{y}-\boldsymbol{s})\}}{(2\pi)^{-n/2}|\mathsf{C}_Z|^{-1/2}\exp\{-\frac{1}{2}\boldsymbol{y}^\top \mathsf{C}_Z^{-1}\boldsymbol{y}\}} \\ &= \exp\left\{-\frac{1}{2}[\boldsymbol{y}^\top \mathsf{C}_Z^{-1}\boldsymbol{s} + \boldsymbol{s}^\top \mathsf{C}_Z^{-1}\boldsymbol{y} - \boldsymbol{s}^\top \mathsf{C}_Z^{-1}\boldsymbol{s}]\right\} \\ &= \exp\left\{\boldsymbol{s}^\top \mathsf{C}_Z^{-1}\boldsymbol{y} - \frac{1}{2}\boldsymbol{s}^\top \mathsf{C}_Z^{-1}\boldsymbol{s}\right\}, \end{aligned} \tag{5.21}$$

where the last line holds because the inverse covariance matrix C_Z^{-1} is symmetric. The optimal test (in the Bayes, minimax, or NP sense) compares $L(\boldsymbol{y})$ to a threshold η. Taking logarithms in (5.21), we obtain the LLRT

$$\boldsymbol{s}^\top \mathsf{C}_Z^{-1}\boldsymbol{y} \underset{H_0}{\overset{H_1}{\gtrless}} \tau \triangleq \ln\eta + \frac{1}{2}\boldsymbol{s}^\top \mathsf{C}_Z^{-1}\boldsymbol{s}. \tag{5.22}$$

We recover the detector of (5.11) if $\boldsymbol{Z}$ is i.i.d. $\mathcal{N}(0, \sigma^2)$. Then $\mathsf{C}_Z = \sigma^2 \mathsf{I}_n$, and the LRT is the correlation test

$$\boldsymbol{s}^\top \boldsymbol{y} \underset{H_0}{\overset{H_1}{\gtrless}} \tau = \sigma^2 \ln\eta + \frac{1}{2}\boldsymbol{s}^\top\boldsymbol{s}. \tag{5.23}$$

We can relate this special case to the general case by defining the pseudo-signal

$$\tilde{\boldsymbol{s}} \triangleq \mathsf{C}_Z^{-1}\boldsymbol{s} \quad \in \mathbb{R}^n \tag{5.24}$$

and writing (5.22) as a correlation test where the correlation is taken with the pseudo-signal $\tilde{\boldsymbol{s}}$ instead of the original signal $\boldsymbol{s}$:

$$\tilde{\boldsymbol{s}}^\top \boldsymbol{y} \underset{H_0}{\overset{H_1}{\gtrless}} \tau = \ln\eta + \frac{1}{2}\tilde{\boldsymbol{s}}^\top\boldsymbol{s}. \tag{5.25}$$

The structure of this detector is shown in Figure 5.8.

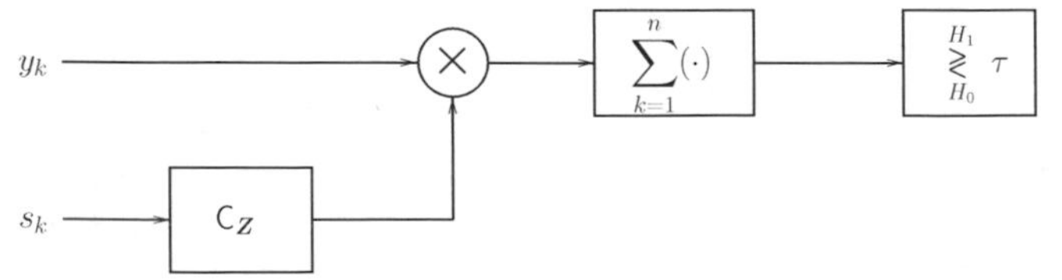

Figure 5.8 Optimal detector structure for known signal in correlated Gaussian noise

5.4.1 Reduction to i.i.d. Noise Case

Another widely applicable approach when dealing with correlated Gaussian noise is to reduce the problem to a detection problem involving i.i.d. Gaussian noise. This is done by applying a special linear transform to the observations. Specifically, given observations $\boldsymbol{Y} \in \mathbb{R}^n$, design a matrix $\mathsf{B} \in \mathbb{R}^{n\times n}$ and define:

- the transformed observations $\tilde{\boldsymbol{Y}} \triangleq \mathsf{B}\boldsymbol{Y}$;
- the transformed signal $\hat{\boldsymbol{s}} \triangleq \mathsf{B}\boldsymbol{s}$;
- the transformed noise $\tilde{\boldsymbol{Z}} \triangleq \mathsf{B}\boldsymbol{Z}$.

We use the notation $\hat{\boldsymbol{s}}$ instead of $\tilde{\boldsymbol{s}}$ to avoid confusion with the pseudo-signal $\tilde{\boldsymbol{s}}$ of (5.24). If B is invertible, the hypothesis test is equivalent to

$$\begin{aligned} H_0 &: \tilde{\boldsymbol{Y}} = \tilde{\boldsymbol{Z}}, \\ H_1 &: \tilde{\boldsymbol{Y}} = \hat{\boldsymbol{s}} + \tilde{\boldsymbol{Z}}. \end{aligned} \tag{5.26}$$

The matrix B is designed such that the transformed noise is i.i.d. $\mathcal{N}(0, 1)$:

$$\mathsf{C}_{\tilde{Z}} = \mathsf{C}_{\mathsf{B}Z} = \mathsf{B}\mathsf{C}_Z\mathsf{B}^\top = \mathsf{I}_n. \tag{5.27}$$

This is the so-called *whitening condition*, and we now consider several ways of choosing B.

(a) *Eigenvector decomposition of* C_Z: let

$$\mathsf{C}_Z = \mathsf{V}\Lambda\mathsf{V}^\top = \sum_{k=1}^{n} \lambda_k \boldsymbol{v}_k \boldsymbol{v}_k^\top, \tag{5.28}$$

where

$$\mathsf{V} = [\boldsymbol{v}_1, \boldsymbol{v}_2, \ldots, \boldsymbol{v}_n]$$

is the matrix of eigenvectors, and

$$\Lambda = \mathrm{diag}(\lambda_1, \ldots, \lambda_n)$$

is the diagonal matrix whose entry $\Lambda_{kk} = \lambda_k$ is the eigenvalue associated with eigenvector $\boldsymbol{v}_k$. The eigenvalues are positive because C_Z was assumed to have full rank and is thus positive definite. The eigenvector matrix V is orthonormal. Substituting (5.28) into (5.27), we obtain

$$\mathsf{I}_n = \mathsf{B}\mathsf{V}\Lambda\mathsf{V}^\top\mathsf{B}^\top = (\mathsf{B}\mathsf{V}\Lambda^{1/2})(\mathsf{B}\mathsf{V}\Lambda^{1/2})^\top,$$

which holds (among other choices) if

$$\mathsf{B}\mathsf{V}\Lambda^{1/2} = \mathsf{I}_n,$$

hence

$$\mathsf{B} = \Lambda^{-1/2}\mathsf{V}^\top. \tag{5.29}$$

For each $k = 1, \ldots, n$, component k of the transformed observation vector is given by

$$\tilde{Y}_k = (\mathsf{B}\boldsymbol{Y})_k = \frac{1}{\sqrt{\lambda_k}} \boldsymbol{v}_k^\top \boldsymbol{Y},$$

which is the projection of $\boldsymbol{Y}$ onto the kth eigenvector $\boldsymbol{v}_k$, normalized by $\lambda_k^{-1/2}$. Hence $\tilde{\boldsymbol{Y}}$ is the representation of $\boldsymbol{Y}$ in the eigenvector basis for the covariance noise matrix, followed by scaling of individual component by the reciprocal of the square root of the associated eigenvalues. Likewise, we have

$$\hat{s}_k = \frac{1}{\sqrt{\lambda_k}} \boldsymbol{v}_k^\top \boldsymbol{s} \quad \text{and} \quad \tilde{Z}_k = \frac{1}{\sqrt{\lambda_k}} \boldsymbol{v}_k^\top \boldsymbol{Z},$$

where $\tilde{Z}_k,\ k = 1, \ldots, n$, are i.i.d. $\mathcal{N}(0, 1)$. Writing the test (5.26) as

$$\begin{cases} H_0 & : \ \tilde{Y}_k = \tilde{Z}_k \\ H_1 & : \ \tilde{Y}_k = \hat{s}_k + \tilde{Z}_k, \quad 1 \le k \le n \end{cases}$$

we see that the *informative components* of the observations are those corresponding to large signal components: they contribute an amount $\hat{s}_k^2$ to the *generalized signal-to-noise ratio* (GSNR)

$$\text{GSNR} \triangleq \|\hat{\boldsymbol{s}}\|^2 = \|\mathsf{B}\boldsymbol{s}\|^2 = \boldsymbol{s}^\top \mathsf{B}^\top \mathsf{B}\boldsymbol{s} = \boldsymbol{s}^\top \mathsf{C}_{\boldsymbol{Z}}^{-1} \boldsymbol{s}. \tag{5.30}$$

(b) *Positive Square Root of* $\mathsf{C}_{\boldsymbol{Z}}^{-1}$: choose $\mathsf{B} = \mathsf{C}_{\boldsymbol{Z}}^{-1/2} = \mathsf{V}\Lambda^{-1/2}\mathsf{V}^\top$. This is V multiplied by the whitening matrix of (5.29).

(c) *Choleski decomposition of* $\mathsf{C}_{\boldsymbol{Z}}$: this decomposition takes the form $\mathsf{C}_{\boldsymbol{Z}} = \mathsf{G}\mathsf{G}^\top$ where G is a lower triangular matrix. Then $\mathsf{B} = \mathsf{G}^{-1}$ is a whitening transform, and B is also lower triangular. We obtain $\tilde{y}_k = \sum_{\ell=1}^{k} \mathsf{B}_{k\ell}\, y_l$ for $1 \le k \le n$. Hence this whitening transform can be implemented by the application of a causal, linear time-varying filter to the observed sequence $\{y_k\}_{k=1}^n$.

5.4.2 Performance Analysis

The LRT statistic of (5.25),

$$T \triangleq \tilde{\boldsymbol{s}}^\top \boldsymbol{Y} = \sum_{k=1}^{n} \tilde{s}_k Y_k,$$

is a linear combination of jointly Gaussian random variables and is therefore also Gaussian. The distributions of T under H_0 and H_1 are determined by the means and variances of T:

$$\begin{aligned}
\mathbb{E}_0[T] &= \mathbb{E}_0[\tilde{\boldsymbol{s}}^\top \boldsymbol{Y}] = \tilde{\boldsymbol{s}}^\top \mathbb{E}_0[\boldsymbol{Y}] = 0, \\
\mathbb{E}_1[T] &= \mathbb{E}_1[\tilde{\boldsymbol{s}}^\top \boldsymbol{Y}] = \tilde{\boldsymbol{s}}^\top \mathbb{E}_1[\boldsymbol{Y}] = \tilde{\boldsymbol{s}}^\top \boldsymbol{s} = \boldsymbol{s}^\top \mathsf{C}_{\boldsymbol{Z}}^{-1} \boldsymbol{s} \triangleq \mu, \\
\mathrm{Var}_0[T] &= \mathbb{E}_0[T^2] = \mathbb{E}_0[(\tilde{\boldsymbol{s}}^\top \boldsymbol{Y})^2] = \mathbb{E}_0[\tilde{\boldsymbol{s}}^\top \boldsymbol{Y}\boldsymbol{Y}^\top \tilde{\boldsymbol{s}}] \\
&= \tilde{\boldsymbol{s}}^\top \mathbb{E}_0[\boldsymbol{Y}\boldsymbol{Y}^\top]\tilde{\boldsymbol{s}} = \tilde{\boldsymbol{s}}^\top \mathsf{C}_{\boldsymbol{Z}} \tilde{\boldsymbol{s}} = \boldsymbol{s}^\top \mathsf{C}_{\boldsymbol{Z}}^{-1} \boldsymbol{s} = \mu, \\
\mathrm{Var}_1[T] &= \mathrm{Var}_0[T].
\end{aligned}$$

Therefore the test (5.25) can be expressed as

$$\begin{cases} H_0 & : T \sim \mathcal{N}(0, \mu) \\ H_1 & : T \sim \mathcal{N}(\mu, \mu) \end{cases}$$

with $\mu = \boldsymbol{s}^\top \mathsf{C}_Z^{-1} \boldsymbol{s}$. The normalized distance between the two distributions is the ratio of the difference between the two means to the standard deviation:

$$d = \sqrt{\mu} = \sqrt{\boldsymbol{s}^\top \mathsf{C}_Z^{-1} \boldsymbol{s}}, \tag{5.31}$$

which is the square root of the GSNR of (5.30). We note that d is a special case of the *Mahalanobis distance* [2], which is defined as follows:

DEFINITION 5.1 The Mahalanobis distance between two n-dimensional vectors $\boldsymbol{x}_1$ and $\boldsymbol{x}_2$, corresponding to a positive definite matrix A, is given by

$$d_{\mathsf{A}}(\boldsymbol{x}_1, \boldsymbol{x}_2) = \sqrt{(\boldsymbol{x}_2 - \boldsymbol{x}_1)^\top \mathsf{A} (\boldsymbol{x}_2 - \boldsymbol{x}_1)}.$$

The probability of false alarm and the probability of correct detection of the LRT (5.25) are respectively given by

$$\mathrm{P_F} = \mathrm{P_0}\{T \geq \tau\} = Q\left(\frac{\tau}{d}\right), \tag{5.32}$$

$$\mathrm{P_D} = \mathrm{P_1}\{T \geq \tau\} = Q\left(\frac{\tau - d^2}{d}\right). \tag{5.33}$$

Hence the ROC is given by

$$\mathrm{P_D} = Q\left(Q^{-1}(\alpha) - d\right). \tag{5.34}$$

If the threshold is set to $\tau = \frac{\mu}{2}$ (as occurs in the Bayesian setting with equal priors, or in the minimax setting), the error probabilities are given by

$$\mathrm{P_F} = \mathrm{P_M} = \mathrm{P_e} = Q\left(\frac{d}{2}\right). \tag{5.35}$$

For the case of i.i.d. noise ($\mathsf{C}_Z = \sigma^2 \mathsf{I}_n$), we have $d^2 = \|\boldsymbol{s}\|^2/\sigma^2$, which represents the signal-to-noise ratio (SNR). For general C_Z, the GSNR $d^2 = \boldsymbol{s}^\top \mathsf{C}_Z^{-1} \boldsymbol{s}$ determines the performance of the LRT.

5.5 m-ary Signal Detection

Consider the following detection problem with $m > 2$ hypotheses:

$$H_j : \boldsymbol{Y} = \boldsymbol{s}_j + \boldsymbol{Z}, \quad j = 1, \ldots, m, \tag{5.36}$$

where $\boldsymbol{Z} \sim \mathcal{N}(0, \mathsf{C}_Z)$, and the signals $\boldsymbol{s}_i$, $i = 1, \ldots, m$ are known.

5.5.1 Bayes Classification Rule

Assume a prior distribution π on the hypotheses. Denoting by $\pi(i|\boldsymbol{y})$ the posterior probability of H_i given $\boldsymbol{Y} = \boldsymbol{y}$, the MPE classification rule is the MAP rule

$$\begin{aligned}\delta_{\text{MAP}}(\boldsymbol{y}) &= \arg\max_{1\le i\le m} \pi(i|\boldsymbol{y}) \\ &= \arg\max_{1\le i\le m} \frac{\pi_i\, p_i(\boldsymbol{y})}{p(\boldsymbol{y})} \\ &= \arg\max_{1\le i\le m} \frac{\pi_i\, p_i(\boldsymbol{y})}{p_0(\boldsymbol{y})},\end{aligned}$$

where we have introduced the pdf $p_0 = \mathcal{N}(0, \mathsf{C}_{\boldsymbol{Z}})$ corresponding to a dummy noise-only hypothesis

$$H_0 \,:\, \boldsymbol{Y} = \boldsymbol{Z}.$$

Therefore,

$$\begin{aligned}\delta_{\text{MAP}}(\boldsymbol{y}) &= \arg\max_{1\le i\le m} \left(\ln \pi_i + \ln \frac{p_i(\boldsymbol{y})}{p_0(\boldsymbol{y})} \right) \\ &= \arg\max_{1\le i\le m} \left(\ln \pi_i + \boldsymbol{s}_i^\top \mathsf{C}_{\boldsymbol{Z}}^{-1} \boldsymbol{y} - \frac{1}{2} \boldsymbol{s}_i^\top \mathsf{C}_{\boldsymbol{Z}}^{-1} \boldsymbol{s}_i \right),\end{aligned}$$

where the last line follows from (5.21). In the special case where $\pi_i = \frac{1}{m}$ (equal priors) and $\mathsf{C}_{\boldsymbol{Z}} = \sigma^2 \mathsf{I}_n$, we obtain

$$\delta_{\text{MAP}}(\boldsymbol{y}) = \arg\max_{1\le i\le m} \left(\frac{1}{\sigma^2} \boldsymbol{s}_i^\top \boldsymbol{y} - \frac{1}{2\sigma^2} \|\boldsymbol{s}_i\|^2 \right).$$

Further assuming equal signal energy, which means $\|\boldsymbol{s}_i\|^2 = E_s \ \forall i$, the MAP rule reduces to the maximum-correlation rule

$$\delta_{\text{MAP}}(\boldsymbol{y}) = \arg\max_{1\le i\le m} \boldsymbol{s}_i^\top \boldsymbol{y}. \tag{5.37}$$

This is also equivalent to the *minimum-distance* detection rule

$$\delta_{\text{MD}}(\boldsymbol{y}) = \arg\min_{1\le i\le m} \|\boldsymbol{y} - \boldsymbol{s}_i\|^2$$

because $\|\boldsymbol{y} - \boldsymbol{s}_i\|^2 = \|\boldsymbol{y}\|^2 + E_s - 2\boldsymbol{s}_i^\top \boldsymbol{y}$.

5.5.2 Performance Analysis

Assume equal priors, equal-energy signals, and $\mathsf{C}_{\boldsymbol{Z}} = \sigma^2 \mathsf{I}_n$. The error probability of the LRT is given by

$$\mathrm{P_e} = \frac{1}{m} \sum_{i=1}^{m} \mathrm{P}\{\text{Error}|H_i\}. \tag{5.38}$$

We now analyze the conditional error probability $\mathrm{P}\{\mathrm{Error}|H_i\}$ which is somewhat unwieldy but can be upper bounded in terms of error probabilities for binary hypothesis tests. Specifically, define the correlation statistics $T_i = \mathbf{s}_i^\top \mathbf{Y}$ for $i = 1, \dots, m$. By (5.6) and (5.11), an optimal statistic for testing H_i versus H_j (for any $i \neq j$) is $(\mathbf{s}_i - \mathbf{s}_j)^\top \mathbf{Y} = T_i - T_j$. If hypotheses H_i and H_j have the same prior probability, the error probability for the optimal binary test between H_i and H_j is given by

$$\mathrm{P}_i\{T_j > T_i\} = \mathrm{P}_j\{T_i > T_j\} = Q\left(\frac{\|\mathbf{s}_i - \mathbf{s}_j\|}{2\sigma}\right). \tag{5.39}$$

The maximum-correlation rule (5.37) takes the form $\delta_{\mathrm{MAP}}(\mathbf{Y}) = \arg\max_{1 \le i \le m} T_i$. Therefore,

$$\begin{aligned}\forall i : \quad \mathrm{P}\{\mathrm{Error}|H_i\} &= P_i\{\exists j \neq i : T_j > T_i\} \\ &= P_i\{\cup_{j \neq i}\{T_j > T_i\}\} \\ &\le \sum_{j \neq i} P_i\{T_j > T_i\} \\ &= \sum_{j \neq i} Q\left(\frac{\|\mathbf{s}_i - \mathbf{s}_j\|}{2\sigma}\right), \end{aligned} \tag{5.40}$$

where the inequality follows from the union-of-events bound, and the last equality from (5.39).

In many problems, the power (energy per sample) of the signals is roughly independent of n, and so is the power of the difference between any two signals. Then the normalized Euclidean distance $d_{ij} \triangleq \|\mathbf{s}_i - \mathbf{s}_j\|/\sigma$ between signals $\mathbf{s}_i$ and $\mathbf{s}_j$ increases linearly with $\sqrt{n}$, and the argument of the Q functions is large. This is referred to as the *high-SNR regime*. While the Q function does not admit a closed-form expression, the upper bound $Q(x) \le e^{-x^2/2}$ holds for all $x \ge 0$, and the asymptotic equality

$$Q(x) \sim \frac{e^{-x^2/2}}{\sqrt{2\pi}\,x}, \quad \text{as } x \to \infty \tag{5.41}$$

is already accurate for moderately large x. Since the function $Q(x)$ decreases exponentially fast, the sum of the Q functions in (5.40) is dominated by the largest term (corresponding to the smallest distance). Let $d_{\min} \triangleq \min_{i \neq j} d_{ij}$. We obtain the upper bound

$$\mathrm{P_e} \le \frac{1}{m} \sum_{i=1}^{m} \sum_{j \neq i} Q\left(\frac{d_{ij}}{2}\right) \le (m-1)\, Q\left(\frac{d_{\min}}{2}\right), \tag{5.42}$$

which is typically fairly tight in the high-SNR regime, when m is relatively small.

5.6 Signal Selection

Having derived expressions for error probability, applicable to signal detection in correlated Gaussian noise, we now ask how the signals can be designed to minimize error probability, a typical problem in digital communications and radar applications. For simplicity we consider the binary case, $m = 2$. The hypothesis test is given by (5.6):

$$\begin{cases} H_0 : \boldsymbol{Y} = \boldsymbol{s}_0 + \boldsymbol{Z} \\ H_1 : \boldsymbol{Y} = \boldsymbol{s}_1 + \boldsymbol{Z}, \end{cases}$$

where $\boldsymbol{Z} \sim \mathcal{N}(0, \mathsf{C}_{\boldsymbol{Z}})$. We first consider the case of i.i.d. noise, then the general case of correlated noise.

5.6.1 i.i.d. Noise

For i.i.d. noise of variance σ^2, we have $\mathsf{C}_{\boldsymbol{Z}} = \sigma^2 \mathsf{I}_n$. Let $d = \|\boldsymbol{s}_1 - \boldsymbol{s}_0\|/\sigma$. For Bayesian detection with equal priors, the problem is to minimize

$$\mathrm{P_e} = Q\left(\frac{d}{2}\right) = Q\left(\frac{\|\boldsymbol{s}_1 - \boldsymbol{s}_0\|}{2\sigma}\right)$$

over all signals $\boldsymbol{s}_0$ and $\boldsymbol{s}_1$. The solution would be trivial (arbitrarily small $\mathrm{P_e}$) in the absence of any constraint on $\boldsymbol{s}_0$ and $\boldsymbol{s}_1$, since the distance $\|\boldsymbol{s}_1 - \boldsymbol{s}_0\|$ between them could be made arbitrarily large. For physical reasons, we assume that the signals are subject to the energy constraints $\|\boldsymbol{s}_0\|^2 \le E_s$ and $\|\boldsymbol{s}_1\|^2 \le E_s$. Minimizing $\mathrm{P_e}$ is equivalent to maximizing the squared distance

$$\begin{aligned} \|\boldsymbol{s}_1 - \boldsymbol{s}_0\|^2 &= \|\boldsymbol{s}_0\|^2 + \|\boldsymbol{s}_1\|^2 - 2\boldsymbol{s}_0^\top \boldsymbol{s}_1 \\ &\overset{(a)}{\le} \|\boldsymbol{s}_0\|^2 + \|\boldsymbol{s}_1\|^2 + 2\|\boldsymbol{s}_0\|\,\|\boldsymbol{s}_1\| \\ &\overset{(b)}{\le} E_s + E_s + 2E_s \\ &= 4E_s, \end{aligned} \tag{5.43}$$

where inequality (a) holds by the Cauchy–Schwarz inequality, and (b) holds because both $\boldsymbol{s}_0$ and $\boldsymbol{s}_1$ are subject to the energy constraint E_s. The upper bound (5.43) holds with equality if and only if $\boldsymbol{s}_1 = -\boldsymbol{s}_0$ and $\|\boldsymbol{s}_0\| = \|\boldsymbol{s}_1\| = \sqrt{E_s}$. Hence the optimal solution is *antipodal signaling* with maximum power. The SNR is

$$d^2 = \frac{4E_s}{\sigma^2},$$

and the error probability is $\mathrm{P_e} = Q\left(\frac{\sqrt{E_s}}{\sigma}\right)$.

5.6.2 Correlated Noise

For general positive definite $\mathsf{C}_{\boldsymbol{Z}}$, the signal selection problem takes the form

$$\min_{\boldsymbol{s}_0, \boldsymbol{s}_1} \mathrm{P_e} = Q\left(\frac{d}{2}\right),$$

where

$$d^2 = (\boldsymbol{s}_1 - \boldsymbol{s}_0)^\top \mathsf{C}_{\boldsymbol{Z}}^{-1} (\boldsymbol{s}_1 - \boldsymbol{s}_0)$$

is the GSNR of (5.30) or squared Mahalanobis distance (see Definition 5.1) between $\boldsymbol{s}_0$ and $\boldsymbol{s}_1$, and the signals $\boldsymbol{s}_0$ and $\boldsymbol{s}_1$ are subject to the energy constraints $\|\boldsymbol{s}_0\|^2 \le E_s$ and $\|\boldsymbol{s}_1\|^2 \le E_s$. Since the Q function is decreasing, the signal selection problem is equivalent to maximizing d^2 subject to the energy constraints.

To solve this problem, consider the eigenvector decomposition $\mathsf{C}_Z = \mathsf{V}\Lambda\mathsf{V}^\top$ of the noise covariance matrix. For $j = 0, 1$, denote by $\hat{\boldsymbol{s}}_j = \mathsf{V}^\top \boldsymbol{s}_j$ the representation of the signal $\boldsymbol{s}_j$ in the eigenvector basis.

We may write

$$
\begin{aligned}
d^2 &= (\boldsymbol{s}_1 - \boldsymbol{s}_0)^\top \mathsf{V}\Lambda^{-1}\mathsf{V}^\top (\boldsymbol{s}_1 - \boldsymbol{s}_0) \\
&= (\hat{\boldsymbol{s}}_1 - \hat{\boldsymbol{s}}_0)^\top \Lambda^{-1} (\hat{\boldsymbol{s}}_1 - \hat{\boldsymbol{s}}_0) \\
&= \sum_{k=1}^{n} \frac{(\hat{s}_{1k} - \hat{s}_{0k})^2}{\lambda_k}.
\end{aligned}
$$

Since V is orthonormal, we have $\|\hat{\boldsymbol{s}}_i\|^2 = \|\boldsymbol{s}_i\|^2$ for $i = 0, 1$, and thus $\|\hat{\boldsymbol{s}}_1 - \hat{\boldsymbol{s}}_0\|^2 \le 4E_s$, as in (5.43). Defining $\lambda_{\min} = \min_{1\le k\le n} \lambda_k$, we obtain the upper bound

$$
\begin{aligned}
\sum_{k=1}^{n} \frac{(\hat{s}_{1k} - \hat{s}_{0k})^2}{2\lambda_k} &\le \frac{1}{2\lambda_{\min}} \|\hat{\boldsymbol{s}}_1 - \hat{\boldsymbol{s}}_0\|^2 \\
&\le \frac{2E_s}{\lambda_{\min}}.
\end{aligned}
$$

Equality is achieved if and only if:

(i) $\hat{\boldsymbol{s}}_1 = -\hat{\boldsymbol{s}}_0$,
(ii) $\|\hat{\boldsymbol{s}}_0\|^2 = E_s$, and
(iii) $\hat{s}_{0k} = 0$ for all k such that $\lambda_k > \lambda_{\min}$.

Hence the optimal signals satisfy $\boldsymbol{s}_1 = \sqrt{E_s}\boldsymbol{v}_{\min} = -\boldsymbol{s}_0$ where $\boldsymbol{v}_{\min}$ is an eigenvector corresponding to $\lambda_{\min}$. Moreover

$$
d^2 - \frac{4E_s}{\lambda_{\min}}.
$$

Example 5.1 Consider $n = 2$ and

$$
\mathsf{C}_Z = \sigma^2 \begin{pmatrix} 1 & \rho \\ \rho & 1 \end{pmatrix},
$$

where $|\rho| < 1$. The eigenvector decomposition of C_Z yields

$$
\Lambda = \sigma^2 \begin{pmatrix} 1+\rho & 0 \\ 0 & 1-\rho \end{pmatrix} \tag{5.44}
$$

and

$$
\mathsf{V} = \begin{pmatrix} \frac{1}{\sqrt{2}} & -\frac{1}{\sqrt{2}} \\ \frac{1}{\sqrt{2}} & \frac{1}{\sqrt{2}} \end{pmatrix}. \tag{5.45}
$$

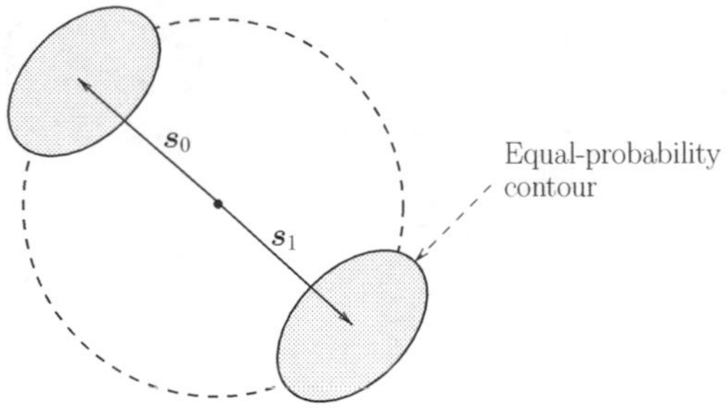

Figure 5.9 Signal selection for detection in correlated Gaussian noise

Hence $\lambda_{\min} = \sigma^2(1 - |\rho|)$ and

$$d^2 = \frac{4E_s}{\sigma^2(1 - |\rho|)}$$

is strictly larger than the value obtained for i.i.d. noise ($\rho = 0$) of the same variance. Hence we can exploit the correlation of the noise by designing antipodal signals along the direction of the minimum eigenvector. The corresponding error probability is $P_e = Q\left(\frac{\sqrt{E_s}}{\sigma\sqrt{1-|\rho|}}\right)$ as illustrated in Figure 5.9. The worst design would be to choose signals in the direction of the maximum eigenvector. □

5.7 Detection of Gaussian Signals in Gaussian Noise

So far we have studied the problem of detecting known signals in noise. However, in various signal processing and communications applications, the signals are not known precisely. Only some general characteristics of the signals may be known, for instance, their probability distribution.

Consider the following model for detection of Gaussian signals in Gaussian noise:

$$\begin{cases} H_0: \ \boldsymbol{Y} = \boldsymbol{S}_0 + \boldsymbol{Z} \\ H_1: \ \boldsymbol{Y} = \boldsymbol{S}_1 + \boldsymbol{Z}, \end{cases} \tag{5.46}$$

where $\boldsymbol{S}_j$ and $\boldsymbol{Z}$ are jointly Gaussian, for $j = 0, 1$. The problem in (5.46) is a special case of the following Gaussian hypothesis testing problem:

$$\begin{cases} H_0: \ \boldsymbol{Y} \sim \mathcal{N}(\boldsymbol{\mu}_0, \mathsf{C}_0) \\ H_1: \ \boldsymbol{Y} \sim \mathcal{N}(\boldsymbol{\mu}_1, \mathsf{C}_1). \end{cases} \tag{5.47}$$

The log-likelihood ratio for this pair of hypotheses is given by

$$\begin{aligned} \ln L(\boldsymbol{y}) = & \frac{1}{2}\boldsymbol{y}^\top \left(\mathsf{C}_0^{-1} - \mathsf{C}_1^{-1}\right)\boldsymbol{y} + \left(\boldsymbol{\mu}_1^\top \mathsf{C}_1^{-1} - \boldsymbol{\mu}_0^\top \mathsf{C}_0^{-1}\right)\boldsymbol{y} \\ & + \frac{1}{2}\ln\frac{|\mathsf{C}_0|}{|\mathsf{C}_1|} + \frac{1}{2}\boldsymbol{\mu}_0^\top \mathsf{C}_0^{-1}\boldsymbol{\mu}_0 - \frac{1}{2}\boldsymbol{\mu}_1^\top \mathsf{C}_1^{-1}\boldsymbol{\mu}_1. \end{aligned} \tag{5.48}$$

The first term on the right hand side is quadratic in $\boldsymbol{y}$, and the second term is linear in $\boldsymbol{y}$. The remaining three terms are constant in $\boldsymbol{y}$. We denote their sum by

$$\kappa = \frac{1}{2}\ln\frac{|\mathsf{C}_0|}{|\mathsf{C}_1|} + \frac{1}{2}\boldsymbol{\mu}_0^\top \mathsf{C}_0^{-1}\boldsymbol{\mu}_0 - \frac{1}{2}\boldsymbol{\mu}_1^\top \mathsf{C}_1^{-1}\boldsymbol{\mu}_1. \tag{5.49}$$

If $\mathsf{C}_0 = \mathsf{C}_1 = \mathsf{C}$ and $\boldsymbol{\mu}_0 \neq \boldsymbol{\mu}_1$, (5.48) reduces to

$$\ln L(\boldsymbol{y}) = \left(\boldsymbol{\mu}_1 - \boldsymbol{\mu}_0\right)^\top \mathsf{C}^{-1}\boldsymbol{y} + \kappa \tag{5.50}$$

and the optimum detector is a *purely* linear detector (with κ being absorbed into the threshold), which corresponds to detection of deterministic signals in Gaussian noise, as studied in Section 5.4.

If $\boldsymbol{\mu}_0 = \boldsymbol{\mu}_1$ (which can be taken to equal $\mathbf{0}$ without loss generality) and $\mathsf{C}_0 \neq \mathsf{C}_1$, (5.48) reduces to

$$\ln L(\boldsymbol{y}) = \frac{1}{2}\boldsymbol{y}^\top\left(\mathsf{C}_0^{-1} - \mathsf{C}_1^{-1}\right)\boldsymbol{y} + \kappa \tag{5.51}$$

and the optimum detector is a *purely* quadratic detector.

5.7.1 Detection of a Gaussian Signal in White Gaussian Noise

We now consider the following special case of (5.46):

$$\begin{cases} H_0: \boldsymbol{Y} = \boldsymbol{Z} \\ H_1: \boldsymbol{Y} = \boldsymbol{S} + \boldsymbol{Z}, \end{cases} \tag{5.52}$$

where $\boldsymbol{S} \sim \mathcal{N}(\mu_{\boldsymbol{S}}, \mathsf{C}_{\boldsymbol{S}})$ is independent of $\boldsymbol{Z} \sim \mathcal{N}(0, \sigma^2\mathsf{I}_n)$. The signal covariance matrix $\mathsf{C}_{\boldsymbol{S}}$ need not be full rank. The test (5.52) can be written equivalently as

$$\begin{cases} H_0: \boldsymbol{Y} \sim \mathcal{N}(0, \sigma^2\mathsf{I}_n) \\ H_1: \boldsymbol{Y} \sim \mathcal{N}(\mu_{\boldsymbol{S}}, \mathsf{C}_{\boldsymbol{S}} + \sigma^2\mathsf{I}_n). \end{cases} \tag{5.53}$$

Specializing (5.48) to (5.53), we get

$$\ln L(\boldsymbol{y}) = \frac{1}{2}\boldsymbol{y}^\top\left(\frac{1}{\sigma^2}\mathsf{I}_n - (\mathsf{C}_{\boldsymbol{S}} + \sigma^2\mathsf{I}_n)^{-1}\right)\boldsymbol{y} + \mu_{\boldsymbol{S}}^\top(\mathsf{C}_{\boldsymbol{S}} + \sigma^2\mathsf{I}_n)^{-1}\boldsymbol{y} + \kappa \tag{5.54}$$

with

$$\kappa = -\frac{1}{2}\mu_{\boldsymbol{S}}{}^\top(\mathsf{C}_{\boldsymbol{S}} + \sigma^2\mathsf{I}_n)^{-1}\mu_{\boldsymbol{S}} - \frac{1}{2}\ln|\mathsf{C}_{\boldsymbol{S}} + \sigma^2\mathsf{I}_n| + \frac{n}{2}\ln\sigma^2.$$

Denote by

$$\begin{aligned} \mathsf{Q} &= \frac{1}{\sigma^2}\mathsf{I}_n - (\mathsf{C}_{\boldsymbol{S}} + \sigma^2\mathsf{I}_n)^{-1} \\ &= \left(\frac{1}{\sigma^2}(\mathsf{C}_{\boldsymbol{S}} + \sigma^2\mathsf{I}_n) - I_n\right)(\mathsf{C}_{\boldsymbol{S}} + \sigma^2\mathsf{I}_n)^{-1} \\ &= \frac{1}{\sigma^2}\mathsf{C}_{\boldsymbol{S}}(\mathsf{C}_{\boldsymbol{S}} + \sigma^2\mathsf{I}_n)^{-1} \end{aligned} \tag{5.55}$$

the difference between the inverse covariance matrices under H_0 and H_1, and note that $\mathsf{Q} \geq 0$ (nonnegative definite matrix). With this notation, (5.54) becomes

$$\ln L(\boldsymbol{y}) = \frac{1}{2}\boldsymbol{y}^\top \mathsf{Q}\boldsymbol{y} + \boldsymbol{\mu}_S^\top(\mathsf{C}_S + \sigma^2\mathsf{I}_n)^{-1}\boldsymbol{y} + \kappa. \tag{5.56}$$

We first consider the case of a signal with zero mean: $\boldsymbol{\mu}_S = 0$. Then the log-likelihood ratio in (5.56) simplifies to

$$\ln L(\boldsymbol{y}) = \frac{1}{2}\boldsymbol{y}^\top \mathsf{Q}\boldsymbol{y} + \kappa$$

and the optimum test has the form

$$\delta_{\text{OPT}}(\boldsymbol{y}) = \begin{cases} 1 : & \boldsymbol{y}^\top\mathsf{Q}\boldsymbol{y} \quad \geq \tau, \\ 0 : & \qquad\quad < \end{cases} \tag{5.57}$$

This test admits an interesting interpretation. Since the linear minimum mean squared-error (MMSE) estimator of $\boldsymbol{S}$ given $\boldsymbol{Y}$ (see Section 11.7.4) is

$$\hat{\boldsymbol{S}} = \mathsf{C}_S\mathsf{C}_{\boldsymbol{Y}}^{-1}\boldsymbol{Y} = \mathsf{C}_S(\mathsf{C}_S + \sigma^2\mathsf{I}_n)^{-1}\boldsymbol{Y} = \sigma^2\mathsf{Q}\boldsymbol{Y},$$

the test of (5.57) can also be written as

$$\delta_{\text{OPT}}(\boldsymbol{y}) = \begin{cases} 1 : & \boldsymbol{y}^\top\hat{\boldsymbol{s}} \quad \geq \sigma^2\tau, \\ 0 : & \qquad < \end{cases} \tag{5.58}$$

which is analogous to the correlator-detector of Section 5.3.1, using the MMSE estimate $\hat{\boldsymbol{s}}$ in place of $\boldsymbol{s}$. Such a detector is also known as an *estimator-correlator*.

5.7.2 Detection of i.i.d. Zero-Mean Gaussian Signal

Assume the signal $\boldsymbol{S}$ is i.i.d. with mean zero and variance σ_S^2, then

$$\mathsf{C}_S = \sigma_S^2\mathsf{I}_n \tag{5.59}$$

and $\mathsf{Q} = \frac{\sigma_S^2}{\sigma^2(\sigma_S^2+\sigma^2)}\mathsf{I}_n$. The quadratic detector of (5.57) reduces to the *energy detector*

$$\delta_{\text{OPT}}(\boldsymbol{y}) = \begin{cases} 1 : & \|\boldsymbol{y}\|^2 \quad \geq \tau\frac{\sigma^2(\sigma_S^2+\sigma^2)}{\sigma_S^2} \triangleq \tau'. \\ 0 : & \qquad < \end{cases} \tag{5.60}$$

Note that the hypothesis test may be written as

$$\begin{cases} H_0 : & Y_i \sim \text{i.i.d. } \mathcal{N}(0, \sigma^2) \\ H_1 : & Y_i \sim \text{i.i.d. } \mathcal{N}(0, \sigma^2 + \sigma_S^2), \quad 1 \leq i \leq n. \end{cases} \tag{5.61}$$

The equal-probability contours under both H_0 and H_1 are spherically symmetric, and so is the decision boundary. This is illustrated in Figure 5.10 for the case $n = 2$.

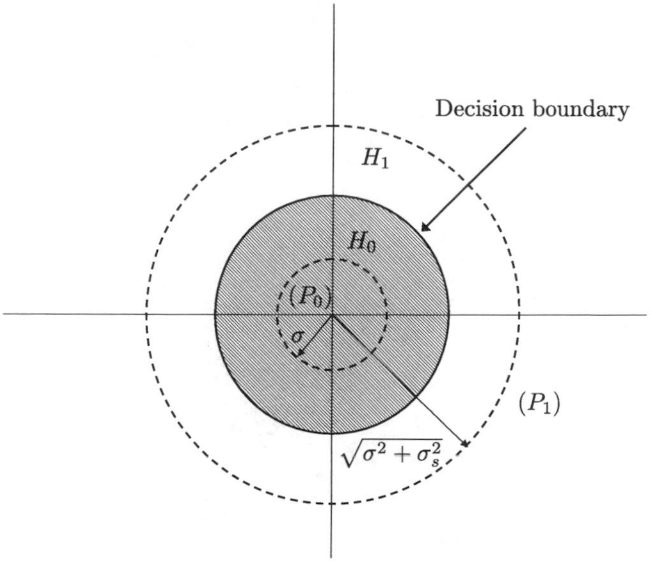

Figure 5.10 Equal probability contours and decision regions for i.i.d. Gaussian signals

Performance Analysis The test statistic $\|\boldsymbol{Y}\|^2$ in (5.60) is σ^2 times a χ_n^2 random variable under H_0, and $(\sigma^2 + \sigma_S^2)$ times a χ_n^2 random variable under H_1. Hence the probabilities of false alarm and correct detection are respectively given by

$$\mathrm{P_F} = \mathrm{P}\left\{\chi_n^2 > \tau \frac{\sigma^2 + \sigma_S^2}{\sigma_S^2}\right\} \quad \text{and} \quad \mathrm{P_D} = \mathrm{P}\left\{\chi_n^2 > \tau \frac{\sigma^2}{\sigma_S^2}\right\}. \tag{5.62}$$

5.7.3 Diagonalization of Signal Covariance

The derivations in Section 5.7.2 were simple because both the signal covariance matrix and the noise covariance matrix were multiples of I_n. To handle a general signal covariance matrix C_S, we use the same type of whitening technique that was applied to the noise in Section 5.4.1. Consider the eigenvector decomposition

$$\mathsf{C}_S = \mathsf{V}\Lambda\mathsf{V}^\top, \tag{5.63}$$

where $\mathsf{V} = [\boldsymbol{v}_1, \ldots, \boldsymbol{v}_n]$ is the matrix of eigenvectors, $\Lambda = \mathrm{diag}\{\lambda_k\}_{k=1}^n$, and λ_k is the eigenvalue corresponding to eigenvector $\boldsymbol{v}_k$. Consequently, $\mathsf{C}_S + \sigma^2\mathsf{I}_n = \mathsf{V}(\Lambda + \sigma^2\mathsf{I}_n)\mathsf{V}^\top$ and $\mathsf{Q} = \mathsf{V}\mathsf{D}\mathsf{V}^\top$ where $\mathsf{D} = \mathrm{diag}\left\{\frac{\lambda_k}{\sigma^2(\lambda_k+\sigma^2)}\right\}_{k=1}^n$. Now define

$$\tilde{\boldsymbol{y}} = \mathsf{V}^\top \boldsymbol{y}, \tag{5.64}$$

i.e., $\tilde{y}_k = \boldsymbol{v}_k^\top \boldsymbol{y}$ is the component of the observations in the direction of $\boldsymbol{v}_k$, the kth eigenvector. The hypothesis test (5.53) may be written in the equivalent form

$$\begin{cases} H_0 : \tilde{\boldsymbol{Y}} \sim \mathcal{N}(0, \sigma^2\mathsf{I}_n) \\ H_1 : \tilde{\boldsymbol{Y}} \sim \mathcal{N}(0, \Lambda + \sigma^2\mathsf{I}_n) \end{cases} \tag{5.65}$$

and the decision rule (5.57) reduces to the diagonal form

$$y^\top Q y = \tilde{y}^\top D \tilde{y} = \sum_{k=1}^{n} \frac{\lambda_k}{\sigma^2(\lambda_k + \sigma^2)} \tilde{y}_k^2 \underset{H_0}{\overset{H_1}{\gtrless}} \tau. \tag{5.66}$$

Note that if $\lambda_k \ll \sigma^2$, component k of the rotated observation vector $\tilde{\boldsymbol{y}}$ receives low weight.

Example 5.2 Let $n = 2$ and

$$C_S = \begin{pmatrix} 1 & \rho \\ \rho & 1 \end{pmatrix} \sigma_S^2, \tag{5.67}$$

where $|\rho| \le 1$. The eigenvector decomposition of this matrix is given in (5.44) (5.45). Here V is the 45° rotation matrix and the eigenvalues λ_1, λ_2 are $(1 \pm \rho)\sigma_S^2$. The rotated signal is given by

$$\tilde{\boldsymbol{Y}} = \frac{1}{\sqrt{2}} \begin{bmatrix} Y_1 + Y_2 \\ Y_1 - Y_2 \end{bmatrix} = \frac{1}{\sqrt{2}} \begin{bmatrix} S_1 + S_2 \\ S_1 - S_2 \end{bmatrix} + \tilde{\boldsymbol{Z}},$$

where the rotated noise components $\tilde{Z}_1$ and $\tilde{Z}_2$ are i.i.d. $\mathcal{N}(0, \sigma^2)$. The equal-probability contours for $\boldsymbol{Y}$ are elliptic, and so is the decision boundary (see Figure 5.11).

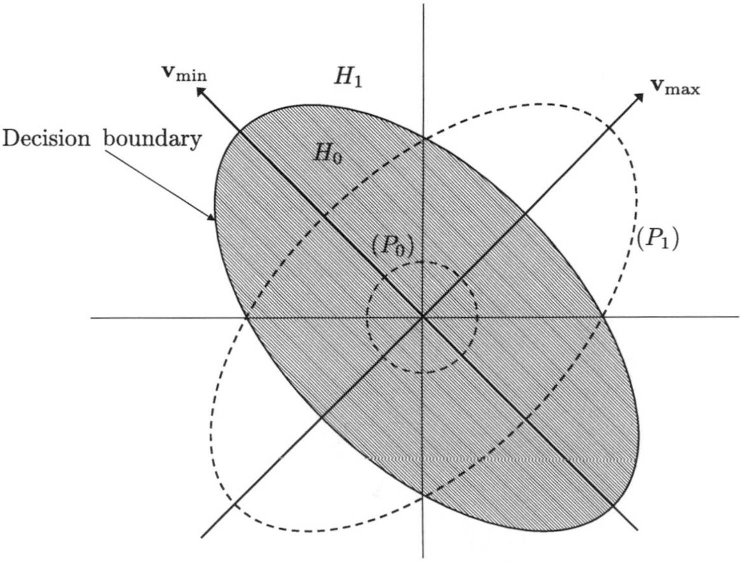

Figure 5.11 Equal-probability contours for $\boldsymbol{Y}$ and decision regions for the 2×2 signal covariance matrix of (5.67)

5.7.4 Performance Analysis

Note that the rotated vector $\tilde{\boldsymbol{Y}} = \mathsf{V}^\top \boldsymbol{Y}$ of (5.64) is Gaussian with covariance matrix equal to $\sigma^2 \mathsf{I}_n$ under H_0 and to $\Lambda + \sigma^2 \mathsf{I}_n$ under H_1. Hence

$$\begin{cases} H_0 : \tilde{Y}_k \sim \mathcal{N}(0, \sigma^2) \\ H_1 : \tilde{Y}_k \sim \mathcal{N}(0, \lambda_k + \sigma^2), \quad 1 \le k \le n, \end{cases}$$

where $\{\tilde{Y}_k\}_{k=1}^n$ are independent. We have

$$\mathrm{P_F} = \mathrm{P}_0 \left\{ \sum_{k=1}^n W_k > \tau \right\},$$

$$\mathrm{P_D} = \mathrm{P}_1 \left\{ \sum_{k=1}^n W_k > \tau \right\},$$

where

$$W_k \triangleq \frac{\lambda_k}{\sigma^2(\lambda_k + \sigma^2)} \tilde{Y}_k^2, \quad k = 1, 2, \ldots, n$$

are independent and have the following variances under H_0 and H_1:

$$\sigma_{0k}^2 = \frac{\lambda_k}{\lambda_k + \sigma^2}, \quad \sigma_{1k}^2 = \frac{\lambda_k}{\sigma^2}, \quad k = 1, 2, \ldots, n.$$

Hence W_k follows a gamma distribution (see Appendix C) under H_j:

$$p_{j,W_k}(x) = \frac{1}{\sqrt{2\pi x \sigma_{jk}^2}} \exp\left\{ -\frac{x}{2\sigma_{jk}^2} \right\}, \tag{5.68}$$

i.e., $W_k \sim \mathrm{Gamma}\left(\frac{1}{2}, \frac{1}{2\sigma_{jk}^2}\right)$.

The case of i.i.d. Gaussian signals ($\lambda_k \equiv \sigma_S^2$) was studied in Section 5.7.2. Then $\{W_k\}_{k=1}^n$ are i.i.d., and the sum of n i.i.d. Gamma(a, b) random variables is Gamma(na, b), which is a scaled χ_n^2 random variable. In the correlated-signal scenario, there is no simple way to evaluate $\mathrm{P_F}$ and $\mathrm{P_D}$. We outline a few approaches in the remainder of Section 5.7.4.

Characteristic Function Approach Denote by

$$\Phi_{j,W_k}(\omega) = \mathbb{E}_j\left(e^{i\omega W_k}\right), \quad \omega \in \mathbb{R}, \quad i = \sqrt{-1},$$

the characteristic function of W_k under H_j. If $T_n = \sum_{k=1}^n W_k$, then under H_j, T_n has characteristic function:

$$\Phi_{j,T_n}(\omega) = \prod_{k=1}^n \Phi_{j,W_k}(\omega) = \prod_{k=1}^n \frac{1}{\sqrt{1 - 2i\omega\sigma_{jk}^2}}.$$

In principle one could apply the inverse Fourier Transform and obtain the pdf p_{T_n} numerically. This is, however, not straightforward because the inverse Fourier transform is an integral, and evaluating it by means of the discrete Fourier transform requires truncation and sampling of the original Φ_{T_n}. Unfortunately the resulting inversion method is unstable and therefore unsuitable for computing small tail probabilities.

Saddlepoint Approximation If each W_k has a moment-generating function (mgf)

$$M_{j,W_k}(\sigma) = \mathbb{E}_j\left(e^{\sigma W_k}\right), \quad \sigma \in \mathbb{R},$$

under H_j, then so does T_n:

$$M_{j,T_n}(\sigma) = \prod_{k=1}^{n} M_{W_k}(\sigma), \quad \sigma \in \mathbb{R},$$

and the definition of the mgf can be extended to the complex plane:

$$M_{j,T_n}(u) = \mathbb{E}_j\left(e^{uT_n}\right), \quad u \in \mathbb{C}.$$

Thus $M_{T_n}(i\omega) = \Phi_{j,T_n}(\omega)$ for $\omega \in \mathbb{R}$.

To alleviate the instability problem, one may compute p_{j,T_n} by evaluating the inverse Laplace transform of M_{j,T_n}. This is done by integrating along a vertical line in the complex plane, intersecting the real axis at a location $u^* \in \mathbb{R}$. (The inverse Fourier transform is obtained when $u^* = 0$.) Then

$$p_{j,T_n}(t) = \frac{1}{2\pi i}\int_{u^*-i\infty}^{u^*+i\infty} M_{j,T_n}(\omega)e^{-ut}\,du.$$

The optimal value of u^* (for which the inversion process is most stable) is the saddlepoint of the integrand on the real axis, which satisfies [3]

$$0 = \frac{d}{du}[-ut + \ln M_{j,T_n}(u)]\bigg|_{u=u^*}.$$

Tail probabilities $\mathrm{P}_j\{T_n > t\}$ can also be approximated using saddlepoint methods, as developed by Lugannani and Rice [4]. For a detailed account of these methods, see [5, Ch. 14].

Large Deviations Large-deviations methods (as developed in Chapter 8) can be used to upper bound the tail probabilities $\mathrm{P_F}$ and $\mathrm{P_M}$; this method is particularly useful for small $\mathrm{P_F}$ and $\mathrm{P_M}$. In fact these methods are closely related to the saddlepoint approximation methods.

5.7.5 Gaussian Signals With Nonzero Mean

Up to this point, we assumed that the mean of $\boldsymbol{S}$ is zero. Now let $\mu_S \neq 0$. Whitening the observations again, we write (5.56) as

$$\ln L(\tilde{\boldsymbol{y}}) = \frac{1}{2}\tilde{\boldsymbol{y}}^\top \mathsf{D}\tilde{\boldsymbol{y}} + \tilde{\boldsymbol{y}}^\top(\Lambda + \sigma^2\mathsf{I}_n)^{-1}\tilde{\boldsymbol{\mu}}_S - \frac{1}{2}\tilde{\boldsymbol{\mu}}_S(\Lambda + \sigma^2\mathsf{I}_n)^{-1}\tilde{\boldsymbol{\mu}}_S + \kappa,$$

where $\tilde{\boldsymbol{\mu}}_S = \{\tilde{\mu}_k\}_{k=1}^n = \mathsf{V}^\top \boldsymbol{\mu}$. Hence

$$\sum_{k=1}^{n} \frac{1}{2} \frac{\lambda_k}{\sigma^2(\lambda_k+\sigma^2)} \tilde{y}_k^2 + \frac{\tilde{y}_k \tilde{\mu}_k}{\lambda_k+\sigma^2} - \frac{\tilde{\mu}_k^2}{2(\lambda_k+\sigma^2)} \underset{H_0}{\overset{H_1}{\gtrless}} \tau.$$

Performance analysis is greatly simplified in the special case of a white signal ($\mathsf{C}_S = \sigma_S^2$, hence $\lambda_k \equiv \sigma_S^2$). Then $\mathsf{V} = \mathsf{I}_n$ and the test becomes

$$\sum_{i=1}^{n} (y_i - u_i)^2 \underset{H_0}{\overset{H_1}{\gtrless}} \tau' = \frac{2\sigma^2(\sigma_S^2+\sigma^2)}{\sigma_S^2},$$

where

$$u_i = -\frac{\sigma^2}{\sigma_S^2} \tilde{\mu}_{S,i}, \quad 1 \le k \le n.$$

After elementary algebraic manipulations, we obtain that the optimal test has the form

$$T \triangleq \sum_{i=1}^{n} (y_i - u_i)^2 \underset{H_0}{\overset{H_1}{\gtrless}} \tau' = \frac{\sigma^2(\sigma_S^2+\sigma^2)}{\sigma_S^2} \left(2\tau + \frac{\|\mu\|^2}{\sigma_S^2} \right). \tag{5.69}$$

Under both H_0 and H_1, the test statistic T is a scaled noncentral χ_n^2 random variable. The scale factor is σ under H_1 and $\sqrt{\sigma_S^2+\sigma^2}$ under H_1. We have

$$\begin{aligned} \mathrm{P_F}(\delta) &= \mathrm{P}\{\chi_n^2(\lambda_0) > \sigma^2\tau'\}, \\ \mathrm{P_D}(\delta) &= \mathrm{P}\{\chi_n^2(\lambda_1) > (\sigma_S^2+\sigma^2)\tau'\}, \end{aligned} \tag{5.70}$$

where the noncentrality parameters are respectively

$$\lambda_0 = \|\boldsymbol{\mu}\|^2 \frac{\sigma^2}{\sigma_S^4} \quad \text{and} \quad \lambda_1 = \|\boldsymbol{\mu}\|^2 \frac{\sigma^4}{\sigma_S^4(\sigma_S^2+\sigma^2)}.$$

5.8 Detection of Weak Signals

In many applications, the signal to be detected is weak and is known only up to a scaling constant. The detection problem may be formulated as

$$\begin{cases} H_0 : \boldsymbol{Y} = \boldsymbol{Z} \\ H_1 : \boldsymbol{Y} = \theta \boldsymbol{s} + \boldsymbol{Z}, \quad \theta > 0, \end{cases} \tag{5.71}$$

where $\boldsymbol{s}$ is known but θ is not. We assume that the noise is i.i.d. with pdf p_Z. A *locally optimum* detector for this problem can be obtained by approximating the log-likelihood ratio as a linear function of θ in the vicinity of $\theta = 0$:

$$\ln L_\theta(\boldsymbol{y}) = \sum_{k=1}^{n} [\ln p_Z(y_k - \theta s_k) - \ln p_Z(y_k)] = \theta \sum_{k=1}^{n} \psi(y_k) s_k + o(\theta),$$

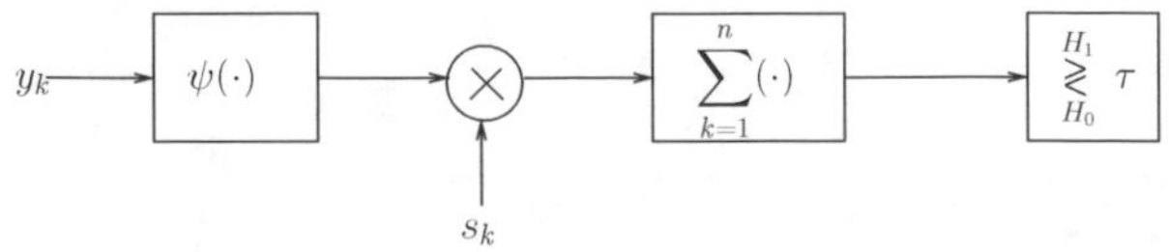

Figure 5.12 Locally optimal detector for detection of weak signals in i.i.d. noise

where we have defined[1]

$$\psi(z) = -\frac{d \ln p_Z(z)}{dz}.$$

Hence the locally optimum test takes the form

$$\sum_{k=1}^{n} \psi(y_k)s_k \overset{H_1}{\underset{H_0}{\gtrless}} \tau, \tag{5.72}$$

where the threshold τ is usually selected to achieve a target probability of false alarm. The test statistic is reminiscent of the correlation statistic for detection in i.i.d. Gaussian noise; however, each observation y_i is passed through a nonlinearity $\psi(\cdot)$ prior to performing the correlation. The locally optimum test is a nonlinear correlator-detector, and its structure is diagrammed in Figure 5.12.

For Gaussian noise $Z \sim \mathcal{N}(0, \sigma^2)$, we have $\psi(z) = z/\sigma^2$, and so the test statistic is the same linear correlation statistic that was derived in Section 5.3.1.

Example 5.3 Let p_Z be the Laplacian pdf of (5.13). Then $\psi(z) = a\,\mathrm{sgn}(z)$, and the test statistic is a times the correlation of the signal $\boldsymbol{s}$ with the sequence of signs of the observations. Further assume that the signal is constant, $s_i \equiv 1$, hence the test statistic $T_n = a\sum_{i=1}^{n}\psi(Y_i)$. We now derive the distributions of the test statistic. Under P_θ, $\psi(Y_i) = -1$ iff $Z_i < -\theta$, hence $P_\theta\{\psi(Y_i) = -1\} = e^{-\theta}/2$, and T_n/a is a scaled binomial random variable: $T_n/a \sim 2\mathrm{Bi}(n, 1 - e^{-\theta}/2) - n$, which has mean $n(1 - e^{-\theta})$ and variance $n(1 - e^{-\theta}/2)e^{-\theta}/2$. Therefore

$$\mathrm{P_F}(\delta_{\mathrm{LO}}) = \mathrm{P}\{2\mathrm{Bi}(n, 1/2) - n > \tau/a\}, \tag{5.73}$$

$$\mathrm{P_D^\theta}(\delta_{\mathrm{LO}}) = \mathrm{P}\{2\mathrm{Bi}(n, 1 - e^{-\theta}/2) - n > \tau/a\}, \quad \theta > 0. \tag{5.74}$$

In order to achieve a probability of false alarm approximately equal to α, one can use the Gaussian approximation of Section 5.3.4 to the distribution of the test statistic and set the value of the test threshold to $\tau = a\sqrt{n}Q^{-1}(\alpha)$.

5.9 Detection of Signal with Unknown Parameters in White Gaussian Noise

We now study detection of a signal that is known up to some parameter θ. The signal is denoted by $\boldsymbol{s}(\theta) \in \mathbb{R}^n$. For detection in i.i.d. Gaussian noise, the problem is formulated as

[1] The pdf p_Z might have points where the derivative does not exist, as in the Laplacian example 5.3; however, the probability that a sample Y_i takes one of these values is zero.

$$\begin{cases} H_0 : \boldsymbol{Y} = \boldsymbol{Z} \\ H_1 : \boldsymbol{Y} = \boldsymbol{s}(\theta) + \boldsymbol{Z}, \quad \theta \in \mathcal{X}_1, \end{cases} \tag{5.75}$$

where the noise $\boldsymbol{Z}$ is i.i.d. $\mathcal{N}(0, \sigma^2)$. This model finds a variety of applications including communications (where θ is an unknown phase, as in Example 5.4), template matching (as in Example 5.5), and radar signal processing (where θ includes target location and velocity parameters). In *linear models*, the signal depends linearly on the unknown parameter:

$$\boldsymbol{s}(\theta) = \mathsf{L}\theta, \tag{5.76}$$

where L is a known $n \times d$ matrix, and the unknown $\theta \in \mathcal{X}_1 = \mathbb{R}^d$. Without loss of generality, we assume that the columns of L are orthonormal vectors, i.e., $\mathsf{L}^\top \mathsf{L} = \mathsf{I}_d$.[2]

5.9.1 General Approach

The Gaussian noise model (5.75) is a special case of the following general composite binary hypothesis test (Chapter 4) with vector observations:

$$\begin{cases} H_0 : \boldsymbol{Y} \sim \mathrm{P}(\cdot|H_0) \\ H_1 : \boldsymbol{Y} \sim \mathrm{P}_\theta, \quad \theta \in \mathcal{X}_1. \end{cases} \tag{5.77}$$

Let $L_\theta(\boldsymbol{y}) \triangleq \frac{p_\theta(\boldsymbol{y})}{p(\boldsymbol{y}|H_0)}$. For the Gaussian problem of (5.75), we have

$$L_\theta(\boldsymbol{y}) = \exp\left\{ \frac{1}{\sigma^2}\boldsymbol{s}(\theta)^\top \boldsymbol{y} - \frac{\|\boldsymbol{s}(\theta)\|^2}{2\sigma^2} \right\}, \quad \theta \in \mathcal{X}_1. \tag{5.78}$$

If θ were known, the optimal test would be a Bayes rule $\delta_{\mathrm{B}}^\theta$:

$$\delta_{\mathrm{B}}^\theta(\boldsymbol{y}) = \begin{cases} 1 & : \ L_\theta(\boldsymbol{y}) \ge \eta_\theta, \\ 0 & : \qquad\quad < \end{cases} \tag{5.79}$$

where the threshold η_θ is selected according to the desired performance criterion (Bayes, minimax, or Neyman–Pearson). The corresponding probabilities of false alarm and miss are

$$\begin{aligned} \mathrm{P}_{\mathrm{F}}(\delta_{\mathrm{B}}^\theta) &= \mathrm{P}\{L_\theta(\boldsymbol{Y}) \ge \eta_\theta \mid H_0\}, \\ \mathrm{P}_{\mathrm{M}}^\theta(\delta_{\mathrm{B}}^\theta) &= \mathrm{P}_\theta\{L_\theta(\boldsymbol{Y}) < \eta_\theta\}. \end{aligned} \tag{5.80}$$

For unknown θ, we consider two strategies. First the GLRT setting, where the test takes the form

$$\delta_{\mathrm{GLRT}}(\boldsymbol{y}) = \begin{cases} 1 : & \max_{\theta \in \mathcal{X}_1} L_\theta(\boldsymbol{y}) \ge \eta, \\ 0 : & \qquad\quad < \end{cases}$$

Second, the Bayesian setting where θ is viewed as a realization of a random variable Θ with prior π, and costs are uniform over each hypothesis (as studied in Section 4.2.1). The likelihood ratio is given by

[2] If this was not the case, a linear reparameterization of θ could be used to obtain $\mathsf{L}^\top \mathsf{L} = \mathsf{I}_d$.

$$L(\boldsymbol{y}) = \int_{\mathcal{X}_1} L_\theta(\boldsymbol{y})\pi(\theta)d\nu(\theta)$$

and the Bayes test takes the form

$$\delta_{\mathrm{B}}(\boldsymbol{y}) = \begin{cases} 1: & L(\boldsymbol{y}) \geq \eta. \\ 0: & \quad < \end{cases}$$

5.9.2 Linear Gaussian Model

To gain insight into the GLRT, we first consider the linear model (5.75), (5.76) for detection in additive white Gaussian noise. Then

$$\|\mathsf{L}\theta\|^2 = \theta^\top \underbrace{\mathsf{L}^\top\mathsf{L}}_{=\mathsf{I}_d} \theta = \|\theta\|^2,$$

and the LLR (5.78) becomes

$$\ln L_\theta(\boldsymbol{y}) = \frac{1}{\sigma^2}\theta^\top\tilde{\boldsymbol{y}} - \frac{\|\theta\|^2}{2\sigma^2}, \quad \theta \in \mathbb{R}^d, \tag{5.81}$$

where we have defined $\tilde{\boldsymbol{y}} = \mathsf{L}^\top \boldsymbol{y} \in \mathbb{R}^d$. Maximizing (5.81) over θ yields the solution $\hat{\theta}_{\mathrm{ML}} = \tilde{\boldsymbol{y}}$, and

$$\ln L_G(\boldsymbol{y}) = \max_\theta \ln L_\theta(\boldsymbol{y}) = \frac{\|\hat{\theta}_{\mathrm{ML}}\|^2}{2\sigma^2} = \frac{\|\tilde{\boldsymbol{y}}\|^2}{2\sigma^2},$$

which represents one half of the estimated "signal-to-noise ratio." Hence the GLRT takes the form

$$\ln L_G(\boldsymbol{y}) = \frac{\|\tilde{\boldsymbol{y}}\|^2}{2\sigma^2} \underset{H_0}{\overset{H_1}{\gtrless}} \ln \eta.$$

Performance Analysis Under H_0, we have $\boldsymbol{Y} \sim \mathcal{N}(0, \sigma^2\mathsf{I}_n)$. Since $\mathsf{L}^\top\mathsf{L} = \mathsf{I}_d$, we obtain $\tilde{\boldsymbol{Y}} \sim \mathcal{N}(0, \sigma^2\mathsf{I}_d)$. Thus $\ln L_G(\boldsymbol{Y})$ is half a χ^2 random variable with d degrees of freedom. The probability of false alarm is given by

$$\mathrm{P_F}(\delta_{\mathrm{GLRT}}) = \mathrm{P}\{\ln L_G(\boldsymbol{Y}) > \ln \eta | H_0\} = \mathrm{P}\{\chi_d^2 > 2\ln\eta\}.$$

Under H_1, we have $\boldsymbol{Y} \sim \mathcal{N}(\mathsf{L}\theta, \sigma^2\mathsf{I}_n)$, hence $\tilde{\boldsymbol{Y}} \sim \mathcal{N}(\theta, \sigma^2\mathsf{I}_d)$. Thus $\ln L_G(\boldsymbol{Y})$ is half a noncentral χ^2 random variable with d degrees of freedom and noncentrality parameter $\lambda = \|\theta\|^2/\sigma^2$. The probability of correct detection is given by

$$\mathrm{P_D^\theta}(\delta_{\mathrm{GLRT}}) = \mathrm{P}_\theta\{\ln L_G(\boldsymbol{Y}) > \ln\eta\} = \mathrm{P}\{\chi_d^2(\lambda) > \ln\eta\}.$$

5.9.3 Nonlinear Gaussian Model

Performance analysis in the case the signal depends nonlinearly on θ is more complicated than in the linear case. We present a well-known example for which the analysis is tractable.

Example 5.4 *Sinusoid Detection.* Assume the Gaussian noise model of (5.75) with $\mathcal{X}_1 = [0, 2\pi]$, $\omega = 2\pi \frac{k}{n}$ where $k \in \{1, 2, \ldots, n\}$, and $s_i(\theta) = \sin(\omega i + \theta)$ for $i = 1, 2, \ldots, n$. Thus $\boldsymbol{s}(\theta)$ is a sinusoid with known angular frequency ω and unknown phase θ. Moreover $\|\boldsymbol{s}(\theta)\|^2 = n/2$ for all θ. Assume equal priors on H_0 and H_1, hence $\eta = 1$. Using the identity $\sin(\omega i + \theta) = \sin(\omega i)\cos\theta + \cos(\omega i)\sin\theta$, we write the GLR statistic as

$$T_{\text{GLR}} = \max_{0 \le \theta \le 2\pi} \boldsymbol{s}(\theta)^\top \boldsymbol{Y} = \max_{0 \le \theta \le 2\pi} [Y_s \cos\theta + Y_c \sin\theta] = \sqrt{Y_s^2 + Y_c^2} \triangleq R,$$

where $Y_s = \sum_{i=1}^n \sin(\omega i) Y_i$ and $Y_c = \sum_{i=1}^n \cos(\omega i) Y_i$ are the projections of the observation vector onto the sine vector and the cosine vector with angular frequency ω and zero phase. We have

$$\delta_{\text{GLRT}}(\boldsymbol{Y}) = \begin{cases} 1 & : R \ge \tau \\ 0 & : R < \tau, \end{cases} \tag{5.82}$$

where $\tau = \sigma^2 \ln \eta + n/4 = n/4$. Likewise define the noise vector projections $Z_s \triangleq \sum_{i=1}^n \sin(\omega i) Z_i$ and $Z_c \triangleq \sum_{i=1}^n \cos(\omega i) Z_i$. Since $Z_i,\ 1 \le i \le n$ are i.i.d. $\mathcal{N}(0, \sigma^2)$, it follows from the trigonometric identities

$$\sum_{i=1}^n \sin^2(\omega i) = \sum_{i=1}^n \cos^2(\omega i) = \frac{n}{2} \quad \text{and} \quad \sum_{i=1}^n \sin(\omega i)\cos(\omega_i) = \frac{1}{2}\sum_{i=1}^n \sin(2\omega i) = 0$$

that the random variables Z_s and Z_c are i.i.d. $\mathcal{N}(0, \frac{n\sigma^2}{2})$. Therefore, under H_0, $R^2 = Z_s^2 + Z_c^2$ follows an exponential distribution with mean equal to $n\sigma^2$. Then we obtain a closed-form solution for the false-alarm probability of the GLRT:

$$\begin{aligned} \mathrm{P_F}(\delta_{\text{GLRT}}) &= \mathrm{P}\{R^2 \ge \tau^2 \,|\, H_0\} \\ &= \exp\left\{-\frac{\tau^2}{n\sigma^2}\right\} = \exp\left\{-\frac{n}{16\sigma^2}\right\}. \end{aligned}$$

Under H_1, the random variables Y_s and Y_c are independent with respective means

$$\mathbb{E}_\theta(Y_s) = \sum_{i=1}^n \sin(\omega i)\sin(\omega i + \theta) = \frac{n}{2}\cos\theta,$$

$$\mathbb{E}_\theta(Y_c) = \sum_{i=1}^n \cos(\omega i)\sin(\omega i + \theta) = \frac{n}{2}\sin\theta,$$

and the same variances as under H_0. Hence $Y_s \sim \mathcal{N}(\frac{n}{2}\cos\theta, \frac{n\sigma^2}{2})$ and $Y_c \sim \mathcal{N}(\frac{n}{2}\sin\theta, \frac{n\sigma^2}{2})$, and $R \sim \text{Rice}(\frac{n}{2}, \frac{n\sigma^2}{2})$ (for all θ). (See Appendix C for the definition of the Ricean distribution.) The probability of correct detection is obtained using (C.7):

$$\begin{aligned} \mathrm{P_D^\theta}(\delta_{\text{GLRT}}) &= \mathrm{P}\{R \ge \tau\} \\ &= \int_\tau^\infty \frac{2r}{n\sigma^2} \exp\left\{-\frac{r^2 + (n/2)^2}{n\sigma^2}\right\} I_0\left(\frac{r}{\sigma^2}\right) dr \\ &= Q\left(\sqrt{\frac{n}{2\sigma^2}}, \tau\sqrt{\frac{n}{2\sigma^2}}\right), \quad \forall \theta \in \mathcal{X}_1, \end{aligned}$$

where $Q(\rho, t) \triangleq \int_t^\infty x \exp\{-\frac{1}{2}(x^2 + \rho^2)\} I_0(\rho x)\, dx$ is Marcum's Q function.

It is also instructive to derive the Bayes rule assuming a uniform prior on Θ in this case. The likelihood ratio is then given by

$$\begin{aligned} L(\boldsymbol{Y}) &= \frac{1}{2\pi} \int_0^{2\pi} \exp\left\{ \frac{1}{\sigma^2} \boldsymbol{s}(\theta)^\top \boldsymbol{Y} - \frac{n}{4\sigma^2} \right\} d\theta \\ &= \exp\left\{ -\frac{n}{4\sigma^2} \right\} \frac{1}{2\pi} \int_0^{2\pi} \exp\left\{ \frac{Y_c \sin\theta + Y_s \cos\theta}{\sigma^2} \right\} d\theta \\ &= \exp\left\{ -\frac{n}{4\sigma^2} \right\} I_0\left(\frac{\sqrt{Y_c^2 + Y_s^2}}{\sigma^2} \right), \end{aligned}$$

where I_0 is the modified Bessel function of the first kind. Hence an optimal test statistic is $\sqrt{Y_c^2 + Y_s^2}$, which is identical to the GLR statistic T_{GLR}, and the Bayes test δ_{B} coincides with δ_{GLRT}.

5.9.4 Discrete Parameter Set

Next we turn our attention to the case of discrete $\mathcal{X}_1$, for which performance bounds can be conveniently derived. Again we consider the case of known θ first. By (5.32), (5.33), the probabilities of false alarm and miss for δ_{B}^θ are given by

$$\text{P}_{\text{F}}(\delta_{\text{B}}^\theta) = Q\left(\frac{\tau_\theta}{\sigma \|\boldsymbol{s}(\theta)\|} \right), \quad \text{P}_{\text{M}}(\delta_{\text{B}}^\theta) = Q\left(\frac{\|\boldsymbol{s}(\theta)\|^2 - \tau_\theta}{\sigma \|\boldsymbol{s}(\theta)\|} \right). \tag{5.83}$$

In the special case of equal-energy signals, we have $\|\boldsymbol{s}(\theta)\|^2 = nE_S$ for each $\theta \in \mathcal{X}_1$. For equal priors $\pi(H_0) = \pi(H_1) = \frac{1}{2}$, the test threshold is given by $\tau_\theta = \frac{1}{2} n E_S$, and thus

$$\text{P}_{\text{F}}(\delta_{\text{B}}^\theta) = \text{P}_{\text{M}}^\theta(\delta_{\text{B}}^\theta) = Q\left(\frac{\sqrt{nE_S}}{2\sigma} \right). \tag{5.84}$$

Performance Bounds for GLRT For unknown θ, we derive upper bounds on $\text{P}_{\text{F}}(\delta_{\text{GLRT}})$ and $\text{P}_{\text{M}}(\delta_{\text{GLRT}})$, as well as a lower bound on $\text{P}_{\text{F}}(\delta_{\text{GLRT}})$. The first part of the derivation does not require a Gaussian observation model. The false-alarm probability for the GLRT with threshold η is given by

$$\begin{aligned} \text{P}_{\text{F}}(\delta_{\text{GLRT}}) &= \text{P}\left\{ \max_{\theta \in \mathcal{X}_1} L_\theta(\boldsymbol{Y}) \geq \eta \,\middle|\, H_0 \right\} \qquad (5.85) \\ &= \text{P}\left\{ \bigcup_{\theta \in \mathcal{X}_1} \{L_\theta(\boldsymbol{Y}) \geq \eta\} \,\middle|\, H_0 \right\} \\ &\overset{(a)}{\leq} |\mathcal{X}_1| \max_{\theta \in \mathcal{X}_1} \text{P}\{L_\theta(\boldsymbol{Y}) \geq \eta \mid H_0\}, \qquad (5.86) \end{aligned}$$

where the inequality follows from the union-of-events bound. Also, for each $\theta \in \mathcal{X}_1$ the probability of miss is given by

$$\begin{aligned} P_M^\theta(\delta_{\mathrm{GLRT}}) &= P_\theta \left\{ \max_{\theta' \in \mathcal{X}_1} L_{\theta'}(\boldsymbol{Y}) < \eta \right\} \\ &\le P_\theta \left\{ L_\theta(\boldsymbol{Y}) < \eta \right\}, \end{aligned} \tag{5.87}$$

where the last inequality in (5.87) uses the fact that $L_\theta(\boldsymbol{Y})$ is a lower bound on $\max_{\theta' \in \mathcal{X}_1} L_{\theta'}(\boldsymbol{Y})$.

Specializing (5.86) to the Gaussian problem of (5.75), and using (5.78), we obtain

$$\begin{aligned} P_F(\delta_{\mathrm{GLRT}}) &\le |\mathcal{X}_1| \max_{\theta \in \mathcal{X}_1} P\left\{ \boldsymbol{s}(\theta)^\top \boldsymbol{Z} \ge \tau \right\} \\ &\le |\mathcal{X}_1| \max_{\theta \in \mathcal{X}_1} Q\left(\frac{\tau}{\sigma \|\boldsymbol{s}(\theta)\|} \right). \end{aligned} \tag{5.88}$$

Applying (5.87), we also obtain

$$\begin{aligned} P_M^\theta(\delta_{\mathrm{GLRT}}) &\le P_\theta \left[\boldsymbol{s}(\theta)^\top \boldsymbol{Y} < \tau \right] \\ &= Q\left(\frac{\|\boldsymbol{s}(\theta)\|^2 - \tau}{\sigma \|\boldsymbol{s}(\theta)\|} \right). \end{aligned} \tag{5.89}$$

In the special case of equal-energy signals, we have

$$P_F(\delta_{\mathrm{GLRT}}) \le |\mathcal{X}_1| Q\left(\frac{\tau}{\sigma \sqrt{n E_S}} \right), \quad P_M^\theta(\delta_{\mathrm{GLRT}}) \le Q\left(\frac{n E_S - \tau}{\sigma \sqrt{n E_S}} \right) \quad \forall \theta \in \mathcal{X}_1. \tag{5.90}$$

The arguments of the Q functions are equal if $\tau = \frac{1}{2} n E_S$, in which case

$$P_F(\delta_{\mathrm{GLRT}}) \le |\mathcal{X}_1| Q\left(\frac{\sqrt{n E_S}}{2\sigma} \right), \quad P_M^\theta(\delta_{\mathrm{GLRT}}) \le Q\left(\frac{\sqrt{n E_S}}{2\sigma} \right), \quad \forall \theta \in \mathcal{X}_1. \tag{5.91}$$

The Q function vanishes exponentially fast with n, hence comparing (5.91) and (5.84) we conclude that performance of the GLRT is nearly as good as that of δ_B^θ: for the same choice of the threshold τ, we have $P_M^\theta(\delta_{\mathrm{GLRT}}) \le P_M(\delta_B^\theta)$ and $P_F(\delta_{\mathrm{GLRT}}) \le |\mathcal{X}_1| P_F(\delta_B^\theta)$.

From (5.85), using again the fact that $L_\theta(\boldsymbol{Y})$ is a lower bound on $\max_{\theta' \in \mathcal{X}_1} L_{\theta'}(\boldsymbol{Y})$ for all $\theta \in \mathcal{X}_1$, we have the lower bound

$$P_F(\delta_{\mathrm{GLRT}}) \ge \max_{\theta \in \mathcal{X}_1} P\left\{ L_\theta(\boldsymbol{Y}) \ge \eta \mid H_0 \right\},$$

which is within a factor $|\mathcal{X}_1|$ of the upper bound in (5.86).

Performance Bounds for Bayesian Detection In the Bayesian setup, the likelihood ratio for finite $\mathcal{X}_1$ is given by

$$L(\boldsymbol{Y}) = \sum_{\theta \in \mathcal{X}_1} \pi(\theta) L_\theta(\boldsymbol{Y}),$$

hence

$$\begin{aligned} \mathrm{P_F}(\delta_{\mathrm{B}}) &= \mathrm{P}\left\{ \sum_{\theta \in \mathcal{X}_1} \pi(\theta) L_\theta(\boldsymbol{Y}) \geq \eta \,\Bigg|\, H_0 \right\} \\ &\leq \mathrm{P}\left\{ \max_{\theta \in \mathcal{X}_1} L_\theta(\boldsymbol{Y}) \geq \eta \,\Bigg|\, H_0 \right\} = \mathrm{P_F}(\delta_{\mathrm{GLRT}}), \end{aligned}$$

where the upper bound holds because the average of $L_\theta(\boldsymbol{Y})$ over $\mathcal{X}_1$ cannot exceed the maximum. Also

$$\begin{aligned} \mathrm{P}_{\mathrm{M}}^{\theta}(\delta_{\mathrm{B}}) &= \mathrm{P}_\theta \left\{ \sum_{\theta' \in \mathcal{X}_1} \pi(\theta') L_{\theta'}(\boldsymbol{Y}) < \eta \right\} \\ &\leq \mathrm{P}_\theta \left\{ \pi(\theta) L_\theta(\boldsymbol{Y}) < \eta \right\}, \quad \forall \theta \in \mathcal{X}_1, \end{aligned}$$

which is the same upper bound as (5.87) for δ_{GLRT}, except that the threshold η is replaced by $\eta / \pi(\theta)$.

Hence, for the Gaussian problem of (5.75), we obtain the same upper bound as in (5.89), except that a term $\sigma^2 \ln \pi(\theta)$ must be subtracted from τ:

$$\mathrm{P}_{\mathrm{M}}^{\theta}(\delta_{\mathrm{B}}) \leq Q\left(\frac{\|\boldsymbol{s}(\theta)\|^2 - \tau + \sigma^2 \ln \pi(\theta)}{\sigma \|\boldsymbol{s}(\theta)\|} \right), \quad \forall \theta \in \mathcal{X}_1.$$

We can also derive the following lower bounds (see Exercise 5.12):

$$\mathrm{P}_{\mathrm{F}}^{\theta}(\delta_{\mathrm{B}}) \geq \max_{\theta \in \mathcal{X}_1} \mathrm{P}\left\{ \pi(\theta) L_\theta(\boldsymbol{Y}) \geq \eta | H_0 \right\}, \tag{5.92}$$

$$\mathrm{P}_{\mathrm{M}}^{\theta}(\delta_{\mathrm{B}}) \geq \mathrm{P}_{\mathrm{M}}^{\theta}(\delta_{\mathrm{GLRT}}). \tag{5.93}$$

Assuming equal-energy signals again and $\tau = \frac{1}{2} n E_S$, we obtain

$$\mathrm{P_F}(\delta_{\mathrm{B}}) \leq Q\left(\frac{\sqrt{n E_S}}{2\sigma} \right), \quad \mathrm{P}_{\mathrm{M}}^{\theta}(\delta_{\mathrm{B}}) \leq Q\left(\frac{\sqrt{n E_S}}{2\sigma} + \frac{\sigma \ln \pi(\theta)}{\sqrt{n E_S}} \right), \quad \forall \theta \in \mathcal{X}_1, \tag{5.94}$$

which is the same as (5.84), except for the presence of a vanishing extra term in the argument of the second Q function. To see the effect of this extra term, let $u = \frac{\sqrt{n E_S}}{2\sigma}$ and $\epsilon = \frac{\sigma \ln \pi(\theta)}{\sqrt{n E_S}}$ in which case the right hand side of (5.94) may be written as $Q(u + \epsilon)$. Using the asymptotic expression (5.41) for the Q function, we have

$$\begin{aligned} Q(u + \epsilon) &\sim \frac{\exp\left\{ -\frac{1}{2}(u + \epsilon)^2 \right\}}{\sqrt{2\pi}(u + \epsilon)} \\ &\sim e^{-\epsilon u} \, Q(u) = e^{-\frac{1}{2} \ln \pi(\theta)} \, Q(u) = \frac{1}{\sqrt{\pi(\theta)}} \, Q(u) \quad \text{as } n \to \infty, \end{aligned}$$

hence (5.94) becomes

$$\mathrm{P_F}(\delta_{\mathrm{B}}) \leq Q\left(\frac{\sqrt{n E_S}}{2\sigma} \right), \quad \mathrm{P}_{\mathrm{M}}^{\theta}(\delta_{\mathrm{B}}) \leq \frac{1}{\sqrt{\pi(\theta)}} \, Q\left(\frac{\sqrt{n E_S}}{2\sigma} \right) [1 + o(1)], \quad \forall \theta \in \mathcal{X}_1. \tag{5.95}$$

This upper bound differs from $\mathrm{P}_\mathrm{M}^\theta(\delta_\mathrm{B}^\theta)$ by a multiplying constant $\frac{1}{\sqrt{\pi(\theta)}} \geq 1$. For a uniform prior, this constant is $\sqrt{|\mathcal{X}_1|}$, which is smaller than the multiplying constant $|\mathcal{X}_1|$ for the GLRT.

Example 5.5 *Template Matching.* Assume $\boldsymbol{s}(\theta) = \Gamma_\theta \boldsymbol{u}$ where $\boldsymbol{u}$ is a known signal and Γ_θ denotes a cyclic shift of the samples by $\theta \in \mathcal{X}_1 = \{0, 1, \dots, n-1\}$:

$$(\Gamma_\theta \boldsymbol{u})_i \triangleq u_{i-\theta \bmod n}, \quad i \in \{1, 2, \dots, n\}.$$

The energy of the signal is given by $\|\boldsymbol{s}(\theta)\|^2 = \|\boldsymbol{u}\|^2$ for all $\theta \in \mathcal{X}_1$. Since $|\mathcal{X}_1| = n$, the performance of the GLRT is obtained from (5.91) as

$$\mathrm{P}_\mathrm{F}(\delta_\mathrm{GLRT}) \leq n\, Q\left(\frac{\|\boldsymbol{u}\|}{2\sigma}\right), \quad \mathrm{P}_\mathrm{M}^\theta(\delta_\mathrm{GLRT}) \leq Q\left(\frac{\|\boldsymbol{u}\|}{2\sigma}\right) \quad \forall \theta \in \mathcal{X}_1.$$

The performance of the Bayesian detector, assuming a uniform prior on Θ, is obtained from (5.95) as

$$\mathrm{P}_\mathrm{F}(\delta_\mathrm{B}) \leq Q\left(\frac{\|\boldsymbol{u}\|}{2\sigma}\right), \quad \mathrm{P}_\mathrm{M}^\theta(\delta_\mathrm{B}) \leq \sqrt{n}\, Q\left(\frac{\|\boldsymbol{u}\|}{2\sigma}\right) \quad \forall \theta \in \mathcal{X}_1. \qquad \square$$

Example 5.6 *Impulse Detection.* In this example we show that the union bound used to derive (5.86) can be tight. Consider the template matching example (Example 5.5), with $u_i = \sqrt{n}\, \mathbb{1}\{i = 1\}$ for $1 \leq i \leq n$; hence $\boldsymbol{s}(\theta)$ is an impulse with energy n at location $1 + \theta$. Let $\tau = \frac{n}{2}$. In this case (5.86) may be written as

$$\begin{aligned}
\mathrm{P}_\mathrm{F}(\delta_\mathrm{GLRT}) &= \mathrm{P}\left[\bigcup_{\theta \in \mathcal{X}_1} \{\boldsymbol{s}(\theta)^\top \boldsymbol{Z} \geq \tau\}\right] \\
&= \mathrm{P}\left[\bigcup_{\theta \in \mathcal{X}_1} \left\{\sqrt{n}\, Z_{1+\theta} \geq \frac{n}{2}\right\}\right] \\
&\overset{(a)}{=} 1 - \prod_{\theta \in \mathcal{X}_1} \mathrm{P}\left[\sqrt{n}\, Z_{1+\theta} < \frac{n}{2}\right] \\
&\overset{(b)}{=} 1 - \left[1 - Q\left(\frac{\sqrt{n}}{2\sigma}\right)\right]^n \\
&\overset{(c)}{=} nQ\left(\frac{\sqrt{n}}{2\sigma}\right)[1 + o(1)],
\end{aligned} \tag{5.96}$$

where (a) holds because the events $\{\sqrt{n}\, Z_{1+\theta} \geq \frac{n}{2}\}$, $\theta \in \mathcal{X}_1$ are mutually independent, (b) because each of these events has probability $Q(\frac{\sqrt{n}}{2\sigma})$, and (c) by binomial expansion. $\square$

5.10 Deflection-Based Detection of Non-Gaussian Signal in Gaussian Noise

Thus far we have considered the detection of Gaussian signals in Gaussian noise, and deterministic signals with unknown parameters in Gaussian noise. We end this chapter with a discussion of an approach to the difficult problem of detecting random signals in noise, i.e., the model in (5.2). We study a special case of this problem where the hypotheses are given by

$$\begin{cases} H_0 : \boldsymbol{Y} = \boldsymbol{Z} \\ H_1 : \boldsymbol{Y} = \boldsymbol{S} + \boldsymbol{Z}, \end{cases}$$

where $\boldsymbol{Z} \sim \mathcal{N}(\mathbf{0}, \mathsf{C}_Z)$ and $\boldsymbol{S}$ is a zero-mean, possibly non-Gaussian signal with covariance matrix C_S. The more general case where there is a signal under both hypotheses with possibly nonzero mean is studied in [7].

We do not make any assumptions about the distribution of $\boldsymbol{S}$ other than it being zero mean with covariance matrix C_S. As such, the optimum (likelihood ratio) test cannot be evaluated. We note that even if we had a characterization of the distribution of $\boldsymbol{S}$, the LRT may be difficult to implement and its performance may be difficult to analyze. To make progress with this problem, we soften the optimality criterion to a signal-to-noise-based criterion called *deflection*, introduced by Baker [6].

To define the deflection criterion, we restrict our attention to threshold tests of the form

$$\tilde{\delta}(\boldsymbol{y}) = \begin{cases} 1 & > \\ 1 \text{ w.p. } \gamma \;:\; T(\boldsymbol{y}) & = \tau, \\ 0 & < \end{cases}$$

where $T(\boldsymbol{y})$ is a one-dimensional (scalar) statistic. If we knew the distribution of $\boldsymbol{S}$, we could choose $T(\boldsymbol{y})$ to be the log-likelihood ratio. The deflection of the statistic $T(\boldsymbol{y})$ is defined as follows.

DEFINITION 5.2 For a scalar test statistic $T(\boldsymbol{y})$, the deflection $D(T)$ is defined as

$$D(T) \triangleq \frac{(\mathbb{E}_1[T(\boldsymbol{Y})] - \mathbb{E}_0[T(\boldsymbol{Y})])^2}{\mathrm{Var}_0(T(\boldsymbol{Y}))}.$$

If $\boldsymbol{S}$ is zero mean and Gaussian, we know from Section 5.7.1 that the log-likelihood ratio is purely quadratic in $\boldsymbol{y}$, i.e., $\ln L(\boldsymbol{y}) = \boldsymbol{y}^\top \mathsf{Q} \boldsymbol{y} + \kappa$, with Q being a symmetric positive definite matrix. This motivates using a purely quadratic statistic

$$T_{\mathsf{Q}}(\boldsymbol{y}) = \boldsymbol{y}^\top \mathsf{Q} \boldsymbol{y} \tag{5.97}$$

even when $\boldsymbol{S}$ is not Gaussian. This leads to the following optimization problem for the deflection based detector:

$$\mathsf{Q}^* = \arg\max_{\mathsf{Q} \in \mathcal{A}} D(T_{\mathsf{Q}}), \tag{5.98}$$

where $\mathcal{A}$ is the set of symmetric $n \times n$ positive definite matrices. The corresponding deflection-based threshold test is given by

$$\delta_{\mathrm{D}}(\boldsymbol{y}) = \begin{cases} 1 \;:\; \boldsymbol{y}^\top \mathsf{Q}^* \boldsymbol{y} & > \tau. \\ 0 & < \end{cases} \tag{5.99}$$

Randomization is not required since the quadratic test statistic does not have any point masses under either hypothesis. Note that the $D(T_{\mathsf{Q}})$ can be computed using only the mean and covariance of $\boldsymbol{S}$, which is another reason for restricting T to the class of quadratic statistics.

To compute $D(T_{\mathsf{Q}})$, we need to find expressions for $\mathbb{E}_1[T_{\mathsf{Q}}(\boldsymbol{Y})]$, $\mathbb{E}_0[T_{\mathsf{Q}}(\boldsymbol{Y})]$, and $\mathrm{Var}_0(T_{\mathsf{Q}}(\boldsymbol{Y}))$:

$$\begin{aligned}\mathbb{E}_j\left[T_{\mathsf{Q}}(\boldsymbol{Y})\right] &= \mathbb{E}_j\left[\boldsymbol{Y}^\top \mathsf{Q}\boldsymbol{Y}\right]\\ &= \mathbb{E}_j\left[\mathrm{Tr}\left(\boldsymbol{Y}^\top \mathsf{Q}\boldsymbol{Y}\right)\right]\\ &= \mathbb{E}_j\left[\mathrm{Tr}\left(\mathsf{Q}\boldsymbol{Y}\boldsymbol{Y}^\top\right)\right]\\ &= \mathrm{Tr}\left(\mathsf{Q}\mathbb{E}_j\left[\boldsymbol{Y}\boldsymbol{Y}^\top\right]\right)\\ &= \begin{cases}\mathrm{Tr}\left(\mathsf{Q}\mathsf{C}_Z\right) & \text{if } j=0\\ \mathrm{Tr}\left(\mathsf{Q}\left(\mathsf{C}_Z+\mathsf{C}_S\right)\right) & \text{if } j=1,\end{cases}\end{aligned} \tag{5.100}$$

where Tr denotes the trace of a square matrix, i.e., the sum of its diagonal components, and we have exploited the fact that if A and B are such that AB is a square matrix, then $\mathrm{Tr}(\mathsf{AB}) = \mathrm{Tr}(\mathsf{BA})$ (see Appendix A).

From (5.100) it follows that

$$\mathbb{E}_1[T_{\mathsf{Q}}(\boldsymbol{Y})] - \mathbb{E}_0[T_{\mathsf{Q}}(\boldsymbol{Y})] = \mathrm{Tr}\left(\mathsf{Q}\mathsf{C}_S\right). \tag{5.101}$$

The term $\mathrm{Var}_0(T_{\mathsf{Q}}(\boldsymbol{Y}))$ involves up to fourth-order moments of the $\boldsymbol{Y}$. We compute it using a general formula for the moments of a zero-mean Gaussian vector (see, e.g., [7]):

$$\begin{aligned}\mathbb{E}_0\left[\left(T_{\mathsf{Q}}(\boldsymbol{Y})\right)^2\right] &= \mathbb{E}_0\left[\boldsymbol{Y}^\top \mathsf{Q}\boldsymbol{Y}\boldsymbol{Y}^\top \mathsf{Q}\boldsymbol{Y}\right]\\ &= \mathbb{E}_0\left[\boldsymbol{Y}^\top \mathsf{Q}\boldsymbol{Y}\right]\mathbb{E}_0\left[\boldsymbol{Y}^\top \mathsf{Q}\boldsymbol{Y}\right]\\ &\quad + \mathrm{Tr}\left(\mathbb{E}_0\left[\mathsf{Q}\boldsymbol{Y}\boldsymbol{Y}^\top\right]\mathbb{E}_0\left[\mathsf{Q}\boldsymbol{Y}\boldsymbol{Y}^\top\right]\right)\\ &\quad + \mathrm{Tr}\left(\mathbb{E}_0\left[\boldsymbol{Y}\boldsymbol{Y}^\top \mathsf{Q}\right]\mathbb{E}_0\left[\boldsymbol{Y}\boldsymbol{Y}^\top \mathsf{Q}\right]\right)\\ &= \left[\mathrm{Tr}\left(\mathsf{Q}\mathsf{C}_Z\right)\right]^2 + 2\mathrm{Tr}\left(\mathsf{Q}\mathsf{C}_Z\mathsf{Q}\mathsf{C}_Z\right).\end{aligned}$$

This implies that

$$\mathrm{Var}_0(T_{\mathsf{Q}}(\boldsymbol{Y})) = \mathbb{E}_0\left[\left(T_{\mathsf{Q}}(\boldsymbol{Y})\right)^2\right] - \left(\mathbb{E}_0[T_{\mathsf{Q}}(\boldsymbol{Y})]\right)^2 = 2\mathrm{Tr}\left(\mathsf{Q}\mathsf{C}_Z\mathsf{Q}\mathsf{C}_Z\right). \tag{5.102}$$

From (5.101) and (5.102) we get that

$$D(T_{\mathsf{Q}}) = \frac{\left[\mathrm{Tr}\left(\mathsf{Q}\mathsf{C}_S\right)\right]^2}{2\mathrm{Tr}\left(\mathsf{Q}\mathsf{C}_Z\mathsf{Q}\mathsf{C}_Z\right)}. \tag{5.103}$$

To find the value of Q that maximizes the right hand side of (5.103), we apply the Cauchy–Schwarz inequality for the inner product space of $n \times n$ matrices, with inner product of $n \times n$ matrices A and B defined to be $\mathrm{Tr}\left(\mathsf{A}\mathsf{B}^\top\right)$. In particular,

$$\left[\mathrm{Tr}\left(\mathsf{AB}^\top\right)\right]^2 \le \mathrm{Tr}\left(\mathsf{AA}^\top\right)\mathrm{Tr}\left(\mathsf{BB}^\top\right) \tag{5.104}$$

with equality if and only if $\mathsf{A} = k\mathsf{B}$ for some $k \in \mathbb{R}$.

Applying (5.104) with $\mathsf{A} = \mathsf{C}_Z^{-\frac{1}{2}}\mathsf{C}_S\mathsf{C}_Z^{-\frac{1}{2}}$ and $\mathsf{B} = \mathsf{C}_Z^{\frac{1}{2}}\mathsf{Q}\mathsf{C}_Z^{\frac{1}{2}}$ yields

$$\left[\mathrm{Tr}\left(\mathsf{QC}_S\right)\right]^2 \le \mathrm{Tr}\left(\mathsf{QC}_Z\mathsf{QC}_Z\right)\mathrm{Tr}\left(\mathsf{C}_S\mathsf{C}_Z^{-1}\mathsf{C}_S\mathsf{C}_Z^{-1}\right),$$

from which we conclude that

$$D(T_\mathsf{Q}) \le \frac{\mathrm{Tr}\left(\mathsf{C}_S\mathsf{C}_Z^{-1}\mathsf{C}_S\mathsf{C}_Z^{-1}\right)}{2}$$

with equality if and only if

$$\mathsf{C}_Z^{-\frac{1}{2}}\mathsf{C}_S\mathsf{C}_Z^{-\frac{1}{2}} = k\mathsf{C}_Z^{\frac{1}{2}}\mathsf{Q}\mathsf{C}_Z^{\frac{1}{2}}.$$

Setting $k = 1$ without loss of generality, we obtain a solution to the optimization problem (5.98):

$$\mathsf{Q}^* = \mathsf{C}_Z^{-1}\mathsf{C}_S\mathsf{C}_Z^{-1}, \tag{5.105}$$

with the corresponding deflection-based detector given by

$$\delta_\mathrm{D}(\boldsymbol{y}) = \begin{cases} 1 : \boldsymbol{y}^\top\mathsf{C}_Z^{-1}\mathsf{C}_S\mathsf{C}_Z^{-1}\boldsymbol{y} & > \tau. \\ 0 & < \end{cases} \tag{5.106}$$

An application of this test in the detection of a sinusoid with drifting phase in white Gaussian noise is given in [8]. See also Exercise 5.15.

REMARK 5.1 The deflection-based test δ_D does not match the optimum (likelihood ratio) test even when the signal $\boldsymbol{S}$ is Gaussian. In particular, the log-likelihood ratio statistic is given by (see (5.51))

$$T_\mathrm{opt}(\boldsymbol{y}) = \boldsymbol{y}^\top\left(\mathsf{C}_Z^{-1} - (\mathsf{C}_S + \mathsf{C}_Z)^{-1}\right)\boldsymbol{y},$$

which is not equivalent to the deflection statistic

$$T_{\mathsf{Q}^*}(\boldsymbol{y}) = \boldsymbol{y}^\top\mathsf{C}_Z^{-1}\mathsf{C}_S\mathsf{C}_Z^{-1}\boldsymbol{y}$$

in general. Even in the special case where the noise is white, i.e., $\mathsf{C}_Z = \sigma^2\mathsf{I}_n$, $T_\mathrm{opt}(\boldsymbol{y})$ is not equivalent to $T_{\mathsf{Q}^*}(\boldsymbol{y})$, except when $\boldsymbol{S}$ is white as well.

Interestingly, when we consider the detection of a zero-mean Gaussian signal with unknown amplitude in white Gaussian noise, the deflection-based detector coincides with the locally most powerful (LMP) detector in the weak signal limit as the signal amplitude goes to zero. See Exercise 5.16.

Exercises

5.1 *Detection of Known Signal in i.i.d. Generalized Gaussian Noise*. Derive the structure of an optimal detector of a known signal corrupted by i.i.d. noise with pdf $p_Z(z) = C_{a,b} \exp\{-a|z|^b\}$ with parameters $a > 0$ and $b \in (0, 2]$. Show that each observation is passed through a nonlinear function, and sketch this function for $b = 1/2$ and for $b = 3/2$.

5.2 *Poisson Observations*. Denote by P_λ the Poisson distribution with parameter $\lambda > 0$. Given two length-n sequences $\boldsymbol{s}$ and $\boldsymbol{\lambda}$ such that $0 < s_i < \lambda_i$ for all $1 \le i \le n$, consider the binary hypothesis test

$$\begin{cases} H_0 & : Y_i \overset{\text{indep.}}{\sim} P_{\lambda_i + s_i} \\ H_1 & : Y_i \overset{\text{indep.}}{\sim} P_{\lambda_i - s_i}, \quad 1 \le i \le n, \end{cases}$$

and assume equal priors. Derive the LRT and show that it reduces to a correlation test.

5.3 *Additive and Multiplicative Noise*. Consider a known signal $\boldsymbol{s}$ and parameter $\theta_0 > 0$ and two i.i.d. $\mathcal{N}(0, 1)$ noise sequences $\boldsymbol{Z}$ and $\boldsymbol{W}$. We test

$$\begin{cases} H_0 & : \theta = 0 \\ H_1 & : \theta = \theta_0, \end{cases}$$

where the observations are given by $Y_i = \sqrt{\theta} s_i W_i + Z_i$ for $1 \le i \le n$.

(a) Derive the NP detector at significance level a.
(b) Let $s_i = a$ for all $1 \le i \le n$. Give an approximation to the threshold of the NP test of part (a) that is valid for large n.
(c) Now assume θ is unknown and we need to test $H_0 : \theta = 0$ versus $H_1 : \theta > 0$. Give a necessary and sufficient condition on $\boldsymbol{s}$ such that a UMP test exists.
(d) Assume that θ is unknown and we need to test $H_0 : \theta = 0$ versus $H_1 : \theta \neq 0$. Derive the GLRT and simplify it in the case $s_i = a$ for all $1 \le i \le n$.

5.4 *CFAR Detector*. Consider the following problem of detecting an unknown signal in exponential noise:

$$\begin{cases} H_0 & : Y_k = Z_k, \quad k = 1, \dots, n \\ H_1 & : Y_k = s_k + Z_k, \quad k = 1, \dots, n, \end{cases}$$

where the signal is unknown except that we know that $s_k > 0$ for all k, and the noise variables $Z_1, \dots, Z_n$ are i.i.d. $\text{Exp}(1/\mu)$ random variables, i.e.,

$$p_{Z_k}(z) = \mu e^{-\mu z} \, \mathbb{1}\{z \ge 0\}, \quad k = 1, \dots, n,$$

with $\mu > 0$. The false-alarm probability of a LRT for this problem is a function of μ. Now suppose μ is unknown, and we wish to design a detector whose false-alarm rate is independent of μ and is still able to detect the signal reasonably well. Such a detector is called a CFAR (constant false-alarm rate) detector.

(a) Let $n \geq 3$. Show that the following detector is CFAR:

$$\delta_\eta(\boldsymbol{y}) = \begin{cases} 1 & : \text{if } y_2 + y_3 + \cdots + y_n \geq \eta\, y_1 \\ 0 & \text{otherwise} \end{cases}, \quad \eta > 0.$$

(b) Suppose n is a large integer. Use the Law of Large Numbers to find a good approximation for η as a function of n and α, for an α-level CFAR test.

(c) Now find a closed-form expression for η as a function of α, for an α-level CFAR test. Compare your answer to the approximation you found in part (b).

5.5 *Signal Detection in Correlated Gaussian Noise*. Derive the minimum error probability for detecting the length-3 signal $\boldsymbol{s} = [1\ 2\ 3]^\top$ in Gaussian noise with mean zero and covariance matrix $\mathsf{K} = \begin{bmatrix} 2 & 0 & 0 \\ 0 & 2 & 1 \\ 0 & 1 & 2 \end{bmatrix}$.

5.6 *Signal Detection in Correlated Gaussian Noise*. Consider the detection problem with

$$\begin{cases} H_0: \ \boldsymbol{Y} = \begin{bmatrix} -a \\ 0 \end{bmatrix} + \boldsymbol{Z} \\ H_1: \ \boldsymbol{Y} = \begin{bmatrix} a \\ 0 \end{bmatrix} + \boldsymbol{Z}, \end{cases}$$

where $\boldsymbol{Z} \sim \mathcal{N}(\mathbf{0}, \mathsf{C}_Z)$ with

$$\mathsf{C}_Z = \begin{bmatrix} 1 & \rho \\ \rho & 1 + \rho^2 \end{bmatrix}.$$

Assume that $a > 0$ and $\rho \in (0, 1)$.

(a) For equal priors show that the minimum-probability-of-error detector is given by

$$\delta_B(\boldsymbol{y}) = \begin{cases} 1 & : \ y_1 - b y_2 \geq \tau, \\ 0 & : \ \qquad\qquad < \end{cases}$$

where $b = \rho/(1 + \rho^2)$ and $\tau = 0$.

(b) Determine the minimum probability of error.

(c) Consider the test of part (a) in the limit as $\rho \to 0$. Explain why the dependence on y_2 goes away in this limit.

(d) Now suppose the observations $\boldsymbol{Y} \sim \mathcal{N}([a\ 0]^\top, \mathsf{C}_Z)$, with $\rho = 1$ but a being an unknown parameter, and we wish to test between the hypotheses:

$$\begin{cases} H_0: \ 0 < a \leq 1 \\ H_1: \ a > 1. \end{cases}$$

Show that a UMP test exists for this problem, and find the UMP test of level $\alpha \in (0, 1)$.

5.7 *Signal Detection in Correlated Gaussian Noise.* Consider the hypothesis testing problem with n-dimensional observations:

$$\begin{cases} H_0: \boldsymbol{Y} = \boldsymbol{Z} \\ H_1: \boldsymbol{Y} = \boldsymbol{s} + \boldsymbol{Z}, \end{cases}$$

where the components of $\boldsymbol{Z}$ are zero-mean correlated Gaussian random variables with

$$\mathbb{E}[Z_k Z_\ell] = \sigma^2 \rho^{|k-\ell|}, \quad \text{for all } 1 \le k, \ell \le n$$

where $|\rho| < 1$.

(a) Show that the NP test for this problem has the form

$$\delta_\tau(\boldsymbol{y}) = \begin{cases} 1 & : \sum_{k=1}^n b_k x_k \ge \tau, \\ 0 & : \qquad\qquad < \end{cases}$$

where $b_1 = s_1/\sigma$, $x_1 = y_1/\sigma$, and

$$b_k = \frac{s_k - \rho s_{k-1}}{\sigma\sqrt{1-\rho^2}}, \quad x_k = \frac{y_k - \rho y_{k-1}}{\sigma\sqrt{1-\rho^2}}, \quad k = 2, \ldots, n.$$

Hint: Note that $\mathsf{C}_{\boldsymbol{Z}}^{-1} = \mathsf{A}/(\sigma^2(1-\rho^2))$, where A is a tridiagonal matrix with main diagonal $(1 \;\; 1+\rho^2 \;\; 1+\rho^2 \;\; \cdots \;\; 1+\rho^2 \;\; 1)$ and superdiagonal and subdiagonal entries all being $-\rho$.

(b) Find an α-level NP test.

(c) Find the ROC for the detector of Part (b).

5.8 *Signal with Unknown Scale Factor.* Consider the Bayesian binary composite hypothesis test

$$\begin{cases} H_0: \; Y_i = Z_i \\ H_1: \; Y_i = \theta s_i + Z_i, \quad i = 1, 2, \ldots, n, \end{cases}$$

where the hypotheses are equally likely, $(Y_1, Y_2, \ldots, Y_n)$ are the observations, $(s_1, s_2, \ldots, s_n)$ is a known signal, $\theta > 0$ is an unknown scale factor, and $(Z_1, Z_2, \ldots, Z_n)$ are i.i.d. random variables with distribution P and support $[-1, 1]$. The locally optimal detector takes the form

$$\sum_{i=1}^n \sqrt{i|Y_i|} \underset{H_0}{\overset{H_1}{\gtrless}} n.$$

Identify the signal (up to a scale factor) and the distribution P.

5.9 *Detection of* m *Known Signals in i.i.d. Gaussian Noise.* Consider the detection of m known orthogonal signals $\boldsymbol{s}_1, \ldots, \boldsymbol{s}_m$ with equal energies. Show that the minimum probability of error for detection in i.i.d. $\mathcal{N}(0, \sigma^2)$ noise is given by

$$\mathrm{P_e} = 1 - (2\pi)^{-1/2} \int_{\mathbb{R}} [Q(x)]^{m-1} e^{-(x-d)^2/2} \, dx,$$

where $d = \|s_1\|/\sigma$.

5.10 *Locally Optimal Test under Generalized Gaussian Noise.* Derive the locally optimum test for detection of a weak signal in i.i.d. noise with power-exponential pdf:

$$p_Z(z) = C_{a,b} \exp\{-a|z|^b\}.$$

5.11 *Locally Optimal Test under Laplacian Noise.* Implement the test of Example 5.3 with $n = 100$ and set the threshold τ/a to achieve a probability of false alarm of 0.05 using the Gaussian approximation $\tau/a = \sqrt{100}Q^{-1}(0.05)$. Perform 10^4 Monte-Carlo simulations to estimate the probabilities of false alarm and of miss.

5.12 *GLRT for Linear Gaussian Model.* Consider the signal with samples $s_i(\theta) = a + bi$, $1 \le i \le n$, to be detected in i.i.d. $\mathcal{N}(0, \sigma^2)$ noise. The parameters a and b are unknown real numbers. Derive the GLRT and its probability of false alarm assuming equal priors on H_0 and H_1.

5.13 *Detection of Sinusoid with Unknown Amplitude and Phase.* Consider a variation of Example 5.4, where the signal samples are given by $s_i = a\cos(\omega i + \phi)$, where the amplitude a and the phase ϕ are unknown. Show that this can be formulated as a linear Gaussian model, and derive the GLRT.

5.14 *Lower Bounds on* $\mathrm{P_F}$ *and* $\mathrm{P_M}$ *for Composite Bayesian Signal Detection.* Prove the inequalities (5.92) and (5.93).

5.15 *Baseband Detection of a Sinusoid with Drifting Phase.* Consider the detection problem:

$$\begin{cases} H_0 : \boldsymbol{Y} = \boldsymbol{Z} \\ H_1 : \boldsymbol{Y} = \boldsymbol{S} + \boldsymbol{Z}, \end{cases}$$

where:

- the signal $S_k = \sqrt{2}\cos(\Theta_k + \phi)$ is a sinusoid with randomly varying phase;
- the noise is white Gaussian, i.e., $\boldsymbol{Z} \sim \mathcal{N}(\mathbf{0}, \mathsf{I}_n)$;
- $\Theta_k = \sqrt{-2\ln\beta}\sum_{i=1}^{k} W_i$, where $0 < \beta < 1$ and $\{W_i\}$ are i.i.d. $\mathcal{N}(0, 1)$ independent of $\boldsymbol{Z}$, i.e., the phase Θ_k drifts as a random walk;
- ϕ is uniform on $[0, 2\pi]$, and is independent of $\boldsymbol{Z}$ and $\boldsymbol{W}$.

(a) Prove that $\mathbb{E}[\boldsymbol{S}] = 0$, and that

$$\mathbb{E}[S_k S_\ell] = \beta^{|k-\ell|}, \quad 1 \le k, \ell \le n.$$

Hint: You may want to use the fact that the characteristic function of a $\mathcal{N}(0, \sigma^2)$ random variable X is given by $\mathbb{E}[e^{\iota u X}] = e^{-\sigma^2 u^2/2}$.

Note that the signal is not Gaussian and so it is difficult to compute the LRT.

(b) Give an expression for the quadratic test statistic that maximizes the deflection and show that this maximum deflection is given by

$$D_{\max} \triangleq f(\beta, n) = \frac{n}{2} + \frac{n\beta^2}{1-\beta^2} - \frac{\beta^2(1-\beta^{2n})}{(1-\beta^2)^2}.$$

Hint: Use the fact that for any function g defined for nonnegative integers

$$\sum_{k=1}^{n}\sum_{\ell=1}^{n} g(|k-\ell|) = ng(0) + 2\sum_{i=1}^{n-1}(n-i)g(i).$$

(c) Now consider the standard noncoherent detector (SND) that uses the test statistic

$$T_{\mathrm{SND}}(\boldsymbol{y}) = \left(\sum_{k=1}^{n} y_k\right)^2.$$

Show that the deflection for this statistic is given by

$$D(T_{\mathrm{SND}}) = \frac{2\left[f(\sqrt{\beta}, n)\right]^2}{n^2},$$

where $f(\cdot)$ is defined in part (b).

(d) Compare $D_{\max}$ and $D(T_{\mathrm{SND}})$ for $n = 100$ and $\beta = 0.9, 0.5$, and 0.1. Explain why the performance of the SND degrades rapidly as β goes to zero.

5.16 *LMP Detection of a Gaussian Signal in White Gaussian Noise.* Consider the detection problem:

$$\begin{cases} H_0 : \boldsymbol{Y} = \boldsymbol{Z} \\ H_1 : \boldsymbol{Y} = \sqrt{\theta}\boldsymbol{S} + \boldsymbol{Z}, \end{cases}$$

where $\boldsymbol{Z} \sim \mathcal{N}(\mathbf{0}, \mathsf{I}_n)$, $\boldsymbol{S} \sim \mathcal{N}(\mathbf{0}, \mathsf{C}_S)$, and $\theta > 0$ is an unknown parameter.

(a) Show that the LMP test as $\theta \downarrow 0$ uses the quadratic test statistic:

$$T_{\mathrm{LMP}}(\boldsymbol{y}) = \boldsymbol{y}^\top \mathsf{C}_S \boldsymbol{y}.$$

Hint: Spectral factorization of C_S may be useful here.

(b) Show that the quadratic statistic that maximizes the deflection is independent of θ and equals $T_{\mathrm{LMP}}(\boldsymbol{y})$.

References

[1] W. Feller, *An Introduction to Probability Theory and Its Applications*, Vol. II, Wiley, 1971.

[2] P. C. Mahalanobis, "On the Generalised Distance in Statistics," *Proc. Nat. Inst. Sci. India*, Vol. 2, pp. 49–55, 1936.

[3] H. E. Daniels, "Tail Probability Approximations," *Int. Stat. Review*, Vol. 44, pp. 37–48, 1954.

[4] R. Lugannani and S. Rice, "Saddlepoint Approximation for the Distribution of a Sum of Independent Random Variables," *Adv. Applied Prob.*, Vol. 12, pp. 475–490, 1980.

[5] A. DasGupta, *Asymptotic Theory of Statistics and Probability*, Springer, 2008.

[6] C. R. Baker, "Optimum Quadratic Detection of a Random Vector in Gaussian Noise," *IEEE Trans. Commun. Tech.* Vol. COM-14, No. 6, pp. 802–805, 1966.

[7] B. Picinbono and P. Duvaut, "Optimum Linear Quadratic Systems for Detection and Estimation," *IEEE Trans. Inform. Theory*, Vol. 34, No. 2, pp. 304–311, 1988.

[8] V. V. Veeravalli and H. V. Poor, "Quadratic Detection of Signals with Drifting Phase," *J. Acoust. Soc. Am.* Vol. 89, No. 2, pp. 811–819, 1991.

6 Convex Statistical Distances

As we saw in previous chapters, most notably Chapter 5, deriving exact expressions for the performance of detection algorithms is in general difficult, and we often have to rely on bounds on the probabilities of error. To derive such bounds, we first introduce convex statistical distances between distributions termed f-divergences and Ali–Silvey distances. They include Kullback–Leibler divergence, Chernoff divergences, and total variation distance as special cases. We identify the basic properties and relations between those distances, and derive the fundamental data processing inequality (DPI). Ali–Silvey distances have also been used as criteria for signal selection or feature selection in pattern recognition systems [1, 2, 3, 4, 5, 6, 7, 8], and as surrogate loss functions in machine learning [9, 10].

The most fundamental inequality used in this chapter is Jensen's inequality. For any convex function $f : \mathbb{R} \to \mathbb{R}$ and probability distribution P on $\mathbb{R}$, Jensen's inequality states that $\mathbb{E}[f(X)] \geq f(\mathbb{E}[X])$. We often use the shorthand $\int_{\mathcal{Y}} f \, d\mu = \int_{\mathcal{Y}} f(y) \, d\mu(y)$ to denote the integral of a function $f : \mathcal{Y} \to \mathbb{R}$.

6.1 Kullback–Leibler Divergence

The notion of Kullback–Leibler divergence [11, 12] plays a central role in hypothesis testing and related fields, including probability theory, statistical inference, and information theory.

DEFINITION 6.1 Let P_0 and P_1 be two probability distributions on the same space $\mathcal{Y}$, and assume P_0 is dominated by P_1, i.e., $\mathrm{supp}\{\mathrm{P}_0\} \subseteq \mathrm{supp}\{\mathrm{P}_1\}$. The functional

$$D(p_0 \| p_1) \triangleq \int_{\mathcal{Y}} p_0(y) \ln \frac{p_0(y)}{p_1(y)} \, d\mu(y) \tag{6.1}$$

is known as the *Kullback–Leibler (KL) divergence*, or *relative entropy*, between the distributions P_0 and P_1. If P_0 is not dominated by P_1, then $D(p_0 \| p_1) = \infty$.

In (6.1), we use the notational convention $0 \ln 0 = 0$.

PROPOSITION 6.1 *The KL divergence $D(p_0 \| p_1)$ satisfies the following properties:*

(i) Nonnegativity. $D(p_0 \| p_1) \geq 0$ *with equality if and only if* $p_0(Y) = p_1(Y)$ *a.s.*
(ii) No triangle inequality. *There exist distributions* p_0, p_1, p_2 *such that*

$$D(p_0 \| p_2) > D(p_0 \| p_1) + D(p_1 \| p_2).$$

(iii) Asymmetry. $D(p_0 \| p_1) \neq D(p_1 \| p_0)$ *in general.*

(iv) Convexity. *The functional* $D(p_0 \| p_1)$ *is jointly convex in* (p_0, p_1).

(v) Additivity for product distributions. *If both* p_0 *and* p_1 *are product distributions, i.e., the observations* $\mathbf{y} = \{y_i\}_{i=1}^n$ *and* $p_j(\mathbf{y}) = \prod_{i=1}^n p_{ji}(y_i)$ *for* $j = 0, 1$, *then* $D(p_0 \| p_1) = \sum_{i=1}^n D(p_{0i} \| p_{1i})$.

(vi) Data-processing inequality. *Consider a stochastic mapping from* Y *to* Z, *represented by a conditional pdf* $w(z|y)$. *Define the marginals*

$$q_0(z) = \int w(z|y)\, p_0(y)\, d\mu(y)$$

and

$$q_1(z) = \int w(z|y)\, p_1(y)\, d\mu(y).$$

Then $D(q_0 \| q_1) \le D(p_0 \| p_1)$, *with equality if and only if the mapping from* Y *to* Z *is invertible.*

In view of properties (ii) and (iii) in Proposition 6.1, the KL divergence is not a distance in the usual mathematical sense. The additivity property (v) holds because

$$D(p_0 \| p_1) = \mathbb{E}_0 \left[\ln \frac{p_0(\mathbf{Y})}{p_1(\mathbf{Y})} \right] = \sum_{i=1}^n \mathbb{E}_0 \left[\ln \frac{p_{0i}(Y_i)}{p_{1i}(Y_i)} \right] = \sum_{i=1}^n D(p_{0i} \| p_{1i}).$$

The nonnegativity property (i), the convexity property (iv), and the data-processing inequality (vi) will not be proved here as they are special cases of a more general result to be presented in Section 6.4.

Example 6.1 *Gaussians with Same Variance.* Let $p_0 = \mathcal{N}(0, 1)$ and $p_1 = \mathcal{N}(\theta, 1)$. We obtain

$$\begin{aligned} D(p_0 \| p_1) &= \mathbb{E}_0 \left[\ln \frac{p_0(Y)}{p_1(Y)} \right] \\ &= \mathbb{E}_0 \left[-\frac{Y^2}{2} + \frac{(Y-\theta)^2}{2} \right] \\ &= \mathbb{E}_0 \left[-\theta Y + \frac{\theta^2}{2} \right] \\ &= \frac{\theta^2}{2}. \end{aligned} \tag{6.2}$$

Example 6.2 *Gaussians with Same Mean.* Let $p_0 = \mathcal{N}(0, 1)$ and $p_1 = \mathcal{N}(0, \sigma^2)$. We obtain

$$\begin{aligned} D(p_0 \| p_1) &= \mathbb{E}_0 \left[\ln \frac{p_0(Y)}{p_1(Y)} \right] \\ &= \mathbb{E}_0 \left[-\frac{1}{2} \ln(2\pi) - \frac{Y^2}{2} + \frac{1}{2} \ln(2\pi\sigma^2) + \frac{Y^2}{2\sigma^2} \right] \end{aligned}$$

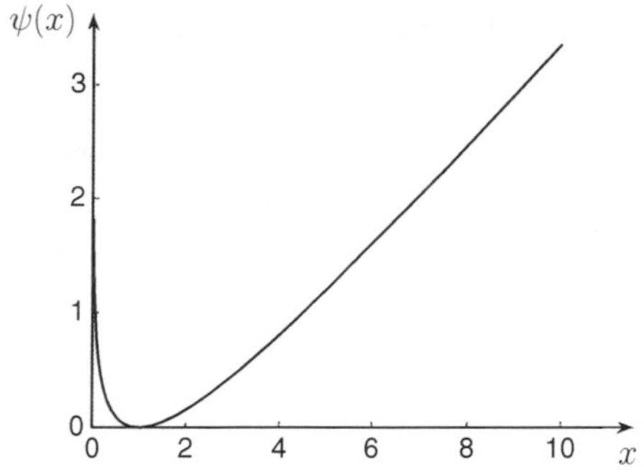

Figure 6.1 The function $\psi(x)$

$$= \frac{1}{2}\left[\ln \sigma^2 - 1 + \frac{1}{\sigma^2}\right]$$
$$= \psi\left(\frac{1}{\sigma^2}\right), \tag{6.3}$$

where we have defined the function

$$\psi(x) = \frac{1}{2}(x - 1 - \ln x), \quad x > 0, \tag{6.4}$$

which is positive, unimodal, and achieves its minimum at $x = 1$. The value of the minimum is $\psi(1) = 0$; see Figure 6.1. □

Example 6.3 *Multivariate Gaussians.* Let $p_0 = \mathcal{N}(\boldsymbol{\mu}_0, \mathsf{K}_0)$ and $p_1 = \mathcal{N}(\boldsymbol{\mu}_1, \mathsf{K}_1)$ where the covariance matrices K_0 and K_1 are $d \times d$ and have full rank. Then (see Exercise 6.2)

$$D(p_0 \| p_1) = \frac{1}{2}\left[\mathrm{Tr}(\mathsf{K}_1^{-1}\mathsf{K}_0) + (\boldsymbol{\mu}_1 - \boldsymbol{\mu}_0)^\top \mathsf{K}_1^{-1}(\boldsymbol{\mu}_1 - \boldsymbol{\mu}_0) - d - \ln \frac{|\mathsf{K}_0|}{|\mathsf{K}_1|}\right]. \tag{6.5}$$

□

Example 6.4 *Two Uniform Random Variables.* Let $\theta > 1$ and p_0 and p_1 be uniform distributions over the intervals $[0, 1]$ and $[0, \theta]$, respectively. We obtain

$$D(p_0 \| p_1) = \int_0^1 \ln \frac{p_0(y)}{p_1(y)}\, dy = \ln \theta.$$

However, $D(p_1 \| p_0) = \infty$. □

6.2 Entropy and Mutual Information

The information-theoretic notions of entropy and mutual information appear in the classification bounds of Section 7.5.3 and can be derived from KL divergence [13].

DEFINITION 6.2 The entropy of a discrete distribution $\pi = (\pi_1, \ldots, \pi_m)$ is

$$H(\pi) \triangleq \sum_{j=1}^{m} \pi_j \ln \frac{1}{\pi_j}. \tag{6.6}$$

Denote by $\mathbb{U}$ the uniform distribution over $\{1, 2, \ldots, m\}$. Then entropy is related to KL divergence as follows:

$$H(\pi) = \ln m - \sum_{j=1}^{m} \pi_j \ln \frac{\pi_j}{1/m} = \ln m - D(\pi \| \mathbb{U}), \tag{6.7}$$

from which we conclude that the function $H(\pi)$ is concave and satisfies the inequalities $0 \le H(\pi) \le \ln m$, where the lower bound is achieved by any degenerate distribution, and the upper bound is achieved by the uniform distribution.

DEFINITION 6.3 The binary entropy function

$$\begin{aligned} h_2(a) &\triangleq -a \ln a - (1-a)\ln(1-a) \\ &= H([a, 1-a]), \quad 0 \le a \le 1, \end{aligned} \tag{6.8}$$

is the entropy of a Bernoulli distribution with parameter a.

If follows from the properties of the entropy function that h_2 is concave, symmetric around $a = 1/2$, and satisfies $h_2(0) = h_2(1) = 0$ and $h_2(1/2) = \ln 2$.

Given a pair (J, Y) with joint distribution $\pi_j\, p_j(y)$ for $j \in \{1, 2, \ldots, m\}$ and $y \in \mathcal{Y}$, denote by

$$\overline{p}(y) \triangleq \sum_j \pi_j p_j(y) \text{ the marginal distribution on } Y,$$

$$(\pi \times p_J)(j, y) \triangleq \pi_j\, p_j(y) \text{ the joint distribution of } (J, Y),$$

$$(\pi \times \overline{p})(j, y) \triangleq \pi_j\, \overline{p}(y) \text{ the product of its marginals.}$$

DEFINITION 6.4 The mutual information between J and Y is the KL divergence between the joint distribution of (J, Y) and the product of its marginals:

$$\begin{aligned} I(J; Y) \triangleq D(\pi \times p_J \| \pi \times \overline{p}) &= \sum_{j=1}^{m} \pi_j \int_{\mathcal{Y}} p_j(y) \ln \frac{p_j(y)}{\overline{p}(y)}\, d\mu(y) \\ &= \sum_{j=1}^{m} \pi_j D(p_j \| \overline{p}). \end{aligned} \tag{6.9}$$

LEMMA 6.1

$$I(J; Y) \le H(\pi) \le \ln m. \tag{6.10}$$

Proof The second inequality was given following (6.7). To show the first inequality, write

$$
\begin{aligned}
I(J;Y) &\overset{(a)}{=} \sum_{j=1}^{m} \pi_j \int_{\mathcal{Y}} p_j(y) \ln \frac{\pi(j|y)}{\pi_j} \, d\mu(y) \\
&= \sum_{j=1}^{m} \pi_j \ln \frac{1}{\pi_j} + \underbrace{\mathbb{E}\left[\ln \pi(J|Y)\right]}_{\leq 0} \\
&\leq \sum_{j=1}^{m} \pi_j \ln \frac{1}{\pi_j} = H(\pi)
\end{aligned}
$$

where equality (a) follows from the definition (6.9) and the identity $\pi_j p_j(y) = \pi(j|y)\overline{p}(y)$. □

6.3 Chernoff Divergence, Chernoff Information, and Bhattacharyya Distance

In addition to KL divergence, the notion of Chernoff divergence also plays a central role in hypothesis testing [14].

DEFINITION 6.5 The *Chernoff divergence* between two probability distributions P_0 and P_1 on a common space is defined by

$$d_u(p_0, p_1) \triangleq -\ln \int_{\mathcal{Y}} p_0^{1-u}(y) p_1^u(y) \, d\mu(y) \geq 0, \quad u \in [0, 1]. \tag{6.11}$$

PROPOSITION 6.2 *The Chernoff divergence $d_u(p_0, p_1)$ satisfies the following properties:*

(i) Nonnegativity. *$d_u(p_0, p_1) \geq 0$ with equality if and only if $p_0(Y) = p_1(Y)$ a.s.*
(ii) No triangle inequality. *$d_u(p_0, p_1)$ does not satisfy the triangle inequality, even for $u = \frac{1}{2}$.*
(iii) Asymmetry. *$d_u(p_0, p_1) \neq d_u(p_1, p_0)$ in general, except for $u = \frac{1}{2}$.*
(iv) Convexity. *The functional $d_u(p_0, p_1)$ is jointly convex in (p_0, p_1) for each $u \in (0, 1)$.*
(v) Additivity for product distributions. *If both p_0 and p_1 are product distributions, i.e., the observations $\mathbf{y} = \{y_i\}_{i=1}^n$ and $p_j(\mathbf{y}) = \prod_{i=1}^n p_{ji}(y_i)$ for $j = 0, 1$, then $d_u(p_0, p_1) = \sum_{i=1}^n d_u(p_{0i}, p_{1i})$.*
(vi) Data-processing inequality. *Consider a stochastic mapping from Y to Z, represented by a conditional pdf $w(z|y)$. Define the marginals*

$$q_0(z) = \int w(z|y) p_0(y) \, d\mu(y), \quad q_1(z) = \int w(z|y) p_1(y) \, d\mu(y).$$

Then $d_u(q_0, q_1) \leq d_u(p_0, p_1)$, with equality if and only if Z is an invertible function of Y.

In view of properties (ii) and (iii) in Proposition 6.2, the Chernoff divergence is not a distance in the usual mathematical sense. Properties (i), (iv), and (vi) are a special case of a more general result to be derived in Section 6.4. Property (v) holds because

$$
\begin{aligned}
d_u(p_0, p_1) &= -\ln \mathbb{E}_0 \left(\frac{p_1(\mathbf{Y})}{p_0(\mathbf{Y})}\right)^u \\
&= -\ln \mathbb{E}_0 \prod_{i=1}^{n} \left(\frac{p_{1i}(Y_i)}{p_{0i}(Y_i)}\right)^u \\
&= -\ln \prod_{i=1}^{n} \mathbb{E}_0 \left(\frac{p_{1i}(Y_i)}{p_{0i}(Y_i)}\right)^u \\
&= \sum_{i=1}^{n} d_u(p_{0i}, p_{1i}),
\end{aligned}
$$

where the third equality holds because the expectation of a product of independent variables is equal to the product of their individual expectations.

In the special case $u = 1/2$, the Chernoff divergence is known as the *Bhattacharyya distance*

$$B(p_0, p_1) \triangleq d_{1/2}(p_0, p_1) = -\ln \int_{\mathcal{Y}} \sqrt{p_0(y)p_1(y)}\, d\mu(y). \tag{6.12}$$

Related quantities of interest are the *Chernoff u-coefficient*

$$\rho_u \triangleq \int_{\mathcal{Y}} p_0^{1-u} p_1^u \, d\mu = e^{-d_u(p_0, p_1)}, \quad u \in [0, 1], \tag{6.13}$$

and the *Bhattacharyya coefficient*

$$\rho_{1/2} = \int_{\mathcal{Y}} \sqrt{p_0 p_1}\, d\mu = e^{-B(p_0, p_1)}. \tag{6.14}$$

Also, maximizing the Chernoff divergence over the parameter u yields a symmetric distance measure called the *Chernoff information*:

$$C(p_0, p_1) \triangleq \max_{u \in [0,1]} d_u(p_0, p_1) = \max_{u \in [0,1]} -\ln \int_{\mathcal{Y}} p_0^{1-u}(y) p_1^u(y) d\mu(y). \tag{6.15}$$

The Bhattacharyya distance and Chernoff information, while both being symmetric in p_0 and p_1, are still not distances in the usual mathematical sense they do not satisfy the triangle inequality.

Scaling Chernoff divergence, one obtains the Rényi divergence [15]

$$D_u(p_0 \| p_1) \triangleq \frac{1}{1-u} d_u(p_0, p_1), \quad u \neq 1, \tag{6.16}$$

which has the property that $\lim_{u \to 1} D_u(p_0 \| p_1) = D(p_0 \| p_1)$.

There is another interesting connection between the Chernoff information and the KL divergence. If we define the "geometric mixture" of p_0 and p_1 by

$$p_u(y) \triangleq \frac{p_0^{1-u}(y) p_1^u(y)}{\int_{\mathcal{Y}} p_0^{1-u}(t) p_1^u(t) d\mu(t)}, \tag{6.17}$$

then it can be shown that (see Exercise 6.4)

$$C(p_0, p_1) = D(p_{u^*} \| p_0) = D(p_{u^*} \| p_1), \tag{6.18}$$

where u^* achieves the maximum in (6.15).

Example 6.5 *Multivariate Gaussians.* Let $p_0 = \mathcal{N}(\boldsymbol{\mu}_0, \mathsf{K}_0)$ and $p_1 = \mathcal{N}(\boldsymbol{\mu}_1, \mathsf{K}_1)$ where the covariance matrices K_0 and K_1 are $m \times m$ and have full rank. Then (see Exercise 6.2):

$$d_u(p_0, p_1) = \frac{u(1-u)}{2}(\boldsymbol{\mu}_1 - \boldsymbol{\mu}_0)^\top \mathsf{K}^{-1}(\boldsymbol{\mu}_1 - \boldsymbol{\mu}_0) + \frac{1}{2}\ln\frac{|(1-u)\mathsf{K}_0 + u\mathsf{K}_1|}{|\mathsf{K}_0|^{1-u}|\mathsf{K}_1|^u}, \tag{6.19}$$

where $\mathsf{K}^{-1} = (1-u)\mathsf{K}_0^{-1} + u\mathsf{K}_1^{-1}$. □

Example 6.6 *Uniform random variables.* Let $\theta > 1$ and p_0 and p_1 be uniform distributions over the intervals $[0, 1]$ and $[0, \theta]$, respectively. We obtain

$$d_u(p_0, p_1) = -\ln\int_0^1 p_0^{1-u}(y)p_1^u(y)\,dy = u\ln\theta.$$ □

6.4 Ali–Silvey Distances

Consider two probability distributions p_0 and p_1 on a common space $\mathcal{Y}$. In Sections 6.1 and 6.3 we have defined the KL divergence

$$D(p_0\|p_1) = \mathbb{E}_0\left[\ln\frac{p_0(Y)}{p_1(Y)}\right]$$

and the Chernoff divergence of order u

$$d_u(p_0, p_1) = -\ln\mathbb{E}_0\left[\left(\frac{p_1(Y)}{p_0(Y)}\right)^u\right], \quad u \in [0, 1].$$

Other statistical "distances" of interest include:

(i) *Total variation* (an actual distance):

$$d_V(p_0, p_1) = \frac{1}{2}\int_{\mathcal{Y}}|p_1(y) - p_0(y)|\,d\mu(y) = \frac{1}{2}\mathbb{E}_0\left|\frac{p_1(Y)}{p_0(Y)} - 1\right|. \tag{6.20}$$

It may be shown that (see Exercise 6.3):

$$d_V(p_0, p_1) = \sup_{\mathcal{Y}_1\subseteq\mathcal{Y}}[P_1(\mathcal{Y}_1) - P_0(\mathcal{Y}_1)] \tag{6.21}$$

(where $\mathcal{Y}_1$ ranges over the Borel subsets of $\mathcal{Y}$) which is actually the most general definition of d_V for arbitrary probability measures. We also write

$$d_V(\pi_0 p_0, \pi_1 p_1) = \frac{1}{2}\int_{\mathcal{Y}}|\pi_1 p_1(y) - \pi_0 p_0(y)|\,d\mu(y) = \sup_{\mathcal{Y}_1\subseteq\mathcal{Y}}[\pi_1 P_1(\mathcal{Y}_1) - \pi_0 P_0(\mathcal{Y}_1)]. \tag{6.22}$$

(ii) *Pearson's χ^2 divergence:*

$$\chi^2(p_1, p_0) = \int_{\mathcal{Y}}\frac{[p_1(y) - p_0(y)]^2}{p_0(y)}\,d\mu(y) = \mathbb{E}_0\left(\frac{p_1(Y)}{p_0(Y)} - 1\right)^2 = \mathbb{E}_0\left(\frac{p_1(Y)}{p_0(Y)}\right)^2 - 1. \tag{6.23}$$

(iii) *Squared Hellinger distance:*

$$H^2(p_0, p_1) = \int_{\mathcal{Y}} [\sqrt{p_0(y)} - \sqrt{p_1(y)}]^2 d\mu(y) = 2(1 - \rho_{1/2}).$$

(iv) *Symmetrized divergence:*

$$d_{\text{sym}}(p_0, p_1) = \frac{1}{2}[d(p_0, p_1) + d(p_1, p_0)], \tag{6.24}$$

for which $d_{\text{sym}}(p_0, p_1) = d_{\text{sym}}(p_1, p_0)$. Note that d_{sym} might not satisfy the triangle inequality and so might not be an actual distance.

(v) *Jensen–Shannon divergence:*

$$d_{\text{JS}}(p_0, p_1) = \frac{1}{2}\left[D\left(p_0 \middle\| \frac{p_0 + p_1}{2}\right) + D\left(p_1 \middle\| \frac{p_0 + p_1}{2}\right)\right], \tag{6.25}$$

which is the square of a distance [16]. It has been used in random graph theory [17] and has recently become popular in machine learning [18].

Denote by $\mathrm{P_e}(\delta)$ the Bayesian error probability for a binary hypothesis test between p_0 and p_1 using decision rule $\delta : \mathcal{Y} \to \{0, 1\}$. There is a remarkable connection between the minimum error probability and total variation. For any deterministic decision rule δ with acceptance region $\mathcal{Y}_1$, we have

$$\begin{aligned}
\mathrm{P_e}(\delta) &= \pi_0 P_0(\mathcal{Y}_1) + \pi_1 P_1(\mathcal{Y}_0) \\
&= \pi_1 - [\pi_1 P_1(\mathcal{Y}_1) - \pi_0 P_0(\mathcal{Y}_1)] \\
&\geq \pi_1 - d_V(\pi_0 p_0, \pi_1 p_1),
\end{aligned}$$

where the last line follows from the definition (6.22). The lower bound is achieved by the MPE rule, hence

$$\mathrm{P_e}(\delta_{\mathrm{B}}) = \pi_1 - d_V(\pi_0 p_0, \pi_1 p_1). \tag{6.26}$$

In the case of a uniform prior, we obtain $P_e(\delta_{\mathrm{B}}) = \frac{1}{2}[1 - d_V(p_0, p_1)]$.

As summarized in Table 6.1, the "distances" in Sections 6.1–6.3 are special cases of Ali–Silvey distances, which are of the general form [2]

$$d(p_0, p_1) = g\left(\mathbb{E}_0\left[f\left(\frac{p_1(Y)}{p_0(Y)}\right)\right]\right), \tag{6.27}$$

where g is an increasing function on $\mathbb{R}$ and f is a convex function of $\mathbb{R}^+$. In the special case where $g(t) = t$, the Ali–Silvey distance is known as a f-divergence, and

$$d(p_0, p_1) = \mathbb{E}_0\left[f\left(\frac{p_1(Y)}{p_0(Y)}\right)\right]. \tag{6.28}$$

The fundamental properties of Ali–Silvey distances are as follows:

(i) *Minimum value.* By Jensen's inequality we have

$$\mathbb{E}_0\left[f\left(\frac{p_1(Y)}{p_0(Y)}\right)\right] \geq f\left(\mathbb{E}_0\left[\frac{p_1(Y)}{p_0(Y)}\right]\right) = f(1) \tag{6.29}$$

with equality if and only if $p_0 = p_1$. Moreover the condition for equality is necessary if f is strictly convex. Since g is increasing, the inequality (6.29) also implies

Table 6.1 Ali–Silvey distances

$g(t)$	$f(x)$	$d(p_0, p_1)$	Name
t	$-\ln x$	$D(p_0\|p_1)$	KL divergence
t	$x\ln x$	$D(p_1\|p_0)$	KL divergence
t	$\frac{x-1}{2}\ln x$	$D_{\text{sym}}(p_0, p_1)$	J divergence
t	$\frac{x}{2}\ln x - \frac{x+1}{2}\ln\frac{x+1}{2}$	$d_{\text{JS}}(p_0, p_1)$	Jensen–Shannon divergence
$-\ln(-t)$	$-x^u$	$d_u(p_0, p_1)$	Chernoff divergence
$-\ln(-t)$	$-\sqrt{x}$	$b(p_0, p_1)$	Bhattacharyya distance
t	$2(1-\sqrt{x})$	$H^2(p_0, p_1)$	Squared Hellinger distance
t	$(x-1)^2$	$\chi^2(p_1, p_0)$	χ^2 divergence
t	$\frac{1}{2}\|x-1\|$	$d_V(p_0, p_1)$	Total variation distance

$d(p_0, p_1) \ge g(f(1))$ with inequality if $p_0 = p_1$. Usually g and f are chosen such that $g(f(1)) = 0$.

(ii) *Data grouping.* Consider a many-to-one mapping $Z = \phi(Y)$ and denote by $\phi^{-1}(z) \triangleq \{y \in \mathcal{Y} : \phi(y) = z\}$ the image of z under the inverse mapping. Define the marginal distributions $q_0(z) = \int_{\phi^{-1}(z)} p_0(y)\, d\mu(y)$ and $q_1(z) = \int_{\phi^{-1}(z)} p_1(y)\, d\mu(y)$. Then $d(q_0\|q_1) \le d(p_0\|p_1)$. This property is a special case of the data processing inequality in (iii).

(iii) *Data processing inequality.* Consider a stochastic mapping from Y to Z, represented by a conditional pdf $w(z|y)$. Define the marginals

$$q_0(z) = \int_{\mathcal{Y}} w(z|y) p_0(y)\, d\mu(y)$$

and

$$q_1(z) = \int_{\mathcal{Y}} w(z|y) p_1(y)\, d\mu(y).$$

Then[1]

$$d(q_0, q_1) \le d(p_0, p_1), \tag{6.30}$$

with equality if the likelihood ratio $\frac{p_1(Y)}{p_0(Y)}$ is a degenerate random variable (under P_0) given Z.

Proof Using the law of iterated expectations we write

$$\mathbb{E}_0\left[f\left(\frac{p_1(Y)}{p_0(Y)}\right)\right] = \mathbb{E}_0\left\{\mathbb{E}_0\left[f\left(\frac{p_1(Y)}{p_0(Y)}\right)\middle| Z\right]\right\}. \tag{6.31}$$

For each z we have, by Jensen's inequality applied to the convex function f,

$$\begin{aligned}\mathbb{E}_0\left[f\left(\frac{p_1(Y)}{p_0(Y)}\right)\middle| Z = z\right] &\ge f\left(\mathbb{E}_0\left[\frac{p_1(Y)}{p_0(Y)}\middle| Z = z\right]\right)\\ &= f\left(\int_{\mathcal{Y}} p(y|H_0, Z = z)\frac{p_1(y)}{p_0(y)} d\mu(y)\right)\end{aligned}$$

[1] For a deterministic mapping from y to z we have $w(z|y) = \mathbb{1}\{z = \phi(y)\}$.

$$= f\left(\int_{\mathcal{Y}} \frac{p_0(y)w(z|y)}{q_0(z)} \frac{p_1(y)}{p_0(y)} d\mu(y)\right)$$
$$= f\left(\frac{q_1(z)}{q_0(z)}\right). \tag{6.32}$$

Equality holds in (6.32) if the likelihood ratio $\frac{p_1(Y)}{p_0(Y)}$ is a degenerate random variable (under P_0) given $Z = z$. Combining (6.31) and (6.32) we obtain

$$\mathbb{E}_0\left[f\left(\frac{p_1(Y)}{p_0(Y)}\right)\right] \geq \mathbb{E}_0\left[f\left(\frac{q_1(Z)}{q_0(Z)}\right)\right].$$

The inequality is preserved by application of the increasing function g, hence (6.30) follows. □

(iv) Let $\theta \in [a, b]$ be a real parameter and $p_\theta(y)$ a parametric family of distributions with the monotone likelihood ratio property in y. That is, the likelihood ratio $p_{\theta_2}(y)/p_{\theta_1}(y)$ is strictly increasing in y for each $a < \theta_1 < \theta_2 < b$. Then if $a < \theta_1 < \theta_2 < \theta_3 < b$, we have

$$d(p_{\theta_1}, p_{\theta_2}) \leq d(p_{\theta_1}, p_{\theta_3}).$$

Proof See [2]. □

(v) *Joint convexity*: f-divergences are jointly convex in (p_0, p_1).

Proof Define the function $g(p, q) \triangleq qf(\frac{p}{q})$ for $0 \leq p, q \leq 1$. For any $\xi \in [0, 1]$ and $0 \leq q, q' \leq 1$, let

$$\lambda = \frac{\xi q}{\xi q + (1-\xi)q'} \quad \Rightarrow \quad 1 - \lambda = \frac{(1-\xi)q'}{\xi q + (1-\xi)q'} \quad \in [0, 1].$$

Then

$$f \text{ is convex} \Leftrightarrow \lambda f\left(\frac{p}{q}\right) + (1-\lambda) f\left(\frac{p'}{q'}\right) \geq f\left(\lambda \frac{p}{q} + (1-\lambda)\frac{p'}{q'}\right)$$
$$= f\left(\frac{\xi p + (1-\xi)p'}{\xi q + (1-\xi)q'}\right),$$
$$\forall \xi, p, p', q, q \in [0, 1]$$
$$\Leftrightarrow \xi q f\left(\frac{p}{q}\right) + (1-\xi)q' f\left(\frac{p'}{q'}\right)$$
$$\geq (\xi q + (1-\xi)q') f\left(\frac{\xi p + (1-\xi)p'}{\xi q + (1-\xi)q'}\right), \quad \forall \xi, p, p', q, q \in [0, 1]$$
$$\Leftrightarrow \xi g(p, q) + (1-\xi) g(p', q') \geq g(\xi p + (1-\xi)p', \xi q + (1-\xi)q')$$
$$\Leftrightarrow g \text{ is convex}$$
$$\Rightarrow d(p_0, p_1) = \mathbb{E}_0\left[f\left(\frac{p_1(Y)}{p_0(Y)}\right)\right]$$
$$= \int_{\mathcal{Y}} g(p_0(y), p_1(y))\, d\mu(y) \text{ is convex.}$$
□

Product Distributions If $p_0 = \prod_{i=1}^n p_{0i}$ and $p_1 = \prod_{i=1}^n p_{1i}$, then for some choices of f, g we have

$$d(p_0, p_1) = \sum_{i=1}^n d(p_{0i}, p_{1i}). \tag{6.33}$$

We have seen in Sections 6.1 and 6.3 that (6.33) holds for KL divergence and Chernoff divergence. The additivity property also holds for $\ln(1 + \chi^2(p_0, p_1))$ but not for $\chi^2(p_0, p_1)$ or for $d_V(p_0, p_1)$.

Bregman Divergences These divergences generalize the notion of distances between pairs of points in a convex set [19]. Let $F : \mathbb{R}^+ \to \mathbb{R}$ be a strictly convex function, and p_0 and p_1 be arbitrary pdfs. The Bregman divergence associated with the generator function F is

$$B_F(p_0, p_1) \triangleq \int_{\mathcal{Y}} F(p_0(y)) - F(p_1(y)) - [p_0(y) - p_1(y)]F'(p_1(y))\, d\mu(y), \tag{6.34}$$

where F' denotes the derivative of F. The integrand is nonnegative by convexity of F, and is zero if and only if $p_0(y) = p_1(y)$. The function $B_F(p_0, p_1)$ is therefore nonnegative and convex in the first argument but not necessarily in the second argument. Examples include:

- $F(x) = x^2$, for which $F'(x) = 2x$ and $B_F(p_0, p_1) = \|p_0 - p_1\|^2$.
- $F(x) = x \ln x$, for which $F'(x) = 1 + \ln x$ and

$$\begin{aligned} B_F(p_0, p_1) &= \int_{\mathcal{Y}} [p_0 \ln p_0 - p_1 \ln p_1 - (p_0 - p_1)(1 + \ln p_1)]\, d\mu \\ &= \int_{\mathcal{Y}} [p_0 \ln \frac{p_0}{p_1} - p_0 + p_1]\, d\mu \\ &= D(p_0 \| p_1). \end{aligned} \tag{6.35}$$

- $F(x) = -\ln x$, for which

$$B_F(p_0, p_1) = \int_{\mathcal{Y}} \left[\frac{p_0}{p_1} - \ln \frac{p_0}{p_1} - 1 \right] d\mu \tag{6.36}$$

  is called the *Itakura–Saito distance* between p_0 and p_1.

The KL divergence is the only f-divergence that is also a Bregman divergence.

6.5 Some Useful Inequalities

THEOREM 6.1 Hoeffding and Wolfowitz [20] *The squared Bhattacharyya coefficient*

$$\rho^2 \geq \exp\{-D(p_0 \| p_1)\}. \tag{6.37}$$

Proof The proof is a direct application of Jensen's inequality, $\ln \mathbb{E}[X] \geq \mathbb{E}[\ln X]$:

$$\ln \rho^2 = 2 \ln \int_{\mathcal{Y}} \sqrt{p_0 p_1} = 2 \ln \mathbb{E}_0 \left[\sqrt{\frac{p_1(Y)}{p_0(Y)}} \right] \geq 2\mathbb{E}_0 \left[\ln \sqrt{\frac{p_1(Y)}{p_0(Y)}} \right] = -D(p_0 \| p_1).$$

COROLLARY 6.1

$$\rho^2 \geq \exp\{-D(p_1 \| p_0)\}, \tag{6.38}$$

$$\rho^2 \geq \exp\{-D_{\text{sym}}(p_0 \| p_1)\}. \tag{6.39}$$

The bound (6.38) is obtained the same way as (6.37). Averaging the two lower bounds on $\ln \rho^2$ yields (6.39). □

THEOREM 6.2

$$D(p_0 \| p_1) \leq \ln(1 + \chi^2(p_0, p_1)) \leq \chi^2(p_0, p_1). \tag{6.40}$$

Proof The first inequality again follows by application of Jensen's inequality:

$$D(p_0 \| p_1) = \mathbb{E}_0 \left[\ln \frac{p_0(Y)}{p_1(Y)} \right] \leq \ln \mathbb{E}_0 \left[\frac{p_0(Y)}{p_1(Y)} \right] = \ln(1 + \chi^2(p_0, p_1)).$$

The second inequality holds because $\ln(1 + x) \leq x$ for all $x > -1$ (by concavity of the log function). □

THEOREM 6.3 (Vajda [21])

$$D(p_0 \| p_1) \geq 2 \ln \frac{1 + d_V(p_0, p_1)}{1 - d_V(p_0, p_1)} - \frac{2d_V(p_0, p_1)}{1 + d_V(p_0, p_1)}. \tag{6.41}$$

This inequality is stronger than the famous Pinsker's inequality (also known as Csiszar–Kemperman–Kullback–Pinsker inequality)

$$D(p_0 \| p_1) \geq 2d_V^2(p_0, p_1). \tag{6.42}$$

In particular, Vajda's inequality is tight when $d_V(p_0, p_1) = 0$ and $d_V(p_0, p_1) = 1$.

Exercises

6.1 *Poisson Distributions.* Derive the KL and Chernoff divergences between two Poisson distributions.

6.2 Prove (6.5) and (6.19).

6.3 Prove (6.21).

6.4 Prove (6.18).
Hint: You may want to use the fact that $p_u(y)$ can be written as:

$$p_u(y) = \frac{p_0(y)\, L(y)^u}{\mathbb{E}_0[L(Y)^u]},$$

where $L(y) = p_1(y)/p_0(y)$ is the likelihood ratio.

6.5 *Jensen–Shannon Divergence.*

(a) Show that Jensen–Shannon divergence (6.25) is an Ali–Silvey distance.
(b) Show that $0 \le d_{JS}(p_0, p_1) \le \ln 2$ where the upper bound is achieved for p_0 and p_1 that have disjoint support sets.
(c) Show that $d_{JS}(p_0, p_1) = H(\frac{p_0+p_1}{2}) - \frac{1}{2}[H(p_0) + H(p_1)]$ when $\mathcal{Y}$ is a finite set.

6.6 *Discard Parity Bit?* Let Z be a random variable taking values in the set of even numbers $\{0, 2, 4, 6, 8\}$, and B a Bernoulli random variable with parameter θ, which is independent of Z. Let $Y = Z + B$, hence B represents a parity bit. Express Z as a function of Y and show that the DPI applied to this function is satisfied with equality.

6.7 *Data Grouping*. Consider the probability vectors $P_0 = [.7, .2, .1]$ and $P_1 = [.1, .6, .3]$ on $\mathcal{Y} = \{1, 2, 3\}$. Find a data grouping function $\phi : \mathcal{Y} \to \mathcal{Z} = \{a, b\}$ such that $d(P_0, P_1) = d(Q_0, Q_1)$ for any f-divergence, where Q_0 and Q_1 are the distributions on $\mathcal{Z}$ induced by the mapping ϕ.

6.8 *DPI and Binary Quantization.* Let Y be distributed as $P_0 = \mathcal{N}(0, 1)$ or some other distribution P_1. Give a nontrivial condition on P_1 such that the DPI applied to the mapping $Z = \operatorname{sgn}(Y)$ is achieved with equality.

6.9 *Binary Quantization*. Let Y be distributed as $P_0 = \mathcal{N}(0, 1)$ or $P_1 = \mathcal{N}(\mu, 1)$.

(a) Give the corresponding distributions Q_0 and Q_1 on $Z = \operatorname{sgn}(Y)$, and express $D(Q_0 \| Q_1)$ as a function of μ.
(b) Graph $D(P_0 \| P_1)$ and $D(Q_0 \| Q_1)$ as a function of μ.
(c) Show that

$$\lim_{\mu \to 0} \frac{D(P_0 \| P_1)}{D(Q_0 \| Q_1)} = \frac{\pi}{2} \quad \text{and} \quad \lim_{|\mu| \to \infty} \frac{D(P_0 \| P_1)}{D(Q_0 \| Q_1)} = 2.$$

6.10 *Ternary Quantization*. Let Y be distributed as $P_0 = \mathcal{N}(0, 1)$ or $P_1 = \mathcal{N}(\mu, 1)$. Consider $t \ge 0$. We quantize Y and represent it by a ternary random variable

$$Z = \begin{cases} 1 & : Y - \mu/2 > t \\ 0 & : |Y - \mu/2| \le t \\ -1 & : Y - \mu/2 < -t. \end{cases}$$

(a) Give the corresponding distributions Q_0 and Q_1 on Z, and express $D(Q_0 \| Q_1)$ as a function of μ and t.
(b) Numerically maximize $D(Q_0 \| Q_1)$ over t, for $\mu \in \{0.1, 0.5, 1, 2, 3, 4, 5\}$. Evaluate the KL divergence ratio $f(\mu) = D(P_0 \| P_1)/\max_t D(Q_0 \| Q_1)$ for $\mu \in \{0.1, 0.5, 1, 2, 3\}$.

6.11 *Linear Dimensionality Reduction*. Consider the hypotheses $H_0 : Y \sim P_0 = \mathcal{N}(0, \mathsf{I}_n)$ and $H_1 : Y \sim P_1 = \mathcal{N}(0, \mathsf{C}_1)$. We wish to select a direction vector $u \in \mathbb{R}^n$ such that the projection $Z = u^\top Y \in \mathbb{R}$ contains as much information for discrimination as possible, in the sense that u maximizes $D(Q_0 \| Q_1)$ where Q_0 and Q_1 are the distributions of Z under H_0 and H_1, respectively.

(a) Show that the maximum is achieved by any eigenvector associated with the largest eigenvalue of C_1, and give an expression for the maximal $D(Q_0\|Q_1)$.
(b) Repeat when $P_0 = \mathcal{N}(0, \mathsf{C}_0)$ where C_0 is an arbitrary full-rank covariance matrix. *Hint*: Exploit the invariance of KL divergence to invertible transformations to relate this problem to that of part (a).
(c) Assume again that $\mathsf{C}_0 = \mathsf{I}_n$ and fix some $d \in \{2, 3, \ldots, n-1\}$. Now we wish to select a $d \times n$ matrix A such that the projection $Z = \mathsf{A}Y \in \mathbb{R}^d$ has maximal $D(Q_0\|Q_1)$ where Q_0 and Q_1 are the distributions of Z under H_0 and H_1, respectively. Argue that there is no loss of optimality in constraining $\mathsf{A}\mathsf{A}^\top = \mathsf{I}_d$ (again exploiting the invariance properties of KL divergence). Show that the maximum is achieved by any A whose columns are eigenvectors associated with the largest eigenvalues of C_1, and give an expression for the maximal $D(Q_0\|Q_1)$ in terms of these eigenvalues.

6.12 *Boolean Function Selection*. This may be viewed as a nonlinear dimensionality reduction problem. Consider two probability distributions P_0 and P_1 for a random variable Y taking values over the set $\{(0,0), (0,1), (1,0), (1,1)\}$. The distribution P_0 is uniform, and $P_1 = [3/16, 9/16, 3/16, 1/16]$. Let Q_0 and Q_1 be the corresponding distributions on $Z = f(Y)$ where f is a Boolean function, i.e., $Z \in \{0, 1\}$.

(a) Evaluate the Bhattacharyya distance $B(P_0, P_1)$.
(b) Find f that maximizes $B(Q_0\|Q_1)$.
Hint: There are 16 Boolean functions on Y but you need to consider only 8 of them.

6.13 *Asymptotics of f-Divergences for Increasingly Similar Distributions*. Assume $\mathcal{Y}$ is a finite set and $p_0(y) > 0$ for all $y \in \mathcal{Y}$. Fix any zero-mean vector $\{h(y)\}_{y\in\mathcal{Y}}$ and let $p_1 = p_0 + \epsilon h$, which is a probability vector for ϵ small enough. Let d be a f-divergence such that $f(x)$ is twice differentiable at $x = 1$. Using a second-order Taylor series expansion of f, show that the following asymptotic equality holds:

$$d(p_0, p_1) \sim \frac{f''(1)}{2}\chi^2(p_0, p_1) \quad \text{as } \epsilon \to 0.$$

6.14 *Divergences*. Fix an arbitrary integer $n > 1$ and evaluate $D(p_0\|p_1)$, $D(p_1\|p_0)$, $\|p_0 - p_1\|_{\mathrm{TV}}$, and $\chi^2(p_0, p_1)$ for the following distributions on $\mathcal{Y} = \{1, 2, 3, \ldots\}$:

$$p_0(y) = 2^{-y}, \quad p_1(y) = \frac{2^{-y}}{1 - 2^{-n}}\,\mathbb{1}\{1 \le y \le n\}, \quad y \in \mathcal{Y}.$$

Give the limits of these expressions as $n \to \infty$.

References

[1] I. Csiszár, "Eine Informationtheoretische Ungleichung and ihre Anwendung auf den Beweis der Ergodizität on Markoffschen Ketten," *Publ. Math. Inst. Hungarian Acad. Sci.* ser. A, Vol. 8, pp. 84–108, 1963.
[2] S. M. Ali and S. D. Silvey, "A General Class of Coefficients of Divergence of One Distribution from Another," *J. Royal Stat. Soc. B*, Vol. 28, pp. 132–142, 1966.
[3] T. Kailath, "The Divergence and Bhattacharyya Distance Measures in Signal Selection," *IEEE Trans. Comm. Technology*, Vol. 15, No. 1, pp. 52–60, 1967.

[4] H. Kobayashi and J.B. Thomas, "Distance Measures and Related Criteria," *Proc. Allerton Conf.*, Monticello, IL, pp. 491–500, 1967.
[5] M. Basseville, "Distance Measures for Signal Processing and Pattern Recognition," *Signal Processing*, Vol. 18, pp. 349–369, 1989.
[6] L. K. Jones and C. L. Byrne, "General Entropy Criteria for Inverse Problems, with Applications to Data Compression, Pattern Classification, and Cluster Analysis," *IEEE Trans. Information Theory*, Vol. 36, No. 1, pp. 23–30, 1990.
[7] A. Jain, P. Moulin, M. I. Miller, and K. Ramchandran, "Information Theoretic Bounds for Degraded Target Recognition," *IEEE Trans. PAMI*, Vol. 24, No. 9, pp. 1153–1166, 2002.
[8] F. Liese and I. Vajda, "On Divergences and Informations in Statistics and Information Theory," *IEEE Trans. Inf. Theory*, Vol. 52, No. 10, pp. 4394–4412, 2006.
[9] X. Nguyen, M. J. Wainwright, and M. I. Jordan, "On Information Divergence Measures, Surrogate Loss Functions and Decentralized Hypothesis Testing," *Proc. 43rd Allerton Conf.*, Monticello, IL, 2005.
[10] X. Nguyen, M. J. Wainwright, and M. I. Jordan, "On Surrogate Loss Functions and f-Divergences," *Annal. Stat.*, Vol. 37, No. 2, pp. 876–904, 2009.
[11] S. Kullback and R. Leibler, "On Information and Sufficiency," *Ann. Math. Stat.*, Vol. 22, pp. 79–86, 1951.
[12] S. Kullback, *Information Theory and Statistics*, Wiley, 1959.
[13] T. M. Cover and J. A. Thomas, *Elements of Information Theory*, 2nd edition, Wiley, 2006.
[14] H. Chernoff, "A Measure of Asymptotic Efficiency for Test of a Hypothesis Based on the Sum of Observations," *Ann. Math. Stat.*, Vol. 23, pp. 493–507, 1952.
[15] A. Rényi, "On Measures of Entropy and Information," *Proc. 4th Berkeley Symp. on Probability Theory and Mathematical Statistics*, pp. 547–561, Berkeley University Press, 1961.
[16] D. M. Endres and J. E. Schindelin, "A New Metric for Probability Distributions," *IEEE Trans. Inf. Theory*, Vol. 49, pp. 1858–1860, 2003.
[17] A. K. C. Wong and M. You, "Entropy and Distance of Random Graphs with Application to Structural Pattern Recognition," *IEEE Trans. Pattern Analysis Machine Intelligence*, Vol. 7, pp. 599–609, 1985.
[18] I. Goodfellow, "Generative Adversarial Networks," *NIPS tutorial*, 2016. Available at `https://arxiv.org/pdf/1701.00160.pdf`.
[19] L. M. Bregman, "The Relaxation Method of Finding the Common Points of Convex Sets and its Application to the Solution of Problems in Convex Programming," *USSR Comput. Math. Math. Phys.*, Vol. 7, No. 3, pp. 200–217, 1967.
[20] W. Hoeffding and J. Wolfowitz, "Distinguishability of Sets of Distributions," *Ann. Math. Stat.*, Vol. 29, pp. 700–718, 1958.
[21] I. Vajda, "Note on Discrimination Information and Variation," *IEEE Trans. Inf. Theory*, Vol. 16, pp. 771–773, 1970.

7 Performance Bounds for Hypothesis Testing

We saw in Chapters 2 and 4 that decision rules for binary hypothesis problems take the form:

$$\tilde{\delta}(y) = \begin{cases} 1 & \\ 1 \text{ w.p. } \gamma & \text{if } T(y) \begin{matrix} > \\ = \\ < \end{matrix} \tau, \\ 0 & \end{cases}$$

where $T(y)$ is an appropriate test statistic, τ is the test threshold, and γ the randomization parameter. For the special case of simple hypothesis testing, an optimal test statistic is the log-likelihood ratio $\ln L(y)$. The performance of the test $\tilde{\delta}$ is characterized in terms of the quantities:

$$\begin{aligned} \mathrm{P_F}(\tilde{\delta}) &= \mathrm{P}_0\{T(Y) > \tau\} + \gamma \mathrm{P}_0\{T(Y) = \tau\}, \\ \mathrm{P_M}(\tilde{\delta}) &= \mathrm{P}_1\{T(Y) < \tau\} + (1-\gamma)\mathrm{P}_1\{T(Y) = \tau\}, \\ \mathrm{P_e}(\tilde{\delta}) &= \pi_0 \mathrm{P_F}(\tilde{\delta}) + (1-\pi_0)\mathrm{P_M}(\tilde{\delta}). \end{aligned}$$

Barring special cases, it is generally impractical or impossible to compute the exact values of $\mathrm{P_F}$, $\mathrm{P_M}$, and $\mathrm{P_e}$. However, it is often sufficient to obtain good lower and upper bounds on these error probabilities. This chapter derives analytically tractable performance bounds for hypothesis testing. To derive such bounds, we exploit the convex statistical distances discussed in Chapter 6. We begin with simple lower bounds before introducing a family of upper bounds based on Chernoff divergence and defining the large-deviations rate function.

7.1 Simple Lower Bounds on Conditional Error Probabilities

Theorem 7.1 presents some simple bounds on conditional error probabilities based on the data-processing inequality.

DEFINITION 7.1 The binary KL divergence function

$$\begin{aligned} d(a\|b) &\triangleq a \ln \frac{a}{b} + (1-a) \ln \frac{1-a}{1-b} \\ &= D([a, 1-a]\|[b, 1-b]), \quad 0 \le a, b \le 1 \end{aligned} \tag{7.1}$$

is the KL divergence between two Bernoulli distributions with parameters a and b.

Again we use the notational convention $0 \ln 0 = 0$. The function $d(a\|b)$ is not symmetric; however, $d(a\|b) = d(1-a\|1-b)$. The function is jointly convex in (a, b) and attains its minimum (0) when $a = b$. For fixed $a \in (0, 1)$, $d(a\|b)$ tends to ∞ as b approaches 0 or 1. The following theorem provides an upper bound (parameterized by α) on the ROC.

THEOREM 7.1 Kullback [1]. *For all tests $\tilde{\delta}$ such that $\mathrm{P_F}(\tilde{\delta}) = \alpha$ and $\mathrm{P_M}(\tilde{\delta}) = \beta$, the following inequality holds:*

$$d(\alpha\|1-\beta) \le D(p_0\|p_1). \tag{7.2}$$

Proof Consider deterministic tests first. Let $\mathcal{Y}_1$ be the acceptance region for the test and define the binary random variable $B = \mathbb{1}\{Y \in \mathcal{Y}_1\}$, which is a deterministic function of Y. The probability of false alarm and the probability of correct detection are respectively equal to

$$\begin{aligned} \alpha &= P_0\{Y \in \mathcal{Y}_1\} = P_0\{B = 1\} = q_0(1) \\ \text{and} \quad 1-\beta &= P_1\{Y \in \mathcal{Y}_1\} = P_1\{B = 1\} = q_1(1), \end{aligned}$$

hence $D(q_0\|q_1) = d(\alpha\|1-\beta)$. Then the claim follows from the data processing inequality (6.30) and the fact that B is a function of Y. For randomized tests, let $B = \mathbb{1}\{\text{say } H_1\}$ which is a stochastic function of Y. Again the data processing inequality applies and (7.2) holds. □

REMARK 7.1 By convexity of the binary divergence function, the set

$$\mathcal{R}_0 \triangleq \{(\alpha, 1-\beta) \, : \, d(\alpha\|1-\beta) \le D(p_0\|p_1)\} \tag{7.3}$$

is a convex region inside $[0, 1]^2$. In particular, the line segment $\alpha = 1-\beta \in [0, 1]$ belongs to $\mathcal{R}_0$, and $\mathcal{R}_0$ satisfies the symmetry condition $(\alpha, 1-\beta) \in \mathcal{R}_0 \Leftrightarrow (1-\alpha, \beta) \in \mathcal{R}_0$. By Theorem 7.1, $\mathcal{R}_0$ is a superset of

$$\mathcal{R} = \{(\alpha, 1-\beta) \, : \, \exists\tilde{\delta} \, : \, \mathrm{P_F}(\tilde{\delta}) = \alpha, \mathrm{P_M}(\tilde{\delta}) = \beta\}. \tag{7.4}$$

Some interesting points in $\mathcal{R}_0$ are given by

$$\begin{aligned} \alpha = 0 \quad &\Rightarrow \quad \beta \ge e^{-D(p_0\|p_1)}, \\ \alpha = \frac{1}{2} \quad &\Rightarrow \quad \beta(1-\beta) \ge \frac{1}{4}e^{-2D(p_0\|p_1)}, \\ \alpha \to 1 \quad &\Rightarrow \quad \beta \ge e^{-D(p_0\|p_1)/(1-\alpha)}. \end{aligned}$$

The last line holds because we must have $\beta \to 0$ for any sequence of $(\alpha, 1-\beta) \in \mathcal{R}_0$ such that $\alpha \to 1$. Hence $d(\alpha\|1-\beta) \sim (1-\alpha)\ln\frac{1}{\beta}$.

REMARK 7.2 The lower bound in the case $\alpha = \frac{1}{2}$ is generally poor in the asymptotic setting with n i.i.d. observations, owing to the extraneous factor of 2 in the exponent (see the Chernoff–Stein Lemma 8.1 in the next chapter).

REMARK 7.3 Similarly to (7.2), a dual bound can be derived using $D(p_1 \| p_0)$. We have

$$d(\beta \| 1-\alpha) \leq D(p_1 \| p_0), \tag{7.5}$$

resulting in the convex region

$$\mathcal{R}_1 \triangleq \{(\alpha, 1-\beta) \,:\, d(\beta \| 1-\alpha) \leq D(p_1 \| p_0)\}, \tag{7.6}$$

and the following bounds:

$$\begin{aligned}
\beta = 0 \quad &\Rightarrow \quad \alpha \geq e^{-D(p_1 \| p_0)}, \\
\beta = \frac{1}{2} \quad &\Rightarrow \quad \alpha(1-\alpha) \geq \frac{1}{4} e^{-2D(p_1 \| p_0)}, \\
\beta \to 1 \quad &\Rightarrow \quad \alpha \geq (1-\beta) e^{-D(p_1 \| p_0)}.
\end{aligned}$$

7.2 Simple Lower Bounds on Error Probability

Consider a binary test $H_0 \,:\, Y \sim p_0$ versus $H_1 \,:\, Y \sim p_1$ with priors π_0 and π_1 on H_0 and H_1, respectively. Denote by $\mathrm{P_e}$ the minimum error probability. Theorem 7.2 provides a simple lower bound on $\mathrm{P_e}$.

THEOREM 7.2 Kailath Lower Bound on $\mathrm{P_e}$ [2] *Let* $\rho = \int_{\mathcal{Y}} \sqrt{p_0 p_1}\, d\mu$ *(Bhattacharyya coefficient). Then*

$$\mathrm{P_e} \geq \frac{1}{2}\left[1 - \sqrt{1 - 4\pi_0\pi_1\rho^2}\right]. \tag{7.7}$$

The following successively weaker lower bounds also apply:

$$\mathrm{P_e} \geq \pi_0\pi_1\rho^2, \tag{7.8}$$

$$\mathrm{P_e} \geq \pi_0\pi_1 \exp\{-D_{sym}(p_0 \| p_1)\}. \tag{7.9}$$

Proof First note the identity $\min(a, b) + \max(a, b) = a + b$. Hence

$$\int_{\mathcal{Y}} \min\{\pi_0 p_0, \pi_1 p_1\}\, d\mu + \int_{\mathcal{Y}} \max\{\pi_0 p_0, \pi_1 p_1\}\, d\mu = \int_{\mathcal{Y}} (\pi_0 p_0 + \pi_1 p_1)\, d\mu = 1. \tag{7.10}$$

We have

$$\begin{aligned}
\pi_0\pi_1\rho^2 &= \left(\int_{\mathcal{Y}} \sqrt{\pi_0\pi_1 p_0 p_1}\, d\mu\right)^2 \\
&= \left(\int_{\mathcal{Y}} \sqrt{\min\{\pi_0 p_0, \pi_1 p_1\} \, \max\{\pi_0 p_0, \pi_1 p_1\}}\, d\mu\right)^2 \\
&\overset{(a)}{\leq} \int_{\mathcal{Y}} \min\{\pi_0 p_0, \pi_1 p_1\}\, d\mu \int_{\mathcal{Y}} \max\{\pi_0 p_0, \pi_1 p_1\}\, d\mu \\
&\overset{(b)}{=} \int_{\mathcal{Y}} \min\{\pi_0 p_0, \pi_1 p_1\}\, d\mu \left(1 - \int_{\mathcal{Y}} \min\{\pi_0 p_0, \pi_1 p_1\}\, d\mu\right) \\
&= \mathrm{P_e}(1 - \mathrm{P_e}),
\end{aligned}$$

where the inequality (a) is obtained by application of the Cauchy–Schwarz inequality, and (b) follows from (7.10). This establishes (7.7). The weaker bound (7.8) follows from the inequality $\sqrt{1+2x} \le 1 + x$ for $x \ge -\frac{1}{2}$ and is tight for small ρ. The even weaker bound (7.9) follows from (6.39). □

7.3 Chernoff Bound

The basic problem is to upper bound a probability $\mathrm{P}\{X \ge a\}$ where X is any real random variable and $a > \mathbb{E}[X]$. Similarly, we may wish to upper bound $\mathrm{P}\{X \le a\}$ where $a < \mathbb{E}[X]$.

7.3.1 Moment-Generating and Cumulant-Generating Functions

The *moment-generating function (mgf)* for a real random variable X is defined as

$$M_X(u) \triangleq \mathbb{E}[e^{uX}] = \int_{\mathbb{R}} p(x) e^{ux}\, d\mu(x), \qquad u \in \mathbb{R}, \tag{7.11}$$

assuming the integral exists. The logarithm of the mgf is the *cumulant-generating function* (cgf)

$$\kappa_X(u) \triangleq \ln \mathbb{E}[e^{uX}]. \tag{7.12}$$

Recall some fundamental properties of the mgf and cgf:

- The kth moment of X is obtained by differentiating the mgf k times at the origin. For $k = 0, 1, 2$, we obtain

$$M_X(0) = 1, \quad M_X'(0) = \mathbb{E}[X], \quad M_X''(0) = \mathbb{E}[X^2].$$

- The kth cumulant of X is obtained by differentiating the cgf k times at the origin. For $k = 0, 1, 2$, we obtain

$$\kappa_X(0) = 0, \quad \kappa_X'(0) = \mathbb{E}[X], \quad \kappa_X''(0) = \mathrm{Var}(X).$$

- Both the mgf and the cgf are convex functions. The mgf is easily seen to be convex because $M_X''(u) = \mathbb{E}[e^{uX} X^2] \ge 0$. For the convexity of the cgf, we apply Hölder's inequality, which states that

$$|\mathbb{E}[XY]| \le \left(E[|X|^p]\right)^{1/p} \left(E[|X|^q]\right)^{1/q}$$

for $p, q \in [1, \infty)$ such that $1/p + 1/q = 1$. In particular, if we consider $\alpha \in (0, 1)$, and set $p = 1/\alpha$ and $q = 1/(1-\alpha)$, then

$$\begin{aligned}
\kappa_X(\alpha u_1 + (1-\alpha) u_2) &= \ln \mathbb{E}\left[e^{\alpha u_1 X + (1-\alpha) u_2 X}\right] \\
&= \ln \mathbb{E}\left[|e^{u_1 X}|^{\alpha}\, |e^{u_2 X}|^{(1-\alpha)}\right] \\
&\le \alpha \ln \mathbb{E}\left[e^{u_1 X}\right] + (1-\alpha) \ln \mathbb{E}\left[e^{u_2 X}\right] \\
&= \alpha \kappa_X(u_1) + (1-\alpha) \kappa_X(u_2).
\end{aligned}$$

- The domain of κ_X is either the entire real line $\mathbb{R}$, or a half-line, or an interval. Except for the case of the degenerate random variable $X = 0$, the supremum of κ_X over its domain is ∞.
- If X and Y are independent random variables, then the cgf for their sum $Z = X + Y$ is the sum of the individual cgfs: $\kappa_Z(u) = \kappa_X(u) + \kappa_Y(u)$.

Example 7.1 *Gaussian Random Variable.* Let $X \sim \mathcal{N}(0, 1)$. Then $M_X(u) = e^{u^2/2}$ and $\kappa_X(u) = u^2/2$ for all $u \in \mathbb{R}$, as shown in Figure 7.1(a).

Example 7.2 *Exponential Random Variable.* Let X be the unit exponential distribution. The mgf is given by

$$M_X(u) = \int_0^\infty e^{-x} e^{ux}\, d\mu(x) = \int_0^\infty e^{-(1-u)x}\, d\mu(x) = \begin{cases} \frac{1}{1-u} & : u < 1 \\ \infty & : u \geq 1. \end{cases}$$

The domain of M_X is $(-\infty, 1)$ and the cgf is $\kappa_X(u) = -\ln(1-u)$ for $u < 1$, as depicted in Figure 7.1(b).

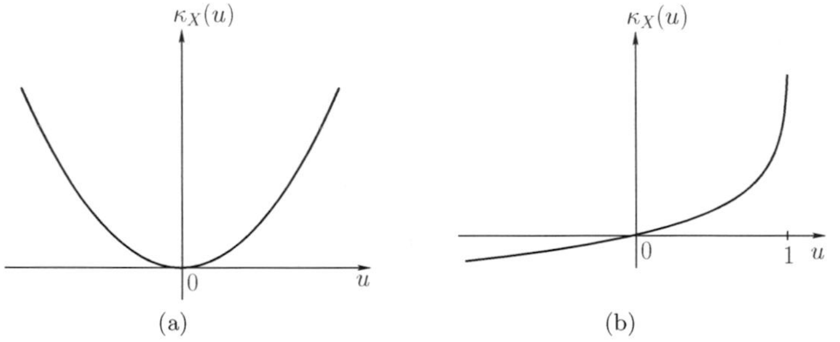

Figure 7.1 cgf for (a) $\mathcal{N}(0, 1)$ and (b) unit exponential distribution

7.3.2 Chernoff Bound

Recall *Markov's inequality*: for any nonnegative random variable Z and any $b > 0$, we have

$$P\{Z \geq b\} \leq \frac{\mathbb{E}[Z]}{b}. \tag{7.13}$$

Chebyshev's inequality

$$P\left\{\frac{|X - \mathbb{E}(X)|}{\sqrt{\mathrm{Var}(X)}} \geq c\right\} \leq c^{-2}, \quad \forall c > 0,$$

follows directly from Markov's inequality with $Z = \frac{(X-\mathbb{E}(X))^2}{\mathrm{Var}(X)}$ and $b = c^2$.

Both Markov's inequality and Chebyshev's inequality are generally loose. Often better upper bounds on $P\{X > a\})$ are obtained by applying Markov's inequality to the nonnegative random variable

$$Z = e^{uX} > 0$$

and letting $b = e^{ua}$, where u is a free parameter to be optimized. For $u > 0$, the function e^{ux} is increasing over $x \in \mathbb{R}$, and we have

$$\mathrm{P}\{X \geq a\} = \mathrm{P}\{e^{uX} \geq e^{ua}\} \leq \frac{\mathbb{E}[e^{uX}]}{e^{ua}}, \quad \forall u > 0,\ a \in \mathbb{R}. \tag{7.14}$$

Combining (7.14) and the definition (7.12) of the cgf, we obtain a *collection of exponential upper bounds* indexed by $u > 0$:

$$\mathrm{P}\{X \geq a\} \leq e^{-[ua - \kappa_X(u)]}, \quad \forall u > 0,\ a \in \mathbb{R}. \tag{7.15}$$

Similarly, to upper bound $\mathrm{P}\{X < a\}$, we observe that for $u < 0$, the function $z = e^{ux}$ is decreasing over $x \in \mathbb{R}$. Hence $\mathrm{P}\{X \leq a\} = \mathrm{P}\{e^{uX} \geq e^{ua}\}$ and

$$\mathrm{P}\{X \leq a\} \leq e^{-[ua - \kappa_X(u)]}, \quad \forall u < 0,\ a \in \mathbb{R}. \tag{7.16}$$

Note the family of bounds in (7.16) can be obtained from those in (7.15) by considering $Y = -X$ in (7.15) and using $\kappa_Y(u) = \kappa_X(-u)$. Therefore only the upper bound in (7.15) needs to be considered. To find the tightest bound in (7.15), we seek $u > 0$ that maximizes the function

$$g_a(u) \triangleq ua - \kappa_X(u), \quad u \in \mathrm{dom}(\kappa_X).$$

Performing this maximization, we obtain the *large-deviation rate function*

$$\Lambda_X(a) \triangleq \sup_{u \in \mathbb{R}}[ua - \kappa_X(u)], \quad a \in \mathbb{R}. \tag{7.17}$$

This may be viewed as a transformation of the function κ_X into Λ_X and is also called the *convex conjugate transform*, or the *Legendre–Fenchel transform*. It is widely used in convex optimization theory and related fields. If the supremum in (7.17) is achieved for $u > 0$, we have

$$\mathrm{P}\{X \geq a\} \leq e^{-\Lambda_X(a)}.$$

If the supremum is achieved for $u < 0$, we have

$$\mathrm{P}\{X \leq a\} \leq e^{-\Lambda_X(a)}.$$

The maximization problem (7.17) satisfies the following properties, which are illustrated in Figure 7.2:

- Since both ua and $-\kappa_X(u)$ are concave functions, so is their sum $g_a(u)$, and finding the maximum u^* of $g_a(u)$ is a simple concave optimization problem.
- We have $g_a(0) = 0$ and $g_a'(0) = a - \mathbb{E}[X]$. Hence $u^* > 0$ for $a > \mathbb{E}[X]$ and $u^* < 0$ for $a < \mathbb{E}[X]$.
- The large-deviation rate function $\Lambda_X(a)$ is the supremum of a collection of linear functions of a and is therefore convex.

In summary, the following upper bounds hold:

$$\mathrm{P}\{X \geq a\} \leq \begin{cases} e^{-[ua - \kappa_X(u)]} & : a > \mathbb{E}[X],\ u > 0 \\ 1 & : a \leq \mathbb{E}[X] \end{cases} \tag{7.18}$$

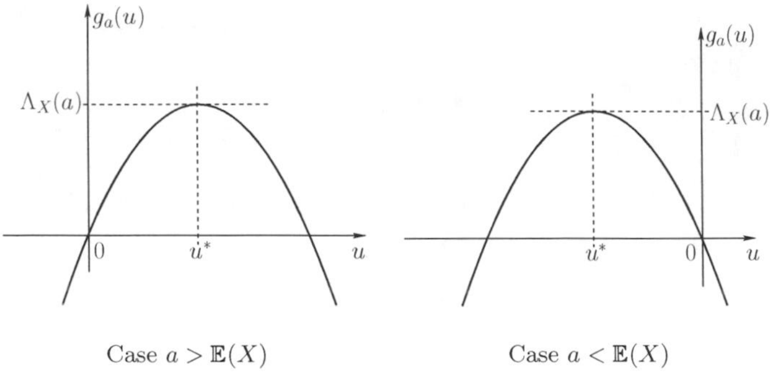

Figure 7.2 Plots of $g_a(u)$ and its maxima

and

$$\mathrm{P}\{X \le a\} \le \begin{cases} e^{-[ua-\kappa_X(u)]} & : a < \mathbb{E}[X],\ u < 0 \\ 1 & : a \ge \mathbb{E}[X]. \end{cases} \tag{7.19}$$

In particular, choosing the maximizing u in (7.17) yields

$$\mathrm{P}\{X \ge a\} \le \begin{cases} e^{-\Lambda_X(a)} & : a > \mathbb{E}[X] \\ 1 & : a \le \mathbb{E}[X] \end{cases} \tag{7.20}$$

and

$$\mathrm{P}\{X \le a\} \le \begin{cases} e^{-\Lambda_X(a)} & : a < \mathbb{E}[X] \\ 1 & : a \ge \mathbb{E}[X]. \end{cases} \tag{7.21}$$

Example 7.3 *Gaussian Random Variable, revisited.* Let $X \sim \mathcal{N}(0, 1)$, which has mean $\mathbb{E}(X) = 0$. The cgf of X is $\kappa_X(u) = u^2/2$ for all $u \in \mathbb{R}$. Hence $g_a(u) = ua - u^2/2$. Since $g_a(u)$ is concave, we obtain its global maximum u^* by setting the derivative of g_a to zero:

$$0 = g_a'(u^*) = a - u^*,$$

hence $u^* = a$ and $\Lambda_X(a) = a^2/2$ for all $a \in \mathbb{R}$. Thus we have the following upper bounds:

$$\mathrm{P}\{X \ge a\} \le e^{-a^2/2}, \quad \forall a > 0,$$
$$\mathrm{P}\{X \le a\} \le e^{-a^2/2}, \quad \forall a < 0.$$

The exact values of both probabilities are $Q(|a|)$. For large a, recall from (5.41) that

$$Q(a) \sim \frac{e^{-a^2/2}}{\sqrt{2\pi}\,a}, \quad a > 0.$$

Hence the upper bound is tight for large a. □

Example 7.4 *Exponential Random Variable, revisited.* Let X follow the unit exponential distribution, which has mean $\mathbb{E}(X) = 1$. The cgf $\kappa_X(u) = -\ln(1 - u)$, $u < 1$, was

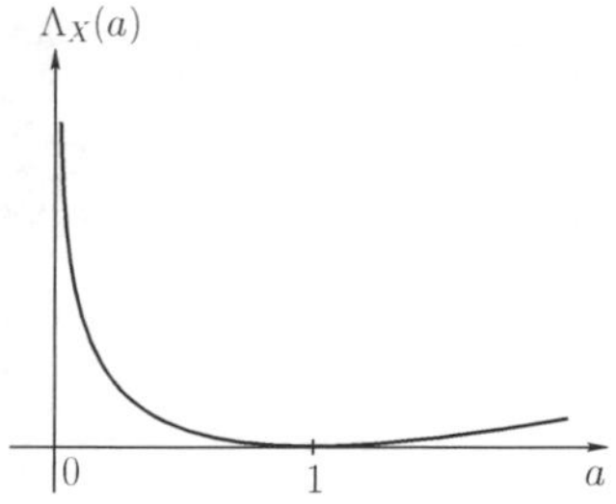

Figure 7.3 Large-deviations function $\Lambda_X(a)$ for unit exponential random variable

derived in Example 7.2. Hence $g_a(u) = ua + \ln(1-u)$ for $u < 1$. Again its global maximum u^* is obtained by setting the derivative of g_a to zero:

$$0 = g_a'(u^*) = a - \frac{1}{1-u^*}.$$

Hence

$$u^* = 1 - \frac{1}{a}$$

and

$$\Lambda_X(a) = g_a(u^*) = a - 1 - \ln a, \quad a > 0,$$

which is depicted in Figure 7.3. For any $a > 1 = \mathbb{E}(X)$, we have $u^* > 0$ and the large-deviations bound yields

$$\mathrm{P}\{X \geq a\} \leq e^{-\Lambda_X(a)} = e^{-a+1+\ln a} = (ae)e^{-a}.$$

Comparing with the exact value $\mathrm{P}\{X \geq a\} = e^{-a}$, we see the bound is especially useful for large a because it captures the exponential decay of $\mathrm{P}\{X \geq a\}$ with a. In contrast, the Chebyshev bound would yield, for $a > 1$,

$$\mathrm{P}\{X \geq a\} = \mathrm{P}\{X - 1 \geq a - 1\} \leq \frac{1}{(a-1)^2},$$

which decays only quadratically with a. □

7.4 Application of Chernoff Bound to Binary Hypothesis Testing

We now apply the large-deviations bounds to the binary hypothesis problem

$$\begin{cases} H_0 & : Y \sim p_0 \\ H_1 & : Y \sim p_1, \end{cases} \tag{7.22}$$

for which decision rules take the form:

$$\tilde{\delta}(y) = \begin{cases} 1 & & > \\ 1 \text{ w.p. } \gamma & \text{if } T(y) & = \tau, \\ 0 & & < \end{cases}$$

where $T(Y)$ is the test statistic. For the case of simple hypothesis testing, an optimal test statistic is the log-likelihood ratio, for which

$$T(Y) \triangleq \ln \frac{p_1(Y)}{p_0(Y)}.$$

7.4.1 Exponential Upper Bounds on $\mathrm{P_F}$ and $\mathrm{P_M}$

We have

$$\mathrm{P_F}(\tilde{\delta}) = \mathrm{P}_0\{T(Y) > \tau\} + \gamma \mathrm{P}_0\{T(Y) = \tau\} \le \mathrm{P}_0\{T(Y) \ge \tau\}, \tag{7.23}$$

$$\mathrm{P_M}(\tilde{\delta}) = \mathrm{P}_1\{T(Y) < \tau\} + (1-\gamma)\mathrm{P}_1\{T(Y) = \tau\} \le \mathrm{P}_1\{T(Y) \le \tau\}. \tag{7.24}$$

Denote by κ_0 and κ_1 the cgfs of T under H_0 and H_1, respectively. Then the large-deviations bounds (7.18) and (7.19) applied to $\mathrm{P_F}$ and $\mathrm{P_M}$ respectively are given by

$$\mathrm{P_F}(\tilde{\delta}) \le e^{-[u\tau - \kappa_0(u)]}, \quad \forall u > 0,$$

$$\mathrm{P_M}(\tilde{\delta}) \le e^{-[v\tau - \kappa_1(v)]}, \quad \forall v < 0.$$

We now derive a relationship between κ_0 and κ_1. We have

$$\begin{aligned}\kappa_0(u) &= \ln \mathbb{E}_0[e^{u \ln L(Y)}] \\ &= \ln \int p_0(y) \left(\frac{p_1(y)}{p_0(y)}\right)^u d\mu(y) \\ &= \ln \int p_0^{1-u}(y) p_1^u(y)\, d\mu(y)\end{aligned}$$

and

$$\begin{aligned}\kappa_1(v) &= \ln \mathbb{E}_1[e^{v \ln L(Y)}] \\ &= \ln \int_{\mathcal{Y}} p_1(y) \left(\frac{p_1(y)}{p_0(y)}\right)^v d\mu(y) \\ &= \ln \int_{\mathcal{Y}} p_0(y) \left(\frac{p_1(y)}{p_0(y)}\right)^{v+1} d\mu(y) \\ &= \kappa_0(v+1).\end{aligned}$$

Hence the second cgf κ_1 is obtained by simply shifting the first cgf κ_0. We further observe that κ_0 is the negative of the Chernoff divergence $d_u(p_0, p_1)$ introduced in (6.11).

Setting $u = v + 1$, we rewrite the upper bounds on $\mathrm{P_F}$ and $\mathrm{P_M}$ as

$$\mathrm{P_F}(\tilde{\delta}) \le e^{-[u\tau - \kappa_0(u)]}, \quad \forall u > 0, \tag{7.25}$$

$$\mathrm{P_M}(\tilde{\delta}) \le e^{-[(u-1)\tau - \kappa_0(u)]}, \quad \forall u < 1. \tag{7.26}$$

The functions $u\tau - \kappa_0(u)$ and $(u-1)\tau - \kappa_0(u)$ differ only by a constant (τ) and their maximum over $u \in \mathbb{R}$ is therefore achieved by the same value u^*. Define

$$g(u) \stackrel{\Delta}{=} u\tau - \kappa_0(u). \tag{7.27}$$

Then by the convexity of κ_0, g is concave and

$$u^* = \arg\max_{u \in \mathbb{R}} g(u) \tag{7.28}$$

satisfies $g'(u^*) = 0$, hence $\tau = \kappa_0'(u^*)$. The derivative of the function κ_0 is given by

$$\begin{aligned}
\kappa_0'(u) &= \frac{d}{du} \ln \int_{\mathcal{Y}} p_0^{1-u}(y) p_1^u(y)\, d\mu(y) \\
&= e^{-\kappa_0(u)} \frac{d}{du} \int_{\mathcal{Y}} p_0(y) \left(\frac{p_1(y)}{p_0(y)} \right)^u d\mu(y) \\
&= e^{-\kappa_0(u)} \int_{\mathcal{Y}} p_0^{1-u}(y)\, p_1^u(y) \ln \frac{p_1(y)}{p_0(y)}\, d\mu(y).
\end{aligned}$$

The large-deviations rate functions for $T(Y)$ under P_0 and P_1 are respectively given by

$$\Lambda_0(\tau) = u^*\tau - \kappa_0(u^*), \tag{7.29}$$

$$\Lambda_1(\tau) = (u^* - 1)\tau - \kappa_0(u^*) = \Lambda_0(\tau) - \tau. \tag{7.30}$$

Recalling the definition of KL divergence in (6.1), the statistic $T(Y)$ has mean

$$\mathbb{E}_0[T(Y)] = \int_{\mathcal{Y}} p_0(y) \ln \frac{p_1(y)}{p_0(y)}\, d\mu(y) = -D(p_0 \| p_1) = \kappa_0'(0) \tag{7.31}$$

under H_0, and mean

$$\mathbb{E}_1[T(Y)] = \int_{\mathcal{Y}} p_1(y) \ln \frac{p_1(y)}{p_0(y)}\, d\mu(y) = D(p_1 \| p_0) = \kappa_1'(0) \tag{7.32}$$

under H_1. We assume that both $D(p_1 \| p_0)$ and $D(p_0 \| p_1)$ are finite. We also recall that $\kappa_0(0) = \kappa_1(0) = \kappa_0(1) = 0$.

Based on these properties of κ_0, we consider the following cases:

Case I: $-D(p_0 \| p_1) < \tau < D(p_1 \| p_0)$ (the case of primary interest);
Case II: $\tau \le -D(p_0 \| p_1)$;
Case III: $\tau \ge D(p_1 \| p_0)$.

The maximization of the function $g(u) = u\tau - \kappa_0(u)$ is illustrated in Figure 7.4. The line with slope τ is tangent to the graph of the cgf κ_0 at u^*. The intercept of that line with the vertical axis is equal to $-[u^*\tau - \kappa_0(u^*)] = -\Lambda_0(\tau)$, and the intercept with the vertical line at $u = 1$ is at $-[(u^* - 1)\tau - \kappa_0(u^*)] = -\Lambda_1(\tau)$. The slopes of $\kappa_0(u)$ at $u = 0$ and $u = 1$ are equal to $-D(p_0 \| p_1)$ and to $D(p_1 \| p_0)$, respectively. The best bounds on P_F and P_M are given in terms of $\Lambda_0(\tau)$ and $\Lambda_1(\tau)$ as in (7.33) and (7.34).

The function g defined in (7.27) has the following properties:

$$\begin{aligned}
g(0) &= \kappa_0(0) = 0, \\
g(1) &= \tau - \kappa_0(1) = \tau - \kappa_1(0) = \tau, \\
g'(0) &= \tau - \kappa_0'(0) = \tau + D(p_0 \| p_1), \\
g'(1) &= \tau - \kappa_0'(1) = \tau - \kappa_1'(0) = \tau - D(p_1 \| p_0).
\end{aligned}$$

Case I: $-D(p_0 \| p_1) < \tau < D(p_1 \| p_0)$. Owing to the first inequality we have $g'(0) > 0$, hence the maximum of $g(\cdot)$ is achieved by $u^* > 0$. Owing to the second inequality we

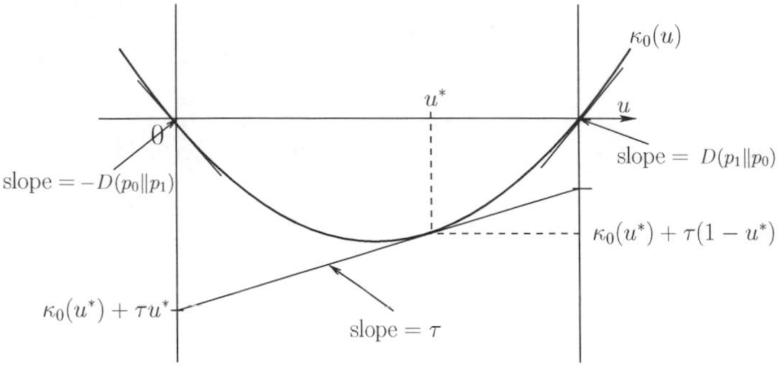

Figure 7.4 Chernoff bounds for Case I, i.e., $-D(p_0\|p_1) < \tau < D(p_1\|p_0)$

have $g'(1) < 0$, hence $u^* < 1$. Thus $u^* \in (0, 1)$. The bounds in (7.25) and (7.26) are both optimized by u^* to yield the tightest Chernoff bounds:

$$\mathrm{P_F}(\tilde{\delta}) \le e^{-\Lambda_0(\tau)}, \tag{7.33}$$

$$\mathrm{P_M}(\tilde{\delta}) \le e^{-\Lambda_1(\tau)} = e^{-(\Lambda_0(\tau)-\tau)}. \tag{7.34}$$

Case II: $\tau \le -D(p_0\|p_1)$. In this case $g'(0) \le 0$, which implies that $u^* \le 0$. Then we have

$$\sup_{u>0} g(u) = g(0) = 0,$$

which implies that the optimized bounds in (7.25) and (7.26) are

$$\mathrm{P_F}(\tilde{\delta}) \le e^{-\sup_{u>0} g(u)} = e^{g(0)} = e^0 = 1,$$
$$\mathrm{P_M}(\tilde{\delta}) \le e^{-\Lambda_1(\tau)},$$

i.e., the bound on $\mathrm{P_F}$ becomes trivial.

Case III: $\tau \ge D(p_1\|p_0)$. In this case $g'(1) \ge 0$, which implies that $u^* \ge 1$. Then we have

$$\sup_{u<1} g(u) = g(1) = \tau,$$

which implies that the optimized bounds in (7.25) and (7.26) are

$$\mathrm{P_F}(\tilde{\delta}) \le e^{-\Lambda_0(\tau)},$$
$$\mathrm{P_M}(\tilde{\delta}) \le e^{-\sup_{u<1}(g(u)-\tau)} = e^{-(g(1)-\tau)} = e^0 = 1,$$

i.e., the bound on $\mathrm{P_M}$ becomes trivial.

7.4.2 Bayesian Error Probability

In the Bayesian setting with uniform costs, assuming that $-D(p_0\|p_1) < \tau < D(p_1\|p_0)$, the bounds in (7.33) and (7.34) can be combined to yield an exponential upper bound on the Bayesian error probability:

$$\mathrm{P_e} = \pi_0 \mathrm{P_F} + \pi_1 \mathrm{P_M} \le (\pi_0 + \pi_1 e^{\tau})\, e^{-\Lambda_0(\tau)} = 2\pi_0\, e^{-\Lambda_0(\tau)}, \tag{7.35}$$

where the last equality holds because $\tau = \ln \frac{\pi_0}{\pi_1}$. Actually, we can get rid of the factor of 2 in (7.35) as follows.

The following theorem due to Chernoff provides a family (indexed by the parameter u) of simple upper bounds on $\mathrm{P_e}$. Recall the definition of the Chernoff u-coefficient $\rho_u \triangleq \int_{\mathcal{Y}} p_0^{1-u} p_1^u \, d\mu$ in (6.13).

THEOREM 7.3 *Chernoff Upper Bound on* $\mathrm{P_e}$ *[3].*

$$\mathrm{P_e} \le \pi_0^{1-u} \pi_1^u \rho_u, \quad \forall u \in [0, 1]. \tag{7.36}$$

Proof The upper bound follows from a basic inequality: $\min\{a, b\} \le a^{1-u} b^u$, which holds for all $a, b \ge 0$ and $u \in [0, 1]$. Hence

$$\begin{aligned}
\mathrm{P_e} &= \int_{\mathcal{Y}} \min\{\pi_0 p_0, \pi_1 p_1\}\, d\mu \\
&\le \int_{\mathcal{Y}} (\pi_0 p_0)^{1-u} (\pi_1 p_1)^u \, d\mu \\
&= \pi_0^{1-u} \pi_1^u \rho_u, \quad \forall u \in [0, 1].
\end{aligned}$$

□

Since $\tau = \ln \frac{\pi_0}{\pi_1}$ and $\kappa_0(u) = \ln \rho_u$, Theorem 7.3 implies

$$\mathrm{P_e} \le \pi_0\, e^{-[u\tau - \kappa_0(u)]} \quad \forall u \in (0, 1). \tag{7.37}$$

Assuming that $-D(p_0 \| p_1) < \tau < D(p_1 \| p_0)$, we can optimize the bound over $u \in (0, 1)$ to get

$$\mathrm{P_e} \le \pi_0\, e^{-[u^* \tau - \kappa_0(u^*)]} = \pi_0 e^{-\Lambda_0(\tau)}, \tag{7.38}$$

where u^* is the solution to $\tau = \kappa_0'(u^*)$.

The *Bhattacharyya bound* on error probability is obtained by setting $u = 1/2$ in (7.36):

$$\mathrm{P_e} \le \sqrt{\pi_0 \pi_1} \int_{\mathcal{Y}} \sqrt{p_0 p_1}\, d\mu = \sqrt{\pi_0 \pi_1}\, e^{-B(p_0, p_1)}, \tag{7.39}$$

where $B(p_0, p_1)$ is the Bhattacharyya distance defined in (6.12). This bound is generally looser but is often useful in practice.

For the special case where $\pi_0 = \pi_1 = 1/2$, we have that $\tau = 0$, which means that the condition $-D(p_0 \| p_1) < \tau < D(p_1 \| p_0)$ is satisfied, and we can simplify (7.37) as

$$\mathrm{P_e} \le \frac{1}{2} e^{\kappa_0(u)} \quad \forall u \in (0, 1).$$

Optimizing over u, we obtain

$$\mathrm{P_e} \le \frac{1}{2} \min_{u \in (0,1)} e^{\kappa_0(u)} = \frac{1}{2} e^{-C(p_0, p_1)}, \tag{7.40}$$

where $C(p_0, p_1)$ is the Chernoff information defined in (6.15).

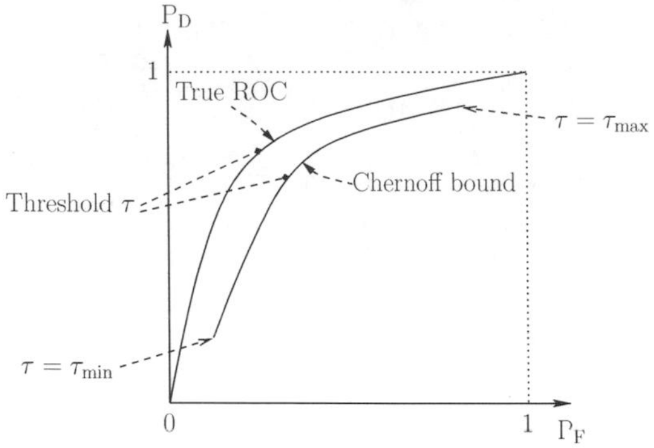

Figure 7.5 ROC curve and Chernoff bound

The upper bounds (7.33) and (7.34) on P_F and P_M coincide when $\tau = 0$, corresponding to the horizontal tangent in Figure 7.4. The common value of the bounds equals the Chernoff information, i.e., $\Lambda_0(0) = \Lambda_1(0) = C(p_0, p_1)$.

7.4.3 Lower Bound on ROC

The ROC curve for the LRT for the hypothesis testing problem of (7.22) is shown in Figure 7.5. Each point on the curve corresponds to a value of the LLRT threshold τ. For each τ, the upper bounds (7.25) and (7.26) on P_F and P_M yield a corresponding value on the ROC bound. Note that $\tau = +\infty$ corresponds to point $(0, 0)$ and $\tau = -\infty$ corresponds to $(1, 1)$ on the ROC curve. However, τ should be in the range $(\tau_{\min}, \tau_{\max}) = (-D(p_0 \| p_1),\ D(p_1 \| p_0))$ for the upper bounds to hold.

7.4.4 Example

Consider the binary hypothesis testing problem for which

$$p_1(y) = \begin{cases} |y| & \text{if } y \in [-1, 1] \\ 0 & \text{otherwise} \end{cases} \quad \text{and} \quad p_0(y) = \begin{cases} \frac{1}{2} & \text{if } y \in [-1, 1] \\ 0 & \text{otherwise.} \end{cases}$$

Assuming equal priors we can find the MPE rule as:

$$\delta_B(y) = \begin{cases} 1 & \text{if } \frac{1}{2} \le |y| \le 1 \\ 0 & \text{if } |y| \le \frac{1}{2} \end{cases}$$

and

$$P_e(\delta_B) = \frac{1}{2} P_0 \left\{ \frac{1}{2} \le |Y| \le 1 \right\} + \frac{1}{2} P_1 \left\{ |Y| \le \frac{1}{2} \right\} = \frac{1}{4} + \frac{1}{8} = \frac{3}{8}.$$

It is interesting to compare this value with the Chernoff bound. For equal priors, we have from (7.40) that

$$P_e \le \frac{1}{2} e^{-C(p_0, p_1)},$$

where

$$C(p_0, p_1) = -\min_{u \in (0,1)} \kappa_0(u),$$

with

$$\kappa_0(u) = \ln \int_{-1}^{1} p_0^{1-u}(y) p_1^u(y) dy = \ln \int_{-1}^{1} \frac{1}{2^{1-u}} |y|^u dy$$
$$= \ln \int_0^1 2^u y^u dy = \ln \frac{2^u}{u+1} = u \ln 2 - \ln(u+1).$$

Since we know that κ_0 is convex, we can minimize by taking the derivative and setting to zero, which yields

$$u^* = \frac{1}{\ln 2} - 1 \approx 0.443,$$

which can be seen to belong to (0, 1) and therefore

$$C(p_0, p_1) = -\kappa_0(u^*) = -(1 - \ln 2 + \ln \ln 2)$$

and

$$\mathrm{P_e} \leq \frac{1}{2} e^{-C(p_0,p_1)} = \frac{e \ln 2}{4} \approx 0.4704 > \frac{3}{8} = 0.375.$$

The Bhattacharyya distance can be computed as

$$B(p_0, p_1) = -\kappa_0(0.5) = -\frac{\ln 2}{2} + \ln(1.5) = \ln(1.5) - \ln \sqrt{2}$$

and the corresponding Bhattacharya bound is given by

$$\mathrm{P_e} \leq \frac{1}{2} e^{-B(p_0,p_1)} = \frac{\sqrt{2}}{3} \approx 0.4714,$$

which is very close to the best bound based on the Chernoff information given earlier.

7.5 Bounds on Classification Error Probability

We now consider m-ary Bayesian hypothesis testing. There are m hypotheses H_j, $j = 1, 2, \ldots, m$, with respective prior probabilities π_j, $j = 1, 2, \ldots, m$. The observational model is $Y \sim p_j$ under H_j. Denote by $\mathrm{P_e}$ the minimum error probability, which is achieved by the MAP decoding rule.

7.5.1 Upper and Lower Bounds in Terms of Pairwise Error Probabilities

Evaluating the minimum error probability is usually difficult for $m > 2$, even when the conditional distributions are "nice" – e.g., Gaussians with the same covariance. However, upper and lower bounds on $\mathrm{P_e}$ can be derived in terms of the $\frac{m(m-1)}{2}$ error probabilities for all pairwise binary hypothesis tests. Let $\mathrm{P_e}(i, j)$ denote the Bayesian error probability for a binary test between H_i and H_j with priors $\frac{\pi_i}{\pi_i+\pi_j}$ and $\frac{\pi_j}{\pi_i+\pi_j}$. The

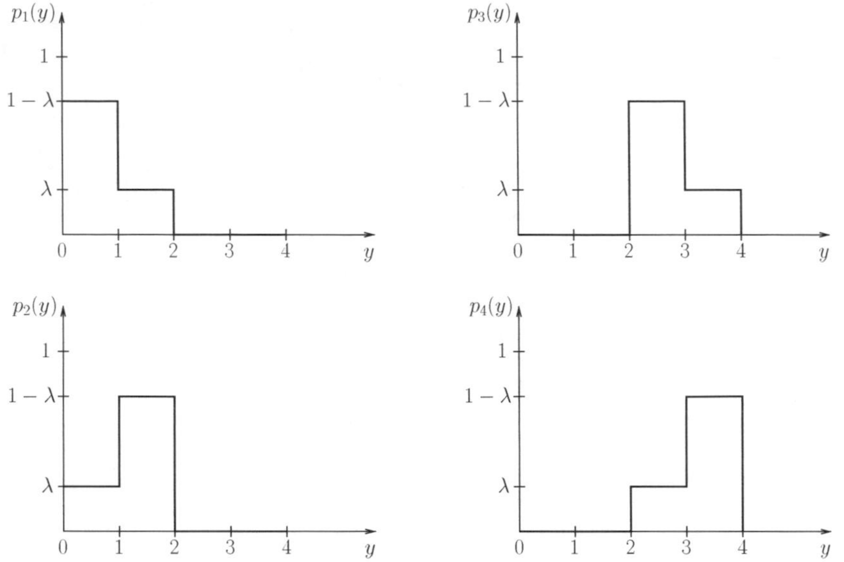

Figure 7.6 Conditional distributions for Example 7.5

lower bound on $\mathrm{P_e}$ in the following theorem is due to Lissack and Fu [4]; an alternate derivation may be found in [5]. The upper bound is due to Chu and Chueh [6].

THEOREM 7.4 *The following lower and upper bounds on the minimum error probability* $\mathrm{P_e}$ *hold:*

$$\frac{2}{m}\sum_{i\neq j}\pi_i \mathrm{P_e}(i,j) \le \mathrm{P_e} \le \sum_{i\neq j}\pi_i \mathrm{P_e}(i,j). \tag{7.41}$$

The lower bound is achieved if the posterior probabilities satisfy

$$\pi(i|y) = \pi(j|y) \quad \forall i,j,y \neq (\mathcal{Y}_i \cup \mathcal{Y}_j),$$

where $\mathcal{Y}_j$ *is the Bayes acceptance region for hypothesis* H_j, $1 \le j \le m$. *The upper bound is achieved if the decision regions* $\mathcal{Y}_j$, $1 \le j \le m$ *are the same as for the binary hypothesis tests.*

Observe that the lower and upper bounds coincide for $m = 2$, and that the gap is relatively small for moderate m, if $\mathrm{P_e}$ is small. The proof of the theorem is given in Section 7.6. We now provide two examples in which the upper and lower bounds of (7.41) are tight, respectively. As the first example shows, the sufficient condition given in the theorem for achieving the upper bound is not necessary.

Example 7.5 *Tight Upper Bound.* Let $\mathcal{Y} = [0,4]$, $m = 4$, uniform prior, $\lambda \in [0, 0.5)$. Consider the following conditional distributions, which are shown in Figure 7.6:

$$p_j(y) = \begin{cases} 1-\lambda & : j-1 \le y < j \\ \lambda & : j \le y \le j+1, \end{cases} \qquad j = 1,3,$$

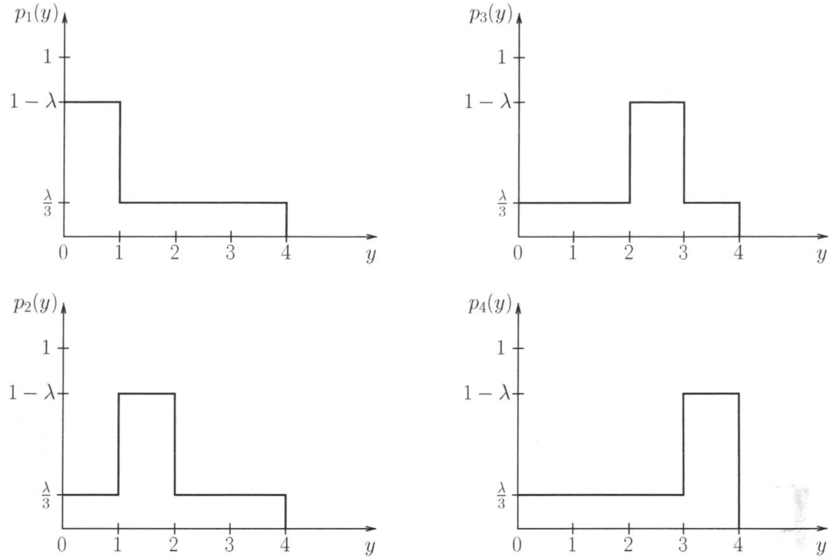

Figure 7.7 Conditional distributions for Example 7.6

$$p_j(y) = \begin{cases} \lambda & : j-2 \le y < j-1 \\ 1-\lambda & : j-1 \le y \le j, \end{cases} \qquad j = 2, 4.$$

The conditional pdfs p_1 and p_2 have common support $[0, 2]$, and p_3 and p_4 have common support $[2, 4]$. The marginal for Y is uniform, hence $\pi_j/p(y) \equiv 1$, and $\pi(j|y) = p_j(y)$. The decision regions are given by $\mathcal{Y}_j = [j-1, j)$, and $\mathrm{P_e} = \mathbb{E}[1 - \max_j \pi(j|Y)] = \lambda$. For the binary hypothesis tests, we have $\mathrm{P_e}(i, j) = \frac{1}{2}\int_{\mathcal{Y}} \min\{p_i, p_j\}\, d\mu = \lambda \mathbb{1}\{j = i+1\}$ for odd i and $\mathrm{P_e}(i, j) = \lambda \mathbb{1}\{j = i-1\}$ for even i. Hence the upper bound of (7.41) is λ, and is achieved. The lower bound is $\frac{1}{2}\lambda$.

Example 7.6 *Tight Lower Bound.* Let $\mathcal{Y} = [0, 4]$, $m = 4$, uniform prior, $\lambda \in [0, 3/4)$, and

$$p_j(y) = \begin{cases} 1-\lambda & : j-1 \le y \le j \\ \lambda/3 & : \text{else}, \end{cases} \qquad j = 1, 2, 3, 4.$$

The conditional pdfs are shown in Figure 7.7. They are obtained by shifts of p_1 and have common support $[0, 4]$. The marginal for Y is uniform, hence $\pi_j/p(y) \equiv 1$, and $\pi(j|y) = p_j(y)$. The decision regions are given by $\mathcal{Y}_j = [j-1, j)$, and the error probability is $\mathrm{P_e} = \mathbb{E}[1 - \max_j \pi(j|Y)] = \lambda$. For the binary hypothesis tests, we have $\mathrm{P_e}(i, j) = \frac{1}{2}\int_{\mathcal{Y}} \min\{p_i, p_j\}\, d\mu = \frac{2}{3}\lambda$ for any $i \neq j$. The lower bound of (7.41) is $\lambda = \mathrm{P_e}$ and is therefore achieved. The upper bound is 2λ.

In many problems, the upper bound of Theorem 7.4 is reasonably tight because one worst-pair (i, j) dominates the sum. This is especially true in an asymptotic setting where the pairwise error probabilities vanish exponentially with the length n of the observation sequence, and the pair (i, j) corresponding to the worst exponent dominates; see the Gaussian example of Section 5.5.

7.5.2 Bonferroni's Inequalities

The union bound is a standard tool for deriving an upper bound on the probability of a union of events $\mathcal{E}_1, \dots, \mathcal{E}_m$:

$$\mathrm{P}\left(\cup_{j=1}^m \mathcal{E}_j\right) \le \sum_{j=1}^m \mathrm{P}\left(\mathcal{E}_j\right) \triangleq S_1.$$

In particular, we applied this bound to the error probability for the m-ary signal detection problem of Section 5.5. However, a lower bound on that probability is sometimes useful too. Let

$$S_2 = \sum_{i<j} \mathrm{P}\left(\mathcal{E}_i \cap \mathcal{E}_j\right).$$

Then we have the lower bound

$$\mathrm{P}\left(\cup_{j=1}^m \mathcal{E}_j\right) \ge S_1 - S_2. \tag{7.42}$$

7.5.3 Generalized Fano's Inequality

Another popular way to obtain lower bounds on classification error probability is via Fano's inequality, which plays a central role in information theory [7]. While Fano's inequality requires a uniform prior on the hypotheses, it can be generalized to the case of arbitrary priors [8]. We state the main result and then specialize it to the classical case of a uniform prior. The proof of the generalized Fano's inequality is obtained by application of the data processing inequality.

Denote by $\pi_{\mathrm{e}} = 1 - \max_j \pi_j$ the minimum error probability in the absence of observations. In this case the MPE rule picks the *a priori* most likely symbol. The generalized Fano's inequality is stated in Theorem 7.5.

THEOREM 7.5 Generalized Fano's Inequality.

(i) The minimum error probability P_{e} *satisfies*

$$\mathrm{P}_{\mathrm{e}} \ge P_{e,LB}, \tag{7.43}$$

where $P_{e,LB}$ *is the unique solution to the equation* $d(\cdot\|\pi_e) = I(J; Y)$ *on* $(0, \pi_e]$.

(ii) Equality is achieved in (7.43) *if and only if there exist a Bayes rule* δ_{B} *and constants* a_0 *and* a_1 *such that*

$$\pi_j p_j(y) = \begin{cases} a_0 \pi_j \overline{p}(y) & : \ \delta_{\mathrm{B}}(y) = j \\ a_1 \pi_j \overline{p}(y) & : \ \delta_{\mathrm{B}}(y) \ne j. \end{cases} \tag{7.44}$$

Proof Consider the MPE rule $\delta_{\mathrm{B}} : \mathcal{Y} \to \{1, 2, \dots, m\}$ and define the Bernoulli random variable $B \triangleq \mathbb{1}\{\delta_{\mathrm{B}}(Y) \neq J\}$ which is a function of (J, Y). We have $\mathrm{P_e} = \mathrm{P}\{B = 1\}$ where P is the joint distribution $\pi \times p_J$. Similarly, $\pi_{\mathrm{e}} = \mathrm{P}\{B = 1\}$ where P is the product $\pi \times \overline{p}$. By the data-processing inequality we have

$$I(J; Y) = D(\pi \times p_J \| \pi \times \overline{p}) \geq d(\mathrm{P_e} \| \pi_{\mathrm{e}}). \tag{7.45}$$

Since $\mathrm{P_e} \leq \pi_{\mathrm{e}}$ and the function $d(\cdot \| \pi_{\mathrm{e}})$ is monotone decreasing on $(0, \pi_{\mathrm{e}}]$, this proves (i). Next, we apply the condition for equality in the data processing equality which is given following (6.30). The condition, applied to our problem, is that the likelihood ratio $\overline{p}(Y)/p_J(Y)$ be a degenerate random variable (under $\pi \times p_J$) given B, i.e.,

$$\frac{p_j(y)}{\overline{p}(y)} = \begin{cases} a_0 & : \delta_{\mathrm{B}}(y) = j \\ a_1 & : \delta_{\mathrm{B}}(y) \neq j. \end{cases}$$

This is equivalent to (7.44). □

THEOREM 7.6 Fano's Inequality *Assume the prior π is uniform over $\{1, 2, \dots, m\}$. For $m > 2$ we have $\mathrm{P_e} \geq P_{e,LB}$ where $P_{e,LB}$ is the unique solution of*

$$\mathrm{P_e} = \frac{\ln m - I(J; Y) - h_2(\mathrm{P_e})}{\ln(m \quad 1)} \triangleq f(\mathrm{P_e}) \tag{7.46}$$

over $[0, 1/2]$. For $m = 2$ the lower bound $P_{e,LB}$ of (7.43) is the unique solution to the equation $h_2(\cdot) = \ln 2 - I(J; Y)$ on $[0, \frac{1}{2}]$.

Proof For a uniform prior we have $\pi_{\mathrm{e}} = 1 - \frac{1}{m}$ and thus

$$\begin{aligned} d(\mathrm{P_e} \| \pi_{\mathrm{e}}) &= \mathrm{P_e} \ln \frac{\mathrm{P_e}}{\pi_{\mathrm{e}}} + (1 - \mathrm{P_e}) \ln \frac{1 - \mathrm{P_e}}{1 - \pi_{\mathrm{e}}} \\ &= \mathrm{P_e} \ln \frac{\mathrm{P_e}}{1 - \frac{1}{m}} + (1 - \mathrm{P_e}) \ln[m(1 - \mathrm{P_e})] \\ &= -h_2(\mathrm{P_e}) + \mathrm{P_e} \ln \frac{m}{m - 1} + (1 - \mathrm{P_e}) \ln m. \end{aligned} \tag{7.47}$$

Combining (7.45) and (7.47), we obtain

$$\begin{aligned} I(J; Y) &\geq -h_2(\mathrm{P_e}) + \mathrm{P_e} \ln \frac{m}{m - 1} + (1 - \mathrm{P_e}) \ln m \\ &= \ln m - h_2(\mathrm{P_e}) - \mathrm{P_e} \ln(m - 1) \\ \Rightarrow \quad \mathrm{P_e} &\geq f(\mathrm{P_e}). \end{aligned}$$

Since $f(\mathrm{P_e})$ is decreasing over $[0, 1/2]$ and increasing over $[1/2, 1]$, this proves the claim. When $m = 2$, we have $\pi_{\mathrm{e}} = 1/2$, hence $I(J; Y) \geq d(\mathrm{P_e} \| \pi_{\mathrm{e}}) = \ln 2 - h_2(\mathrm{P_e})$. □

REMARK 7.4 Fano's inequality (7.46) is sometimes weakened by replacing the denominator of (7.46) with $\ln m$ and by replacing $I(J; Y)$ with the upper bound

$$I(J; Y) \leq \sum_{i \neq j} \pi_i \pi_j D(p_i \| p_j), \tag{7.48}$$

which follows from (6.9) and the fact that $D(p_j \| \overline{p}) \le \sum_i \pi_i D(p_j \| p_i)$ by convexity of KL divergence. However, in many classification problems $\pi_i \pi_j D(p_i \| p_j)$ is larger than $\ln m$ for at least one pair (i, j), and the inequality (7.48) is loose.

Example 7.7 We apply Fano's inequality to Example 7.5. The mutual information is given by $I(J; Y) = \sum_j \pi_j D(p_j \| \overline{p}) = D(p_1 \| \overline{p})$ by symmetry of the conditional pdfs $\{p_j,\ j = 1, 2, 3, 4\}$. Thus

$$I(J; Y) = D(p_1 \| \overline{p}) = (1 - \lambda) \ln \frac{1 - \lambda}{1/4} + \lambda \ln \frac{\lambda}{1/4} + 0 = -h_2(\lambda) + \ln 4.$$

Fano's inequality (7.46) yields

$$\mathrm{P_e} \ge \frac{h_2(\lambda) - h_2(\mathrm{P_e})}{\ln 3},$$

which is satisfied by $\mathrm{P_e} = \lambda$ but also by values below λ. Hence Fano's inequality is loose for this problem.

Example 7.8 We apply Fano's inequality to Example 7.6. Here we have

$$I(J; Y) = D(p_1 \| \overline{p}) = (1 - \lambda) \ln \frac{1 - \lambda}{1/4} + 3 \frac{\lambda}{3} \ln \frac{\lambda/3}{1/4} = -h_2(\lambda) + \ln 4 - \lambda \ln 3.$$

Fano's inequality yields

$$\mathrm{P_e} \ge \frac{h_2(\lambda) + \lambda \ln 3 - h_2(\mathrm{P_e})}{\ln 3},$$

which is satisfied with equality by $\mathrm{P_e} = \lambda$. Hence Fano's inequality is tight for this problem. This result could also have been predicted from the condition (7.44) for equality in the data-processing inequality: we have $\delta_{\mathrm{B}}(y) = \lceil y \rceil$ and

$$\pi_j p_j(y) = \begin{cases} \frac{\lambda}{4} = 4\lambda\, \pi_j \overline{p}(y) & : \ \delta_{\mathrm{B}}(y) = j \\ \frac{\lambda}{12} = \frac{4}{3} \lambda \pi_j \overline{p}(y) & : \ \delta_{\mathrm{B}}(y) \neq j. \end{cases}$$

7.6 Appendix: Proof of Theorem 7.4

The minimum error probability is achieved by a deterministic MAP rule. Without loss of generality we assume that ties among the posterior probabilities are broken in favor of the lower index. Hence for the binary test between H_i and H_j with $i < j$ with respective priors $\frac{\pi_i}{\pi_i + \pi_j}$ and $\frac{\pi_j}{\pi_i + \pi_j}$, the Bayes rule returns i in the event $\pi_i p_i(Y) = \pi_j p_j(Y)$, and

$$\begin{aligned} \mathrm{P_e}(i, j) = &\frac{\pi_i}{\pi_i + \pi_j} P_i\{\pi_i p_i(Y) < \pi_j p_j(Y)\} \\ &+ \frac{\pi_j}{\pi_i + \pi_j} P_j\{\pi_i p_i(Y) \ge \pi_j p_j(Y)\}. \end{aligned} \qquad (7.49)$$

For the m-ary test, recall that $\mathcal{Y}_i$ denotes the acceptance region for hypothesis H_i, $1 \le i \le m$.

Upper Bound By our tie-breaking convention, for any $i < j$ we have

$$\begin{aligned} y \in \mathcal{Y}_i \quad &\Rightarrow \quad \pi_i p_i(y) \ge \pi_j p_j(y) \quad \Rightarrow \quad P_j(\mathcal{Y}_i) \le P_j\{\pi_i p_i(Y) \ge \pi_j p_j(Y)\}, \\ y \in \mathcal{Y}_j \quad &\Rightarrow \quad \pi_i p_i(y) < \pi_j p_j(y) \quad \Rightarrow \quad P_i(\mathcal{Y}_j) \le P_i\{\pi_i p_i(Y) < \pi_j p_j(Y)\}. \end{aligned}$$

Hence from (7.49) we have the inequality

$$\mathrm{P_e}(i, j) \ge \frac{\pi_i}{\pi_i + \pi_j} P_i(\mathcal{Y}_j) + \frac{\pi_j}{\pi_i + \pi_j} P_j(\mathcal{Y}_i). \tag{7.50}$$

Equality holds if the decision regions for the binary test are $\mathcal{Y}_i$ and $\mathcal{Y}_j$. Therefore

$$\begin{aligned} \mathrm{P_e} &= \sum_{i \ne j} \pi_j P_j(\mathcal{Y}_i) \\ &= \sum_{i<j} [\pi_j P_j(\mathcal{Y}_i) + \pi_i P_i(\mathcal{Y}_j)] \\ &\overset{(a)}{\le} \sum_{i<j} (\pi_i + \pi_j)\mathrm{P_e}(i, j) \\ &\overset{(b)}{=} \sum_{i \ne j} \pi_i \mathrm{P_e}(i, j), \end{aligned}$$

where inequality (a) follows from (7.50), and equality (b) holds because $\mathrm{P_e}(i, j) = \mathrm{P_e}(j, i)$.

Lower Bound For the lower bound of (7.41), we show the following identity:

$$\mathrm{P_e} = \frac{2}{m} \sum_{i \ne j} \pi_i \mathrm{P_e}(i, j) + \frac{1}{m} \sum_{i<j} \int_{(\mathcal{Y}_i \cup \mathcal{Y}_j)^c} |\pi_i p_i(y) - \pi_j p_j(y)| \, d\mu(y). \tag{7.51}$$

Since each integral is nonnegative, the lower bound of (7.41) follows. To show (7.51), observe that for each i, j, we have

$$\begin{aligned} &\pi_i P_i(\mathcal{Y}_i) - \pi_i P_i(\mathcal{Y}_j) + \pi_j P_j(\mathcal{Y}_j) - \pi_j P_j(\mathcal{Y}_i) \\ &= \int_{\mathcal{Y}_i} [\pi_i p_i(y) - \pi_j p_j(y)] \, d\mu(y) + \int_{\mathcal{Y}_j} [\pi_j p_j(y) - \pi_i p_i(y)] \, d\mu(y) \\ &= \int_{\mathcal{Y}_i \cup \mathcal{Y}_j} |\pi_i p_i(y) - \pi_j p_j(y)| \, d\mu(y), \end{aligned} \tag{7.52}$$

where the last equality holds because the integrands are nonnegative over their respective domains. Summing over all (i, j), we obtain

$$\sum_{i,j}[\pi_i P_i(\mathcal{Y}_i) - \pi_i P_i(\mathcal{Y}_j) + \pi_j P_j(\mathcal{Y}_j) - \pi_j P_j(\mathcal{Y}_i)]$$
$$= 2\sum_{i<j}([\pi_i P_i(\mathcal{Y}_i) - \pi_i P_i(\mathcal{Y}_j)] + [\pi_j P_j(\mathcal{Y}_j) - \pi_j P_j(\mathcal{Y}_i)])$$
$$= 2\sum_{i<j}\int_{\mathcal{Y}_i \cup \mathcal{Y}_j} |\pi_i p_i(y) - \pi_j p_j(y)| \, d\mu(y), \tag{7.53}$$

where the last equality follows from (7.52). But direct evaluation of the same sum by decomposition into four terms yields

$$\sum_{i,j}[\pi_i P_i(\mathcal{Y}_i) - \pi_i P_i(\mathcal{Y}_j) + \pi_j P_j(\mathcal{Y}_j) - \pi_j P_j(\mathcal{Y}_i)] = 2m(1 - \mathrm{P_e}) - 2. \tag{7.54}$$

Equating (7.53) and (7.54), we obtain

$$2\sum_{i<j}\int_{\mathcal{Y}_i \cup \mathcal{Y}_j} |\pi_i p_i(y) - \pi_j p_j(y)| \, d\mu(y) = 2(m-1) - 2m\mathrm{P_e}$$

and therefore

$$\mathrm{P_e} = \frac{1}{m}\left[m - 1 - \sum_{i<j}\int_{\mathcal{Y}_i \cup \mathcal{Y}_j} |\pi_i p_i(y) - \pi_j p_j(y)| \, d\mu(y)\right]$$
$$= \frac{1}{m}\sum_{i<j}\left[\pi_i + \pi_j - \int_{\mathcal{Y}_i \cup \mathcal{Y}_j} |\pi_i p_i(y) - \pi_j p_j(y)| \, d\mu(y)\right], \tag{7.55}$$

where the last line holds because $\sum_{i<j}[\pi_i + \pi_j] = \sum_{i \neq j} \pi_i = m - 1$. Now

$$(\pi_i + \pi_j)\mathrm{P_e}(i, j) = \int_{\mathcal{Y}} \min\{\pi_i p_i(y), \pi_j p_j(y)\} \, d\mu(y)$$
$$= \frac{1}{2}\left[\pi_i + \pi_j - \int_{\mathcal{Y}} |\pi_i p_i(y) - \pi_j p_j(y)| \, d\mu(y)\right] \quad \forall i \neq j,$$

where the last equality follows from the identity $\min\{a, b\} = \frac{1}{2}(a + b - |a - b|)$. Substituting into (7.55), we obtain

$$\mathrm{P_e} = \frac{1}{m}\sum_{i<j}\left[2(\pi_i + \pi_j)\mathrm{P_e}(i, j) + \int_{(\mathcal{Y}_i \cup \mathcal{Y}_j)^c} |\pi_i p_i(y) - \pi_j p_j(y)| \, d\mu(y)\right],$$

which proves (7.51). For each (i, j), the integral in the previous equation may be written as

$$\int_{(\mathcal{Y}_i \cup \mathcal{Y}_j)^c} |\pi(i|y) - \pi(j|y)| p(y) \, d\mu(y),$$

which is zero under the condition stated following (7.41).

Exercises

7.1 *Acquiring Observations Cannot Increase Bayes Risk*. Consider a binary hypothesis testing problem H_0 versus H_1 with costs $C_{10} = 2$, $C_{01} = 1$, and $C_{00} = C_{11} = 0$. Give the prior Bayes risk r_0, which is defined as the Bayes risk without observations. Use Jensen's inequality to show that the availability of observations cannot increase that risk. Under what condition on the observational model does the risk actually remain r_0?

7.2 *Chernoff Upper Bound and Kailath Lower Bound on* $\mathrm{P_e}$ *(I)*. Evaluate the Chernoff upper bound (7.36) (for $u \in \{.25, .5, .75\}$) and the Kailath lower bound (7.7) for $P_0 = [.1, .1, .8]$ and $P_1 = [0, .5, .5]$ and a uniform prior.

7.3 *Chernoff Upper Bound and Kailath Lower Bound on* $\mathrm{P_e}$ *(II)*. Evaluate the Chernoff upper bound (7.36) (for $0 < u < 1$) and the Kailath lower bound (7.7) for the following distributions on $[0, 1]$:

$$p_0(y) = 2y, \quad p_1(y) = 3y^2$$

assuming a uniform prior. Evaluate the Chernoff information $C(p_0, p_1)$ and the Chernoff bound in (7.40).

7.4 *Chernoff and Bhattacharyya Bounds*. Consider the binary hypothesis testing problem with

$$p_0(y) = \frac{1}{2}e^{-|y|} \quad \text{and} \quad p_1(y) = e^{-2|y|}.$$

Assume equal priors. Find the Chernoff and Bhattacharya bounds on $\mathrm{P_e}$.

7.5 *Detection in Neural Networks*. Consider the following Bayesian detection problem with *equal priors*:

$$\begin{cases} H_0 : Y_k = -m + Z_k, & k = 1, \dots, n \\ H_1 : Y_k = m + Z_k, & k = 1, \dots, n, \end{cases}$$

where $m > 0$ and $Z_1, Z_2, \dots, Z_n$ are i.i.d. continuous-valued random variables with a symmetric distribution around zero, i.e., $p_{Z_k}(x) = p_{Z_k}(-x)$ for all $x \in \mathbb{R}$.

Suppose we quantize each observation to one bit to form

$$U_k = g(Y_k) = \begin{cases} 1 & Y_k \geq 0 \\ 0 & Y_k < 0. \end{cases}$$

(a) Show that the minimum-error-probability detector based on $U_1, U_2, \dots, U_n$ is a majority-logic detector, i.e.,

$$\delta_{\text{opt}}(\boldsymbol{u}) = \begin{cases} 1 & \text{if } \sum_{k=1}^{n} u_k \geq \frac{n}{2} \\ 0 & \text{otherwise.} \end{cases}$$

(b) Now let us specialize to case where $m = \ln 2$ and Z_k has the Laplacian pdf

$$p_{Z_k}(x) = \frac{1}{2} e^{-|x|}.$$

Find the Chernoff bound on the probability of error of the detector of part (a).

7.6 *Bounds on ROC*. The binary Bhattacharyya distance

$$\begin{aligned} d_{1/2}(a, b) &\triangleq -\ln(\sqrt{ab} + \sqrt{(1-a)(1-b)}) \\ &= B([a, 1-a], [b, 1-b]), \quad 0 \le a, b \le 1 \end{aligned}$$

is the Bhattacharyya distance between two Bernoulli distributions with parameters a and b. Let α and β be the probabilities of false alarm and miss for a binary hypothesis test between p_0 and p_1.

(a) Show that $d_{1/2}(\alpha, 1-\beta) \le B(p_0, p_1)$.
(b) Give a lower bound on β when $\alpha = 0$.
(c) Give a lower bound on β when $\alpha = \beta$.
(d) Give a lower bound on β when $\alpha = \frac{1}{2}$ and $B(p_0, p_1) \ge \ln \sqrt{2}$.
(e) Give a lower bound on α when $\beta = 0$.

7.7 *Ternary Gaussian Hypothesis Testing*. Consider the hypotheses $H_j : Y \sim \mathcal{N}(\mu_j, \mathrm{I}_2)$, $j = 1, 2, 3$ with uniform prior. The mean vectors $\mu_1 = [0, 1]^\top$, $\mu_2 = [\sqrt{3}/2, -1/2]^\top$, and $\mu_3 = [-\sqrt{3}/2, -1/2]^\top$ are the vertices of a simplex. Evaluate the bounds of (7.41). Do you expect $\mathrm{P_e}$ to be closer to the lower or to the upper bound? Justify your answer.

7.8 *9-ary Gaussian Hypothesis Testing*. Assume a uniform prior on the hypotheses $H_j : Y \sim \mathcal{N}(3j - 15, 1)$, $1 \le j \le 9$. Evaluate (7.41). Do you expect $\mathrm{P_e}$ to be closer to the lower or to the upper bound? Justify your answer.

7.9 *Signal Design*. Denote by P_λ the Poisson distribution with parameter $\lambda > 0$. Consider the sequence $\boldsymbol{\lambda} = \{\lambda_i\}_{i=1}^{20}$ whose element λ_i equals 10 for $1 \le i \le 10$ and 20 for $11 \le i \le 20$. We wish to design a sequence $\boldsymbol{s}$ and transmit either $\boldsymbol{s}$ or $-\boldsymbol{s}$ to a receiver which observes a sequence $\boldsymbol{Y}$ at the output of an optical communication channel. The corresponding probability distributions P_0 and P_1 are as follows:

$$\begin{cases} H_0 & : Y_i \overset{\text{indep.}}{\sim} P_{\lambda_i + s_i} \\ H_1 & : Y_i \overset{\text{indep.}}{\sim} P_{\lambda_i - s_i}, \quad 1 \le i \le n. \end{cases}$$

Find a sequence $\boldsymbol{s}$ that maximizes $D(P_0 \| P_1)$ subject to the constraint $\sum_i s_i^2 = 1$. You may use the approximation $\ln(1 + \epsilon) \approx \epsilon - \epsilon^2/2$ for small ϵ.

7.10 *Fano's Inequality*. Let $\mathcal{X} = \mathcal{Y} = \{1, 2, 3, 4\}$. The following matrix gives $p_j(y)$ for each row j:

$$\begin{bmatrix} 1 & 0 & 0 & 0 \\ 0 & 1 & 0 & 0 \\ 0 & 0 & 1 & 0 \\ 1/4 & 1/4 & 1/4 & 1/4 \end{bmatrix}.$$

(a) Derive the MPE and Fano's lower bound on it, assuming a uniform prior.
(b) Derive the MPE and the generalized Fano lower bound on it, assuming a prior $\pi = (1/6, 1/6, 1/6, 1/2)$.

References

[1] S. Kullback and R. Leibler, "On Information and Sufficiency," *Ann. Math. Stat.*, Vol. 22, pp. 79–86, 1951.

[2] T. Kailath, "The Divergence and Bhattacharyya Distance Measures in Signal Selection," *IEEE Trans. Comm. Technology*, Vol. 15, No. 1, pp. 52–60, 1967.

[3] H. Chernoff, "A Measure of Asymptotic Efficiency for Test of a Hypothesis Based on the Sum of Observations," *Ann. Math. Stat.*, Vol. 23, pp. 493–507, 1952.

[4] T. Lissack and K.-S. Fu, "Error Estimation in Pattern recognition via L^α Distance Between Posterior Density Functions," *IEEE Trans. Inf. Theory*, Vol. 22, No. 1, pp. 34–45, 1976.

[5] F. D. Garber and A. Djouadi, "Bounds on Bayes Error Classification Based on Pairwise Risk Functions," *IEEE Trans. Pattern Anal. Mach. Intell.*, Vol. 10, No. 2, pp. 281–288, 1988.

[6] J. T. Chu and J. C. Chueh, "Error Probability in Decision Functions for Character Recognition," *Journal ACM*, Vol. 14, pp. 273–280, 1967.

[7] T. M. Cover and J. A. Thomas, *Elements of Information Theory*, 2nd edition, Wiley, 2006.

[8] T. S. Han and S. Verdú, "Generalizing the Fano Inequality," *IEEE Trans. Inf. Theory*, Vol. 40, No. 4, pp. 1247–1251, 1994.

8 Large Deviations and Error Exponents for Hypothesis Testing

In this chapter, we apply large-deviations theory, whose basis is the Chernoff bound studied in Section 7.3, to derive bounds on hypothesis testing with a large number of i.i.d. observations under each hypothesis [1, 2]. We study the asymptotics of these methods and present tight approximations that are based on the method of exponential tilting and can easily be evaluated.

8.1 Introduction

Consider the following hypothesis testing problem as motivation. The hypotheses are:

$$\begin{cases} H_0 & : Y_k \sim \text{i.i.d. } p_0 \\ H_1 & : Y_k \sim \text{i.i.d. } p_1, \quad 1 \le k \le n. \end{cases} \tag{8.1}$$

The LLRT is given by

$$\ln L(\boldsymbol{Y}) = \sum_{k=1}^{n} L_k \underset{<}{\overset{>}{=}} \tau,$$

where

$$L_k \triangleq \ln \frac{p_1(Y_k)}{p_0(Y_k)}, \quad k = 1, \ldots, n. \tag{8.2}$$

Since Y_k are i.i.d., so are L_k for $k = 1, \ldots, n$ under both H_0 and H_1. For instance, the energy detector of Section 5.7 satisfies

$$L_k = Y_k^2 \sim \text{Gamma}\left(\frac{1}{2}, \frac{1}{2\sigma_j^2}\right) \quad \text{under } P_j, \; j = 0, 1. \tag{8.3}$$

We would like to upper bound the false-alarm and miss probabilities

$$\mathrm{P_F} \le \mathrm{P_0}\left[\sum_{k=1}^{n} L_k \ge \tau\right], \tag{8.4}$$

$$\mathrm{P_M} \le \mathrm{P_1}\left[\sum_{k=1}^{n} L_k \le \tau\right]. \tag{8.5}$$

Both inequalities (8.4) and (8.5) hold with equality if randomization is not needed. The two probabilities on the right hand side can be upper bounded using the tools developed in this chapter, resulting in

$$\mathrm{P_F} \le \mathrm{P_F^{UB}} = e^{-n e_{\mathrm{F}}}, \qquad \mathrm{P_M} \le \mathrm{P_M^{UB}} = e^{-n e_{\mathrm{M}}}, \tag{8.6}$$

where e_{F} and e_{M} are two error exponents that we shall see are easy to compute. We will also use refinements of large-deviations theory to derive good asymptotic approximations $\hat{\mathrm{P}}_{\mathrm{F}}$ and $\hat{\mathrm{P}}_{\mathrm{M}}$ in the sense that

$$\mathrm{P_F} \sim \hat{\mathrm{P}}_{\mathrm{F}} \quad \text{and} \quad \mathrm{P_M} \sim \hat{\mathrm{P}}_{\mathrm{M}} \quad \text{as } n \to \infty, \tag{8.7}$$

where the symbol $\sim$ means that the ratio of left and right hand sides tends to 1 as $n \to \infty$.

8.2 Chernoff Bound for Sum of i.i.d. Random Variables

Consider a sequence of n i.i.d. random variables X_i, $1 \le i \le n$ with common cgf κ_X and large-deviations function $\Lambda_X(a)$. We are interested in upper bounding the probability

$$\mathrm{P}\{S_n \ge na\}, \quad a > \mathbb{E}(X), \tag{8.8}$$

where we have defined the sum

$$S_n = \sum_{i=1}^{n} X_i. \tag{8.9}$$

PROPOSITION 8.1 *The large-deviations bound takes the form*

$$\mathrm{P}\{S_n \ge na\} \le e^{-n\Lambda_X(a)}, \quad a > \mathbb{E}(X). \tag{8.10}$$

Proof Since $X_i, i = 1, \dots, n$ are i.i.d., the cgf of S_n is $\kappa_{S_n}(u) = n\kappa_X(u)$. Hence

$$\mathrm{P}\{S_n \ge na\} \le e^{-n[ua - \kappa_X(u)]} \quad \forall a > \mathbb{E}(X),\ u > 0. \tag{8.11}$$

Optimizing u according to (7.17) proves the claim (8.10). □

In comparison, Chebyshev's bound yields

$$\mathrm{P}\{S_n \ge na\} = \mathrm{P}\{S_n^2 > (na)^2\} \le \frac{n\mathbb{E}(X^2)}{(na)^2} = \frac{\mathbb{E}(X^2)}{na^2},$$

which is inversely proportional to n. Hence the Chebyshev bound is loose for large n.

8.2.1 Cramér's Theorem

The following theorem states that the upper bound in Proposition 8.1 is tight in the exponent, by providing a lower bound on $\mathrm{P}\{S_n \ge na\}$.

THEOREM 8.1 Cramér [3]. *For $a > \mathbb{E}[X]$,*

$$\mathrm{P}\{S_n \ge na\} \le e^{-n\Lambda_X(a)}$$

and given $\epsilon > 0$*, there exists* n_ϵ *such that*

$$\mathrm{P}\{S_n \geq na\} \geq e^{-n(\Lambda_X(a)+\epsilon)} \quad \textit{for all } n > n_\epsilon. \tag{8.12}$$

That is:

$$\lim_{n\to\infty} -\frac{1}{n} \ln \mathrm{P}\{S_n \geq na\} = \Lambda_X(a), \quad \forall a > \mathbb{E}[X]. \tag{8.13}$$

A proof of the lower bound (8.12) is given in [3] (also see Exercises 8.12 and 8.13), and is strengthened in Theorem 8.4 under a finiteness assumption on the third absolute moment of X.

Cramér's theorem says that the tail probability $\mathrm{P}\{S_n \geq na\}$ decays exponentially with n, and the exponent $\Lambda_X(a)$ in the upper bound (8.10) cannot be improved. Note, however, that the ratio r_n of the exact tail probabilities to the upper bound $e^{-n\Lambda_X(a)}$ does not necessarily converge to 1. In fact it generally converges to 0, as seen in the case where X_i are i.i.d. $\mathcal{N}(0, 1)$, in which case $S_n \sim \mathcal{N}(0, n)$ and

$$\mathrm{P}\{S_n \geq na\} = Q(\sqrt{n}\,a) \sim \frac{e^{-na^2/2}}{\sqrt{2\pi na}} = \frac{e^{-n\Lambda_X(a)}}{\sqrt{2\pi na}}, \qquad \text{as } n \to \infty.$$

Hence the ratio r_n decays as $(2\pi na^2)^{-1/2}$. The large-deviations bound (8.10) is said to be *tight in the exponent*.

Cramér's theorem can be generalized slightly in the following way, by exploiting the continuity of the rate function Λ_X (see Exercise 8.1).

THEOREM 8.2 *Suppose* $a_n \to a$ *as* $n \to \infty$ *and* $a > \mathbb{E}[X]$*, then*

$$\lim_{n\to\infty} -\frac{1}{n} \ln \mathrm{P}\{S_n \geq na_n\} = \Lambda_X(a). \tag{8.14}$$

8.2.2 Why is the Central Limit Theorem Inapplicable Here?

The partial sum S_n of (8.9) has mean $n\mathbb{E}(X)$ and variance $n\mathrm{Var}(X)$. The normalized random variable

$$T_n \triangleq \frac{S_n - n\mathbb{E}[X]}{\sqrt{n\mathrm{Var}(X)}}$$

has zero mean and unit variance. By the Central Limit Theorem (CLT), T_n converges in distribution to the normal random variable $\mathcal{N}(0, 1)$. Assume $a > \mathbb{E}[X]$. Since

$$\mathrm{P}\{S_n \geq na\} = \mathrm{P}\left\{T_n > \sqrt{n}\,\frac{a - \mathbb{E}[X]}{\sqrt{\mathrm{Var}(X)}}\right\} \tag{8.15}$$

it seems tempting to use the "Gaussian approximation"

$$\mathrm{P}\{S_n \geq na\} \approx Q\left(\sqrt{n}\,\frac{a - \mathbb{E}[X]}{\sqrt{\mathrm{Var}(X)}}\right).$$

Using the bound $Q(x) \leq e^{-x^2/2}$ for $x > 0$ yields the approximation

$$\mathrm{P}\{S_n \geq na\} \approx \exp\left\{-n\frac{(a - \mathbb{E}[X])^2}{2\mathrm{Var}(X)}\right\}. \tag{8.16}$$

Unfortunately (8.16) disagrees with (8.10), except when X is Gaussian to begin with! In general the exponent in (8.16) could be larger than $\Lambda_X(a)$ or smaller. What went wrong?

The problem lies in an incorrect application of the CLT. Convergence of T_n in distribution merely says that

$$\lim_{n\to\infty} \mathrm{P}\{T_n > z\} = Q(z)$$

for any *fixed* $z \in \mathbb{R}$. However, in (8.15) z is not fixed, in fact it grows as $\sqrt{n}$ as $n \to \infty$.

Unless X is Gaussian, the tails of the distribution of T_n are not Gaussian. For instance, if X is uniformly distributed over the unit interval $[-\frac{1}{2}, \frac{1}{2}]$, we have $-\frac{n}{2} \le S_n \le \frac{n}{2}$. Hence

$$\mathrm{P}\left\{S_n > \frac{n}{2}\right\} = 0, \quad \forall n \ge 1,$$

again showing the inadequacy of the Gaussian approximation (8.16). The Gaussian approximation holds in the *small-deviations regime* (also known as the central regime). It still provides the correct exponent in the so-called *moderate-deviations regime* but fails in the large-deviations regime.[1]

8.3 Hypothesis Testing with i.i.d. Observations

We return to our detection problem with i.i.d. observations (8.1) and use large-deviations methods to upper bound $\mathrm{P_F}$ and $\mathrm{P_M}$. The case of i.i.d. observations is a special case of (7.22) with Y replaced by an i.i.d. sequence $\boldsymbol{Y} = (Y_1, \dots, Y_n)$ and p_0 and p_1 replaced by product distributions p_0^n and p_1^n. For each n, LLRTs take the form

$$\tilde{\delta}^{(n)}(y) = \begin{cases} 1 & & > \\ 1 \text{ w.p. } \gamma_n & \text{if } S_n & = n\tau_n, \\ 0 & & < \end{cases} \tag{8.17}$$

where

$$S_n = \sum_{k=1}^{n} \ln \frac{p_1(Y_k)}{p_0(Y_k)} \tag{8.18}$$

is the test statistic, and $n\tau_n$ is the test threshold. Note that S_n has means under H_0 and H_1 that are respectively given by

$$\mathbb{E}_0[S_n] = -nD(p_0\|p_1), \quad \mathbb{E}_1[S_n] = nD(p_1\|p_0).$$

Assume that $-D(p_0\|p_1) < \tau_n < D(p_1\|p_0)$. (Other values of τ_n can be used as discussed in Section 7.4.1 but result in a trivial upper bound on either $\mathrm{P_F}$ or $\mathrm{P_M}$.) The cgf

[1] Moderate deviations refers to the evaluation of tail probabilities such as $\mathrm{P}[S_n > n^\beta a]$ where $\beta \in (\frac{1}{2}, 1)$ is a fixed constant. Then it may be shown that

$$\lim_{n\to\infty} -\frac{1}{n^{2\beta-1}} \ln \mathrm{P}\{S_n > n^\beta a\} = \frac{(a - \mathbb{E}[X])^2}{2\mathrm{Var}[X]}.$$

for S_n is n times the cgf for the random variable $L = \ln \frac{p_1(Y)}{p_0(Y)}$. Applying (7.33) and (7.34), we obtain the following bounds on $\mathrm{P_F}$ and $\mathrm{P_M}$:

$$\mathrm{P_F}(\tilde{\delta}^{(n)}) \le e^{-n\Lambda_0(\tau_n)}, \tag{8.19}$$

$$\mathrm{P_M}(\tilde{\delta}^{(n)}) \le e^{-n[\Lambda_0(\tau_n)-\tau_n]}. \tag{8.20}$$

If the sequence τ_n is bounded away from $-D(p_0\|p_1)$ and $D(p_1\|p_0)$, we see that both error probabilities vanish exponentially fast as $n \to \infty$. If the sequence τ_n converges to a limit $\tau \in (-D(p_0\|p_1), D(p_1\|p_0))$, then we may apply Theorem 8.2 along with the fact that (due to possible randomization)

$$\mathrm{P_0}\{S_n > n\tau_n\} \le \mathrm{P_F}(\tilde{\delta}^{(n)}) \le \mathrm{P_0}\{S_n \ge n\tau_n\}$$

to conclude that the error exponent for $\mathrm{P_F}$ is given by

$$\mathrm{e_F}(\tau) = -\lim_{n\to\infty} \ln \mathrm{P_F}(\tilde{\delta}^{(n)}) = \Lambda_0(\tau) = u^*\tau - \kappa_0(u^*). \tag{8.21}$$

Similarly we can show that the error exponent for $\mathrm{P_M}$ is given by

$$\mathrm{e_M}(\tau) = -\lim_{n\to\infty} \ln \mathrm{P_M}(\tilde{\delta}^{(n)}) = \Lambda_1(\tau) = -\tau + \mathrm{e_F}(\tau). \tag{8.22}$$

It is clear from (8.21) and (8.22) that randomization (i.e., γ_n) plays no role in the error exponents.

8.3.1 Bayesian Hypothesis Testing with i.i.d. Observations

The Bayes LLRT for i.i.d. observations takes the form

$$\delta_{\mathrm{B}}^{(n)}(y) = \begin{cases} 1 & \text{if } \dfrac{S_n}{n} \ge \tau_n = \frac{1}{n}\ln\frac{\pi_0}{\pi_1} \\ 0 & \text{otherwise.} \end{cases}$$

Note that $\tau_n \to 0$ as $n \to \infty$, and therefore from (8.21) and (8.22) it follows that

$$\mathrm{e_F} = \lim_{n\to\infty} -\frac{1}{n} \ln \mathrm{P_F}(\delta_{\mathrm{B}}^{(n)}) = \Lambda_0(0) = C(p_0, p_1)$$

and that

$$\mathrm{e_M} = \lim_{n\to\infty} -\frac{1}{n} \ln \mathrm{P_M}(\delta_{\mathrm{B}}^{(n)}) = \Lambda_1(0) = \Lambda_0(0) = C(p_0, p_1).$$

Combining these results, we get that $\mathrm{P_e} = \pi_0\mathrm{P_F} + \pi_1\mathrm{P_M}$ has an error exponent given by

$$\lim_{n\to\infty} -\frac{1}{n} \ln \mathrm{P_e}(\delta_{\mathrm{B}}^{(n)}) = C(p_0, p_1). \tag{8.23}$$

Finally, the error exponent for minimax hypothesis testing can be shown to be the same as that for Bayesian hypothesis testing, and is given by the Chernoff information $C(p_0, p_1)$ (see Exercise 8.5).

8.3.2 Neyman–Pearson Hypothesis Testing with i.i.d. Observations

The Neyman–Pearson LLRT for i.i.d. observations takes the form

$$\tilde{\delta}_{\text{NP}}^{(n)}(y) = \begin{cases} 1 & > \\ 1 \text{ w.p. } \gamma_n \quad \text{if} \quad \dfrac{S_n}{n} & = \tau_n, \\ 0 & < \end{cases} \tag{8.24}$$

where γ_n and τ_n are chosen so that $\mathrm{P_F}(\tilde{\delta}_{\text{NP}}^{(n)}) = \alpha$, $\alpha \in (0, 1)$. Since $\mathrm{P_F}$ is fixed at α for all n, we are interested in characterizing the error exponent for $\mathrm{P_M}$. This is given in the following theorem, which is proved in Section 8.5.

LEMMA 8.1 Chernoff–Stein[2]. *For every* $\alpha \in (0, 1)$,

$$\mathrm{e_M} = \lim_{n\to\infty} -\frac{1}{n} \ln \mathrm{P_M}(\tilde{\delta}_{\text{NP}}^{(n)}) = D(p_0 \| p_1).$$

An intuitive argument that explains the Chernoff–Stein lemma is as follows. Under H_0, the LLRT statistic S_n of (8.18) is the sum of n i.i.d. random variables with common mean $-D(p_0\|p_1)$ and variance $\sigma^2 = \mathrm{Var}_0\left[\ln \frac{p_1(Y)}{p_0(Y)}\right]$. Assume $0 < \sigma^2 < \infty$. By the Central Limit Theorem, the scaled LLRT statistic

$$\frac{S_n + nD(p_0\|p_1)}{\sqrt{n}}$$

converges in distribution to $\mathcal{N}(0, \sigma^2)$. To achieve a false-alarm probability $\mathrm{P_F} \approx \alpha$, we need an LLRT threshold

$$\tau_n \sim -D(p_0\|p_1) + \frac{1}{\sqrt{n}} \sigma Q^{-1}(\alpha).$$

Hence we must have $\lim_n \tau_n = -D(p_0\|p_1)$. Then by (8.22)

$$\mathrm{e_M} = -(-D(p_0\|p_1)) + \mathrm{e_F} = D(p_0\|p_1)$$

since $\mathrm{e_F} = 0$ due to the constraint $\mathrm{P_F}(\tilde{\delta}_{\text{NP}}^{(n)}) = \alpha$.

REMARK 8.1 By reversing the roles of H_0 and H_1 in the Chernoff–Stein lemma, it follows that if we fix $\mathrm{P_M}(\tilde{\delta}_{\text{NP}}^{(n)}) = \beta$, $\beta \in (0, 1)$, then

$$\mathrm{e_F} = \lim_{n\to\infty} -\frac{1}{n} \ln \mathrm{P_F}(\tilde{\delta}_{\text{NP}}^{(n)}) = D(p_1 \| p_0).$$

8.3.3 Hoeffding Problem

Related to the Neyman–Pearson criterion is the following asymptotic criterion for hypothesis testing, first introduced in a paper by Hoeffding [4]. If $\tilde{\delta}^{(n)}$ is a sequence

[2] This result first appeared in a paper by Chernoff [1], who credits it to an unpublished manuscript by Stein.

of tests between the product distributions p_0^n and p_1^n, $n = 1, 2, \ldots$, the criterion can be described as

$$\text{maximize } \lim_{n\to\infty} -\frac{1}{n} \ln \mathrm{P_M}(\tilde{\delta}^{(n)}), \text{ subject to } \lim_{n\to\infty} -\frac{1}{n} \ln \mathrm{P_F}(\tilde{\delta}^{(n)}) \geq \gamma \tag{8.25}$$

for $\gamma \in (0, D(p_1 \| p_0))$. This criterion can be interpreted as the asymptotic version of the Neyman–Pearson criterion.

We know from Lemma 8.1 (reversing the roles of H_0 and H_1) that the maximum value that the error exponent can take under H_0 is equal to $D(p_1 \| p_0)$ and the minimum value is 0, and therefore Hoeffding's criterion (8.25) is well defined only if $\gamma \in [0, D(p_1 \| p_0)]$. The end points of this interval correspond to the cases covered by Lemma 8.1, and so we further restrict γ to the open interval as in (8.25).

To avoid technicalities in solving (8.25), we restrict our attention to tests of the form

$$\delta^{(n)}(y) = \begin{cases} 1 & \text{if } S_n \geq n\tau_n \\ 0 & \text{otherwise,} \end{cases}$$

where τ_n converges to a limit $\tau \in (-D(p_0 \| p_1), D(p_1 \| p_0))$. Note that we do not consider randomization since we already showed that randomization does not play a role in the error exponents for the constant threshold test (see (8.21) and (8.22)). In addition, we have from (8.21) and (8.22) that

$$\mathrm{e_F}(\tau) = \lim_{n\to\infty} -\frac{1}{n} \ln \mathrm{P_F}(\delta_\tau^{(n)}) = \Lambda_0(\tau)$$

and

$$\mathrm{e_M}(\tau) = \lim_{n\to\infty} -\frac{1}{n} \ln \mathrm{P_M}(\delta_\tau^{(n)}) = \Lambda_1(\tau) = \Lambda_0(\tau) - \tau.$$

From Figure 7.4 it is clear that choosing τ such that $\Lambda_0(\tau) = \gamma$ maximizes $\Lambda_1(\tau)$. Thus, the solution to (8.25) is given by τ_{H} that satisfies

$$\Lambda_0(\tau_{\mathrm{H}}) = \gamma. \tag{8.26}$$

The direct way to compute τ_{H} is to compute for each τ, $\Lambda_0(\tau) = u^*\tau - \kappa_0(u^*)$, with u^* being the solution to $\kappa_0'(u^*) = \tau$, and then solving (8.26). An alternative approach that has a convenient geometric interpretation is as follows.

Consider the geometric mixture between p_0 and p_1 defined in (6.17)

$$p_u(y) = \frac{p_0^{1-u}(y) p_1^u(y)}{\int_{\mathcal{Y}} p_0^{1-u}(t) p_1^u(t) d\mu(t)} = \frac{p_0(y) L(y)^u}{\mathbb{E}_0 [L(Y)^u]} = \frac{p_0(y) L(y)^u}{\exp(\kappa_0(u))}, \quad u \in (0, 1).$$

Then,

$$D(p_u \| p_0) = \mathbb{E}_u \left[\ln \frac{p_u(Y)}{p_0(Y)} \right] = u \mathbb{E}_u [\ln L(Y)] - \kappa_0(u). \tag{8.27}$$

Similarly,

$$D(p_u \| p_1) = (u - 1) \mathbb{E}_u [\ln L(Y)] - \kappa_0(u). \tag{8.28}$$

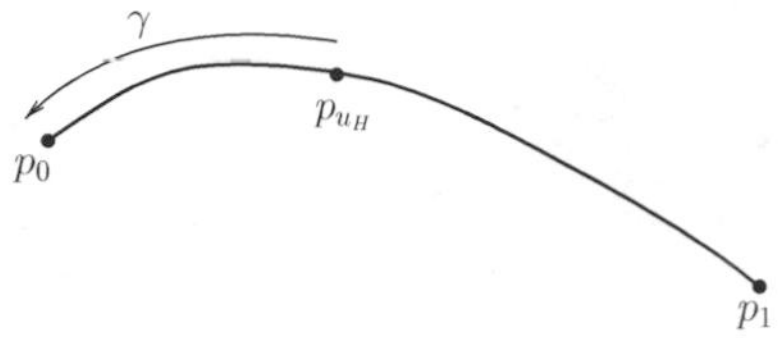

Figure 8.1 Solution to the Hoeffding problem

Now

$$\mathbb{E}_u [\ln L(Y)] = \frac{\mathbb{E}_0 \left[L(Y)^u \ln L(Y)\right]}{\mathbb{E}_0 [L(Y)^u]} = \frac{d}{du} \ln \mathbb{E}_0 \left[L(Y)^u\right] = \kappa_0'(u).$$

Noting that $\kappa_0'(u^*) = \tau$, and plugging $\kappa_0'(u)$ into (8.27) and (8.28), we get

$$\begin{aligned} D(p_{u^*} \| p_0) &= \tau u^* - \kappa_0(u^*) = \Lambda_0(\tau), \\ D(p_{u^*} \| p_1) &= \tau(u^* - 1) - \kappa_0(u^*) = \Lambda_1(\tau), \end{aligned} \tag{8.29}$$

and therefore

$$\tau = D(p_{u^*} \| p_0) - D(p_{u^*} \| p_1). \tag{8.30}$$

In light of (8.29), the solution to (8.25) is obtained by finding u_{H} such that

$$D(p_{u_{\mathrm{H}}} \| p_0) = \gamma.$$

The LLRT threshold that solves the Hoeffding problem is then given by

$$\tau_{\mathrm{H}} = \kappa_0'(u_{\mathrm{H}}). \tag{8.31}$$

The corresponding maximal error exponent under H_1 is given by $e_{\mathrm{M}}(\tau_{\mathrm{H}}) = D(p_{u_{\mathrm{H}}} \| p_1)$. This is illustrated in Figure 8.1. Note that $p_{u_{\mathrm{H}}} = p_0$ and $p_{u_{\mathrm{H}}} = p_1$ recover the solutions to the Neyman–Pearson detection problem as described in Lemma 8.1 and Remark 8.1. Furthermore, the Bayesian setting corresponds to having $\tau = 0$, for which

$$D(p_{u^*} \| p_0) = D(p_{u^*} \| p_1) = \Lambda_0(0) = \Lambda_1(0) = C(p_0, p_1).$$

In this case p_{u^*} lies in the "middle" of p_0 and p_1.

8.3.4 Example

Example 8.1 *Testing between i.i.d. exponentials.* Consider the hypothesis test

$$\begin{cases} H_0 & : Y_k \sim \text{i.i.d. } p_0 \\ H_1 & : Y_k \sim \text{i.i.d. } p_1, \quad 1 \le k \le n \end{cases} \tag{8.32}$$

where

$$p_j(y) = \lambda_j \exp(-\lambda_j y)\, \mathbb{1}\{y \ge 0\}, \; j = 0, 1,$$

with $\lambda_1 > \lambda_0$. Define the ratio

$$\xi = \frac{\lambda_1}{\lambda_0}. \tag{8.33}$$

Computing the cgf $\kappa_0(u)$ The likelihood ratio is given by

$$L(y) = \frac{\lambda_1}{\lambda_0} \exp(-(\lambda_1 - \lambda_0)y) \, \mathbb{1}\{y \geq 0\}.$$

Then

$$\kappa_0(u) = \ln \mathbb{E}_0 \left[L(Y)^u \right] = u \ln \frac{\lambda_1}{\lambda_0} + \ln \mathbb{E}_0 \left[\exp(-(\lambda_1 - \lambda_0)Yu) \right]$$

with

$$\begin{aligned} \mathbb{E}_0 \left[\exp(-(\lambda_1 - \lambda_0)Yu) \right] &= \lambda_0 \int_0^\infty \exp(-(\lambda_1 - \lambda_0)yu) \exp(-\lambda_0 y) dy \\ &= \frac{\lambda_0}{\lambda_1 u + \lambda_0(1-u)}. \end{aligned}$$

Thus

$$\begin{aligned} \kappa_0(u) &= u \ln \frac{\lambda_1}{\lambda_0} + \ln \frac{\lambda_0}{\lambda_1 u + \lambda_0(1-u)} \\ &= u \ln \xi - \ln(\xi u + 1 - u). \end{aligned} \tag{8.34}$$

Hence

$$\kappa_0'(u) = \ln \xi - \frac{\xi - 1}{\xi u + 1 - u} \tag{8.35}$$

and the error exponents for the Chernoff–Stein problem (Lemma 8.1) are given by

$$\begin{aligned} D(p_0 \| p_1) &= -\kappa_0'(0) = -\ln \xi + \xi - 1 = 2\psi(\xi), \\ D(p_1 \| p_0) &= \kappa_0'(1) = \ln \xi + \frac{1}{\xi} - 1 = 2\psi\left(\frac{1}{\xi}\right), \end{aligned}$$

where ψ is defined in (6.4); see also Figure 6.1.

Computing the Error Exponents e_F and e_M for an LLRT Consider an LLRT of the form (8.17) with threshold τ_n converging to $\tau \in (-D(p_0\|p_1), D(p_1\|p_0))$. Note that $e_F = \Lambda_0(\tau)$ and $e_M = \Lambda_1(\tau)$, where $\Lambda_0(\tau)$ and $\Lambda_1(\tau)$ can be computed from the solution u^* to the equation $\kappa_0'(u^*) = \tau$, i.e.,

$$\ln \xi - \frac{\xi - 1}{\xi u^* + 1 - u^*} = \tau.$$

This yields

$$u^* = \frac{1}{\ln \xi - \tau} - \frac{1}{\xi - 1}. \tag{8.36}$$

And after some straightforward algebra, we can compute

$$e_F(\tau) = \Lambda_0(\tau) = \tau u^* - \kappa_0(u^*) = 2\psi\left(\frac{\ln \xi - \tau}{\xi - 1}\right) \tag{8.37}$$

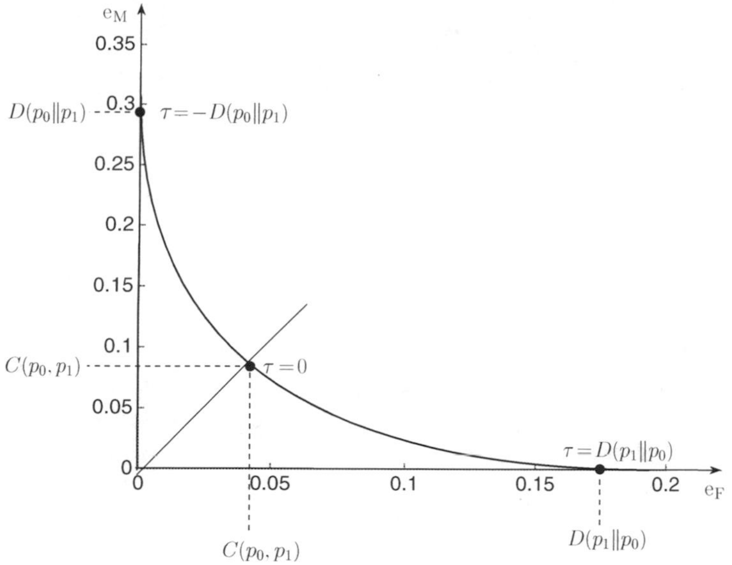

Figure 8.2 Tradeoff between the false-alarm and miss exponents for the hypothesis test of (8.32). Here $\xi = 2$, $D(p_0 \| p_1) = 2\psi(2)$, and $D(p_1 \| p_0) = 2\psi(1/2)$

and

$$e_M(\tau) = \Lambda_1(\tau) = \Lambda_0(\tau) - \tau = 2\psi\left(\frac{\ln \xi - \tau}{\xi - 1}\right) - \tau. \tag{8.38}$$

Figure 8.2 shows the tradeoff between e_M and e_F as τ is varied between the limits of $-D(p_0 \| p_1)$ and $D(p_1 \| p_0)$.

Computing the Chernoff Information $C(p_0, p_1)$ The error exponent for Bayesian hypothesis testing $C(p_0, p_1)$ can be computed using (8.37) as

$$C(p_0, p_1) = \Lambda_0(0) = 2\psi\left(\frac{\ln \xi}{\xi - 1}\right).$$

Hoeffding Problem Finally, we consider the solution to the Hoeffding problem (8.25). To this end, we first compute the geometric mixture pdf

$$\begin{aligned} p_u(y) &= \frac{p_0(y) L(y)^u}{\exp(\kappa_0(u))} \\ &= \lambda_0(\xi u + (1-u)) \exp\left(-\lambda_0(\xi u + (1-u)) y\right) \mathbb{1}\{y \geq 0\}. \end{aligned}$$

Then

$$\begin{aligned} D(p_u \| p_0) &= \mathbb{E}_u\left[\ln \frac{p_u(Y)}{p_0(Y)}\right] \\ &= \ln\left(\xi u + (1-u)\right) + \frac{1}{\xi u + (1-u)} - 1 \\ &= 2\psi\left(\frac{1}{\xi u + (1-u)}\right). \end{aligned}$$

We now need to solve for u_H that satisfies $D(p_{u_H} \| p_0) = \gamma$, which can done numerically through the function ψ, keeping in mind that $u_H \in (0, 1)$.

Finally we can compute the LLRT threshold for the Hoeffding test as (see (8.26))

$$\tau_H = \kappa_0'(u_H) = \ln \xi - \frac{\xi - 1}{\xi u_H + 1 - u_H}.$$

8.4 Refined Large Deviations

We now study the problem stated in (8.7): find exact asymptotic expressions for P_F and P_M. To do so, we first revisit the generic large-deviations problem of Section 7.3.2 and introduce the method of exponential tilting. Then we consider sums of i.i.d. random variables as in Section 8.2 and derive the desired asymptotics. The results are then applied to binary hypothesis testing.

8.4.1 The Method of Exponential Tilting

Let p be a pdf on X (continuous, discrete, or mixed) with respect to a measure μ. Fix some $u > 0$ and define the **tilted distribution** (also known as *exponentially tilted distribution*, or *twisted distribution*)

$$\tilde{p}(x) \triangleq \frac{e^{ux} p(x)}{\int e^{ux} p(x) \, d\mu(x)} = e^{ux - \kappa(u)} p(x), \quad u \in \mathbb{R}. \tag{8.39}$$

This equation defines a mapping from p to $\tilde{p}$ named the *Esscher transform* [5]. Observe that $\tilde{p}(x) = p(x)$ in the case $u = 0$. The first two cumulants of $\tilde{p}$ are given by

$$\mathbb{E}_{\tilde{p}}(X) = \int x \tilde{p}(x) \, d\mu(x) = \frac{\int x e^{ux} p(x) d\mu(x)}{\int e^{ux} p(x) \, d\mu(x)} = \kappa'(u),$$

$$\mathrm{Var}_{\tilde{p}}(X) = \kappa''(u).$$

(This proves that $\kappa''(u) \geq 0$, hence that the cgf is convex.)

Consider $a > \mathbb{E}(X)$. Using the definition (8.39), we derive the identity

$$\begin{aligned} \mathrm{P}\{X \geq a\} &= \int_a^\infty p(x) \, d\mu(x) \\ &= \int_a^\infty e^{-[ux - \kappa(u)]} \, \tilde{p}(x) \, d\mu(x) \\ &= e^{-[ua - \kappa(u)]} \int_a^\infty e^{-u(x-a)} \, \tilde{p}(x) \, d\mu(x) \qquad (8.40) \\ &= e^{-[ua - \kappa(u)]} \, \mathbb{E}_{\tilde{p}} \left[e^{-u(X-a)} \mathbb{1}\{X \geq a\} \right]. \end{aligned}$$

Observe that the integral in (8.40) is the expectation of the function $e^{-u(x-a)} \, \mathbb{1}\{x \geq a\} \leq 1$ when the random variable X is distributed as $\tilde{p}$. For $u \geq 0$, the value of the integral is at most 1, hence the large-deviations upper bound (7.15) follows directly and without applying Markov's inequality. Now we choose $u = u^*$ satisfying $\kappa'(u^*) = a$.

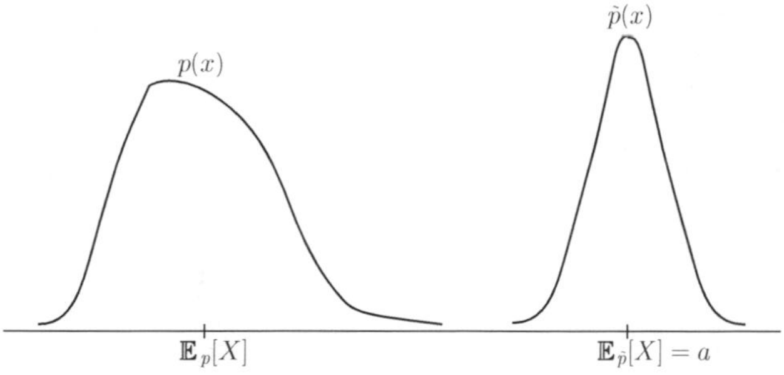

Figure 8.3 Original and tilted pdfs

The tilted distribution has mean a, as illustrated in Figure 8.3. Define the normalized random variable

$$V = \frac{X - \mathbb{E}_{\tilde{p}}(X)}{\sqrt{\mathrm{Var}_{\tilde{p}}(X)}} = \frac{X - \kappa'(u^*)}{\sqrt{\kappa''(u^*)}} = \frac{X - a}{\sqrt{\kappa''(u^*)}}, \tag{8.41}$$

which has zero mean, unit variance, and cdf F_V. Using this notation and changing variables, we write (8.40) as

$$\mathrm{P}\{X \geq a\} = e^{-[u^* a - \kappa(u^*)]} \int_0^\infty e^{-u^* \sqrt{\kappa''(u^*)}\, v}\, dF_V(v). \tag{8.42}$$

The problem is now to evaluate the integral, which in many problems is much simpler than evaluating the original tail probability $\mathrm{P}\{X \geq a\}$.

8.4.2 Sum of i.i.d. Random Variables

Now let $X_i, i \geq 1$ be an i.i.d. sequence drawn from a distribution P. The method of exponential tilting can be applied to the sum $\sum_{i=1}^n X_i$. The cgf for $\sum_{i=1}^n X_i$ is n times the cgf κ for X. Fix $a > \mathbb{E}[X]$ and let u^* be the solution to $\kappa'(u^*) = a$. Analogously to (8.41), the normalized sum

$$V_n \triangleq \frac{\sum_{i=1}^n X_i - na}{\sqrt{n\kappa''(u^*)}} \tag{8.43}$$

has zero mean, unit variance, and cdf F_n. By application of (8.42), the following identity, due to Esscher, holds [5]:

$$\mathrm{P}\left\{\sum_{i=1}^n X_i \geq na\right\} = e^{-n[u^* a - \kappa(u^*)]} \int_0^\infty e^{-u^* \sqrt{n\kappa''(u^*)}\, v}\, dF_n(v). \tag{8.44}$$

As $n \to \infty$, it follows from the Central Limit Theorem that V_n converges in distribution to a normal distribution, as illustrated in Figure 8.4. This property can be exploited to derive exact asymptotics of (8.44) as $n \to \infty$, as well as lower bounds.

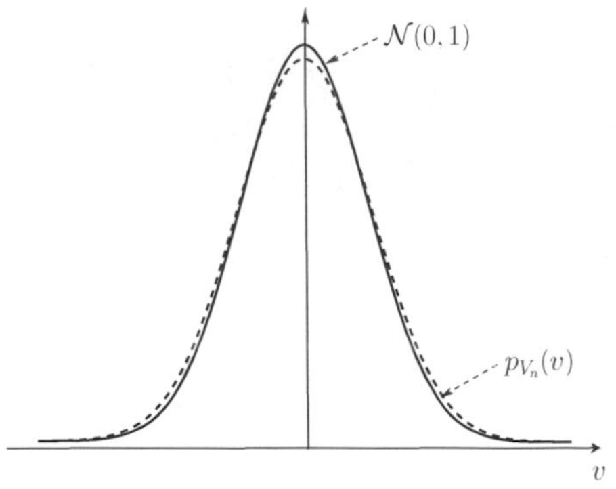

Figure 8.4 Convergence of V_n to a $\mathcal{N}(0, 1)$ random variable

The main result on exact asymptotics is the Bahadur–Rao theorem [6, 7] [3, Ch. 3.7] which applies to the case where X is a continuous random variable, or a discrete random variable of the *nonlattice type*.[3]

THEOREM 8.3 Bahadur–Rao. *Let X be a random variable of the nonlattice type with cgf κ. Fix $a > \mathbb{E}[X]$ and let u^* be the solution to $\kappa'(u^*) = a$. Let $X_i, i \geq 1$ be an i.i.d. sequence of such random variables. Then the following asymptotic equality holds:*

$$\mathrm{P}\left\{\sum_{i=1}^{n} X_i \geq na\right\} \sim \frac{\exp\{-n[u^*a - \kappa(u^*)]\}}{u^*\sqrt{2\pi\kappa''(u^*)\,n}} = \frac{\exp\{-n\Lambda_X(a)\}}{u^*\sqrt{2\pi\kappa''(u^*)\,n}} \quad \text{as } n \to \infty. \tag{8.45}$$

We have used the standard notation $f(n) \sim g(n)$ for asymptotic equality of two functions f and g as $n \to \infty$, i.e., $\lim_{n\to\infty} \frac{f(n)}{g(n)} = 1$. Owing to the term $\sqrt{n}$ in the denominator of (8.45), we now see that the upper bound of (8.11) is increasingly **loose** as $n \to \infty$. (However, this is sometimes not a concern because an upper bound that is tight in the exponent may be satisfactory.)

Sketch of Proof of Theorem 8.3 For simplicity the proof is sketched in the case X is a continuous random variable. Since X is a continuous random variable, so is V_n for each n. The density of V_n is $p_{V_n}(v) = dF_n(v)/dv$, and

$$\lim_{n\to\infty} p_{V_n}(0) = \frac{1}{\sqrt{2\pi}}. \tag{8.46}$$

While convergence might be slow in the tails of p_{V_n}, we now analyze (8.44) and show that only the value of p_{V_n} at $v = 0$ matters. Indeed let

$$\epsilon = \frac{1}{u^*\sqrt{n\kappa''(u^*)}}, \tag{8.47}$$

[3] A real random variable X is said to be of the lattice type if there exist numbers d and x_0 such that X belongs to the lattice $\{x_0 + kd,\ k \in \mathbb{Z}\}$ with probability 1. The largest d for which this holds is called the *span* of the lattice, and x_0 is the *offset*.

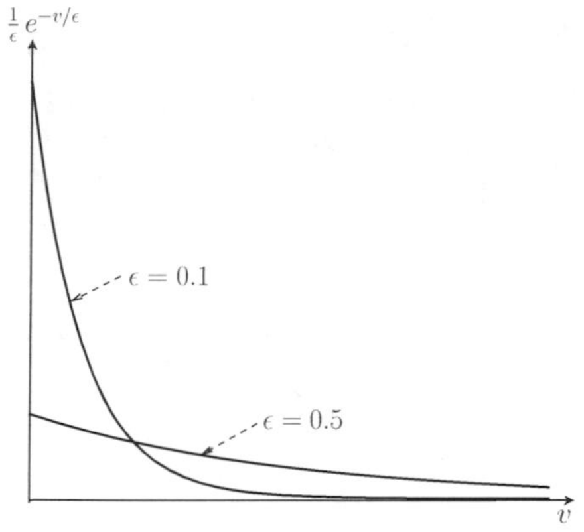

Figure 8.5 Functions $\frac{1}{\epsilon}\, e^{-v/\epsilon},\ v \geq 0$

which tends to zero as $n^{-1/2}$. The function $\frac{1}{\epsilon}\, e^{-v/\epsilon},\ v \geq 0$ integrates to 1, and concentrates at $v = 0$ in the limit as $\epsilon \downarrow 0$ (see Figure 8.5). Hence the integral in (8.44) is of the form

$$\int_0^\infty e^{-v/\epsilon}\, p_{V_n}(v)\, dv = \epsilon \int_0^\infty \underbrace{\left(\frac{1}{\epsilon}\, e^{-v/\epsilon}\right)}_{\text{“Dirac-like”}} p_{V_n}(v)\, dv \overset{(a)}{\sim} \epsilon\, p_{V_n}(0) \overset{(b)}{\sim} \frac{\epsilon}{\sqrt{2\pi}} \quad \text{as } \epsilon \to 0, \tag{8.48}$$

where asymptotic equality (a) holds because p_{V_n} is continuous at $v = 0$, and (b) because of (8.46). A formal proof of this claim (taking into account that ϵ is actually a function of n) can be found e.g., in [3]. Combining (8.44), (8.47), and (8.48), we obtain the desired asymptotic equality (8.45). □

If X is a discrete random variable of the lattice type, the constant $(2\pi)^{-1/2}$ is replaced by a bounded oscillatory function of n, as derived by Blackwell and Hodges [6].

Special Case If $X_i,\ 1 \leq i \leq n$ are i.i.d. $\mathcal{N}(0, 1)$, then V_n is also $\mathcal{N}(0, 1)$ for each n. To evaluate (8.45), we note that $\kappa(u) = u^2/2$ and that $u^* = a$, as shown in Example 7.1. Thus (8.45) becomes

$$\mathrm{P}\left\{\sum_{i=1}^n X_i \geq na\right\} \sim \frac{e^{-na^2/2}}{a\sqrt{2\pi n}} \quad \text{as } n \to \infty. \tag{8.49}$$

This expression could have been derived in a more direct way for this special case. Since $\sum_{i=1}^n X_i \sim \mathcal{N}(0, n)$, we have

$$\mathrm{P}\left\{\sum_{i=1}^n X_i \geq na\right\} = Q(a\sqrt{n}).$$

Using the asymptotic formula (5.41) for the Q function with $x = a\sqrt{n}$ results immediately in the asymptotic expression (8.49).

8.4.3 Lower Bounds on Large-Deviations Probabilities

One can derive lower bounds on the large-deviations probability $\mathrm{P}\left\{\sum_{i=1}^n X_i \ge na\right\}$ using appropriate probability inequalities. Shannon and Gallager derived a lower bound in which the leading exponential term $\exp\{-n\Lambda_X(a)\}$ is multiplied by an $\exp\{O(\sqrt{n})\}$ term. Theorem 8.4 provides a sharper lower bound in case the random variables have finite absolute third moment.

THEOREM 8.4 *Let X be a random variable with cgf κ. Fix $a > \mathbb{E}(X)$ and let u^* be the solution to $\kappa'(u^*) = a$. Denote by $\rho = \mathbb{E}(|X|^3)/\mathrm{Var}(X)$ the normalized absolute third moment of X under the tilted distribution. Let X_i, $i \ge 1$ be an i.i.d. sequence of such random variables. Then*

$$\mathrm{P}\left\{\sum_{i=1}^n X_i \ge na\right\} \ge \frac{c}{\sqrt{n}} e^{-n\Lambda_X(a)} \quad \forall n \ge \left(\frac{1}{u^*\sqrt{\kappa''(u^*)}} + \rho\sqrt{2\pi e}\right)^2, \tag{8.50}$$

where the constant

$$c = \frac{\exp\{-1 - \rho u^* \sqrt{2\pi e \kappa''(u^*)}\}}{u^* \sqrt{2\pi e \kappa''(u^*)}}.$$

Comparing with the Bahadur–Rao Theorem 8.3, we see that the lower bound (8.50) yields the same $O(1/\sqrt{n})$ subexponential correction, with a lower asymptotic constant.

Proof of Theorem 8.4 The starting point is the Esscher identity (8.44), which can be restated as

$$\mathrm{P}\left\{\sum_{i=1}^n X_i \ge na\right\} = e^{-n\Lambda_X(a)}\, I_n, \tag{8.51}$$

where

$$I_n \triangleq \int_0^\infty e^{-u^*\sqrt{n\kappa''(u^*)}\,v}\, dF_n(v). \tag{8.52}$$

We now derive a lower bound on I_n. By the Central Limit Theorem, the normalized random variable V_n converges in distribution to $\mathcal{N}(0, 1)$. Since X_i have normalized absolute third moment, the deviation of the cdf of V_n from the Gaussian cdf can be bounded using the Berry–Esséen theorem [8]:

$$\forall n \ge 1: \quad \sup_{v\in\mathbb{R}} |\Phi(v) - F_n(v)| \le \frac{\rho}{2\sqrt{n}},$$

where F_n denotes the cdf of V_n. Fix any $b > 0$, let $h = (b+1)\rho\sqrt{2\pi e}$, and assume $n \ge h^2$. Since the integrand of (8.52) is nonnegative, we can lower bound I_n by truncating the integral:

$$\begin{aligned} I_n &\ge \int_0^{h/\sqrt{n}} e^{-u^*\sqrt{n\kappa''(u^*)}\,v}\, dF_n(v) \\ &\ge e^{-hu^*\sqrt{\kappa''(u^*)}} \int_0^{h/\sqrt{n}} dF_n(v). \end{aligned}$$

Now

$$\int_0^{h/\sqrt{n}} dF_n(v) = F_n\left(\frac{h}{\sqrt{n}}\right) - F_n(0)$$
$$\geq \Phi\left(\frac{h}{\sqrt{n}}\right) - \Phi(0) - \frac{\rho}{\sqrt{n}}$$
$$\geq \frac{h}{\sqrt{n}}\phi\left(\frac{h}{\sqrt{n}}\right) - \frac{\rho}{\sqrt{n}}$$
$$\geq \frac{h\phi(1) - \rho}{\sqrt{n}} = \frac{b\rho}{\sqrt{n}},$$

where the first inequality follows from the Berry–Esséen theorem, the second from the fact that the Gaussian pdf $\phi(t)$ is decreasing for $t > 0$, and the third holds because $n > h^2$. Hence

$$I_n \geq \frac{b\rho}{\sqrt{n}} e^{-(b+1)\rho\sqrt{2\pi e}u^*\sqrt{\kappa''(u^*)}}. \tag{8.53}$$

This lower bound holds for all $b > 0$ and $n > 2\pi e\rho^2(b+1)^2$. To obtain the best bound, we maximize it over b and obtain

$$b = \frac{1}{\rho u^*\sqrt{2\pi e\kappa''(u^*)}}.$$

Substituting into (8.53), we obtain $I_n \geq c/\sqrt{n}$, which proves the claim. □

8.4.4 Refined Asymptotics for Binary Hypothesis Testing

Consider

$$\begin{cases} H_0 & : \boldsymbol{Y} \sim p_0^n \\ H_1 & : \boldsymbol{Y} \sim p_1^n. \end{cases} \tag{8.54}$$

The LLRT is given by

$$\sum_{k=1}^{n} L_k \underset{H_0}{\overset{H_1}{\gtrless}} n\tau,$$

where τ is the normalized threshold and $L_k = \ln p_1(Y_k)/p_0(Y_k)$ are i.i.d. random variables.

Assuming that $-D(p_0\|p_1) < \tau < D(p_1\|p_0)$, the probability of false alarm $\mathrm{P_F}$ and the probability of miss $\mathrm{P_M}$ are upper bounded by (8.19) and (8.20), respectively. The refined asymptotics of the previous section can be applied directly to yield

$$\mathrm{P_F} \leq P_0^n\left\{\sum_{k=1}^n L_k \geq n\tau\right\} \sim \frac{e^{-n[u^*\tau - \kappa_0(u^*)]}}{u^*\sqrt{2\pi\kappa_0''(u^*)n}}, \tag{8.55}$$

$$\mathrm{P_M} \leq P_1^n\left\{\sum_{k=1}^n L_k \leq n\tau\right\} \sim \frac{e^{-n[(u^*-1)\tau - \kappa_0(u^*)]}}{(1-u^*)\sqrt{2\pi\kappa_0''(u^*)n}}. \tag{8.56}$$

Example 8.2 We revisit Example 8.1. The hypotheses are given by (8.32). Differentiating (8.35), we obtain

$$\kappa_0''(u) = \frac{(\xi - 1)^2}{[1 + u(\xi - 1)]^2}. \tag{8.57}$$

Replacing in (8.55) and (8.56) yields

$$\mathrm{P_F} \sim \frac{\exp\{-n[u^*\tau - \kappa_0(u^*)]\}}{\frac{u^*(\xi-1)}{1+u^*(\xi-1)}\sqrt{2\pi n}}, \tag{8.58}$$

$$\mathrm{P_M} \sim \frac{\exp\{-n[(u^* - 1)\tau - \kappa_0(u^*)]\}}{\frac{(1-u^*)(\xi-1)}{1+u^*(\xi-1)}\sqrt{2\pi n}}, \tag{8.59}$$

with u^* given by (8.36) and the threshold τ in the range $(-2\psi(\xi), 2\psi(1/\xi))$.

The ROC using the upper bounds of (8.6), (8.37), and (8.38) on $\mathrm{P_F}$ and $\mathrm{P_M}$ is compared with the asymptotic approximation to the ROC using (8.58) and (8.59) in Figure 8.6 for $\xi = 1$ and $n = 80$. Since the error probabilities are small, the curve $-\frac{1}{n}\ln \mathrm{P_M}$ versus $-\frac{1}{n}\ln \mathrm{P_F}$ is given instead of the ROC. The asymptotic approximation is very accurate as illustrated by the comparison with Monte Carlo simulations of Figure 8.6.

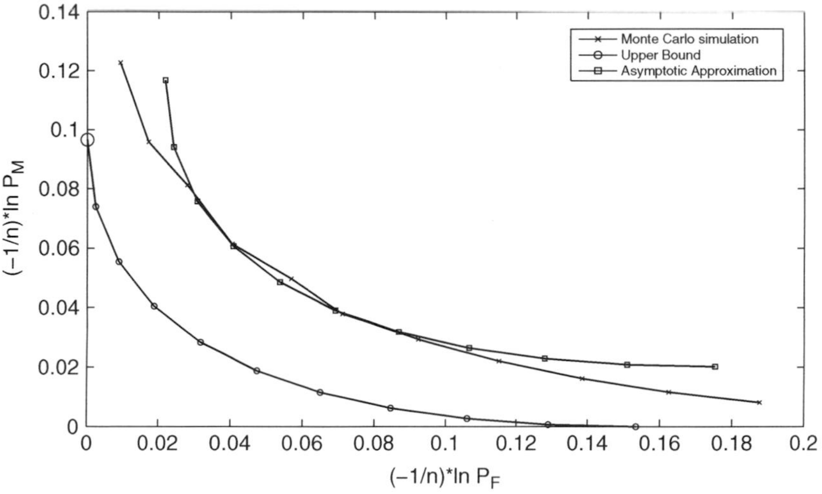

Figure 8.6 Monte Carlo simulation of the curve $-\frac{1}{n}\ln \mathrm{P_M}$ versus $-\frac{1}{n}\ln \mathrm{P_F}$ for $\xi = 1$ and $n = 80$, compared with Chernoff lower bound on ROC and with asymptotic approximation

□

8.4.5 Non-i.i.d. Components

A variation of the hypothesis testing problem (8.54) involves non-i.i.d. data:

$$\begin{cases} H_0 & : \boldsymbol{Y} \sim \prod_{k=1}^n p_{0k} \\ H_1 & : \boldsymbol{Y} \sim \prod_{k=1}^n p_{1k}. \end{cases} \tag{8.60}$$

The log-likelihood ratio is still a sum of n independent random variables $\{L_k\}$ but they are no longer identically distributed:

$$\ln \frac{p_1(\boldsymbol{Y})}{p_0(\boldsymbol{Y})} = \sum_{k=1}^{n} L_k,$$

where

$$L_k = \ln \frac{p_{1k}(Y_k)}{p_{0k}(Y_k)}, \qquad 1 \le k \le n.$$

Denote by $\kappa_{0k}(u)$ the cgf for L_k under H_0. Since $\{L_k\}$ are mutually independent, the cgf for $\sum_{k=1}^{n} L_k$ under H_0 is given by

$$\kappa_0(u) = \sum_{k=1}^{n} \kappa_{0k}(u).$$

Using the same methods as in Section 8.3, one can derive the upper bounds

$$\mathrm{P_F} \le \left(\prod_{k=1}^{n} P_{0k}\right)\left\{\sum_{k=1}^{n} L_k \ge \tau\right\} \le e^{-[u^*\tau - \kappa_0(u^*)]},$$
$$\mathrm{P_M} \le \left(\prod_{k=1}^{n} P_{1k}\right)\left\{\sum_{k=1}^{n} L_k \le \tau\right\} \le e^{-[(u^*-1)\tau - \kappa_0(u^*)]}.$$

The Gärtner–Ellis theorem [3], which is a generalization of Cramér's theorem, can be used to show that these upper bounds are tight in the exponent, if both the normalized cgf $\frac{1}{n}\sum_{k=1}^{n} \kappa_{0k}(u)$ and the normalized threshold τ/n converge to a finite limit as $n \to \infty$.

One may also ask whether refined asymptotics may apply. To answer this question, we revisit the core problem of evaluating $\mathrm{P}\{\sum_{i=1}^{n} X_i \ge na\}$, where X_i, $i \ge 1$ are independent but not identically distributed random variables with respective cgfs $\kappa_i(u)$, $i \ge 1$. Define the normalized sum of cgfs

$$\overline{\kappa}_n(u) = \frac{1}{n}\sum_{k=1}^{n} \kappa_k(u).$$

(In the i.i.d. case, $\overline{\kappa}_n(u)$ is independent of n.) Let u_n^* be the solution of $\overline{\kappa}_n'(u_n^*) = a$. Analogously to (8.43), define

$$V_n \triangleq \frac{\sum_{i=1}^{n} X_i - na}{\sqrt{n\overline{\kappa}_n'(u_n^*)}}. \tag{8.61}$$

From (8.44), we have

$$\mathrm{P}\left\{\sum_{i=1}^{n} X_i \ge na\right\} = e^{-n[u_n^* a - \overline{\kappa}_n(u_n^*)]} \int_0^{\infty} e^{-u_n^*\sqrt{\overline{\kappa}_n''(u_n^*)n}\, v}\, p_{V_n}(v)\, dv. \tag{8.62}$$

The question is whether the function $\overline{\kappa}_n$ converges to a finite limit as $n \to \infty$ (in which case u_n^* also converges to a limit u^*) as in the Gärtner–Ellis theorem, and whether V_n of (8.61) converges in distribution to a normal distribution. If so, we have

$$\mathrm{P}\left\{\sum_{i=1}^{n} X_i \geq na\right\} \sim \frac{e^{-n[u_n^* a - \overline{\kappa}_n(u_n^*)]}}{u_n^* \sqrt{2\pi \overline{\kappa}_n''(u_n^*) n}} \quad \text{as } n \to \infty. \tag{8.63}$$

Conditions for convergence of V_n to a normal distribution are given by the Lindeberg conditions [8, p. 262], which are necessary and sufficient. Let $s_n^2 = \sum_{i=1}^n \mathrm{Var}[X_i]$ and denote by F_i the cdf of the centered random variable $X_i - \mathbb{E}(X_i)$. The Lindeberg condition is

$$\forall \epsilon > 0: \quad \lim_{n\to\infty} \frac{1}{s_n^2} \sum_{i=1}^{n} \int_{-\epsilon s_n}^{\epsilon s_n} t^2 dF_i(t) = 1.$$

This implies that no component of the sum dominates the others:

$$\forall \epsilon > 0, \exists n(\epsilon) : \forall n > n(\epsilon) : \quad \max_{1\leq i\leq n} \mathrm{Var}[X_i] < \epsilon s_n^2.$$

Refined large-deviations results can be obtained in even more general settings involving dependent random variables [9].

8.5 Appendix: Proof of Lemma 8.1

For any LLRT of the form in (8.17), we may bound $\mathrm{P_M}(\tilde{\delta}^{(n)})$ using a change of measure argument as

$$\begin{aligned} \mathrm{P_M}(\tilde{\delta}^{(n)}) &\leq \mathrm{P}_1\{S_n \leq n\tau_n\} \\ &= \mathbb{E}_1\left[\mathbb{1}\{S_n \leq n\tau_n\}\right] \\ &= \mathbb{E}_0\left[\mathbb{1}\{S_n \leq n\tau_n\}\, e^{S_n}\right] \leq e^{n\tau_n}, \end{aligned} \tag{8.64}$$

since e^{S_n} is the likelihood ratio of the observations up to time n. This yields the upper bound

$$\frac{1}{n} \ln \mathrm{P_M}(\tilde{\delta}^{(n)}) \leq \tau_n. \tag{8.65}$$

Given any $\epsilon > 0$, consider the LLRT of the form in (8.17) that uses the constant threshold $\tau = -D(p_0\|p_1) + \epsilon$. Then, by (8.19), it follows that $\mathrm{P_F}(\tilde{\delta}_\tau^{(n)}) \leq \alpha$ for n sufficiently large. But the NP solution for level α must have a smaller $\mathrm{P_M}$ than any other test of level α, and therefore

$$\frac{1}{n} \ln \mathrm{P_M}(\tilde{\delta}_{\mathrm{NP}}^{(n)}) \leq \frac{1}{n} \ln \mathrm{P_M}(\tilde{\delta}_\tau^{(n)}) \leq \tau = -D(p_0\|p_1) + \epsilon \tag{8.66}$$

for n sufficiently large, where the last equality follows from (8.65).

Let $\tau_n(\alpha)$ denote the threshold of the α-level test $\tilde{\delta}_{\mathrm{NP}}^{(n)}$. It may be assumed that

$$\liminf_{n\to\infty} \tau_n(\alpha) \geq -D(p_0\|p_1).$$

Otherwise since

$$\mathrm{P_F}(\tilde{\delta}_{\mathrm{NP}}^{(n)}) \geq \mathrm{P}_0\left\{\frac{S_n}{n} > \tau_n(\alpha)\right\}$$

and $\frac{S_n}{n} \to -D(p_0\|p_1)$ as $n \to \infty$ by the law of large numbers, we have that

$$\limsup_{n\to\infty} \mathrm{P_F}(\tilde{\delta}_{\mathrm{NP}}^{(n)}) = 1$$

and the constraint of $\alpha \in (0, 1)$ will not be met for some large n. Furthermore, by the weak law of large numbers, we have that, for any given $\epsilon > 0$,

$$\mathrm{P}_0\left\{-D(p_0\|p_1) - \epsilon < \frac{S_n}{n} < \tau_n(\alpha)\right\} \geq 1 - \alpha, \quad \text{for sufficiently large } n.$$

Now, by an argument similar to that in (8.64),

$$\begin{aligned}\frac{1}{n}\ln \mathrm{P_M}(\tilde{\delta}_{\mathrm{NP}}^{(n)}) &\geq \frac{1}{n}\ln \mathbb{E}_0\left[\mathbb{1}\{n(-D(p_0\|p_1) - \epsilon) < S_n < n\tau_n\}\, e^{S_n}\right] \\ &\geq -D(p_0\|p_1) - \epsilon + \frac{1}{n}\ln \mathrm{P}_0\{n(-D(p_0\|p_1) - \epsilon) < S_n < n\tau_n\} \\ &\geq -D(p_0\|p_1) - 2\epsilon \quad \text{for sufficiently large } n.\end{aligned} \tag{8.67}$$

Combining (8.66) and (8.67), and using the fact that $\epsilon > 0$ was chosen arbitrarily, we have the result. □

Exercises

8.1 *Slight Generalization of Cramér's Theorem.* Prove Theorem 8.2.
Hint: Use the bounds in Theorem 8.1 and the continuity of Λ_X.

8.2 *Hypothesis Testing for Geometric Distributions.* Let $p_\theta(y) = (\theta - 1)y^{-\theta}$ for $\theta > 1$ and $y \geq 1$.

(a) Derive an expression for the Chernoff divergence of order $u \in (0, 1)$ between p_α and p_β, valid for all $\alpha, \beta > 1$.
(b) Fix $\alpha > \beta$ and consider the binary hypothesis test with $\mathrm{P}\{H_0\} = 0.5$.

$$\begin{cases} H_0 & : Y_i \sim \text{ i.i.d. } p_\alpha \\ H_1 & : Y_i \sim \text{ i.i.d. } p_\beta, \quad 1 \leq i \leq n. \end{cases}$$

Give the LRT in its simplest form.
(c) Show that the Bayes error probability for the problem of part (b) vanishes exponentially with n, and give the error exponent.

8.3 *Hypothesis Testing for Bernoulli Random Variables.* Let N_i, $1 \leq i \leq n$, be i.i.d. Bernoulli random variables with $\mathrm{P}\{N = 1\} = \theta \leq \frac{1}{2}$. Consider the following binary hypothesis test:

$$\begin{cases} H_0 : & Y_i = N_i \\ H_1 : & Y_i = 1 - N_i, \quad 1 \leq i \leq n, \end{cases}$$

(a) Write the LRT.

(b) Give a large-deviations upper bound on the Bayes error probability of the test, assuming uniform prior.
(c) Give a good approximation to the threshold of the NP test at fixed significance level α.
(d) Give a good approximation to the threshold of the NP test at significance level $\exp\{-n^{1/3}\}$.

8.4 *Large Deviations for Poisson Random Variables.* Denote by P_λ the Poisson distribution with parameter $\lambda > 0$. The cumulant-generating function for P_λ is $\kappa(u) = \lambda(e^u - 1)$, $u \in \mathbb{R}$.

(a) Derive the large-deviations function for P_λ.
(b) Let Y_i, $1 \le i \le n$ be i.i.d. P_λ. Let $S_n = \sum_{i=1}^n Y_i$. Derive the large-deviations bound on $\mathrm{P}\{S_n \ge ne\lambda\}$.
(c) Derive an exact asymptotic expression for $\mathrm{P}\{S_n \ge ne\lambda\}$.

8.5 *Error Exponent for Minimax Hypothesis Testing.* Prove that the error exponent for binary minimax hypothesis testing with i.i.d. observations is the same as that for Bayesian hypothesis testing, and is given by the Chernoff information $C(p_0, p_1)$.

8.6 *Optical Communications.* The following problem is encountered in optical communications. A device counts N photons during T seconds and decides whether the observations are due to background noise or to signal plus noise. The hypothesis test is of the form

$$H_0 : N \sim \text{Poisson}(T) \quad \text{versus} \quad H_1 : N \sim \text{Poisson}(e\,T).$$

We would like to design the detector to minimize the probability of miss $\mathrm{P_M}$ subject to an upper bound $\mathrm{P_F} \le e^{-\alpha T}$ on the probability of alarm. Here $\alpha > 0$ is a tradeoff parameter. When T is large, we can use large-deviations bounds to approximate the optimal test.

Derive a large-deviations bound on $\mathrm{P_M}$, and determine the range of α in which your bound is useful. Sketch the corresponding curve $-\frac{1}{T}\ln \mathrm{P_M}$ versus $-\frac{1}{T}\ln \mathrm{P_F}$ and identify its endpoints.

8.7 *Hypothesis Testing for Poisson Random Variables.* Denote by P_λ the Poisson distribution with parameter $\lambda > 0$. Consider the binary hypothesis test

$$\begin{cases} H_0 & : Y_i \overset{\text{indep.}}{\sim} P_{1/\ln i} \\ H_1 & : Y_i \overset{\text{indep.}}{\sim} P_{e/\ln i}, \quad 2 \le i \le n. \end{cases}$$

(a) Assume equal priors on the hypotheses. Derive the LRT.
(b) Derive a tight upper bound on the probability of false alarm.
Hint: A decreasing function $x(t)$ is said to be slowly decreasing if

$$\lim_{t\to\infty} \frac{x(at)}{x(t)} = 1 \text{ for all } a > 0.$$

For such a function, the following asymptotic equality holds:

$$\sum_{i=1}^{n} x(i) \sim n x(n) \text{ as } n \to \infty.$$

8.8 *Signal Detection in Sparse Noise*. The random variable Z is equal to 0 with probability $1-\epsilon$, and to a $\mathcal{N}(0,1)$ random variable with probability ϵ. Given a length-n i.i.d. sequence $\boldsymbol{Z}$ generated from this distribution and a known signal $\boldsymbol{s}$, consider the binary hypothesis test

$$\begin{cases} H_0 & : \boldsymbol{Y} = -\boldsymbol{s} + \boldsymbol{Z} \\ H_1 & : \boldsymbol{Y} = \boldsymbol{s} + \boldsymbol{Z}. \end{cases}$$

Derive an upper bound on the probability of error for that test; the bound should be tight in the exponent as $n \to \infty$.

8.9 *Hypothesis Testing for Exponential Distributions*. Let X_i, $1 \le i \le n$, be i.i.d. random variables with unit exponential distribution.

(a) Use a large-deviations method to derive an upper bound on $\mathrm{P}\{\sum_{i=1}^{n} X_i > n^{\alpha}\}$, where $\alpha > 1$.
(b) Use the Central Limit Theorem to approximate $\mathrm{P}\{\sum_{i=1}^{n} X_i > n + n^{\alpha}\}$, where $\frac{1}{2} \le \alpha < 1$.
(c) Use the large-deviations method of part (a) to derive an upper bound on the tail probability of part (b). You may use asymptotic approximations, e.g., $\ln(1+\epsilon) \sim \epsilon - \frac{1}{2}\epsilon^2$ as $\epsilon \to 0$. Compare your result with that of part (b).

8.10 *Error Exponents for Hypothesis Testing with Gaussian Observations (I).*
Consider the detection problem where the observations, $\{Y_k, k \ge 1\}$ are i.i.d. $\mathcal{N}(\mu_j, \sigma^2)$ random variables under H_j, $j = 0, 1$.

(a) Evaluate the cgf $\kappa_0(u)$.
(b) Use $\kappa_0(u)$ to find $D(p_0 \| p_1)$ and $D(p_1 \| p_0)$.
(c) Find the rate functions $\Lambda_0(\tau)$ and $\Lambda_1(\tau)$.
(d) Find the Chernoff information $C(p_0, p_1)$.
(e) Solve the Hoeffding problem when the constraint on the false-alarm exponent is equal to γ, for $\gamma \in (0, D(p_1 \| p_0))$.

8.11 *Error Exponents for Hypothesis Testing with Gaussian Observations (II).*
Repeat Exercise 8.10 for the case where the observations $\{Y_k, k \ge 1\}$ are i.i.d. $\mathcal{N}(0, \sigma_j^2)$ random variables under H_j, $j = 0, 1$.
Hint: This problem can be reduced to the one studied in Example 8.1. In particular,

$$\kappa_0(u) = \frac{1-u}{2} \ln \frac{\sigma_1^2}{\sigma_0^2} - \frac{1}{2} \ln \left(1 + (1-u)\left(\frac{\sigma_1^2}{\sigma_0^2} - 1\right)\right).$$

8.12 *Lower Bound on Large-Deviations Probability (I).* Here we derive a version of Theorem 8.4 which does not assume finiteness of the absolute third moment of X. Start from the Esscher identity (8.51) and show that

$$\lim_{n\to\infty} \frac{1}{\sqrt{n}} \left(\ln \mathrm{P}\left\{ \sum_{i=1}^{n} X_i \ge na \right\} + n\Lambda_X(a) \right) = 0$$

or equivalently, $\mathrm{P}\left\{\sum_{i=1}^{n} X_i \ge na\right\} = \exp\{-n\Lambda_X(a) + o(\sqrt{n})\}$.
Hint: The integral I_n of (8.52) is upper bounded by 1, and lower bounded by $\int_0^{\epsilon} e^{-u^* \sqrt{n\kappa''(u^*)}\, v}\, dF_n(v)$ for any $\epsilon > 0$. Invoke the Central Limit Theorem to claim that $\lim_{n\to\infty}[F_n(\epsilon) - F_n(0)] = \Phi(\epsilon) - \Phi(0)$. Note that while ϵ can be as small as desired, ϵ may not depend on n in this analysis.

8.13 *Lower Bound on Large-Deviations Probability (II).* Here we derive a version of Theorem 8.4 which does not assume finiteness of Var(X). Start from the Esscher identity (8.51) and show that

$$\lim_{n\to\infty} \frac{1}{n} \left(\ln \mathrm{P}\left\{ \sum_{i=1}^{n} X_i \ge na \right\} + n\Lambda_X(a) \right) = 0$$

or equivalently, $\mathrm{P}\left\{\sum_{i=1}^{n} X_i \ge na\right\} = \exp\{-n\Lambda_X(a) + o(n)\}$.
Hint: Write I_n of (8.52) as

$$I_n = \mathbb{E}_{\tilde{p}}\left[e^{-u(\sum_i X_i - na)} \mathbb{1}\left\{ \sum_i X_i \ge na \right\} \right],$$

which is upper bounded by 1 and lower bounded by

$$I_n = \mathbb{E}_{\tilde{p}}\left[e^{-u(\sum_i X_i - na)} \mathbb{1}\left\{ na \le \sum_i X_i \le n(a+\epsilon) \right\} \right]$$

for any $\epsilon > 0$. Invoke the Weak Law of Large Numbers to claim that there exists $\delta > 0$ such that $\lim_{n\to\infty} \mathrm{P}\{na \le \sum_i X_i \le n(a+\epsilon)\} > \delta$. Note that while ϵ and δ can be as small as desired, they may not depend on n in this analysis.

8.14 *Bernoulli Random Variables.* Denote by Be(θ) the Bernoulli distribution with success probability θ.

(a) Derive the cumulant-generating function for Be(θ).
(b) Show that the large-deviations function for Be(θ) is $\Lambda(a) = d(a\|\theta),\ a \in [0,1]$.
(c) Use refined large deviations to give a precise approximation to $\mathrm{P}\left\{S < \frac{10}{e+1}\right\}$ where S is the sum of 10 i.i.d. Be$\left(\frac{e}{e+1}\right)$ random variables.

References

[1] H. Chernoff, "Large Sample Theory: Parametric Case," *Ann. Math. Stat.*, Vol. 27, pp. 1–22, 1956.

[2] H. Chernoff, "A Measure of Asymptotic Efficiency for Tests of a Hypothesis Based on a Sum of Observations," *Ann. Math. Stat.*, Vol. 23, pp. 493–507, 1952.

[3] A. Dembo and O. Zeitouni, *Large Deviations Techniques and Applications*, Springer, 1998.
[4] W. Hoeffding, "Asymptotically Optimal Tests for Multinomial Distributions," *Ann. Math. Stat.*, Vol. 36, No. 2, pp. 369–401, 1965.
[5] F. Esscher, "On the Probability Function in the Collective Theory of Risk," *Skandinavisk Aktuarietidskrift*, Vol. 15, pp. 175–195, 1932.
[6] D. Blackwell and J. L. Hodges, "The Probability in the Extreme Tail of a Convolution," *Ann. Math. Stat.*, Vol. 30, pp. 1113–1120, 1959.
[7] R. Bahadur and R. Ranga Rao, "On Deviations of the Sample Mean," *Ann. Math. Stat.*, Vol. 31, pp. 1015–1027, 1960.
[8] W. Feller, *An Introduction to Probability Theory and Its Applications*, Wiley, 1971.
[9] N. R. Chaganty and J. Sethuraman, "Strong Large Deviation and Local Limit Theorems," *Ann. Prob.*, Vol. 21, No. 3, pp. 1671–1690, 1993.

9 Sequential and Quickest Change Detection

In this chapter, we continue our study of the problem of hypothesis testing with multiple observations, but in a sequential setting where we are allowed to choose when to stop taking observations before making a decision. We first study the sequential binary hypothesis testing problem, introduced by Wald [1], where the goal is to choose a stopping time at which a binary decision about the hypothesis is made. We then study the quickest change detection problem, where the observations undergo a change in distribution at some unknown time and the goal is to detect the change as quickly as possibly, subject to false-alarm constraints. More details on the sequential detection problem are provided in [2, 3], and on the quickest change detection problem are provided in [3, 4, 5].

9.1 Sequential Detection

In the detection problems we studied in Chapter 5, we are given a sample of n observations with which we need to make a decision about the hypothesis. The size of the sample n is assumed to be fixed beforehand, i.e., before the observations are taken, and therefore this setting is referred to as the fixed sample size (FSS) setting. In contrast, in the problems that we study in this chapter, the sample size is determined on-line, i.e., the observations come from a potentially infinite stream of observations and we are allowed to choose when to stop taking observations based on a desired level of accuracy for decision-making about the hypothesis.

9.1.1 Problem Formulation

We will restrict our attention to binary hypothesis testing and i.i.d. observation models here. In particular we have:

$$\begin{cases} H_0 & : \; Y_k \sim \text{i.i.d. } p_0 \\ H_1 & : \; Y_k \sim \text{i.i.d. } p_1, \quad k = 1, 2, \ldots. \end{cases} \tag{9.1}$$

A sequential test has two components: (i) a *stopping rule*, which is used to determine when to stop taking additional observations and make a decision, and (ii) a *decision rule*, which is used to make a decision about the hypothesis using the observations up to the stopping time. There are two goals in sequential detection: (i) to make an accurate decision regarding the hypothesis (which might require a large number of observations);

and (ii) to use as few observations as possible. These are conflicting goals and hence the optimization problem we need to solve is a *tradeoff* problem.

9.1.2 Stopping Times and Decision Rules

The number of observations taken by a sequential test is a random variable called a *stopping time*, which is defined formally as follows.

DEFINITION 9.1 A *stopping time* on a sequence of random variables $Z_1, Z_2, \ldots$ is a random variable N such that $\{N = n\} \subset \sigma(Z_1, \ldots, Z_n)$, where $\sigma(Z_1, \ldots, Z_n)$ denotes the σ-algebra generated by the random variables $Z_1, \ldots, Z_n$. Equivalently, $\mathbb{1}\{N = n\}$ is a function of only the observations $Z_1, \ldots, Z_n$, i.e., it is a causal function of the observations.

A sequential test stops taking observations and makes a decision about the hypothesis at a stopping time N on the observations sequence $Y_1, Y_2, \ldots$. We have the following definition.

DEFINITION 9.2 A *sequential test* is a pair[1] $(N; \delta)$, such that at the stopping time N on the observation sequence, the decision rule δ is used to make a decision about the hypothesis based on $Y_1, \ldots, Y_N$.

9.1.3 Two Formulations of the Sequential Hypothesis Testing Problem

As in the case of fixed sample size hypothesis testing, there are Bayesian and non-Bayesian approaches to sequential hypothesis testing. We begin with a Bayesian approach, for which we assume:

- uniform decision costs;[2]
- known priors on the hypotheses π_0, π_1;
- cost per observation c (to penalize the use of observations).

We may then associate a risk with the sequential test $(N; \delta)$.

DEFINITION 9.3 The *Bayes risk* associated with a sequential test $(N; \delta)$ is given by

$$r(N; \delta) = \underbrace{c\left[\pi_0 \mathbb{E}_0[N] + \pi_1 \mathbb{E}_1[N]\right]}_{\text{Average penalty for stopping time } N} + \underbrace{\pi_0 \mathrm{P_F}(N; \delta) + \pi_1 \mathrm{P_M}(N; \delta)}_{\text{Average probability of error}}, \tag{9.2}$$

where

$$\mathrm{P_F}(N; \delta) = \mathrm{P}_0\{\delta(Y_1, \ldots, Y_N) = 1\}, \tag{9.3}$$

$$\mathrm{P_M}(N; \delta) = \mathrm{P}_1\{\delta(Y_1, \ldots, Y_N) = 0\}. \tag{9.4}$$

[1] Including the stopping time N (which is a random variable) in the definition of the test is slightly inconsistent with the way in which we have defined decision rules in the previous chapters. We could have defined a stopping rule η, which chooses the stopping time N, and use η in place of N in defining the sequential test. However, such a stopping rule η and N would be equivalent. We therefore adopt the more standard way to define the sequential test using N as introduced by Wald [1].

[2] This can easily be generalized to nonuniform decision costs.

A *Bayesian* sequential test is one that minimizes the risk in (9.2), i.e.,

$$(N;\delta)_B = \underset{(N;\delta)}{\operatorname{argmin}}\, r(N;\delta). \tag{9.5}$$

An alternative to the Bayesian formulation is a Neyman–Pearson-like non-Bayesian formulation. Here no priors are assumed for the hypotheses. Furthermore, we do not have any explicit costs on the decisions or the observations. In particular, the non-Bayesian formulation of the sequential hypothesis testing problem is

$$\begin{aligned} &\text{Minimize } \mathbb{E}_0[N] \text{ and } \mathbb{E}_1[N] \\ &\text{subject to } \mathrm{P_F}(N;\delta) \le \alpha \text{ and } \mathrm{P_M}(N;\delta) \le \beta, \text{ with } \alpha,\beta \in (0,1). \end{aligned} \tag{9.6}$$

Note that we require a simultaneous minimization of $\mathbb{E}_0[N]$ and $\mathbb{E}_1[N]$ under constraints on both of the error probabilities. One of the most celebrated results in sequential analysis, due to Wald and Wolfowitz [6], is that optimization problem (9.6) has a solution whose structure is the same as that of the solution to Bayesian optimization problem given in (9.5). We discuss this test structure next.

9.1.4 Sequential Probability Ratio Test

The log-likelihood ratio of the observations $Y_1, \ldots, Y_n$ is given by the sum:

$$S_n = \sum_{k=1}^{n} \ln L(Y_k), \tag{9.7}$$

where as usual $L(y) = p_1(y)/p_0(y)$.

Let $a < 0 < b$. The stopping time and decision rule of the *sequential probability ratio test (SPRT)* are given by:

$$N_{\mathrm{SPRT}} = \inf\{n \ge 1 : S_n \notin (a,b)\} \tag{9.8}$$

and

$$\delta_{\mathrm{SPRT}} = \begin{cases} 1 & \text{if } S_{N_{\mathrm{SPRT}}} \ge b \\ 0 & \text{if } S_{N_{\mathrm{SPRT}}} \le a. \end{cases} \tag{9.9}$$

The SPRT $(N_{\mathrm{SPRT}}; \delta_{\mathrm{SPRT}})$ can be summarized as:

- stop and decide H_1 if $S_n \ge b$;
- stop and decide H_0 if $S_n \le a$;
- continue observing if $a < S_n < b$.

Typical sample paths of the S_n, along with the decision boundaries are depicted in Figure 9.1.

As mentioned previously, the SPRT test structure can be shown to be optimal for both the Bayesian and non-Bayesian optimization problems introduced in Section 9.1.3 (see [1] and [6]). Therefore, an optimal solution to (9.5) is obtained by finding thresholds a and b that minimize the Bayes risk (9.2), and an optimal to solution to (9.6) is obtained by finding a and b that satisfy the constraints on the error probabilities in (9.6) with equality. (Note that randomization might be required for the latter problem.)

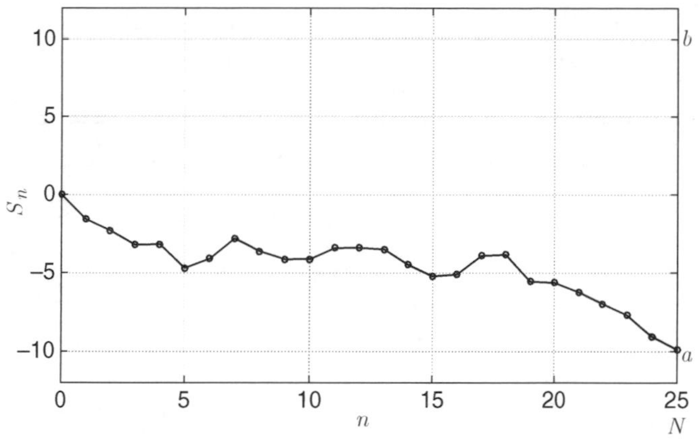

(a)

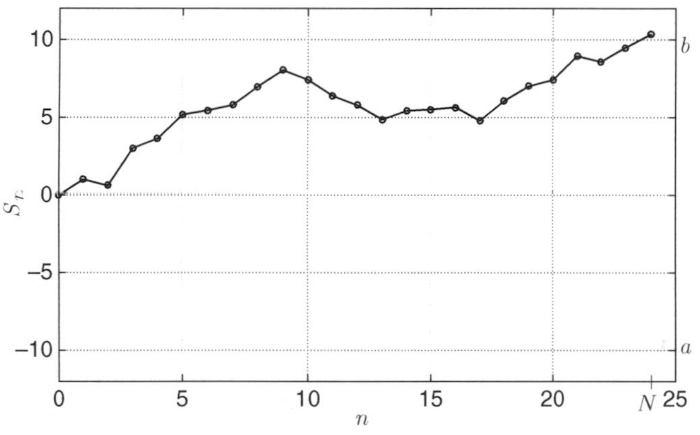

(b)

Figure 9.1 Typical evolution of the SPRT statistic S_n with $p_0 \sim \mathcal{N}(0, 1)$ and $p_1 \sim \mathcal{N}(1, 1)$: (a) H_0 is the true hypothesis, (b) H_1 is the true hypothesis

We now provide some intuition as to why the SPRT structure provides a good trade-off between decision errors and sample size. The log-likelihood ratio sequence S_n that defines the SPRT is a random walk with components $\ln L(Y_k)$ that have the following means under H_0 and H_1, respectively:

$$\mathbb{E}_0[\ln L(Y_k)] = \mathbb{E}_0\left[\ln \frac{p_1(Y_k)}{p_0(Y_k)}\right] = -D(p_0 \| p_1) < 0 \tag{9.10}$$

and

$$\mathbb{E}_1[\ln L(Y_k)]] = \mathbb{E}_1\left[\ln \frac{p_1(Y_k)}{p_0(Y_k)}\right] = D(p_1 \| p_0) > 0. \tag{9.11}$$

This means that the random walk S_n has a positive drift under H_1 and a negative drift under H_0. Therefore, under H_0, S_n is more likely to cross the lower threshold a than the upper threshold b, and the opposite is true under H_1. Furthermore, the decision accuracy can be improved by making the threshold b larger and the threshold a smaller.

Example 9.1 *Gaussian Observations.* Consider testing the hypotheses:

$$\begin{aligned} H_0 &: Y_1, Y_2, \ldots \quad i.i.d. \sim \mathcal{N}(\mu_0, \sigma^2), \\ H_1 &: Y_1, Y_2, \ldots \quad i.i.d. \sim \mathcal{N}(\mu_1, \sigma^2), \end{aligned}$$

with $\mu_1 > \mu_0$.

In this case,

$$S_n = \frac{\mu_1 - \mu_0}{\sigma^2} \sum_{k=1}^{n} Y_k - n\left(\frac{\mu_1^2 - \mu_0^2}{2\sigma^2}\right).$$

It is easy to see that $S_n \in (a, b)$ is equivalent to $\sum_{k=1}^{n} Y_k \in (a'_n, b'_n)$ where

$$\begin{aligned} a'_n &= \frac{\sigma^2}{\mu_1 - \mu_0} a + n \frac{\mu_1 + \mu_0}{2}, \\ b'_n &= \frac{\sigma^2}{\mu_1 - \mu_0} b + n \frac{\mu_1 + \mu_0}{2}. \end{aligned}$$

Therefore, we can write our SPRT as:

- stop and decide H_1 if $\sum_{k=1}^{n} Y_k \geq b'_n$;
- stop and decide H_0 if $\sum_{k=1}^{n} Y_k \leq a'_n$;
- continue observing otherwise.

9.1.5 SPRT Performance Evaluation

The performance of the SPRT is characterized by values of $\mathrm{P_M}, \mathrm{P_F}, \mathbb{E}_0[N], \mathbb{E}_1[N]$ achieved by the SPRT. An exact analysis of these quantities is impossible except in special cases, and therefore we rely on bounds and approximations. We begin by studying Wald's bounds and approximations for the error probabilities $\mathrm{P_F}$ and $\mathrm{P_M}$.

The likelihood ratio of $Y_1, \ldots, Y_n$ can be written as

$$X_n \triangleq e^{S_n} = \prod_{k=1}^{n} L(Y_k) = \prod_{k=1}^{n} \frac{p_1(Y_k)}{p_0(Y_k)}.$$

Error Probability Bounds For the SPRT

$$\mathrm{P_F} = \mathrm{P}_0\{S_N \geq b\} = \mathrm{P}_0\{X_N \geq e^b\},$$

the event $\{X_N \geq e^b\}$ can be rewritten as the following union of disjoint events:

$$\{X_N \geq e^b\} = \bigcup_{n=1}^{\infty} \left(\{N = n\} \cap \{X_n \geq e^b\}\right).$$

Thus

$$\begin{aligned}
\mathrm{P}_0\{X_N \ge e^b\} &= \sum_{n=1}^{\infty} \mathrm{P}_0(\{N=n\}\cap\{X_n \ge e^b\}) \\
&= \sum_{n=1}^{\infty} \int_{\{N=n\}\cap\{X_n\ge e^b\}} \prod_{k=1}^{n} p_0(y_k)\, d\mu(y_1) d\mu(y_2)\dots d\mu(y_n) \\
&= \sum_{n=1}^{\infty} \int_{\{N=n\}\cap\{X_n\ge e^b\}} \frac{\prod_{k=1}^{n} p_1(y_k)}{X_n}\, d\mu(y_1) d\mu(y_2)\dots d\mu(y_n) \\
&\le \frac{1}{e^b}\sum_{n=1}^{\infty} \int_{\{N=n\}\cap\{X_n\ge e^b\}} \prod_{k=1}^{n} p_1(y_k)\, d\mu(y_1) d\mu(y_2)\dots d\mu(y_n) \\
&= e^{-b}\,\mathrm{P}_1\{X_N \ge e^b\} = e^{-b}\,(1-\mathrm{P_M}).
\end{aligned}$$

That is,

$$\mathrm{P_F} \le e^{-b}(1-\mathrm{P_M}) \le e^{-b}. \tag{9.12}$$

An analogous derivation for $\mathrm{P_M}$ noting that

$$\mathrm{P_M} = \mathrm{P}_1\{X_N \le e^a\}$$

yields

$$\mathrm{P_M} \le e^{a}(1-\mathrm{P_F}) \le e^{a}. \tag{9.13}$$

Together, (9.12) and (9.13) are known as *Wald's bounds* on the error probabilities.

Now, we consider setting the thresholds a and b to meet constraints $\mathrm{P_F} \le \alpha$ and $\mathrm{P_M} \le \beta$. An obvious way to do this is to set $b = -\ln\alpha$ and $a = \ln\beta$. An alternative suggested by Wald is to treat (9.12) and (9.13) as approximations, with the understanding that if the decision is in favor of H_1, $X_n \approx e^b$, and if the decision is in favor of H_0, $X_n \approx e^a$, i.e.,

$$\mathrm{P_F} \approx e^{-b}(1-\mathrm{P_M}), \quad \text{and } \mathrm{P_M} \approx e^{a}(1-\mathrm{P_F}).$$

Then setting $\mathrm{P_F} = \alpha$ and $\mathrm{P_M} = \beta$ yields the following approximations for the thresholds:

$$a = \ln\frac{\beta}{1-\alpha}, \quad \text{and } b = \ln\frac{1-\beta}{\alpha}.$$

Although this choice of thresholds does not necessarily result in error probabilities that meet the constraints $\mathrm{P_F} \le \alpha$ and $\mathrm{P_M} \le \beta$, we can show using (9.12) and (9.13) that (see Exercise 9.1):

$$\mathrm{P_F} + \mathrm{P_M} \le \alpha + \beta. \tag{9.14}$$

Thus, setting thresholds based on assuming equality in the Wald bounds guarantees that the sum of the false-alarm probability and missed detection probabilities is constrained by the sum of their individual constraints. It is important to note that we did not assume anything about the distributions of the observations when deriving the bounds. Nevertheless, the SPRT itself is a function of the distributions through the log-likelihood ratio statistic S_n. More accurate approximations for the error probabilities derived using renewal theory do indeed depend on the distributions P_0 and P_1 [2].

Expected Stopping Time Approximations In order to calculate approximations for $\mathbb{E}_0[N]$ and $\mathbb{E}_1[N]$, we will need a few preliminary lemmas.

LEMMA 9.1 *If Z is a nonnegative integer valued random variable*

$$\mathbb{E}[Z] = \sum_{k=0}^{\infty} \mathrm{P}\{Z > k\}.$$

(If Z is a nonnegative continuous random variable, we replace the sum with an integral.)

Proof

$$\begin{aligned}
\mathbb{E}[Z] &= \sum_{k=0}^{\infty} k\mathrm{P}\{Z = k\} = \sum_{k=1}^{\infty} k\mathrm{P}\{Z = k\} \\
&= \sum_{k=1}^{\infty} k\left(\mathrm{P}\{Z \geq k\} - \mathrm{P}\{Z \geq k+1\}\right) \\
&= \sum_{k=1}^{\infty} \mathrm{P}\{Z \geq k\} = \sum_{k=0}^{\infty} \mathrm{P}\{Z > k\}.
\end{aligned}$$

□

LEMMA 9.2 *Assume $P_0 \neq P_1$. Then, the stopping time of the SPRT is finite w.p. 1 under both hypotheses. Further, $\mathbb{E}_j[N^m] < \infty$ for $j = 0, 1$, for all integer $m \geq 1$.*

Proof For $k \geq 0$, using Markov's inequality,

$$\begin{aligned}
\mathrm{P}_0\{N > k\} &\leq \mathrm{P}_0\{X_k \in (e^a, e^b)\} \\
&\leq \mathrm{P}_0\{X_k > e^a\} \\
&= \mathrm{P}_0\{\sqrt{X_k} > \sqrt{e^a}\} \\
&\leq \frac{\mathbb{E}_0[\sqrt{X_k}]}{\sqrt{e^a}} \\
&= \frac{\rho^k}{\sqrt{e^a}},
\end{aligned}$$

where $\rho = \int_{\mathcal{Y}} \sqrt{p_0(y)p_1(y)}d\mu(y)$ is the Bhattacharyya coefficient between P_0 and P_1 (see also (6.14)). Since $\rho < 1$, whenever $P_0 \neq P_1$, $P_0\{N > k\}$ goes to zero exponentially fast with k, thus establishing that N is finite with probability one under H_0. A similar argument with b replacing the role of a establishes the corresponding result under H_1.

Now, we apply Lemma 9.1 to obtain that for integer $m \geq 1$,

$$\mathbb{E}_0[N^m] = \sum_{k=0}^{\infty} \mathrm{P}_0\{N^m > k\} = \sum_{k=0}^{\infty} \mathrm{P}_0\{N > k^{1/m}\} \leq \frac{1}{\sqrt{e^a}} \sum_{k=0}^{\infty} \rho^{k^{1/m}} < \infty,$$

since $\rho < 1$. A similar argument with b replacing the role of a establishes the corresponding result under H_1. □

LEMMA 9.3 Wald's Identity. *Let $Z_1, Z_2, \ldots$ be i.i.d with mean μ. Let N be a stopping time with respect to $Z_1, Z_2, \ldots$ such that N is finite w.p. 1. Then*

$$\mathbb{E}\left[\sum_{k=1}^{N} Z_k\right] = \mathbb{E}[N]\mu.$$

Proof (Note: Since N is a function of the sequence $\{Z_k\}$, this statement is not obvious. If N were independent of $\{Z_k\}$, then a simple application of Law of Iterated Expectation after first conditioning on N would lead to the result.)

We begin by noting that

$$\mathbb{1}\{N \geq k\} = 1 - \mathbb{1}\{N \leq k-1\}.$$

By the definition of a stopping time (see Definition 9.1), the indicator function on the right hand side is a function of only $Z_1, \ldots, Z_{k-1}$, and is thus independent of Z_k. Therefore, we get

$$\mathbb{E}\left[\sum_{k=1}^{N} Z_k\right] = \mathbb{E}\left[\sum_{k=1}^{\infty} Z_k \mathbb{1}\{k \leq N\}\right] = \sum_{k=1}^{\infty} \mathbb{E}\left[Z_k\right]\mathbb{E}\left[\mathbb{1}\{k \leq N\}\right]$$
$$= \sum_{k=1}^{\infty} \mu \mathrm{P}\{N \geq k\} = \mu\mathbb{E}[N],$$

where the last step follows from Lemma 9.1. □

We now apply Lemmas 9.2 and 9.3 to the analysis of the expected stopping time of the SPRT. At the stopping time N, the test statistic S_N has just crossed the threshold b from below or just crossed the threshold a from above. Therefore, $S_N \approx a$ or $S_N \approx b$. Then, under H_0,

$$S_N \approx \begin{cases} a & \text{w.p. } 1 - \mathrm{P_F} \\ b & \text{w.p. } \mathrm{P_F} \end{cases}$$

and therefore

$$\mathbb{E}_0[S_N] \approx a(1 - \mathrm{P_F}) + b\mathrm{P_F}. \tag{9.15}$$

Similarly, under H_1

$$S_N \approx \begin{cases} a & \text{w.p. } \mathrm{P_M} \\ b & \text{w.p.} 1 - \mathrm{P_F} \end{cases}$$

and therefore

$$\mathbb{E}_1[S_N] \approx a\mathrm{P_M} + b(1 - \mathrm{P_M}). \tag{9.16}$$

Now recall that

$$S_N = \sum_{k=1}^{N} \ln L(Y_k).$$

Since the stopping time of the SPRT N is finite w.p. 1 under both H_0 and H_1, we can apply Wald's identity to S_N to obtain

$$\mathbb{E}_0[S_N] = \mathbb{E}_0[N]\,\mathbb{E}_0\,[\ln L(Y_K)] = -\mathbb{E}_0[N] D(p_0 \| p_1)$$

and

$$\mathbb{E}_1[S_N] = \mathbb{E}_1[N]\,\mathbb{E}_1\,[\ln L(Y_K)] = \mathbb{E}_1[N] D(p_1 \| p_0).$$

Plugging these expressions into (9.15) and (9.16), respectively, we get the Wald approximations for the expected stopping time:

$$\mathbb{E}_0[N] \approx \frac{a(1 - \mathrm{P_F}) + b\mathrm{P_F}}{-D(p_0 \| p_1)}, \tag{9.17}$$

$$\mathbb{E}_1[N] \approx \frac{a\mathrm{P_M} + b(1 - \mathrm{P_M})}{D(p_1 \| p_0)}. \tag{9.18}$$

Example 9.2 *Comparison of Fixed Sample Size Hypothesis Test with SPRT.* Consider the hypotheses:

$$\begin{cases} H_0 : & Y_1, Y_2, \ldots \text{ i.i.d.} \sim \mathcal{N}(0, \sigma^2) \\ H_1 : & Y_1, Y_2, \ldots \text{ i.i.d.} \sim \mathcal{N}(\mu, \sigma^2), \end{cases}$$

where $\mu > 0$.

Our goal is to be efficient in terms of number of observations, while meeting error probability constraints $\mathrm{P_F} \le \alpha$ and $\mathrm{P_M} \le \beta$. First, we consider the fixed sample size (FSS) case. Here we design a sample size n before we take observations such that the constraints are met. By the optimality of the Neyman–Pearson test, it is clear that for any fixed n, an α-level Neyman–Pearson test results in the smallest $\mathrm{P_M}$, while meeting the constraint on $\mathrm{P_F}$. Therefore the most efficient FSS sample size test that meets the constraints on $\mathrm{P_M}$ and $\mathrm{P_F}$ must be an α-level Neyman–Pearson test, for which $\mathrm{P_M}$ for a fixed n can be computed using (5.34).

$$\mathrm{P_M} = 1 - \mathrm{P_D} = 1 - Q(Q^{-1}(\alpha) - d),$$

where (using $\mathbf{1}$ to denote the vector of all ones of length n)

$$d^2 = \mu \mathbf{1}^T (\sigma^2 I)^{-1} \mu \mathbf{1} = \frac{1}{\sigma^2}\mu^2 n.$$

Therefore

$$\mathrm{P_M} = 1 - Q(Q^{-1}(\alpha) - \sqrt{n}\frac{\mu}{\sigma}).$$

Thus, to meet the $\mathrm{P_M} \le \beta$ constraint, we pick the sample size

$$n_{\mathrm{FSS}} = \left\lceil \left(\frac{\sigma}{\mu} \left(Q^{-1}(1 - \beta) - Q^{-1}(\alpha) \right) \right)^2 \right\rceil. \tag{9.19}$$

For the sequential test, using the Wald approximations we set

$$b = \ln\left(\frac{1 - \beta}{\alpha}\right), \qquad a = \ln\left(\frac{\beta}{1 - \alpha}\right)$$

and compute the expected sample sizes using Wald's approximations

$$\mathbb{E}_0[N] \approx \frac{(1-\mathrm{P_F})a + \mathrm{P_F}b}{-D(p_0\|p_1)} \approx \frac{(1-\alpha)\ln\frac{\beta}{1-\alpha} + \alpha\ln\frac{1-\beta}{\alpha}}{-D(p_0\|p_1)},$$

$$\mathbb{E}_1[N] \approx \frac{\mathrm{P_M}a + (1-\mathrm{P_M})b}{D(p_1\|p_0)} \approx \frac{\beta\ln\frac{\beta}{1-\alpha} + (1-\beta)\ln\frac{1-\beta}{\alpha}}{D(p_1\|p_0)},$$

with

$$D(p_1\|p_0) = D(p_0\|p_1) = \frac{\mu^2}{2\sigma^2}.$$

If $\alpha = \beta = 0.1$ and $\frac{\mu}{\sigma} = 1$, then we can compute $n_{\mathrm{FSS}} = 22$ and $\mathbb{E}_0[N] \approx \mathbb{E}_1[N] \approx 9$. Therefore the SPRT takes fewer than half the number of observations as the FSS test on *average*.

While the average number of observation taken is lower for the sequential test, there is no guarantee that sequential test will take fewer observations for every realization of the observation sequence. Therefore, in practice, truncated sequential procedures are used, where the test stops at $\min\{N, k\}$ for some fixed integer k, so that at most k observations are used. The performance analysis of truncated sequential procedures is considerably more difficult than for the SPRT, but with a sufficiently large truncation parameter k, one can continue to use Wald's approximations for the performance.

9.2 Quickest Change Detection

In the quickest change detection (QCD) problem, we have a sequence of observations $\{Y_k, k \geq 1\}$, which initially have distribution P_0. At some point of time λ, due to some event, the distribution of the random variables changes to P_1. We restrict our attention to models where the observations are i.i.d. in the pre- and post-change regimes. Thus

$$Y_k \sim \text{i.i.d. } p_0 \text{ for } k < \lambda,$$
$$Y_k \sim \text{i.i.d. } p_1 \text{ for } k \geq \lambda.$$

We assume that p_0 and p_1 are known but λ is unknown. We call λ the *change point.*

Our goal is to detect the change in distribution as quickly as possible (after it happens), subject to false-alarm constraints. To this end, a procedure for quickest change detection is constructed as a stopping time N (see Definition 9.1) on the observation sequence $\{Y_k\}$, with the understanding that at the stopping time it is declared that the change has happened. The notion of quick detection is then captured by the requirement that we wish to detect the change with minimum possible delay, i.e., minimize some metric that captures the average of $(N-\lambda)^+$ (where we use the notation $x^+ = \max\{x, 0\}$). Clearly, selecting $N = 1$ always minimizes this quantity; however, if $\lambda > 1$, we have a false-alarm event. In practice we want to avoid false alarms as much as possible. Thus, the QCD problem is to find a stopping time on $\{Y_k\}$ that results in an optimal tradeoff between detection delay and false-alarm rate.

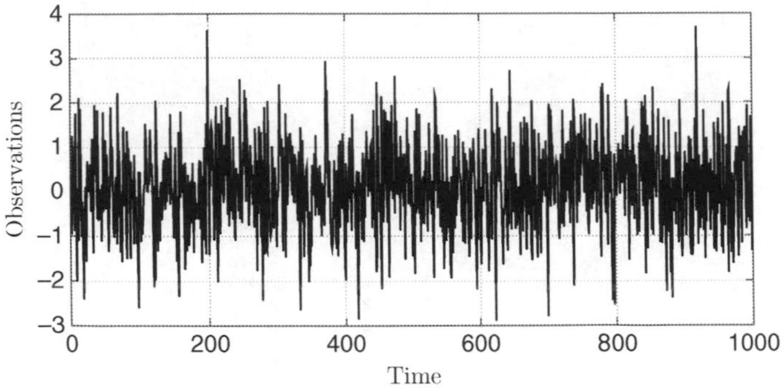

Figure 9.2 Stochastic sequence with observations from $p_0 = \mathcal{N}(0, 1)$ before the change (time slot 200), and with observations from $p_1 = \mathcal{N}(0.1, 1)$ after the change

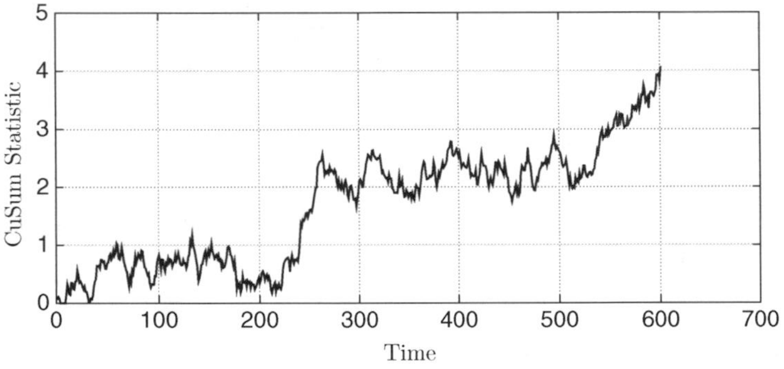

Figure 9.3 Evolution of the CuSum algorithm when applied to the samples given in Figure 9.2. We see that the change is detected with a delay of roughly 400 time slots

To motivate the need for quickest change detection algorithms, in Figure 9.2 we plot a sample path of a stochastic sequence whose samples are distributed as $\mathcal{N}(0, 1)$ before the change, and distributed as $\mathcal{N}(0.1, 1)$ after the change. For illustration, we choose time slot 200 as the change point. As is evident from the figure, the change cannot be detected through manual inspection.

In Figure 9.3, we plot the evolution of the *CuSum* test statistic (discussed in detail in Section 9.2.1), computed using the observations shown in Figure 9.2. As seen in Figure 9.3, the value of the test statistic stays close to zero before the change point, and starts to grow after the change point. The change is detected by using a threshold of 4.

We also see from Figure 9.3 that it takes around 500 samples to detect the change after it occurs. Can we do better than that, at least on an average? Clearly, declaring change before the change point (time slot 200) will result in zero delay, but it will cause a false alarm. The theory of quickest change detection deals with finding algorithms that have provable optimality properties, in the sense of minimizing the average detection delay under a false-alarm constraint.

9.2.1 Minimax Quickest Change Detection

We begin by considering a non-Bayesian setting, where we do not assume a prior on the change point λ. We focus on first constructing good algorithms for this problem, and then later comment on their optimality properties.

A fundamental fact that is used in the construction of these algorithms is that the mean of the log-likelihood ratio $\ln L(Y_k)$ in the post-change regime, i.e., for $k \geq \lambda$ equals $D(p_1 \| p_0) > 0$, and its mean before change, i.e., for $k < \lambda$ is $-D(p_0 \| p_1) < 0$. A simple way to use this fact was first proposed by Shewhart [7], in which a statistic based on the current observation is compared with a threshold to make a decision about the change. That is the Shewhart test is defined as

$$N_{\text{Shew}} = \inf\{n \geq 1 : \ln L(Y_n) > b\}.$$

Shewhart's test is widely employed in practice due to its simplicity; however, significant gain in performance can be achieved by making use of past observations (in addition to the current observation) to make the decision about the change.

Page [8] proposed such an algorithm that uses past observations, which he called the Cumulative Sum (CuSum) algorithm. The idea behind the CuSum algorithm is based on the behavior of the cumulative log-likelihood ratio sequence:

$$S_n = \sum_{k=1}^{n} \ln L(Y_k).$$

Before the change occurs, the statistic S_n has a negative "drift" due to the fact that $\mathbb{E}_0[\ln L(Y_k)] < 0$ (see (9.10)) and diverges towards $-\infty$. At the change point, the drift of the statistic S_n changes from negative to positive due to the fact that $\mathbb{E}_1[\ln L(Y_k)] > 0$ (see (9.11)), and beyond the change point, S_n starts growing towards ∞. Therefore S_n roughly attains a minimum at the change point (see Figure 9.4). The CuSum algorithm is constructed to detect this change in drift, and stop the first time the growth of S_n after change in the drift is large enough (to avoid false alarms).

Specifically, the stopping time for the CuSum algorithm is defined as

$$N_C = \inf\{n \geq 1 : W_n \geq b\}, \tag{9.20}$$

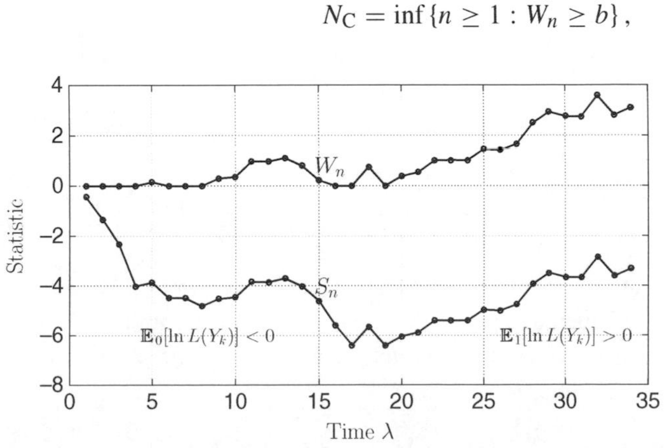

Figure 9.4 Motivation for the CuSum test statistic

where

$$W_n = S_n - \min_{0 \le k \le n} S_k = \max_{0 \le k \le n} S_n - S_k \tag{9.21}$$

$$= \max_{0 \le k \le n} \sum_{\ell=k+1}^{n} \ln L(Y_\ell) = \max_{1 \le k \le n+1} \sum_{\ell=k}^{n} \ln L(Y_\ell), \tag{9.22}$$

with the understanding that $\sum_{\ell=n+1}^{n} \ln L(Y_\ell) = 0$. It is easily shown that W_n can be computed iteratively as follows (see Exercise 9.6):

$$W_n = (W_{n-1} + \ln L(Y_n))^+, \quad W_0 = 0. \tag{9.23}$$

This recursion is useful from the viewpoint of implementing the CuSum test in practice.

The CuSum test also has a generalized likelihood ratio test (GLRT) interpretation. We can consider the quickest change detection (QCD) problem as a dynamic decision making problem, where at each time n, we are faced with a composite binary hypothesis testing problem:

$$\begin{cases} H_0^n : & \lambda > n \\ H_1^n : & \lambda \le n. \end{cases}$$

The joint distribution of the observations under H_0^n is given by

$$\prod_{k=1}^{n} p_0(y_k)$$

and the joint distribution under H_1^n is given by

$$p_\lambda(y_1, y_2, \dots, y_n) = \prod_{k=1}^{\lambda-1} p_0(y_k) \prod_{k=\lambda}^{n} p_1(y_k), \ \lambda = 1, \dots, n$$

with the understanding that $\prod_{k+1}^{k} p_i(X_k) = 1$. Then, the GLRT statistic L_G computed at time n is equal to

$$L_G(y_1, \dots, y_n) = \frac{\max_{1 \le \lambda \le n} p_\lambda(y_1, y_2, \dots, y_n)}{\prod_{k=1}^{n} p_0(y_k)} = \max_{1 \le \lambda \le n} \prod_{k=\lambda}^{n} L(y_k). \tag{9.24}$$

Define the statistic

$$C_n \triangleq \ln L_G(Y_1, \dots, Y_n) = \max_{1 \le k \le n} \sum_{\ell=k}^{n} \ln L(Y_\ell). \tag{9.25}$$

This statistic is closely related to the statistic W_n given in (9.22) except that the maximization does not include $k = n + 1$. This means that, unlike W_n, the statistic C_n can take on negative values. In particular, it is easy to show that C_n can be computed iteratively as (see Exercise 9.6)

$$C_n = (C_{n-1})^+ + \ln L(Y_n), \quad C_0 = 0. \tag{9.26}$$

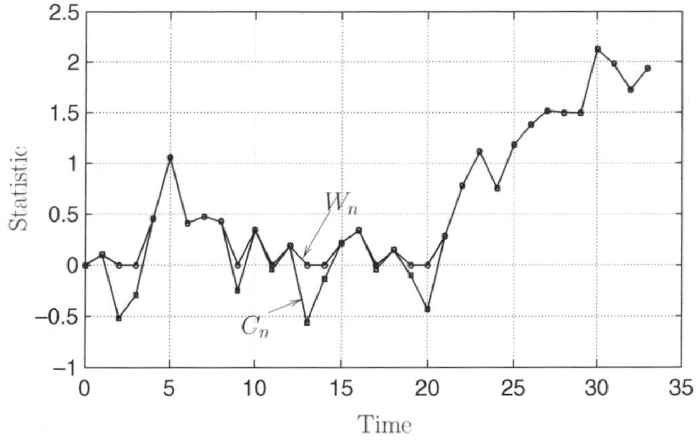

Figure 9.5 Comparison of the test statistics W_n and C_n

Nevertheless it is easy to see that both W_n and C_n cross will cross a positive threshold b at the same time (see Figure 9.5) and hence the CuSum algorithm can be equivalently defined in terms of C_n as

$$N_C = \inf\{n \geq 1 : C_n \geq b\}. \tag{9.27}$$

The GLRT interpretation of the CuSum algorithm has a close connection with another popular algorithm in the literature, called the Shiryaev–Roberts (SR) algorithm. In the SR algorithm, the maximum in (9.25) is replaced by a *sum*. Specifically, let

$$T_n = \sum_{1 \leq \lambda \leq n} \prod_{k=\lambda}^{n} L(Y_k).$$

Then, the SR stopping time is defined as

$$N_{SR} = \inf\{n \geq 1 : T_n > b\}. \tag{9.28}$$

It can be shown that even the SR statistic can be computed iteratively as

$$T_n = (1 + T_{n-1})L(Y_n), \quad T_0 = 0.$$

We now comment on the optimality properties of the CuSum algorithm and the SR algorithm. Without a prior on the change point, a reasonable measure of false alarms is the mean time to false alarm, or its reciprocal, which is the false-alarm rate (FAR):

$$\text{FAR}(N) = \frac{1}{\mathbb{E}_\infty[N]}, \tag{9.29}$$

where $\mathbb{E}_\infty$ is the expectation with respect to the probability measure when the change never occurs. Finding a uniformly powerful test that minimizes the delay over all possible values of γ subject to a FAR constraint is generally not possible. Therefore, it is more appropriate to study the quickest change detection problem in what is known as a minimax setting in this case. There are two important minimax problem formulations, one due to Lorden [9] and the other due to Pollak [10].

In Lorden's formulation, the objective is to minimize the supremum of the average delay conditioned on the worst possible realizations, subject to a constraint on the false-alarm rate. In particular, we define the worst-case average detection delay (WADD):

$$\text{WADD}(N) = \sup_{\lambda \geq 1} \text{ess sup}\, \mathbb{E}_\lambda \left[(N - \lambda + 1)^+ | Y_1, \ldots, Y_{\lambda-1} \right], \tag{9.30}$$

where $\mathbb{E}_\lambda$ denotes the expectation with respect to the probability measure when the change occurs at time λ. We have the following problem formulation.

PROBLEM 9.1 *Lorden.* Minimize $\text{WADD}(N)$ subject to $\text{FAR}(N) \leq \alpha$.

For the i.i.d. setting, Lorden showed that the CuSum algorithm ((9.20) or (9.27)) is asymptotically optimal for Problem 9.1 as $\alpha \to 0$. Specifically, the following result is proved in [9].

THEOREM 9.1 *Setting $b = |\ln \alpha|$ in* (9.20) *or* (9.27) *ensures that*

$$\text{FAR}(N_\text{C}) \leq \alpha,$$

and as $\alpha \to 0$,

$$\inf_{N:\text{FAR}(N) \leq \alpha} \text{WADD}(N) \sim \text{WADD}(N_\text{C}) \sim \frac{|\ln \alpha|}{D(p_1 \| p_0)},$$

where the "$\sim$" notation is used to denote that the ratio of the quantities on the two sides of the "$\sim$" approaches 1 in the limit as $\alpha \to 0$.

It was later shown in [11] that the CuSum algorithm is actually exactly optimal for Problem 9.1. Although the CuSum algorithm enjoys such a strong optimality property under Lorden's formulation, it can be argued that WADD is a somewhat pessimistic measure of delay. A less pessimistic way to measure the delay, the conditional average detection delay (CADD), was suggested by Pollak [10]:

$$\text{CADD}(N) = \sup_{\lambda \geq 1} \mathbb{E}_\lambda [N - \lambda | N \geq \lambda] \tag{9.31}$$

for all stopping times N for which the expectation is well-defined. It can be shown that

$$\text{CADD}(N) \leq \text{WADD}(N). \tag{9.32}$$

We then have the following problem formulation.

PROBLEM 9.2 *Pollak.* Minimize $\text{CADD}(N)$ subject to $\text{FAR}(N) \leq \alpha$.

It was shown by Lai [12] that:

THEOREM 9.2 *As $\alpha \to 0$,*

$$\inf_{N:\text{FAR}(N) \leq \alpha} \text{CADD}(N) \geq \frac{|\ln \alpha|}{D(p_1 \| p_0)} (1 + o(1)). \tag{9.33}$$

From Theorem 9.2 and (9.32), we have the following result:

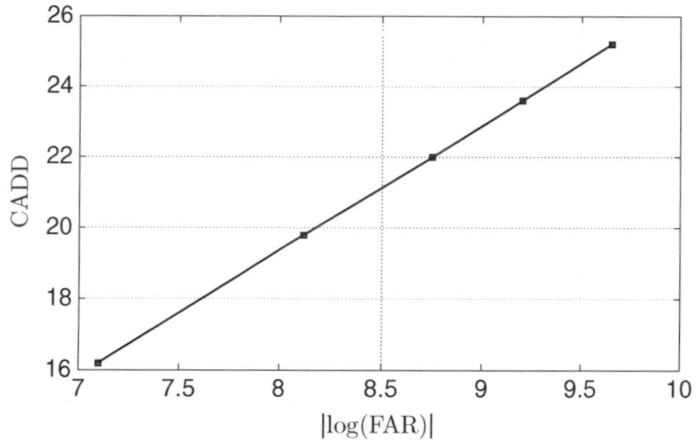

Figure 9.6 Performance of the CuSum algorithm with $p_0 = \mathcal{N}(0, 1)$, $p_1 = \mathcal{N}(0.75, 1)$, and $D(p_1 \| p_0) = 0.2812$

COROLLARY 9.1 *As* $\alpha \to 0$,

$$\inf_{N:\text{FAR}(N) \leq \alpha} \text{CADD}(N) \sim \text{CADD}(N_C) \sim \frac{|\ln \alpha|}{D(p_1 \| p_0)}.$$

Problem 9.2 has been studied in the i.i.d. setting in [10] and [13], where it is shown that the performance of some algorithms based on the Shiryaev–Roberts statistic, specifically, the SR algorithm (see (9.28)), are within a constant of the best possible performance over the set of algorithms meeting an FAR constraint of α, as $\alpha \to 0$. We do not discuss the details here, but those results imply that the SR algorithm is asymptotically optimal, i.e., we have the following theorem.

THEOREM 9.3 *Setting* $b = 1/\alpha$ *in* (9.28) *ensures that*

$$\text{FAR}(N_{\text{SR}}) \leq \alpha,$$

and as $\alpha \to 0$,

$$\inf_{N:\text{FAR}(N) \leq \alpha} \text{CADD}(N) \sim \text{CADD}(N_{\text{SR}}) \sim \frac{|\ln \alpha|}{D(p_1 \| p_0)}.$$

Theorems 9.1–9.3 and Corollary 9.1 show that both the CuSum algorithm and the SR algorithm are asymptotically optimal for both Problem 9.1 and Problem 9.2. The FAR in all cases decreases to zero exponentially with delay (CADD or WADD) with exponent $D(p_1 \| p_0)$, which is the same as the error exponent in the Chernoff–Stein lemma (see Lemma 8.1) with fixed P_M.

In Figure 9.6, we plot tradeoff curve for the CuSum algorithm obtained via simulations for Gaussian observations, i.e., we plot CADD as a function of $-\ln \text{FAR}$. It can be checked that the curve has a slope of approximately $1/D(p_1 \| p_0)$.

9.2.2 Bayesian Quickest Change Detection

In the Bayesian setting it is assumed that the change point is a random variable Λ taking values on the nonnegative integers, with $\pi_\lambda = \text{P}\{\Lambda = \lambda\}$. Define the average detection delay (ADD) and the probability of false alarm (PFA), respectively, as

$$\mathrm{ADD}(N) = \mathbb{E}\left[(N-\Lambda)^+\right] = \sum_{\lambda=1}^{\infty} \pi_\lambda \mathbb{E}_\lambda\left[(N-\lambda)^+\right], \tag{9.34}$$

$$\mathrm{PFA}(N) = \mathrm{P}\{N < \Lambda\} = \sum_{\lambda=0}^{\infty} \pi_\lambda \mathrm{P}_\lambda\{N < \lambda\}. \tag{9.35}$$

Then, the Bayesian quickest change detection problem is to minimize ADD subject to a constraint on PFA, and can be stated formally as follows.

PROBLEM 9.3 Minimize $\mathrm{ADD}(N)$ subject to $\mathrm{PFA}(N) \le \alpha$.

The prior on the change point Λ is usually assumed to be *geometric* with parameter ρ, i.e., for $0 < \rho < 1$,

$$\pi_\lambda = \mathrm{P}\{\Lambda = \lambda\} = \rho(1-\rho)^{\lambda-1}, \quad \lambda \ge 1.$$

The justification for this assumption is that it is a *memoryless* distribution; it also leads to a tractable solution to Problem 9.3 as seen in Theorem 9.4.

Let $Y_1^n = (Y_1, \ldots, Y_n)$ denote the observations up to time n. Also let

$$G_n = \mathrm{P}\{\Lambda \le n \mid Y_1^n\} \tag{9.36}$$

be the *a posteriori* probability at time n that the change has taken place given the observation up to time n. Using Bayes' rule, G_n can be shown to satisfy the recursion (see Exercise 9.8):

$$G_n = \Phi(Y_n, G_{n-1}), \tag{9.37}$$

where

$$\Phi(Y_n, G_{n-1}) = \frac{\tilde{G}_{n-1} L(Y_n)}{\tilde{G}_{n-1} L(Y_n) + (1 - \tilde{G}_{n-1})}, \tag{9.38}$$

$\tilde{G}_{n-1} = G_{n-1} + (1 - G_{n-1})\rho$, and $G_0 = 0$.

THEOREM 9.4 [14] *The optimal solution to Problem 9.3 has the following structure:*

$$N_S = \inf\{n \ge 1 : G_n \ge a\}, \tag{9.39}$$

if $a \in (0, 1)$ can be chosen such that

$$\mathrm{PFA}(N_S) = \alpha. \tag{9.40}$$

The optimal change detection algorithm of (9.39) is referred to as the *Shiryaev algorithm*.

We now discuss some alternative descriptions of the Shiryaev algorithm. Let

$$R_n = \frac{G_n}{(1 - G_n)},$$

and

$$R_{n,\rho} = \frac{G_n}{(1 - G_n)\rho}.$$

We note that R_n is the likelihood ratio between the hypotheses

$$\begin{cases} H_1^{(n)}: & \Lambda \le n \\ H_0^{(n)}: & \Lambda > n \end{cases}$$

averaged over the distribution of the change point. In particular,

$$\begin{aligned} R_n &= \frac{G_n}{(1-G_n)} = \frac{\mathrm{P}(\Lambda \le n | Y_1^n)}{\mathrm{P}(\Lambda > n | Y_1^n)} \\ &= \frac{\sum_{k=1}^n (1-\rho)^{k-1}\rho \prod_{i=1}^{k-1} p_0(Y_i) \prod_{i=k}^n p_1(Y_i)}{(1-\rho)^n \prod_{i=1}^n p_0(Y_i)} \\ &= \frac{1}{(1-\rho)^n} \sum_{k=1}^n (1-\rho)^{k-1} \rho \prod_{i=k}^n L(Y_i). \end{aligned} \tag{9.41}$$

Also, $R_{n,\rho}$ is a scaled version of R_n:

$$R_{n,\rho} = \frac{1}{(1-\rho)^n} \sum_{k=1}^n (1-\rho)^{k-1} \prod_{i=k}^n L(Y_i). \tag{9.42}$$

Like G_n, $R_{n,\rho}$ can also be computed using a recursion (see Exercise 9.8):

$$R_{n,\rho} = \frac{1 + R_{n-1,\rho}}{1-\rho} L(Y_n), \quad R_{0,\rho} = 0. \tag{9.43}$$

We remark here that if we set $\rho = 0$ in (9.42) and (9.43), then the Shiryaev statistic reduces to the SR statistic (see (9.28)) and the Shiryaev recursion reduces to the SR recursion.

It is easy to see that R_n and $R_{n,\rho}$ have one-to-one mappings with the Shiryaev statistic G_n.

ALGORITHM 9.1 *Shiryaev Algorithm.* The following three stopping times are equivalent and define the Shiryaev stopping time:

$$\begin{aligned} N_{\mathrm{S}} &= \inf\{n \ge 1 : G_n \ge a\}, \\ N_{\mathrm{S}} &= \inf\{n \ge 1 : R_n \ge b\}, && (9.44) \\ N_{\mathrm{S}} &= \inf\{n \ge 1 : R_{n,\rho} \ge \frac{b}{\rho}\}, && (9.45) \end{aligned}$$

with $b = \frac{a}{1-a}$.

It is easy to design the Shiryaev algorithm to meet a constraint on PFA. In particular,

$$\begin{aligned} \mathrm{PFA}(N_{\mathrm{S}}) &= \mathrm{P}\{N_{\mathrm{S}} < \Lambda\} = \mathbb{E}[\mathbb{1}\{N_{\mathrm{S}} < \Lambda\}] \\ &= \mathbb{E}\left[\mathbb{E}\left[\mathbb{1}\{N_{\mathrm{S}} < \Lambda\} \mid Y_1, Y_2, \ldots, Y_{N_{\mathrm{S}}}\right]\right] \\ &= \mathbb{E}\left[\mathrm{P}\left\{N_{\mathrm{S}} < \Lambda \mid Y_1, Y_2, \ldots, Y_{N_{\mathrm{S}}}\right\}\right] \\ &= \mathbb{E}[1 - G_{N_{\mathrm{S}}}] \le 1 - a. \end{aligned}$$

Thus, for $a = 1 - \alpha$, we have

$$\mathrm{PFA}(N_{\mathrm{S}}) \le \alpha.$$

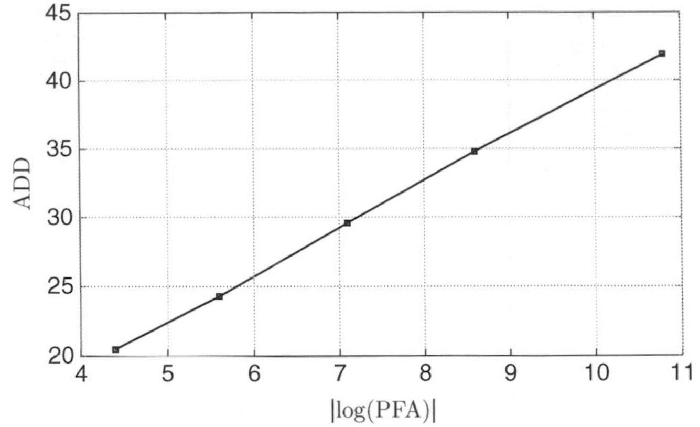

Figure 9.7 Performance of the Shiryaev algorithm with $\rho = 0.01$, $p_0 = \mathcal{N}(0, 1)$, $p_1 = \mathcal{N}(0.75, 1)$, $|\ln(1-\rho)| + D(p_1 \| p_0) = 0.2913$

We now present a first-order asymptotic analysis of the Shiryaev algorithm in the following theorem that is proved in [15].

THEOREM 9.5 *As* $\alpha \to 0$,

$$\inf_{N:\text{PFA}(N)\le\alpha} \mathbb{E}[(N-\Lambda)^+] \sim \mathbb{E}[(N_S-\Lambda)^+] \sim \frac{|\ln\alpha|}{D(p_1\|p_0) + |\ln(1-\rho)|}. \tag{9.46}$$

In Figure 9.7 we plot simulation results for ADD as a function of ln(PFA) for Gaussian observations. For a PFA constraint of α that is small, $b \approx |\ln\alpha|$, and

$$\text{ADD} \approx \mathbb{E}_1[N_S] \approx \frac{|\ln\alpha|}{|\ln(1-\rho)| + D(p_1\|p_0)},$$

giving a slope of roughly $\frac{1}{|\ln(1-\rho)|+D(p_1\|p_0)}$ to the tradeoff curve as we can verify in Figure 9.7.

When $|\ln(1-\rho)| \ll D(p_1\|p_0)$, the observations contain more information about the change than the prior, and the tradeoff slope is roughly $\frac{1}{D(p_1\|p_0)}$, as in the minimax setting. On the other hand, when $|\ln(1-\rho)| \gg D(p_1\|p_0)$, the prior contains more information about the change than the observations, and the tradeoff slope is roughly $\frac{1}{|\ln(1-\rho)|}$. The latter asymptotic slope is achieved by the stopping time that is based only on the prior:

$$N = \inf\{n \ge 1 : \text{P}(\Lambda > n) \le \alpha\}.$$

This is also easy to see from (9.38). With $D(p_1\|p_0)$ small, $L(Y) \approx 1$, and the recursion for G_n reduces to

$$G_n = G_{n-1} + (1 - G_{n-1})\rho, \quad G_0 = 0.$$

Expanding we get $G_n = \rho\sum_{k=0}^{n-1}(1-\rho)^k = 1-(1-\rho)^n$. Setting $G_{N^*} = 1-\alpha$, we get

$$\text{ADD} \approx N^* \approx \frac{|\ln\alpha|}{|\ln(1-\rho)|}.$$

Exercises

9.1 *Wald's Bound on Sum of Error Probabilities.* Prove (9.14).

9.2 *Sequential Hypothesis Testing.* Consider the problem sequential testing between the pmfs:

$$p_1(y) = \begin{cases} \frac{3}{4} & \text{if } y = 1 \\ \frac{1}{4} & \text{if } y = 0 \end{cases}, \qquad p_0(y) = \begin{cases} \frac{1}{4} & \text{if } y = 1 \\ \frac{3}{4} & \text{if } y = 0 \end{cases}$$

using an SPRT with thresholds $a < 0 < b$.

(a) Let $N_1(n)$ denote the number of observations that are equal to 1 among the first n observations. Show that the SPRT can be written in terms of $N_1(n)$.
(b) Suppose $a = -10 \ln 3$, and $b = 10 \ln 3$. Show that Wald's approximations for the error probabilities and expected stopping times are exact in this case.
(c) For the thresholds given in part (b), find the error probabilities P_F and P_M.
(d) For the thresholds given in part (b), find $\mathbb{E}_0[N_{SPRT}]$ and $\mathbb{E}_1[N_{SPRT}]$.
(e) Give a good approximation for the number of samples that a fixed sample size (FSS) test requires to achieve same error probabilities as in part (c).

9.3 *SPRT versus FSS Simulation.* Consider the detection problem:

$$H_1 \quad : \quad Y_1, Y_2, \ldots, \text{ are i.i.d. } \mathcal{N}(0.25, 1),$$

$$H_0 \quad : \quad Y_1, Y_2, \ldots, \text{ are i.i.d. } \mathcal{N}(-0.25, 1).$$

Your goal is to design tests that have error probabilies $\alpha = 0.05$ and $\beta = 0.05$.

(a) Design a fixed sample size test that has the desired error probabilities. That is, find τ and n_{FSS} such that the LRT

$$\delta(y_1, \ldots, y_{n_{FSS}}) = \begin{cases} 1 & \text{if } \sum_{k=1}^{n_{FSS}} y_k \geq \tau \\ 0 & \text{if } \sum_{k=1}^{n_{FSS}} y_k < \tau \end{cases}$$

has the desired error probabilities.
(b) Design an SPRT for the same α and β using Wald's approximations. Also compute the corresponding Wald approximations for $\mathbb{E}_0[N]$ and $\mathbb{E}_1[N]$.
(c) Simulate the performance of the SPRT using at least 1000 trials and compare your answers for the error probabilities and expected stopping times with part (b).
(d) Truncate the SPRT to a maximum of n_{FSS} observations in the simulations of part (c) and find the resulting error probabilities and expected stopping times for this truncated test.

9.4 *A Special Sequential Detection Problem.* Consider the sequential detection problem in which the observations $\{Y_k, k \geq 1\}$ are i.i.d. uniform on the interval $[-\frac{w_0}{2}, \frac{w_0}{2}]$ under H_0, and i.i.d. uniform on the interval $[-\frac{w_1}{2}, \frac{w_1}{2}]$ under H_1. Assume that $w_0 < w_1$. Our goal is to design an SPRT for this problem, with $b > 0 > a$.

(a) Show that the SPRT stopping rule is independent of the upper threshold b; in particular:

$$N = \min\left\{\left\lceil \frac{a}{\ln(w_0/w_1)} \right\rceil, \min\left\{n \geq 1 : \frac{w_0}{2} < |Y_n| < \frac{w_1}{2}\right\}\right\},$$

and we select H_0 if $N = \left\lceil \frac{a}{\ln(w_0/w_1)} \right\rceil$ and $|Y_N| \leq w_0/2$, and H_1 otherwise.

(b) Evaluate the probabilities of false alarm and miss for the rule of part (a). Note that you should be able to evaluate this exactly, without using Wald's approximations.

(c) Evaluate the expected stopping times under H_0 and H_1. Once again you should be able to evaluate these exactly without resorting to Wald's approximations.

9.5 *Sequential Testing with Three Hypotheses.* Suggest a reasonable way to modify the SPRT for sequential testing with three hypotheses. Explain your logic.

9.6 *CuSum Recursion.* Prove that the CuSum statistics W_n and C_n can be computed iteratively as given in this chapter, i.e.,

$$W_n = (W_{n-1} + \ln(L(Y_n))^+, \quad W_0 = 0,$$

and

$$C_n = (C_{n-1})^+ + \ln(L(Y_n)), \quad C_0 = 0.$$

Also use these recursions to conclude that W_n and C_n will cross a positive threshold b at the same time (sample-path wise).

9.7 *CuSum Simulation.* Simulate the performance of the CuSum algorithm with $p_0 = \mathcal{N}(0, 1)$ and $p_1 = \mathcal{N}(0.5, 1)$, and plot your results in a similar manner as in Figure 9.6.

9.8 *Shiryaev and SR Recursions.*

(a) Show that the Shiryaev statistic G_n satisfies the recursion

$$G_n = \Phi(Y_n, G_{n-1}),$$

where

$$\Phi(Y_n, G_{n-1}) = \frac{\tilde{G}_{n-1}L(Y_n)}{\tilde{G}_{n-1}L(Y_n) + (1 - \tilde{G}_{n-1})},$$

$\tilde{G}_{n-1} = G_{n-1} + (1 - G_{n-1})\rho$, and $G_0 = 0$.

(b) Show that the SR statistic can be computed iteratively as

$$T_n = (1 + T_{n-1})L(Y_n), \quad T_0 = 0.$$

9.9 *CuSum Simulation.* Simulate the performance of the Shiryaev algorithm with $\rho = 0.01$, $p_0 = \mathcal{N}(0, 1)$, and $p_1 = \mathcal{N}(0.5, 1)$, and plot your results in a similar manner as in Figure 9.7.

9.10 *ML Detection of Change Point in QCD.* The Cumulative Sum (CuSum) algorithm for quickest change detection uses the statistic:

$$W_n = \max_{1 \le k \le n+1} \sum_{\ell=k}^{n} \ln L(Y_\ell)$$

where $L(y) = \frac{p_1(y)}{p_0(y)}$. Recall that statistic W_n has the recursion

$$W_n = (W_{n-1} + \ln L(Y_n))^+, \quad W_0 = 0.$$

The statistic W_n can be used to detect the change in distribution by comparison with an appropriate threshold.

Now suppose in addition to detecting the occurrence of the change, we are also interested in obtaining a maximum likelihood estimator (MLE) of the actual change point λ at the time n, with the understanding that an MLE value of $n + 1$ corresponds to $\lambda > n$. Let $\hat{\lambda}_n$ denote the MLE of λ at time n. Show that

$$\hat{\lambda}_n = \begin{cases} n+1 & \text{if } W_n = 0 \\ \hat{\lambda}_{n-1} & \text{if } W_n > 0, \end{cases}$$

with $\hat{\lambda}_0 = 1$.

References

[1] A. Wald, *Sequential Analysis*, Wiley, 1947.

[2] D. Siegmund, *Sequential Analysis: Tests and Confidence Intervals*, Springer Science & Business Media, 2013.

[3] A. G. Tartakovsky, I. Nikiforov, and M. Basseville, *Sequential Analysis: Hypothesis Testing and Changepoint Detection*, CRC Press, 2014.

[4] H.V. Poor and O. Hadjiliadis, *Quickest Detection*, Cambridge University Press, 2009.

[5] V. V. Veeravalli and T. Banerjee, "Quickest Change Detection," *Academic Press Library in Signal Processing*, Vol. 3, pp. 209–255, Elsevier, 2014.

[6] A. Wald and J. Wolfowitz, "Optimum Character of the Sequential Probability Ratio test," *Ann. Math. Stat.*, Vol. 19, No. 3, pp. 326–339, 1948.

[7] W. A. Shewhart, "The Application of Statistics as an Aid in Maintaining Quality of a Manufactured Product," *J. Am. Stat. Assoc.*, Vol. 20, pp. 546–548, 1925.

[8] E. S. Page, "Continuous Inspection Schemes," *Biometrika*, Vol. 41, pp. 100–115, 1954.

[9] G. Lorden, "Procedures for Reacting to a Change in Distribution," *Ann. Math. Stat.*, Vol. 42, pp. 1897–1908, 1971.

[10] M. Pollak, "Optimal Detection of a Change in Distribution," *Ann. Stat.*, Vol. 13, pp. 206–227, 1985.

[11] G. V. Moustakides, "Optimal Stopping Times for Detecting Changes in Distributions," *Ann. Stat.*, Vol. 14, pp. 1379–1387, 1986.

[12] T. L. Lai, "Information Bounds and Quick Detection of Parameter Changes in Stochastic Systems," *IEEE Trans. Inf. Theory*, Vol. 44, pp. 2917–2929, 1998.

[13] A. G. Tartakovsky, M. Pollak, and A. Polunchenko, "Third-order Asymptotic Optimality of the Generalized Shiryaev–Roberts Changepoint Detection Procedures," *Theory Prob. Appl.*, Vol. 56, pp. 457–484, 2012.

[14] A. N. Shiryayev, *Optimal Stopping Rules*, Springer-Verlag, 1978.

[15] A. G. Tartakovsky and V. V. Veeravalli, "General Asymptotic Bayesian Theory of Quickest Change Detection," *SIAM Theory Prob. App.*, Vol. 49, pp. 458–497, 2005.

10 Detection of Random Processes

This chapter considers hypothesis testing when the observations are realizations of random processes. We first study discrete-time stationary processes and Markov processes, describe the structure of optimal detectors, and introduce the notions of Kullback–Leibler (KL) and Chernoff divergence rates. Then we consider two continuous-time processes: Gaussian processes, which are studied by means of the Karhunen–Loève transform; and inhomogeneous Poisson processes. The chapter ends with a general approach to detection based on Radon–Nikodym derivatives.

In Chapters 5 and 8 we studied the structure of optimal detectors and their performance when the observations form a length-n vector. We have focused on problems where the samples of the observation vector are *conditionally independent* given the hypothesis, or at least can be reduced to that case (using a whitening transform). The underlying information measures (Kullback–Leibler, Chernoff, etc.) are additive, which facilitates the performance analysis.

This chapter considers several problems where the conditional-independence model does not hold. We show that for several classical models a tractable expression for the optimal tests can still be derived, and performance can still be accurately analyzed. We begin with the case of discrete-time processes (Section 10.1) then move to continuous-time Gaussian processes (Section 10.2) and inhomogeneous Poisson processes (Section 10.3) before concluding with a description of the general approach to such problems (Section 10.4).

10.1 Discrete-Time Random Processes

A random process is an indexed family of random variables $\{X(t), t \in \mathcal{T}\}$ defined in terms of a probability space $(\Omega, \mathscr{F}, P)$ [1]. Then $X(t) = x(t, \omega)$, $\omega \in \Omega$. In m-ary hypothesis testing, there are m probability measures P_j, $1 \le j \le m$, defined over $\mathscr{F}$.

For discrete-time processes, the observations form a length-n sequence $\boldsymbol{Y} = (Y_1, Y_2, \dots, Y_n)$ whose conditional distribution given hypothesis H_j will sometimes be denoted by $p_j^{(n)}$ to emphasize the dependency on n. We will have $\mathcal{T} = \mathbb{Z}$ and $Y_i = X(i)$ for $i = 1, 2, \dots, n$. For many problems encountered in this section, the Kullback–Leibler and Chernoff information measures, normalized by $\frac{1}{n}$, converge to a limit as $n \to \infty$, as defined next.

DEFINITION 10.1 The *Kullback–Leibler divergence rate* between processes P_0 and P_1 is defined as the following limit, when it exists:

$$\overline{D}(\mathrm{P}_0 \| \mathrm{P}_1) \triangleq \lim_{n\to\infty} \frac{1}{n} D(p_0^{(n)} \| p_1^{(n)}). \tag{10.1}$$

DEFINITION 10.2 The *Chernoff divergence rate* of order $u \in (0, 1)$ between processes P_0 and P_1 is defined as the following limit, when it exists:

$$\overline{d}_u(\mathrm{P}_0, \mathrm{P}_1) \triangleq \lim_{n\to\infty} \frac{1}{n} d_u(p_0^{(n)}, p_1^{(n)}). \tag{10.2}$$

The Kullback–Leibler and Chernoff divergence rates can be used to derive the asymptotics of error probability for binary hypothesis tests.

10.1.1 Periodic Stationary Gaussian Processes

A discrete-time stationary Gaussian process $Z = (Z_1, Z_2, \ldots)$ is said to be periodic with period n if its mean function and covariance function are periodic with period n [1]. We shall only be interested in zero-mean processes. The covariance sequence $c_k \triangleq \mathbb{E}[Z_1 Z_{k+1}]$ is symmetric ($c_k = c_{-k}$) and periodic, hence $c_k = c_{n-k}$. The process is represented by a n-vector $\mathbf{Z}$ whose $n \times n$ covariance matrix $\mathsf{K}^{(n)} = \mathbb{E}[\mathbf{Z}\mathbf{Z}^T]$ is circulant Toeplitz, i.e., each row is obtained from a circular shift of the first row $[c_0, c_1, \ldots, c_2, c_1]$:

$$\mathsf{K}^{(n)} = \begin{bmatrix} c_0 & c_1 & c_2 & \ldots & c_2 & c_1 \\ c_1 & c_0 & c_1 & \ldots & c_3 & c_2 \\ \vdots & & & & & \\ c_1 & c_2 & c_3 & \ldots & c_1 & c_0 \end{bmatrix}. \tag{10.3}$$

The matrix is constant along diagonals, symmetric, and nonnegative definite. The discrete Fourier transform (DFT) of the first row is the *power spectral mass function*

$$\lambda_k = \frac{1}{\sqrt{n}} \sum_{j=0}^{n-1} c_j \, e^{-i2\pi jk/n} \geq 0, \quad k = 0, 1, \ldots, n-1 \quad (i = \sqrt{-1}), \tag{10.4}$$

which is symmetric and nonnegative. The inverse DFT transform gives

$$c_j = \frac{1}{\sqrt{n}} \sum_{j=0}^{n-1} \lambda_k \, e^{i2\pi jk/n}, \quad j = 0, 1, \ldots, n-1. \tag{10.5}$$

Hence the process is characterized equivalently by (10.4) or by (10.5).

Denote by $\mathsf{F} \triangleq \{\frac{1}{\sqrt{n}} e^{-i2\pi jk/n}\}_{1 \leq j,k \leq n}$ the $n \times n$ DFT matrix, which is complex-valued, symmetric, and unitary. The eigenvectors of circulant Toeplitz matrices are complex exponentials at harmonically related frequencies. Specifically,

$$\mathsf{K}^{(n)} = \mathsf{F}^{\dagger} \Lambda \mathsf{F},$$

where $\Lambda = \mathrm{diag}[\lambda_0, \lambda_1, \ldots, \lambda_{k-1}]$ is a diagonal matrix comprised of the power spectral mass values, and the superscript $\dagger$ denotes the Hermitian transpose operator: $(\mathsf{F}^{\dagger})_{ij} = \mathsf{F}^*_{ji}$. Since the determinant of a matrix is invariant under unitary transforms, we have

$$|\mathsf{K}^{(n)}| = \prod_{k=0}^{n-1} \lambda_k. \tag{10.6}$$

Now denote by $\tilde{\boldsymbol{Z}} \triangleq \mathsf{F}\boldsymbol{Z}$ the DFT of $\boldsymbol{Z}$, which satisfies the Hermitian-symmetry property $\tilde{Z}_k = \tilde{Z}^*_{n-k}$ for all k. Also

$$\text{Cov}(\tilde{\boldsymbol{Z}}) = \mathbb{E}[\tilde{\boldsymbol{Z}}\tilde{\boldsymbol{Z}}^\dagger] = \mathsf{F}\mathbb{E}[\boldsymbol{Z}\boldsymbol{Z}^T]\mathsf{F}^\dagger = \mathsf{F}\mathsf{K}^{(n)}\mathsf{F}^\dagger = \Lambda \tag{10.7}$$

and

$$\boldsymbol{Z}^T(\mathsf{K}^{(n)})^{-1}\boldsymbol{Z} = \tilde{\boldsymbol{Z}}^T\Lambda^{-1}\tilde{\boldsymbol{Z}} = \sum_{k=0}^{n-1}\frac{|\tilde{Z}_k|^2}{\lambda_k}. \tag{10.8}$$

Binary Hypothesis Testing Consider testing between two zero-mean periodic stationary Gaussian processes with respective power spectral mass functions $\lambda_{0,k}$ and $\lambda_{1,k}$, $k = 0, 1, \ldots, n-1$. Let $\tilde{\boldsymbol{Y}} = \mathsf{F}\boldsymbol{Y}$. The LLR is given by

$$\begin{aligned}\ln\frac{p_1^{(n)}(\boldsymbol{Y})}{p_0^{(n)}(\boldsymbol{Y})} &= \frac{1}{2}\ln\frac{|\mathsf{K}_0^{(n)}|}{|\mathsf{K}_1^{(n)}|} - \frac{1}{2}\boldsymbol{Y}^T(\mathsf{K}_1^{(n)})^{-1}\boldsymbol{Y} + \frac{1}{2}\boldsymbol{Y}^T(\mathsf{K}_0^{(n)})^{-1}\boldsymbol{Y}\\ &= \frac{1}{2}\sum_{k=0}^{n-1}\left[\ln\frac{\lambda_{0,k}}{\lambda_{1,k}} - |\tilde{Y}_k|^2\left(\frac{1}{\lambda_{1,k}} - \frac{1}{\lambda_{0,k}}\right)\right],\end{aligned} \tag{10.9}$$

where the last equality was obtained using the identities (10.6) and (10.8). Thus the LLR is expressed simply in terms of the DFT of the observation vector.

Taking the expectation of (10.9) under H_1 and using $\mathbb{E}_1[|\tilde{Z}_k|^2] = \lambda_{1,k}$ per (10.7), we obtain the Kullback–Leibler divergence

$$D(p_1^{(n)}\|p_0^{(n)}) = \frac{1}{2}\sum_{k=0}^{n-1}\left[-\ln\frac{\lambda_{1,k}}{\lambda_{0,k}} - 1 + \frac{\lambda_{1,k}}{\lambda_{0,k}}\right] = \sum_{k=0}^{n-1}\psi\left(\frac{\lambda_{1,k}}{\lambda_{0,k}}\right) \tag{10.10}$$

and likewise

$$D(p_0^{(n)}\|p_1^{(n)}) = \sum_{k=0}^{n-1}\psi\left(\frac{\lambda_{0,k}}{\lambda_{1,k}}\right), \tag{10.11}$$

where the function $\psi(x) = \frac{1}{2}(x - 1 - \ln x),\ x > 0$, was introduced in (6.4). For testing $\mathrm{P}_0 = \mathcal{N}(0, 1)$ against $\mathrm{P}_1 = \mathcal{N}(0, \xi)$, we obtained $D(p_0\|p_1) = \psi\left(\frac{1}{\xi}\right)$.

The Chernoff divergence of order $u \in (0, 1)$ is obtained from (6.19) as

$$\begin{aligned}d_u(p_0^{(n)}, p_1^{(n)}) &= \frac{1}{2}\sum_{k=0}^{n-1}\ln\frac{(1-u)\lambda_{0,k} + u\lambda_{1,k}}{\lambda_{0,k}^{1-u}\lambda_{1,k}^{u}}\\ &= \frac{1}{2}\sum_{k=0}^{n-1}\left[u\ln\frac{\lambda_{0,k}}{\lambda_{1,k}} + \ln\left(1 - u + u\frac{\lambda_{1,k}}{\lambda_{0,k}}\right)\right].\end{aligned} \tag{10.12}$$

10.1.2 Stationary Gaussian Processes

Consider now testing between two zero-mean stationary Gaussian processes with respective spectral densities $S_0(f)$ and $S_1(f)$, $|f| \le \frac{1}{2}$. The respective covariance matrices are $\mathsf{K}_0^{(n)}$ and $\mathsf{K}_1^{(n)}$. The LLR is given by

$$\begin{aligned}
\ln \frac{p_1^{(n)}(\boldsymbol{Y})}{p_0^{(n)}(\boldsymbol{Y})} &= \frac{1}{2} \ln \frac{|\mathsf{K}_0^{(n)}|}{|\mathsf{K}_1^{(n)}|} - \frac{1}{2}\boldsymbol{Y}^T (\mathsf{K}_1^{(n)})^{-1}\boldsymbol{Y} + \frac{1}{2}\boldsymbol{Y}^T (\mathsf{K}_0^{(n)})^{-1}\boldsymbol{Y} \\
&= \frac{1}{2} \ln \frac{|\mathsf{K}_0^{(n)}|}{|\mathsf{K}_1^{(n)}|} - \frac{1}{2}\mathrm{Tr}\left[\boldsymbol{Y}\boldsymbol{Y}^T[(\mathsf{K}_1^{(n)})^{-1} - (\mathsf{K}_0^{(n)})^{-1}]\right], \qquad (10.13)
\end{aligned}$$

where $\mathrm{Tr}[\mathsf{A}] = \sum_i \mathsf{A}_{ii}$ denotes the trace of a square matrix. (See Appendix A for basic properties of the trace.) In general the LLR is not so easy to evaluate due to the need to evaluate quadratic forms involving the inverse covariance matrices.

Taking the expectation of (10.13) under H_1 and using $\mathbb{E}_1[\boldsymbol{Y}\boldsymbol{Y}^T] = \mathsf{K}_1^{(n)}$, we obtain the Kullback–Leibler divergence

$$D(p_1^{(n)} \| p_0^{(n)}) = \frac{1}{2}\left[\mathrm{Tr}[(\mathsf{K}_0^{(n)})^{-1}\mathsf{K}_1^{(n)}] - n - \ln \frac{|\mathsf{K}_1^{(n)}|}{|\mathsf{K}_0^{(n)}|}\right] \qquad (10.14)$$

and likewise

$$D(p_0^{(n)} \| p_1^{(n)}) = \frac{1}{2}\left[\mathrm{Tr}[(\mathsf{K}_1^{(n)})^{-1}\mathsf{K}_0^{(n)}] - n - \ln \frac{|\mathsf{K}_0^{(n)}|}{|\mathsf{K}_1^{(n)}|}\right] \qquad (10.15)$$

as given in (6.5).

While these expressions are still cumbersome, significant simplifications are obtained in the asymptotic setting where $n \to \infty$. Assume the covariance sequence is absolutely summable: $\sum_k |c_k| < \infty$, which implies that $|c_k|$ decays faster than $\frac{1}{k}$ as $k \to \infty$. Then the KL divergence rate (10.1) exists and can be evaluated using the asymptotics of covariance matrices of stationary processes [2, 3]

$$\lim_{n\to\infty} |\mathsf{K}^{(n)}|^{1/n} = \exp \int_{-1/2}^{1/2} \ln S(f)\, df$$

or equivalently,

$$\lim_{n\to\infty} \frac{1}{n} \ln |\mathsf{K}^{(n)}| = \int_{-1/2}^{1/2} \ln S(f)\, df.$$

Likewise,

$$\lim_{n\to\infty} \frac{1}{n}\mathrm{Tr}[(\mathsf{K}_0^{(n)})^{-1}\mathsf{K}_1^{(n)}] = \int_{-1/2}^{1/2} \frac{S_1(f)}{S_0(f)}\, df.$$

Combining these limits with (10.1) and (10.15), we obtain

$$\bar{D}(\mathrm{P}_0\|\mathrm{P}_1) = \int_{-1/2}^{1/2} \psi\left(\frac{S_1(f)}{S_0(f)}\right) df. \qquad (10.16)$$

For the Chernoff divergence of order $u \in (0, 1)$, starting from (6.19) and applying (10.2), one similarly obtains the Chernoff divergence rate of order u [4]:

$$\begin{aligned}\bar{d}_u(p_0, p_1) &\triangleq \lim_{n\to\infty} \frac{1}{n} d_u(p_0^{(n)}, p_1^{(n)}) \\ &= \frac{1}{2} \lim_{n\to\infty} \frac{1}{n} \ln \frac{|(1-u)\mathsf{K}_0^{(n)} + u\mathsf{K}_1^{(n)}|}{|\mathsf{K}_0^{(n)}|^{1-u} |\mathsf{K}_1^{(n)}|^u} \\ &= \frac{1}{2} \int_{-1/2}^{1/2} \ln \frac{(1-u)S_0(f) + uS_1(f)}{S_0(f)^{1-u} S_1(f)^u} df. \qquad (10.17)\end{aligned}$$

The intuition behind these results is that the $n \times n$ Toeplitz covariance matrix is well approximated by a circulant Toeplitz matrix if n is large enough and $|c_k|$ decays faster than $\frac{1}{k}$. For instance, a tridiagonal Toeplitz covariance matrix K and its circular Toeplitz approximation $\hat{\mathsf{K}}$ are given in (10.18):

$$\mathsf{K} = \begin{bmatrix} c_0 & c_1 & 0 & 0 & \dots & 0 & 0 \\ c_1 & c_0 & c_1 & 0 & \dots & 0 & 0 \\ 0 & c_1 & c_0 & c_1 & \dots & 0 & 0 \\ \vdots & & & & \ddots & & \vdots \\ 0 & 0 & 0 & 0 & \dots & c_0 & c_1 \\ 0 & 0 & 0 & 0 & \dots & c_1 & c_0 \end{bmatrix} \quad \hat{\mathsf{K}} = \begin{bmatrix} c_0 & c_1 & 0 & 0 & \dots & 0 & c_1 \\ c_1 & c_0 & c_1 & 0 & \dots & 0 & 0 \\ 0 & c_1 & c_0 & c_1 & \dots & 0 & 0 \\ \vdots & & & & \ddots & & \vdots \\ 0 & 0 & 0 & 0 & \dots & c_0 & c_1 \\ c_1 & 0 & 0 & 0 & \dots & c_1 & c_0 \end{bmatrix} \qquad (10.18)$$

Only the upper right and lower left corners of the matrix are affected. More generally, replacing each entry c_k of the covariance matrix by $\hat{c}_k = c_k + c_{n-k}$ produces a circulant approximation to the original Toeplitz covariance matrix. If such approximation is valid, then the approach used for periodic stationary Gaussian processes in Section 10.1.1 can be extended to the nonperiodic case, and an additive expression for the KL divergence rate is obtained again.

This approximation method can be also used to construct an approximate LRT, by computing the DFT $\tilde{Y}$ of the observations and applying the LRT (10.9) for periodic stationary Gaussian processes. Here the power spectral values $\lambda_{0,k}$ and $\lambda_{1,k}$ are obtained by taking the DFT of the periodicized correlation sequence $\hat{c}_k$ under H_0 and H_1, respectively.

10.1.3 Markov Processes

Consider a homogeneous Markov chain $(\mathcal{S}, q_0, r_0)$ with finite state space $\mathcal{S}$, initial state probabilities

$$q_0(y) = \mathrm{P}_0\{Y_0 = y\},\ y \in \mathcal{S},$$

and transition probabilities

$$r_0(y|y') = \mathrm{P}_0\{Y_{i+1} = y | Y_i = y'\},$$

for all $i \in \mathbb{N}$ and $y, y' \in \mathcal{S}$ [1]. Consider a second process $(\mathcal{S}, q_1, r_1)$ with same state space $\mathcal{S}$ but potentially different initial state probabilities $q_1(y) = \mathrm{P}_1\{Y_0 = y\}$ and transition probabilities $r_1(y|y') = \mathrm{P}_1\{Y_i = y|Y_{i-1} = y'\}$. The probability of a sequence $\boldsymbol{y} = (y_1, y_2, \dots, y_n) \in \mathcal{S}^n$ under hypothesis $H_j,\ j = 0, 1$ is given by

$$p_j^{(n)}(\boldsymbol{y}) = q_j(y_1) \prod_{i=2}^{n} r_j(y_i|y_{i-1}). \tag{10.19}$$

The LLR is given by

$$\ln \frac{p_1^{(n)}(\boldsymbol{y})}{p_0^{(n)}(\boldsymbol{y})} = \frac{q_1(y_1)}{q_0(y_1)} \prod_{i=2}^{n} \frac{r_1(y_i|y_{i-1})}{r_0(y_i|y_{i-1})} \tag{10.20}$$

and is easily computed owing to the factorization property (10.19) of the two Markov chains.

To compute the KL divergences between $p_0^{(n)}$ and $p_1^{(n)}$ we first need the following definition.

DEFINITION 10.3 Let p and q be two conditional distributions on Y given X, and r a distribution on X. The *conditional Kullback–Leibler divergence* of p and q with reference r is

$$D(p\|q|r) \triangleq \sum_x r(x) \sum_y p(y|x) \ln \frac{p(y|x)}{q(y|x)} = \sum_x r(x) D(p(\cdot|x)\|q(\cdot|x)). \tag{10.21}$$

KL Divergence Rate We have

$$\begin{aligned}
D(p_0^{(n)}\|p_1^{(n)}) &= \mathbb{E}_0\left[\ln \frac{p_0^{(n)}(\boldsymbol{Y})}{p_1^{(n)}(\boldsymbol{Y})}\right] \\
&= \mathbb{E}_0\left[\ln \frac{q_0(Y_1)}{q_1(Y_1)} + \sum_{i=2}^{n} \ln \frac{r_0(Y_i|Y_{i-1})}{r_1(Y_i|Y_{i-1})}\right] \\
&= \mathbb{E}_0\left[\ln \frac{q_0(Y_1)}{q_1(Y_1)}\right] + \sum_{i=2}^{n} \mathbb{E}_0\left[\ln \frac{r_0(Y_i|Y_{i-1})}{r_1(Y_i|Y_{i-1})}\right] \\
&= D(q_0\|q_1) + \sum_{i=2}^{n} D(r_0\|r_1|f_{0,i-1}),
\end{aligned} \tag{10.22}$$

where $f_{0,i} = \{f_{0,i}(y),\ y \in \mathcal{S}\}$ is the distribution of Y_i under H_0. Assume the Markov chain $(\mathcal{S}, q_0, r_0)$ is irreducible (there exists a path connecting any two states) and aperiodic. Then the Markov chain has a unique equilibrium distribution f_0^*, and $f_{0,i}$ converges to f_0^* as $i \to \infty$. Hence $D(r_0\|r_1|f_{0,i-1})$ converges to $D(r_0\|r_1|f_0^*)$. Hence the KL divergence rate (10.1) exists and is equal to

$$\overline{D}(\mathrm{P}_0\|\mathrm{P}_1) = \lim_{n\to\infty} \frac{1}{n} D(p_0^{(n)}\|p_1^{(n)}) = D(r_0\|r_1|f_0^*). \tag{10.23}$$

Note how the log function and the additivity property (10.22) were used to show the existence of a limit and to evaluate that limit.

Chernoff Divergence Rate Define the $|\mathcal{S}| \times |\mathcal{S}|$ matrix R with (nonnegative) elements

$$\mathsf{R}(y, y') = r_0(y|y')^{1-u} r_1(y|y')^u \quad y, y' \in \mathcal{S}, \tag{10.24}$$

the vector $\boldsymbol{v}$ with (nonnegative) elements

$$v(y) = q_0(y)^{1-u} q_1(y)^u, \quad y \in \mathcal{S},\ u \in (0, 1), \tag{10.25}$$

and the vector $\mathbf{1} = (1, \ldots, 1)$ of the same dimension $|\mathcal{S}|$. Both R and $\boldsymbol{v}$ depend on u. Then the Chernoff divergence of order $u \in (0, 1)$ between $p_0^{(n)}$ and $p_1^{(n)}$ is given by

$$\begin{aligned}
d_u(p_0^{(n)}, p_1^{(n)}) &= -\ln \sum_{\boldsymbol{y} \in \mathcal{S}^n} [p_0^{(n)}(\boldsymbol{y})]^{1-u} [p_1^{(n)}(\boldsymbol{y})]^u \\
&= -\ln \sum_{\boldsymbol{y} \in \mathcal{S}^n} [q_0(y_1)^{1-u} q_1(y_1)^u] \prod_{i=2}^{n} [r_0(y_i|y_{i-1})^{1-u} r_1(y_i|y_{i-1})^u] \\
&= -\ln \sum_{\boldsymbol{y} \in \mathcal{S}^n} v(y_1) \prod_{i=2}^{n} \mathsf{R}(y_i, y_{i-1}) \\
&\overset{(a)}{=} -\ln \sum_{y_1 \in \mathcal{S}} v(y_1) \sum_{y_2 \in \mathcal{S}} \mathsf{R}(y_2, y_1) \ldots \sum_{y_n \in \mathcal{S}} \mathsf{R}(y_n, y_{n-1}) \\
&= -\ln \sum_{y_1 \in \mathcal{S}} v(y_1) \prod_{i=2}^{n} \sum_{y_i \in \mathcal{S}} \mathsf{R}(y_i, y_{i-1}) \\
&= -\ln \sum_{y_1 \in \mathcal{S}} v(y_1) \sum_{y_n \in \mathcal{S}} R^{n-1}(y_n, y_1) \\
&= -\ln \boldsymbol{v}^T R^{n-1} \mathbf{1}.
\end{aligned} \tag{10.26}$$

Observe the passage from a sum of products to a product of sums in (a). Since r_0 and r_1 are irreducible, so is the matrix R. By the Perron–Frobenius theorem, the matrix R has a unique largest real eigenvalue, which is positive and will be denoted by λ. Moreover, $\lim_{n\to\infty} \boldsymbol{v}^T \mathsf{R}^{n-1}\mathbf{1}$ exists and is equal to λ.[1] Hence the Chernoff divergence rate (10.2) exists and is equal to

$$\bar{d}_u(\mathrm{P}_0, \mathrm{P}_1) = \lim_{n\to\infty} \frac{1}{n} d_u(p_0^{(n)}, p_1^{(n)}) = -\ln \lambda. \tag{10.27}$$

Example 10.1 Let $\mathcal{S} = \{0, 1\}$. Consider two Markov processes with respective transition probability matrices

$$r_0 = \begin{bmatrix} a & 1-a \\ 1-a & b \end{bmatrix}, \quad \text{and} \quad r_1 = \begin{bmatrix} b & 1-b \\ 1-b & b \end{bmatrix}, \quad a, b \in (0, 1). \tag{10.28}$$

[1] This is reminiscent of the *power method* for finding the largest eigenvalue of a matrix R. Starting from an arbitrary vector $\boldsymbol{v}$ and recursively multiplying by R and normalizing the result yields the corresponding eigenvector in the limit.

Each process admits the uniform stationary distribution $f_0^* = f_1^* = (\frac{1}{2}, \frac{1}{2})$. Typical sample paths of Y are, for $a = 1 - b = 0.1$:

$$H_0 : 10101010100101010\ldots,$$
$$H_1 : 11111111000000000\ldots.$$

By (10.23), the KL divergence rates are given by

$$\overline{D}(\mathrm{P}_0\|\mathrm{P}_1) = d(a\|b) \quad \text{and} \quad \overline{D}(\mathrm{P}_1\|\mathrm{P}_0) = d(b\|a),$$

where d is the binary KL divergence function of (7.1).

By (10.24) and (10.28), the matrix R is given by

$$R = \begin{bmatrix} a^{1-u}b^u & (1-a)^{1-u}(1-b)^u \\ (1-a)^{1-u}(1-b)^u & a^{1-u}b^u \end{bmatrix},$$

which is Toeplitz circulant. Its largest eigenvalue is

$$\lambda = a^{1-u}b^u + (1-a)^{1-u}(1-b)^u = e^{-d_u(a,b)},$$

where

$$d_u(a,b) = -\ln(a^{1-u}b^u + (1-a)^{1-u}(1-b)^u)$$

is the *binary Chernoff divergence* of order u, i.e., the Chernoff divergence between two Bernoulli distributions with parameters a and b respectively. Hence the Chernoff divergence rate of order $u \in (0,1)$ between the processes P_0 and P_1 is equal to $d_u(a,b)$. □

10.2 Continuous-Time Processes

In many signal detection problems, the observations are continuous-time waveforms. These observations are often sampled and processed numerically, which gives rise to the discrete-time detection problems considered so far. In fact, one might even question the need to study detection of continuous-time signals in this context. To motivate such study, consider the following issues:

- Since sampling generally discards information, one would like to know how much information is lost in the sampling process.
- High-resolution sampling produces long observation vectors, and finite-sample performance metrics such as the GSNR of (5.30) might be cumbersome. The use of limits (as sampling resolution increases) often provides considerable insight into the mathematical structure of the problem.
- In particular, problems of singular detection (zero error probability) arise in some continuous-time detection problems, even though their discrete-time counterparts have nonzero error probabilities.

By acquiring n samples of a continuous-time waveform $y(t)$, $0 \le t \le T$, we mean either acquiring values $y(t_i)$ for n distinct time instants t_i, $1 \le i \le n$, or more generally acquiring linear projections $\int_0^T y(t)\phi_i(t)\,dt$ for n suitably designed measurement functions $\phi_i(\cdot)$, $1 \le i \le n$.

Returning to the observed continuous-time waveform $Y = \{Y(t), t \in [0,T]\}$, the main technical difficulty is that Y does not have a density, hence it is not clear at this point what the likelihood ratio should be. A variety of approaches can be developed to approach such detection problems; see [5] for a comprehensive survey. As described by Grenander [6], the most intuitive approach for Gaussian processes is the Karhunen–Loève decomposition, or eigenfunction expansion method.

The Kullback–Leibler and Chernoff information measures can still be defined. When normalized by $\frac{1}{T}$, they often converge to a limit as $T \to \infty$.

Mathematical Preliminaries The space of square-integrable functions over $[0,T]$ is denoted by $L^2([0,T])$. The L^2 norm of a function $f \in L^2([0,T])$ is $\|f\| \triangleq \left(\int_0^T f^2(t)\,dt\right)^{1/2}$. The Dirac impulse is a generalized function, denoted by $\delta(t)$ (not to be confused with a decision rule) and defined via the property $\int f(t)\delta(t)\,dt = f(0)$ for any function f continuous at the origin. A Riemann–Stieltjes integral $\int_0^T f(t)dg(t)$ is the limit of the Riemann sum $\sum_{i=1}^K f(t_i)[g(t_i) - g(t_{i-1})]$ for a discretization $0 = t_0 < t_1 < \cdots < t_K = T$ of the interval $[0,T]$, as the mesh size of the discretization tends to zero. For a random waveform $Y = \{Y(t),\ t \in [0,T]\}$, the stochastic integral $\int_0^T f(t)dY(t)$ is defined as the limit in the mean-square sense of the random sum $\sum_{i=1}^K f(t_i)[Y(t_i) - Y(t_{i-1})]$.

10.2.1 Covariance Kernel

Let $Z = \{Z(t), t \in \mathcal{T}\}$ be a random (not necessarily Gaussian) process with mean zero and covariance kernel

$$C(t,u) \triangleq \mathbb{E}[Z(t)Z(u)], \quad t,u \in \mathcal{T}. \tag{10.29}$$

The kernel is symmetric and nonnegative definite. If $\mathcal{T} = \mathbb{R}$ and the process is weakly stationary (*aka* wide-sense stationary) then $C(t,u) = c(t-u)$ for all t, u, and the process has a *spectral density*

$$S(f) = \int_{\mathbb{R}} c(t)\, e^{i2\pi f t}\, dt, \tag{10.30}$$

which is real-valued, nonnegative, and symmetric.

Classical Gaussian processes include:

Brownian Motion: $C(t,u) = \sigma^2 \min(t,u)$. This is a nonstationary Gaussian process with independent increments. Its sample paths are continuous with probability 1 but are nowhere differentiable, see Figure 10.1 for a sample path.

White Gaussian Noise: $C(t,u) = \sigma^2 \delta(t-u)$ where δ denotes the Dirac impulse. White Gaussian noise is not physically realizable but serves as a powerful mathematical abstraction. It is also the "derivative" of Brownian motion, $W(t) = \frac{dZ(t)}{dt}$,

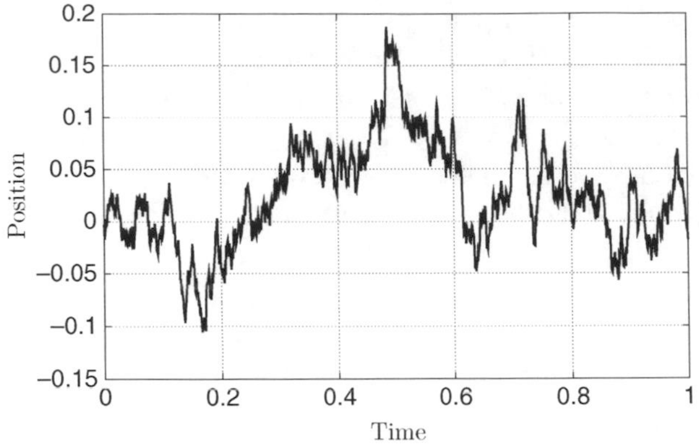

Figure 10.1 Brownian motion with $\sigma = 1$

where the notion of derivative holds in a distributional sense: the stochastic integral $\int f(t)W(t)dt = \int f(t)dZ(t)$ is well defined for every function $f \in L^2([0,T])$.[2] White Gaussian noise is stationary with spectral density $S(f) = \sigma^2$ for all $f \in \mathbb{R}$.

Bandlimited White Noise: This process is weakly stationary with spectral density function $S(f) = \sigma^2 \mathbb{1}\{|f| \le B\}$ for some finite B (the *bandwidth* of the process) and therefore its covariance function is given by $c(t) = 2\sigma^2 B \frac{\sin(2\pi Bt)}{2\pi Bt}$.

Periodic Weakly Stationary Process: Weakly stationary process whose correlation function is periodic with period T. The process admits a power spectral mass function.

Ornstein–Uhlenbeck Process: $C(t,u) = \sigma^2 e^{-\gamma|t-u|}$, where $\gamma > 0$ is the inverse of the "correlation time" of the process. This is a stationary Gauss–Markov process, solution to the stochastic differential equation

$$dZ(t) = -\gamma Z(t) + dW(t),$$

where W is Brownian motion. The spectral density of the process is $S(f) = \frac{2\gamma\sigma^2}{\gamma^2 + (2\pi f)^2}$, $f \in \mathbb{R}$. This is one of the simplest processes with a rational spectrum (i.e., a ratio of polynomials in f^2).

10.2.2 Karhunen–Loève Transform

The Karhunen–Loève transform decomposes a random (not necessarily Gaussian) continuous-time process Z into a countable sequence of *uncorrelated* coefficients. If the process is Gaussian, these coefficients are also mutually independent. Let $\mathcal{T} = [0,T]$ and consider the eigenfunction decomposition of the covariance kernel:

[2] From a mathematical standpoint, white noise can be treated as a generalized random function, in the same way as a Dirac impulse is a generalized function. From an engineering standpoint, deterministic and random generalized functions are meaningful when multiplied by well-behaved functions and integrated. For instance, $\int f(t)\delta(t)\,dt = f(0)$ for any function f that is continuous at 0. The random variable $\int_0^T f(t)W(t)\,dt$ has mean zero and finite variance $\sigma^2 \int_0^T f^2(t)\,dt$.

$$\int_0^T C(t,u)v_k(t)dt = \lambda_k v_k(u), \quad u \in [0,T], \; k = 1,2,\dots, \tag{10.31}$$

where v_k, $k \in \mathbb{N}$, are the eigenfunctions of $C(t,u)$, and $\lambda_k \geq 0$ are the corresponding eigenvalues. The eigenfunctions are orthonormal:

$$\int_0^T v_k(t)v_l(t)\,dt = \mathbb{1}\{k = l\}, \tag{10.32}$$

but do not necessarily form a complete orthonormal basis of $L^2([0,T])$. The coordinates of Z in the eigenfunction basis are

$$\tilde{Z}_k \triangleq \int_0^T v_k(t)Z(t)\,dt, \quad k = 1,2,\dots. \tag{10.33}$$

The random variables $\tilde{Z}_k$, $k \geq 1$ have zero mean, have respective variances λ_k, and are uncorrelated because

$$\begin{aligned}\mathbb{E}[\tilde{Z}_k\tilde{Z}_l] &= \mathbb{E}\left[\int_0^T\int_0^T v_k(t)v_l(u)Z(t)Z(u)\,dtdu\right]\\ &= \int_0^T\int_0^T v_k(t)v_l(u)C(t,u)\,dtdu\\ &= \lambda_k\mathbb{1}\{k=l\},\end{aligned} \tag{10.34}$$

where the last line follows from (10.31) and (10.32). Thus the KLT provides both *geometric orthogonality* of the eigenfunctions $\{v_k\}$ and *statistical orthogonality* of the noise coefficients $\{\tilde{Z}_k\}$. This may be viewed as noise whitening in the continuous-time framework.

For weakly stationary processes one sometimes prefers to work with complex-valued eigenfunctions. Then the coefficients $\tilde{Z}_k$ are complex-valued, and the covariance of (10.34) is replaced by $\mathbb{E}[\tilde{Z}_k\tilde{Z}_l^*] = \lambda_k\mathbb{1}\{k=l\}$ (note the complex conjugation of $\tilde{Z}_l$).

Example 10.2 *Periodic Weakly Stationary Process*. The complex exponentials

$$\frac{1}{\sqrt{T}}e^{i2\pi kt/T}, \quad t \in [0,T], \; k \in \mathbb{Z} \quad (i = \sqrt{-1}) \tag{10.35}$$

are eigenfunctions of $C(t,u)$ and thus

$$\int_0^T c(t-u)e^{i2\pi ku/T}\,du = \left(\int_0^T c(t-u)e^{i2\pi k(u-t)/T}\,du\right)e^{i2\pi kt/T} = \lambda_k\,e^{i2\pi kt/T}$$

with

$$\lambda_k = \int_0^T c(-u)e^{i2\pi ku/T}\,du = \int_0^T c(u)\cos(2\pi ku/T)\,du. \tag{10.36}$$

We obtain real-valued eigenfunctions by taking the real and imaginary parts of (10.35) and normalizing to unit energy. Thus

$$v_0(t) = \frac{1}{\sqrt{T}}, \quad v_{2k-1}(t) = \sqrt{\frac{2}{T}} \sin(2\pi kt/T), \quad v_{2k}(t) = \sqrt{\frac{2}{T}} \cos(2\pi kt/T),$$
$$k = 1, 2, \ldots, \; t \in [0, T]. \quad (10.37)$$

The index k represents a discrete frequency. The coefficients $\tilde{Z}_k$ of (10.33) are the Fourier coefficients of Z, and the sequence of eigenvalues λ_k is the *power spectral mass function* of the process. By the inverse Fourier transform formula, we have

$$c(t) = \frac{\lambda_0}{\sqrt{T}} + \sqrt{\frac{2}{T}} \sum_{k=1}^{\infty} \lambda_{2k} \cos(2\pi kt/T), \quad t \in [0, T]. \quad (10.38)$$

□

Example 10.3 *Brownian Motion.* The eigenfunctions and the corresponding eigenvalues are given by

$$v_k(t) = \sqrt{\frac{2}{T}} \sin \frac{(2k-1)\pi t}{2T}, \quad \lambda_k = \frac{4\sigma^2 T^2}{(2k-1)^2 \pi^2}, \quad k = 1, 2, \ldots. \quad (10.39)$$

The eigenfunctions form a complete orthornormal basis of $L^2([0, T])$. □

Example 10.4 *White Noise.* If $C(t, u) = \sigma^2 \delta(t - u)$ then any complete orthonormal basis of $L^2([0, T])$ is an eigenfunction basis for $C(t, u)$, and the eigenvalues λ_k are all equal to σ^2. □

Example 10.5 *Ornstein–Uhlenbeck process.* It will be convenient here to assume the observation interval is $[-T, T]$. The eigenvalues λ_k are samples of the spectral density:

$$\lambda_k = \frac{2\gamma\sigma^2}{\gamma^2 + b_k^2} = S\left(\frac{b_k}{2\pi}\right), \quad k = 1, 2, \ldots,$$

where the coefficients b_k are the solutions to the nonlinear equation $(\tan(bT) + \frac{b}{\gamma})(\tan(bT) - \frac{\gamma}{b}) = 0$. The eigenfunctions are given by

$$v_k(t) = T^{-1/2} \left(1 + \frac{\sin(2b_k T)}{2b_k T}\right)^{-1} \cos(b_k t), \quad -T \le t \le T \quad : k \text{ odd},$$

$$v_k(t) = T^{-1/2} \left(1 - \frac{\sin(2b_k T)}{2b_k T}\right)^{-1} \sin(b_k t), \quad -T \le t \le T \quad : k \text{ even}.$$

The eigenfunctions are sines and cosines whose frequencies are not harmonically related. However, the coefficients b_k form an increasing sequence, and $b_k \sim (k-1)\frac{\pi}{2T}$ as $k \to \infty$. □

Example 10.6 *Bandlimited White Noise Process.* The eigenvectors of the covariance kernel are *prolate spheroidal wave functions*. The number of significant eigenvalues is approximately $2BT$; the remaining eigenvalues decay exponentially fast. □

Mercer's Theorem [1] Assume that $C(t, u)$ is continuous over $[0, T]^2$. Equivalently, Z is mean-square continuous [1]. Then C admits the representation

$$C(t,u) = \sum_{k=1}^{\infty} \lambda_k v_k(t) v_k(u), \quad 0 \le t, u \le T, \tag{10.40}$$

where the sum converges uniformly over $[0, T]^2$, i.e.,

$$\lim_{n\to\infty} \max_{0\le t,u\le T} \left[C(t,u) - \sum_{k=1}^{n} \lambda_k v_k(t) v_k(u) \right] = 0.$$

Letting $t = u$ in (10.40) and integrating, we obtain $\int_0^T C(t,t)\,dt = \sum_k \lambda_k$. Since this quantity is finite, it follows that the eigenvalues λ_k decay faster than k^{-1}.

Reconstruction of Z The process Z may be represented by

$$Z(t) = \sum_{k=1}^{\infty} \tilde{Z}_k v_k(t), \quad t \in [0, T], \tag{10.41}$$

where the sum converges in the mean-square sense, uniformly over $[0, T]$:

$$\lim_{n\to\infty} \max_{0\le t\le T} \mathbb{E}\left[Z(t) - \sum_{k=1}^{n} \tilde{Z}_k v_k(t) \right]^2 = 0.$$

If Z is white Gaussian noise, its covariance matrix $C(t,u) = \sigma^2 \delta(t-u)$ does not satisfy the conditions of Mercer's theorem, and the sum of (10.41) does not converge in the mean-squared sense. However, (10.40) and (10.41) hold in a weaker sense: for any function $f \in L^2([0, T])$ with coefficients $f_k = \int_0^T f(t) v_k(t)\,dt$, there holds

$$\int_0^T f(u) C(t,u)\,du = \int_0^T f(u) \left[\sum_{k=1}^{\infty} \lambda_k v_k(t) v_k(u) \right] du = \sum_{k=1}^{\infty} \lambda_k f_k v_k(t)$$

and

$$\int_0^T f(t) Z(t)\,dt = \sum_{k=1}^{\infty} f_k \tilde{Z}_k. \tag{10.42}$$

Infinite Observation Interval For $\mathcal{T} = \mathbb{R}$, a spectral representation of weakly stationary processes may be obtained by extension of the spectral representation (10.37) (10.41) of periodic weakly stationary processes. For processes that are mean-square continuous, Cramér's spectral representation theorem [7] gives

$$Z(t) = \int_{\mathbb{R}} e^{i2\pi f t} d\tilde{Z}(f), \quad t \in \mathbb{R}, \tag{10.43}$$

where $\tilde{Z}$ is a process with *orthogonal increments*, i.e., for any integer n and any $f_1 < f_2 < \cdots < f_n$, the increments $\tilde{Z}(f_{i+1}) - \tilde{Z}(f_i)$, $1 \le i \le n$, are uncorrelated.

10.2.3 Detection of Known Signals in Gaussian Noise

Consider the binary hypothesis test

$$\begin{cases} H_0 & : Y(t) = Z(t) \\ H_1 & : Y(t) = s(t) + Z(t), \quad t \in [0, T], \end{cases} \tag{10.44}$$

where $s = \{s(t), t \in [0, T]\}$ is a known signal with finite energy

$$E_s = \int_0^T s^2(t)\, dt \tag{10.45}$$

and $Z = \{Z(t), t \in [0, T]\}$ is a Gaussian random process with mean zero and continuous covariance kernel $C(t, u)$.

Denote the signal and noise coefficients in the kernel eigenfunction basis by

$$\tilde{s}_k = \int_0^T v_k(t) s(t)\, dt, \tag{10.46}$$

$$\tilde{Z}_k = \int_0^T v_k(t) Z(t)\, dt, \quad k = 1, 2, \dots. \tag{10.47}$$

The coefficients of the observed Y are given by

$$\tilde{Y}_k = \int_0^T v_k(t) Y(t)\, dt, \quad k = 1, 2, \dots. \tag{10.48}$$

A representation for Z in terms of $\{\tilde{Z}_k\}_{k \geq 1}$ was given in (10.41).

We also have the signal reconstruction formula

$$s(t) = s_0(t) + \sum_{k=1}^{\infty} \tilde{s}_k v_k(t), \quad t \in [0, T], \tag{10.49}$$

where convergence of the sum is in the mean-square sense. The signal s_0 is the component of s in the null space of the covariance kernel, i.e., $\int_0^T C(t, u) s_0(u)\, du = 0$. We have $\int_0^T s_0^2(t)\, dt + \sum_{k=1}^{\infty} \tilde{s}_k^2 = E_s$. If $s_0 \neq 0$, the detection problem is singular because there is no noise in the direction of s_0, hence perfect detection is possible using the test statistic $\int_0^T s_0(t) Y(t)\, dt$, which equals 0 and $\|s_0\|^2$ with probability 1 under H_0 and H_1, respectively. To avoid such trivial conclusions, we assume that $s_0 \neq 0$ in the following.

The hypothesis test of (10.44) may then be written as

$$\begin{cases} H_0 & : \tilde{Y}_k = \tilde{Z}_k \\ H_1 & : \tilde{Y}_k = \tilde{s}_k + \tilde{Z}_k, \quad k = 1, 2, \dots. \end{cases} \tag{10.50}$$

This is the same formulation as in the discrete-time case, the only difference being that the sequence of observations $\tilde{Y}_k$ is not finite. If we use the first n coefficients to construct a LRT, the LLR is given by

$$\ln L_n(Y) = \sum_{k=1}^{n} \left(\frac{\tilde{s}_k \tilde{Y}_k}{\lambda_k} - \frac{\tilde{s}_k^2}{2\lambda_k} \right), \tag{10.51}$$

similarly to (5.26). For Bayesian hypothesis testing with equal priors, the minimum error probability of the LLRT $\ln L_n(Y) \underset{H_0}{\overset{H_1}{\gtrless}} 0$ is $P_e = Q(d_n/2)$ where

$$d_n^2 = \sum_{k=1}^{n} \frac{\tilde{s}_k^2}{\lambda_k} \tag{10.52}$$

is a nondecreasing sequence.

Now what happens as $n \to \infty$? Assume that the sum of (10.52) converges to a finite limit,

$$d^2 = \sum_{k=1}^{\infty} \frac{\tilde{s}_k^2}{\lambda_k}. \tag{10.53}$$

Then the limit of the log-likelihood ratios,

$$\ln L(Y) \triangleq \sum_{k=1}^{\infty} \left(\frac{\tilde{s}_k \tilde{Y}_k}{\lambda_k} - \frac{\tilde{s}_k^2}{2\lambda_k} \right), \tag{10.54}$$

has finite mean and variance under both H_0 and H_1 (see Exercise 10.6) and therefore the limit exists in the mean-squared sense. For Bayesian hypothesis testing with equal priors, the minimum error probability of the test $\ln L(Y) \underset{H_0}{\overset{H_1}{\gtrless}} 0$ is $P_e = Q(d/2)$.

If only a finite number n of coefficients $\tilde{Y}_k$ are evaluated and used in the log-likelihood ratio, the performance of the LRT is determined by the sum (10.52). The performance loss depends on n as well as on the convergence rate of the sum. For instance, if the individual terms $\tilde{s}_k^2/\lambda_k$ decay as $k^{-\alpha}$ with $\alpha > 1$, then the loss in d^2 is $O(n^{\alpha-1})$.

We now evaluate the limits (10.53) and (10.54) for classical noise processes.

White Noise If $C(t,u) = \sigma^2 \delta(t-u)$ then any orthonormal basis of $L^2([0,T])$ is an eigenfunction basis for $C(t,u)$, with corresponding eigenvalues λ_k all equal to σ^2. In this case (10.54) becomes

$$\begin{aligned} \ln L(Y) &= \sum_{k=1}^{\infty} \left(\frac{\tilde{s}_k \tilde{Y}_k}{\sigma^2} - \frac{\tilde{s}_k^2}{2\sigma^2} \right) \\ &= \frac{1}{\sigma^2} \int_0^T s(t) Y(t)\, dt - \frac{E_s}{2\sigma^2}, \end{aligned} \tag{10.55}$$

where the second inequality follows from (10.42). Also (10.53) becomes

$$d^2 = \frac{E_s}{\sigma^2} < \infty,$$

hence $P_e = Q(d/2) > 0$.[3]

[3] A simpler way to derive (10.55) in the white Gaussian noise case is to choose the first eigenfunction $v_1 = E_s^{-1/2} s$ of the covariance kernel to be aligned with the signal s. Then $\tilde{s}_1 = \sqrt{E_s}$ and $\tilde{s}_k = 0$ for all $k > 1$. The sum of (10.55) then reduces to a single nonzero term.

Colored Noise In general the expression (10.54) does not reduce to a formula as simple as (10.55). In fact, even for finite-energy signals, the sum (10.53) might diverge, in which case $P_e = 0$. For a periodic stationary process, divergence of the sum (10.53) means that the noise variance λ_k at high frequencies decays too fast relative to the signal energy components. The underlying observational model is mathematically valid but does not correspond to a physically plausible situation. The same conclusion applies to aperiodic stationary processes. It is therefore common to model colored noise by the superposition of colored noise with smooth covariance kernel $\tilde{C}(t, u)$ and an independent white noise component with variance σ^2: $C(t, u) = \tilde{C}(t, u) + \sigma_W^2 \delta(t - u)$.

Brownian Motion Detection of a signal in Brownian motion can be analyzed by differentiating the observed signal and reducing the problem to one of detection in white Gaussian noise. Assume the signal s satisfies $s(0) = 0$ and is differentiable on $[0, T]$. Then the process Y may be represented by the sample $Y(0)$ together with the derivative process $Y'(t)$, $t \in [0, T]$. Since $Y(0) = Z(0)$ has the same distribution under H_0 and H_1, we may write the hypotheses as

$$\begin{cases} H_0 & : Y'(t) = W(t) \\ H_1 & : Y'(t) = s'(t) + W(t), \quad t \in [0, T], \end{cases}$$

where W is white Gaussian noise. Therefore $P_e = Q(d/2)$ where

$$d^2 = \frac{1}{\sigma^2} \int_0^T [s'(t)]^2 \, dt. \tag{10.56}$$

Note that d^2 might be infinite (perfect detection) even if the signal s has finite energy. This occurs, for instance, with $s(t) = (T - t)^{-1/3}$, $0 \le t < T$.

The case of a differentiable signal with $s(0) \neq 0$ always results in a singular detection problem. Indeed the test $\delta(Y) = \mathbb{1}\{|Y(0)| > \frac{|s_0|}{2}\}$ has zero error probability because $\mathrm{Var}[Z(0)] = 0$.

10.2.4 Detection of Gaussian Signals in Gaussian Noise

Consider the binary hypothesis test

$$\begin{cases} H_0 & : Y(t) = Z(t) \\ H_1 & : Y(t) = S(t) + Z(t), \quad t \in [0, T], \end{cases} \tag{10.57}$$

where $S = \{S(t), t \in [0, T]\}$ is a Gaussian signal with mean zero and covariance kernel $C_S(t, u) \triangleq \mathbb{E}[S(t)S(u)]$, $t, u \in [0, T]$, and $Z = \{Z(t), t \in [0, T]\}$ is a Gaussian random process with mean zero and covariance kernel $C_Z(t, u) \triangleq \mathbb{E}[Z(t)Z(u)]$, $t, u \in [0, T]$. Assume that S and Z are independent and that the covariance kernels C_S and C_Z have the same eigenfunctions $\{v_k, k = 1, 2, \ldots\}$, and respective eigenvalues $\lambda_{S,k}$ and $\lambda_{Z,k}$, $k = 1, 2, \ldots$. Then the test (10.57) can be equivalently stated as

$$\begin{cases} H_0 & : \tilde{Y}_k = \tilde{Z}_k \\ H_1 & : \tilde{Y}_k = \tilde{S}_k + \tilde{Z}_k, \quad k = 1, 2, \ldots \end{cases} \tag{10.58}$$

or

$$\begin{cases} H_0 & : \tilde{Y}_k \sim \text{indep. } \mathcal{N}(0, \lambda_{Z,k}) \\ H_1 & : \tilde{Y}_k \sim \text{indep. } \mathcal{N}(0, \lambda_{S,k} + \lambda_{Z,k}), \quad k = 1, 2, \ldots . \end{cases} \tag{10.59}$$

Again this is the same formulation as in the discrete-time case (5.65), the only difference being that the sequence of observations $\tilde{Y}_k$ is not finite. It will be notationally convenient to work with the normalized random variables $\hat{Y}_k = \tilde{Y}_k / \sqrt{\lambda_{Z,k}}$ and the variance ratios $\xi_k = \lambda_{S,k} / \lambda_{Z,k}$, in which case the test can be equivalently stated as

$$\begin{cases} H_0 & : \hat{Y}_k \sim \text{indep. } \mathcal{N}(0, 1) \\ H_1 & : \hat{Y}_k \sim \text{indep. } \mathcal{N}(0, 1 + \xi_k), \quad k = 1, 2, \ldots . \end{cases} \tag{10.60}$$

The log-likelihood ratio for n coefficients is denoted by

$$\ln L_n(Y) = \frac{1}{2} \sum_{k=1}^{n} \ln \frac{1}{1 + \xi_k} + \frac{1}{2} \sum_{k=1}^{n} \frac{\xi_k}{1 + \xi_k} \hat{Y}_k^2 . \tag{10.61}$$

Similarly to (5.66), we may write the LLRT for n coefficients as

$$\sum_{k=1}^{n} \frac{\xi_k}{1 + \xi_k} \hat{Y}_k^2 \underset{H_0}{\overset{H_1}{\gtrless}} \tau . \tag{10.62}$$

By (6.3), we have

$$D(p_0^{(n)} \| p_1^{(n)}) = \mathbb{E}_0[\ln L_n(Y)] = \sum_{k=1}^{n} \psi \left(\frac{1}{1 + \xi_k} \right), \tag{10.63}$$

$$D(p_1^{(n)} \| p_0^{(n)}) = \mathbb{E}_1[\ln L_n(Y)] = \sum_{k=1}^{n} \psi \left(1 + \xi_k \right), \tag{10.64}$$

where $\psi(x) = \frac{1}{2}(x - 1 - \ln x)$ was defined in (6.4).

Now consider the asymptotic scenario where $n \to \infty$. First assume the variance ratios $\{\xi_k\}$ are such that the two sequences of KL divergences (10.63) and (10.64) converge to finite limits, respectively denoted by

$$D(\mathrm{P}_0 \| \mathrm{P}_1) \triangleq \mathbb{E}_0[\ln L(Y)] = \sum_{k=1}^{\infty} \psi \left(\frac{1}{1 + \xi_k} \right) \tag{10.65}$$

and

$$D(\mathrm{P}_1 \| \mathrm{P}_0) \triangleq \mathbb{E}_1[\ln L(Y)] = \sum_{k=1}^{\infty} \psi \left(1 + \xi_k \right). \tag{10.66}$$

Then the limit of random sequence (10.61) exists in the mean-squared sense:

$$\ln L(Y) \triangleq \frac{1}{2} \sum_{k=1}^{\infty} \ln \frac{1}{1 + \xi_k} + \frac{1}{2} \sum_{k=1}^{\infty} \frac{\xi_k}{1 + \xi_k} \hat{Y}_k^2 , \tag{10.67}$$

as it has finite mean and variance under both H_0 and H_1 (see Exercise 10.6). Then detection performance saturates as an increasing number n of terms is evaluated.

In many problems, however, the KL divergence sequences of (10.63) and (10.64) do not converge to finite limits. For instance, if the sequence ξ_k does not vanish, the sequences (10.63) and (10.64) are unbounded, and one may show that perfect detection is obtained. This may hold even if the sequence ξ_k vanishes. The second-order Taylor series expansion of $\psi(x)$ near $x = 1$ is given by $\psi(1+\epsilon) = \frac{1}{4}\epsilon^2 + O(\epsilon^3)$. Hence we need $\sum_{k=1}^{\infty} \xi_k^2 < \infty$ for the sums (10.63) and (10.64) to converge, hence ξ_k must decay faster than $\sqrt{1/k}$.

By application of Exercise 8.11, the Chernoff divergence is given by

$$d_u(p_0^{(n)}, p_1^{(n)}) = \frac{1}{2}\sum_{k=1}^{n}[(1-u)\ln(1+\xi_k) - \ln(1+(1-u)\xi_k)], \quad \forall u \in (0,1), \quad (10.68)$$

which can be used to derive upper bounds on $\mathrm{P_F}$, $\mathrm{P_M}$, and $\mathrm{P_e}$ as in (7.25), (7.26), (7.35) and lower bounds as in Theorem 8.4.

Example 10.7 Let $T = 1$ and S be Brownian motion with parameter ϵ. Then we have $\xi_k = \epsilon/\sigma^2$ for all k. The Chernoff divergence of (10.68) tends to ∞ as $n \to \infty$, and perfect detection is possible no matter how small ϵ is! Indeed arbitrarily small error probabilities are achieved using the test (10.62) with an arbitrarily large n. An even simpler test that has the same performance is the following one. Define the random variables $\hat{Y}_k = \sqrt{n}[Y(\frac{k}{n}) - Y(\frac{k-1}{n})]$, $1 \le k \le n$, which are mutually independent given either hypothesis (owing to the independent-increments property of Brownian motion) and have variances σ^2 and $\epsilon + \sigma^2$ under H_0 and H_1, respectively. Since these random variables have the same statistics as $\{\tilde{Y}_k,\ 1 \le k \le n\}$, the LRT based on $\{\hat{Y}_k,\ 1 \le k \le n\}$ has the same performance as (10.62). □

Periodic Stationary Processes If both S and Z are Gaussian periodic stationary processes, the eigenvalues $\lambda_{S,k}$ and $\lambda_{Z,k}$ are samples of the discrete spectral densities of these two processes. Similarly to the known-signal case, colored noise whose spectral mass function falls off at the same rate or faster than the signal's energy results in perfect detection.

10.3 Poisson Processes

Recall the following definitions [1]. The Poisson random variable N with parameter $\lambda > 0$ has pmf

$$p(n) = \frac{\lambda^n}{n!}e^{-\lambda}, \quad n = 0, 1, 2, \ldots \quad (10.69)$$

and cgf $\kappa(u) = \ln \mathbb{E}[e^{uN}] = \lambda(e^u - 1)$. A function f on $\mathbb{R}^+$ is a counting function if $f(0) = 0$, f is nondecreasing and right-continuous, and integer-valued. A random process is a counting process if its sample paths are counting functions with probability

1. A random process $Y(t),\ t \geq 0$ has independent increments if for any integer n and any $0 \leq t_1 \leq t_2 \leq \cdots \leq t_n$, the increments $Y(t_{i+1}) - Y(t_i),\ 1 \leq i < n$, are mutually independent.

DEFINITION 10.4 An inhomogeneous Poisson process with rate function $\lambda(t),\ t \geq 0$ is a counting process $Y(t),\ t \geq 0$ with the following two properties:

(i) Y has independent increments;
(ii) for any $0 \leq t < u$, the increment $Y(u) - Y(t)$ is a Poisson random variable with parameter $\int_t^u \lambda(v)\, dv$.

Denote by N the number of counts in the interval $[0, T]$ and by $T_1 < T_2 < \cdots < T_N$ the arrival times in $[0, T]$. The sample path $Y(t)$ can be represented by the tuple $(N, T_1, T_2, \ldots, T_N)$. See Figure 10.2 for an illustration.

LEMMA 10.1 [8] *The joint pmf for N and density for $T_1 < T_2 < \cdots < T_N$ is given by*

$$p(t_1, \ldots, t_n, N = n) = \exp\left\{\int_0^T \ln \lambda(t) dy(t) - \int_0^T \lambda(t) dt\right\},$$
$$n = 0, 1, 2, \ldots, \quad 0 < t_1 < t_2 < \cdots < t_n < T, \tag{10.70}$$

where $y(t) = \sum_{i=1}^n \mathbb{1}\{t \geq t_i\}$ is the counting function with n jumps located at times $t_1, \ldots, t_n$.

The expression (10.70) is sometimes called *sample-function density* [8].

Now consider the problem of testing between two inhomogeneous Poisson processes P_0 and P_1 with known intensity functions $s_j(t),\ 0 \leq t \leq T,\ j = 0, 1$, given observations $Y = \{Y(t),\ 0 \leq t \leq T\}$:

$$\begin{cases} H_0 & : Y \text{ has intensity function } s_0(t),\ 0 \leq t \leq T \\ H_1 & : Y \text{ has intensity function } s_1(t),\ 0 \leq t \leq T. \end{cases} \tag{10.71}$$

Denote by N the number of counts in the interval $[0, T]$, by $T_1, \ldots, T_N$ the arrival times, and by p_0 and p_1 the sample-function densities under H_0 and H_1, respectively. Using (10.70), we obtain the likelihood ratio as

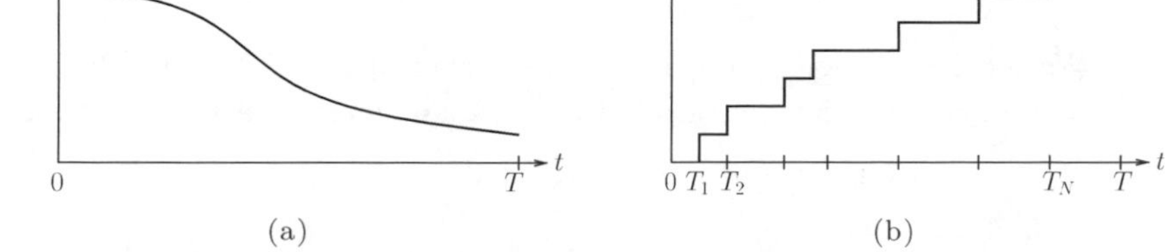

Figure 10.2 (a) Intensity function for an inhomogeneous Poisson process; (b) sample path

$$
\begin{aligned}
L(Y) &= \frac{p_1(T_1, \ldots, T_N, N)}{p_0(T_1, \ldots, T_N, N)} \\
&= \begin{cases} \exp\left\{\sum_{i=1}^N \ln \frac{s_1(T_i)}{s_0(T_i)} - \int_0^T [s_1(t) - s_0(t)]\, dt\right\} & : N > 0 \\ \exp\left\{- \int_0^T [s_1(t) - s_0(t)]\, dt\right\} & : N = 0 \end{cases} \\
&= \exp\left\{\int_0^T \ln \frac{s_1(t)}{s_0(t)} dY(t) - \int_0^T [s_1(t) - s_0(t)]\, dt\right\}.
\end{aligned} \tag{10.72}
$$

Hence the LRT admits the simple implementation

$$
\sum_{i=1}^N \ln \frac{s_1(T_i)}{s_0(T_i)} \underset{H_0}{\overset{H_1}{\gtrless}} \tau = \ln \eta + \int_0^T [s_1(t) - s_0(t)]\, dt, \tag{10.73}
$$

where the left hand side is understood to be zero when $N = 0$.

The KL divergence between the two processes is given by

$$
\begin{aligned}
D(\mathrm{P}_0 \| \mathrm{P}_1) &= -\mathbb{E}_0[\ln L(Y)] \\
&= \int_0^T \left[s_0(t) \ln \frac{s_0(t)}{s_1(t)} + s_1(t) - s_0(t)\right] dt \\
&= \int_0^T s_0(t) \phi\left(\frac{s_0(t)}{s_1(t)}\right) dt,
\end{aligned}
$$

where we have used $\mathbb{E}_0[dY(t)] = s_0(t)\, dt$ in the first line and introduced the function

$$
\phi(x) = \ln x - 1 + \frac{1}{x}, \qquad x > 0,
$$

in the second line, where $\phi(x) = 2\psi(\frac{1}{x})$ and ψ was defined in (6.4).

PROPOSITION 10.1 *Assume the intensity ratio $s_1(t)/s_0(t)$ is bounded away from zero and infinity for all $t \in [0, T]$. For each $u \in (0, 1)$, the Chernoff distance between* P_0 *and* P_1 *is given by*

$$
d_u(\mathrm{P}_0, \mathrm{P}_1) = -\ln \mathbb{E}_0[L(Y)^u] = \int_0^T [(1-u)s_0 + us_1 - s_0^{1-u} s_1^u]. \tag{10.74}
$$

Proof See Section 10.5. □

10.4 General Processes

We conclude this chapter by introducing a general approach to hypothesis testing based on Radon–Nikodym derivatives. This section requires some notions of measure theory.

10.4.1 Likelihood Ratio

Consider a measurable space $(\mathcal{Y}, \mathscr{F})$ consisting of a sample space $\mathcal{Y}$ and a σ-algebra $\mathscr{F}$ of subsets of $\mathcal{Y}$. Let P_0 and P_1 be two probability measures defined on $\mathscr{F}$. Let μ be a probability measure on $\mathscr{F}$ such that both P_0 and P_1 are *absolutely continuous* with respect to μ (denoted by $P_0 \ll \mu$ and $P_1 \ll \mu$), i.e., for any set $\mathcal{A} \in \mathscr{F}$ such that $\mu(\mathcal{A}) = 0$ we must have $P_0(\mathcal{A}) = 0$ and $P_1(\mathcal{A}) = 0$. Both P_0 and P_1 are also said to be *dominated* by μ. This condition is met, for instance, if $\mu = \frac{1}{2}(P_0 + P_1)$.

If there exists a set $\mathcal{E} \in \mathscr{F}$ such that $P_0(\mathcal{E}) = 0$ and $P_1(\mathcal{E}) = 1$ then P_0 and P_1 are said to be *orthogonal probability measures* (denoted by $P_0 \perp P_1$). The detection problem is singular in this case. Indeed the decision rule $\delta(y) = \mathbb{1}\{y \in \mathcal{E}\}$ has zero error probability under both H_0 and H_1. Several of the problems treated in Sections 10.2.3 and 10.2.4 fall in that category.

If P_0 and P_1 are dominated by each other, then P_0 and P_1 are said to be *equivalent* (denoted by $P_0 \equiv P_1$), and perfect detection cannot be achieved. Intermediate cases exist where P_0 and P_1 are neither orthogonal nor equivalent (e.g., $P_0 = \text{Uniform}[0, 2]$ and $P_1 = \text{Uniform}[1, 3]$).

By the Radon–Nikodym theorem [7], there exist functions $p_0(y)$ and $p_1(y)$, called *generalized probability densities* such that

$$P_0(A) = \int_{\mathcal{A}} p_0(y) d\mu(y), \quad P_1(\mathcal{A}) = \int_{\mathcal{A}} p_1(y) d\mu(y), \quad \forall \mathcal{A} \in \mathscr{F}.$$

These functions are also called Radom–Nikodym derivatives and denoted by

$$p_0(y) = \frac{dP_0(y)}{d\mu(y)} \quad \text{and} \quad p_1(y) = \frac{dP_1(y)}{d\mu(y)}.$$

Let $\mathcal{E}_j = \{y : p_j(y) = 0\}$ for $j = 0, 1$, and $\mathcal{E} = \{y : p_0(y) > 0, p_1(y) > 0\}$.

DEFINITION 10.5 The likelihood ratio is defined as

$$L(y) = \begin{cases} \infty & : y \in \mathcal{E}_0 \\ \frac{p_1(y)}{p_0(y)} & : y \in \mathcal{E} \\ 0 & : y \in \mathcal{E}_1, \end{cases} \tag{10.75}$$

where $p_1(y)/p_0(y)$ is the Radon–Nikodym derivative of P_1 relative to P_0.

The expressions (10.54), (10.67), and (10.72) are all Radon–Nikodym derivatives, whether the underlying p_0 and p_1 are densities or generalized densities.

Likelihood Ratio Convergence Theorem [7] We have seen in Section 10.2 that for two signal detection problems in continuous-time Gaussian noise, the sequences of likelihood ratios (10.51) and (10.61) converge to a limit. In a more general setting, we may design a sequence of real-valued observables $\tilde{Y}_k$, $k \geq 1$, that faithfully represents the process Y as an increasing number of observations is taken. Corresponding to the first n observations is a likelihood ratio $L_n(\tilde{Y}_{1:n})$, where $\tilde{Y}_{1:n} \triangleq (\tilde{Y}_1, \ldots, \tilde{Y}_n)$. The likelihood ratio convergence theorem states that, if the distribution of each $\tilde{Y}_{1:n}$ is absolutely continuous with respect to Lebesgue measure in $\mathbb{R}^n$, then

$$\begin{cases} L_n(\tilde{Y}_{1:n}) \to L(Y) & \text{i. p. } (P_0) \\ L_n(\tilde{Y}_{1:n}) \to L(Y) & \text{i. p. } (P_1), \end{cases} \tag{10.76}$$

where i.p. denotes convergence in probability.

10.4.2 Ali–Silvey Distances

Similarly to (6.1), we define the KL divergence between P_0 and P_1 as

$$D(\mathrm{P}_0\|\mathrm{P}_1) = \int_{\mathcal{Y}} p_0(y) \ln \frac{p_0(y)}{p_1(y)} d\mu(y) = \mathbb{E}_0\left[\frac{d\mathrm{P}_0(Y)}{d\mathrm{P}_1(Y)}\right]. \qquad (10.77)$$

f-divergences are likewise defined as

$$D_f(\mathrm{P}_0, \mathrm{P}_1) = \int_{\mathcal{Y}} p_0(y) f\left(\frac{p_1(y)}{p_0(y)}\right) d\mu(y) = \mathbb{E}_0\left[f\left(\frac{d\mathrm{P}_1(Y)}{d\mathrm{P}_0(Y)}\right)\right]. \qquad (10.78)$$

By the data processing inequality, we have

$$\sum_i \mathrm{P}_0(\mathcal{A}_i) f\left(\frac{\mathrm{P}_1(A_i)}{\mathrm{P}_0(A_i)}\right) \le D_f(\mathrm{P}_0, \mathrm{P}_1)$$

for any partition $\{\mathcal{A}_i\}$ of $\mathcal{Y}$. By definition of the Lebesgue integral (10.78),

$$D_f(\mathrm{P}_0, \mathrm{P}_1) = \sup_{\{\mathcal{A}_i\}} \sum_i \mathrm{P}_0(\mathcal{A}_i) f\left(\frac{\mathrm{P}_1(\mathcal{A}_i)}{\mathrm{P}_0(\mathcal{A}_i)}\right).$$

Stationary Processes The general approach can be applied to derive KL divergence rates and Chernoff divergence rates for detection of stationary processes. For instance, let S and Z be two independent continuous-time Gaussian stationary processes with absolutely integrable covariance functions and respective spectral densities $S_S(f)$ and $S_Z(f)$, $f \in \mathbb{R}$. Consider the hypothesis test of (10.57) and denote by $\mathrm{P}_0^{(T)}$ and $\mathrm{P}_1^{(T)}$ the probability measures for $\{Y(t),\ 0 \le t \le T\}$ under H_0 and H_1. Then the eigenvalues $\lambda_{S,k}$, $k \ge 1$ of the covariance kernel $C_S(t,u)$, $0 \le t, u \le T$ converge to samples of the spectral density $S_S(f)$ for increasing T, in the sense that

$$\forall f \in \mathbb{R}: \quad \lambda_{S,\lceil fT \rceil} \sim S_S(f) \quad \text{as } T \to \infty$$

and similarly

$$\forall f \in \mathbb{R}: \quad \lambda_{Z,\lceil fT \rceil} \sim S_Z(f) \quad \text{as } T \to \infty.$$

Then the KL divergence rates are given by

$$\overline{D}(\mathrm{P}_0\|\mathrm{P}_1) = \lim_{T\to\infty} \frac{1}{T} D(\mathrm{P}_0^{(T)}\|\mathrm{P}_1^{(T)}) = \int_{\mathbb{R}} \psi\left(\frac{S_Z(f)}{S_Z(f)+S_S(f)}\right) df,$$
$$\overline{D}(\mathrm{P}_1\|\mathrm{P}_0) = \lim_{T\to\infty} \frac{1}{T} D(\mathrm{P}_1^{(T)}\|\mathrm{P}_0^{(T)}) = \int_{\mathbb{R}} \psi\left(1+\frac{S_S(f)}{S_Z(f)}\right) df,$$

which are the Itakura–Saito distances between the spectral densities S_Z and $S_Z + S_S$ (and vice-versa). The Chernoff divergence rates are given by

$$\begin{aligned}\overline{d}_u(\mathrm{P}_0, \mathrm{P}_1) &= \lim_{T\to\infty} \frac{1}{T} D_u(\mathrm{P}_0^{(T)}, \mathrm{P}_1^{(T)}) \\ &= \frac{1}{2}\int_{\mathbb{R}} \left[(1-u)\ln\left(1+\frac{S_S(f)}{S_Z(f)}\right) - \ln\left(1+(1-u)\frac{S_S(f)}{S_Z(f)}\right)\right] df, \\ &\qquad \forall u \in (0,1).\end{aligned}$$

Other inference problems that can be treated using this approach include continuous-time Markov processes, processes with independent increments, and compound Poisson processes.

10.5 Appendix: Proof of Proposition 10.1

For simplicity, we first prove the claim in case the intensity ratio $s_1(t)/s_0(t)$ is constant over the interval $[0, T]$. Denote this constant by ρ. Under H_0, the random variable $N \triangleq \int_0^T dY(t)$ is Poisson with parameter $\int_0^T s_0$ under H_0. From (10.72) we obtain

$$L(Y) = \exp\left\{N \ln \rho - \int_0^T [s_1(t) - s_0(t)]\, dt\right\}.$$

Then

$$\begin{aligned}
d_u(\mathrm{P}_0, \mathrm{P}_1) &= -\ln \mathbb{E}_0[L(Y)^u] \\
&= -\ln \mathbb{E}_0\left[\exp\left\{uN \ln \rho - u\int_0^T [s_1 - s_0]\right\}\right] \\
&\overset{(a)}{=} -\ln\left[\sum_{n=0}^{\infty} \frac{(\int_0^T s_0)^n}{n!} e^{-\int_0^T s_0} \exp\left\{un \ln \rho - u\int_0^T [s_1 - s_0]\right\}\right] \\
&= \int_0^T [(1-u)s_0 + us_1] - \ln \sum_{n=0}^{\infty} \frac{[(\int_0^T s_0)\rho^u]^n}{n!} \\
&\overset{(b)}{=} \int_0^T [(1-u)s_0 + us_1 - s_0\rho^u],
\end{aligned} \tag{10.79}$$

where (a) uses the fact that N is Poisson with parameter $\int_0^T s_0$ under H_0 and (b) uses the identity $e^x = \sum_n x^n/n!$. The claim (10.74) follows from (10.79) because $s_0\rho^u = s_0^{1-u}s_1^u$.

Next consider the more general case where the intensity ratio $s_1(t)/s_0(t)$ is piecewise constant, taking values ρ_i over regions $\mathcal{T}_i$, $i \in \mathcal{I}$, forming a partition of $[0, T]$:

$$\frac{s_1(t)}{s_0(t)} = \sum_{i \in \mathcal{I}} \rho_i \mathbb{1}\{t \in \mathcal{T}_i\}.$$

By Definition 10.4, the random variables $N_i \triangleq \int_{\mathcal{T}_i} dY(t)$, $i \in \mathcal{I}$ are mutually independent and are Poisson distributed with respective parameters $\int_{\mathcal{T}_i} s_0$ under H_0. It follows that

$$\begin{aligned}
d_u(\mathrm{P}_0, \mathrm{P}_1) &= -\ln \mathbb{E}_0[L(Y)^u] \\
&= -\ln \mathbb{E}_0\left[\exp \sum_{i \in \mathcal{I}} \left\{uN_i \ln \rho_i - u\int_{\mathcal{T}_i} [s_1 - s_0]\right\}\right] \\
&\overset{(a)}{=} \sum_{i \in \mathcal{I}} -\ln \mathbb{E}_0\left[\exp\left\{uN_i \ln \rho_i - u\int_{\mathcal{T}_i} [s_1 - s_0]\right\}\right] \\
&\overset{(b)}{=} \sum_{i \in \mathcal{I}} \int_{\mathcal{T}_i} [(1-u)s_0 + us_1 - s_0\rho_i^u]
\end{aligned} \tag{10.80}$$

where (a) holds because $\{N_i\}$ are mutually independent, and (b) follows from (10.79) with the interval $[0, T]$ replaced by $\mathcal{T}_i$ and ρ by ρ_i. Hence (10.74) holds again.

Finally, in the general case where the intensity ratio $s_1(t)/s_0(t)$ belongs to a finite range $[a, b)$, pick an arbitrarily large integer k and let $\overline{\rho}_i = a + i\frac{b-a}{k}$ for $i = 0, 1, \ldots, k$. The level sets

$$\mathcal{T}_i = \left\{ t \in [0, T] \,:\, \overline{\rho}_{i-1} \le \frac{s_1(t)}{s_0(t)} < \overline{\rho}_i \right\}, \quad i \in \mathcal{I} \triangleq \{1, 2, \ldots, k\}$$

form a partition of $[0, T]$. Define the nonnegative random variables $N_i = \int_{\mathcal{T}_i} dY(t)$, $i \in \mathcal{I}$. Then $d_u(\mathrm{P}_0, \mathrm{P}_1)$ may be sandwiched as follows:

$$\begin{aligned}
&-\ln \mathbb{E}_0 \left[\exp \sum_{i \in \mathcal{I}} \left\{ u N_i \ln \overline{\rho}_i - u \int_{\mathcal{T}_i} [s_1 - s_0] \right\} \right] \\
&\le d_u(\mathrm{P}_0, \mathrm{P}_1) \\
&= -\ln \mathbb{E}_0[L(Y)^u] \\
&< -\ln \mathbb{E}_0 \left[\exp \sum_{i \in \mathcal{I}} \left\{ u N_i \ln \overline{\rho}_{i-1} - u \int_{\mathcal{T}_i} [s_1 - s_0] \right\} \right],
\end{aligned}$$

and the lower and upper bounds evaluated using (10.80). We obtain

$$\sum_{i \in \mathcal{I}} \int_{\mathcal{T}_i} [(1-u)s_0 + u s_1 - s_0 \overline{\rho}_i^u] \le d_u(\mathrm{P}_0, \mathrm{P}_1) \le \sum_{i \in \mathcal{I}} \int_{\mathcal{T}_i} [(1-u)s_0 + u s_1 - s_0 \overline{\rho}_{i-1}^u].$$

Now letting $k \to \infty$, the lower and upper bounds tend to the same limit (10.74). This proves the claim. □

Exercises

10.1 *Detection of Markov Process.* Consider two Markov chains with state space $\mathcal{Y} = \{0, 1\}$. The emission probabilities q_0 and q_1 are uniform, and the probability transition matrices under H_0 and H_1 are respectively given by

$$r_0 = \begin{bmatrix} .5 & .5 \\ .1 & .9 \end{bmatrix} \quad \text{and} \quad r_1 = \begin{bmatrix} .5 & .5 \\ .9 & .1 \end{bmatrix}.$$

(a) Give the Bayes decision assuming uniform prior on the hypotheses and an observed sequence $\boldsymbol{y} = (0, 0, 1, 1, 1)$.

(b) Give the Kullback–Leibler and Rényi divergence rates between the two Markov processes.

10.2 *Detection of Markov Process.* Repeat Exercise 10.1 when

$$r_0 = \begin{bmatrix} .5 & .5 \\ .5 & .5 \end{bmatrix} \quad \text{and} \quad r_1 = \begin{bmatrix} .9 & .1 \\ .1 & .9 \end{bmatrix}.$$

10.3 *Brownian Motion (I).* Consider the noise process $Z(t) = \int_0^t B(u)\,du$, $t \in [0, T]$ where $B(u)$ is Brownian motion with parameter σ^2. Consider the binary hypothesis test

$$\begin{cases} H_0 & : Y(t) = Z(t) \\ H_1 & : Y(t) = s(t) + Z(t), \quad t \in [0, T], \end{cases}$$

where s is a known signal. Assume uniform priors on the hypotheses.

(a) Give the optimal test and the Bayes error probability when $s(t) = t^2$.
(b) Repeat Part (a) when $s(t) = \max\left\{t - \frac{T}{2}, 0\right\}$.

10.4 *Brownian Motion (II).* Consider the following ternary hypothesis test:

$$\begin{cases} H_1 & : Y(t) = t + Z(t) \\ H_2 & : Y(t) = t^2 + t + Z(t) \\ H_3 & : Y(t) = 1 + Z(t), \quad t \in [0, 1], \end{cases}$$

where Z is standard Brownian motion ($\sigma^2 = 1$). Give a Bayes decision rule and the Bayesian error probability assuming a uniform prior.

10.5 *Detection of Poisson Process.* Consider two Poisson processes P_0 and P_1 with respective intensity functions $s_0(t) = e^t$ and $s_1(t) = e^{2t}$ for $t \in [0, \ln 4]$. Assume a uniform prior on the corresponding hypotheses H_0 and H_1.

(a) Give the Bayes decision when counts are observed at times 0.2, 0.9, and 1.2.
(b) Find the optimal exponent u in Proposition 10.1 in terms of the intensity functions s_0 and s_1.
(c) Give an exponential upper bound on the Bayes error probability.

10.6 *Second-Order Statistics of Log-likelihood Ratios.* Derive the mean and variance of the log-likelihood ratios of (10.54) and (10.67).

10.7 *Signal Detection in Correlated Gaussian Noise.* Consider the following detection problem:

$$\begin{cases} H_0 & : Y_t = Z_t \\ H_1 & : Y_t = \cos(4\pi t) + Z_t, \quad 0 \le t \le 1, \end{cases}$$

where Z_t is a zero-mean Gaussian random process with covariance kernel

$$C_Z(t, u) = \cos(2\pi(t - u)), \quad 0 \le t, u \le 1.$$

Assuming equal priors, find the minimum error probability.

10.8 *Composite Hypothesis Testing.* Consider the hypothesis testing problem

$$\begin{cases} H_0 & : Y_t = W_t \\ H_1 & : Y_t = \Theta t + Z_t, \quad 0 \le t \le T, \end{cases}$$

where $\{Z_t, 0 \le t \le T\}$ is the standard Brownian motion. The parameter Θ takes values $+1$ and -1 with equal probabilities and is independent of $\{Z_t, 0 \le t \le T\}$.

(a) Find the likelihood ratio.
(b) Find an α-level Neyman–Pearson test.

References

[1] B. Hajek, *Random Processes for Engineers*, Cambridge University Press, 2015.
[2] U. Grenander and G. Szegö, *Toeplitz Forms and their Applications*, Chelsea, 1958.
[3] R. M. Gray, *Toeplitz and Circulant Matrices: A Review*, NOW Publishers, 2006. Available from `http://ee.stanford.edu/~gray/toeplitz.pdf`
[4] M. Gil, "On Rényi Divergences for Continuous Alphabet Sources," *M. S. thesis*, Dept of Math. and Stat., McGill University, 2011.
[5] T. Kailath and H. V. Poor, "Detection of Stochastic Processes," *IEEE Trans. Inf. Theory*, Vol. 44, No. 6, pp. 2230–2259, 1998.
[6] U. Grenander, "Stochastic processes and Statistical Inference," *Arkiv Matematik*, Vol. 1, No. 17, pp. 195–277, 1950.
[7] U. Grenander, *Abstract Inference*, Wiley, 1981.
[8] D. L. Snyder and M. I. Miller, *Random Point Processes in Time and Space*, 2nd edition, Springer-Verlag, 1991.

Part II

Estimation

11 Bayesian Parameter Estimation

This chapter covers the Bayesian approach to estimation theory, where the unknown parameters are modeled as random. We consider three cost functions, namely squared error, absolute error, and the uniform cost. The cases of scalar- and vector-valued parameter estimation are studied separately to emphasize the similarities and differences between these two cases.

11.1 Introduction

For parameter estimation, the five ingredients of the statistical decision theory framework are as follows:

- $\mathcal{X}$: The set of parameters. A typical parameter is denoted by $\theta \in \mathcal{X}$. For parameter estimation, $\mathcal{X}$ is generally a compact subset of $\mathbb{R}^m$.
- $\mathcal{A}$: The set of estimates of the parameter. A typical estimate (action) is denoted by $a \in \mathcal{A}$.
- $C(a, \theta)$: The cost of action a when the parameter is θ, $C : \mathcal{A} \times \mathcal{X} \mapsto \mathbb{R}^+$.
- $\mathcal{Y}$: The set of observations. The observations could be continuous or discrete.
- $\mathcal{D}$: The set of decision rules, or estimators, typically denoted by $\hat{\theta} : \mathcal{Y} \mapsto \mathcal{A}$.

Often (but not always) we will have $\mathcal{A} = \mathcal{X}$. The set $\mathcal{D}$ of decision rules may be constrained. For instance, $\mathcal{D}$ could be the class of linear estimators, or a class of estimators that satisfy some unbiasedness or invariance property, as will be discussed later. The case of scalar and vector-valued θ will sometimes require a different treatment. We use boldface notation ($\boldsymbol{\theta}$) to denote a vector-valued parameter whenever the vector nature of the parameter is used.

As in the detection problem, we assume that conditional distributions for the observations $\{p_\theta(y), \theta \in \mathcal{X}\}$ are specified, based on which we can compute the *conditional risk* of the estimator $\hat{\theta}$:

$$R_\theta(\hat{\theta}) = \mathbb{E}_\theta\left[C(\hat{\theta}(Y), \theta)\right] = \int_{\mathcal{Y}} C(\hat{\theta}(y), \theta) p_\theta(y) d\mu(y), \quad \theta \in \mathcal{X}.$$

11.2 Bayesian Parameter Estimation

To find optimal solutions to the parameter estimation problem, we first consider the Bayesian setting in which we assume that the parameter $\theta \in \mathcal{X}$ is a realization of a random parameter Θ with prior $\pi(\theta)$. Using the general approach to finding Bayes solutions that we introduced earlier in Section 1.6, we obtain the Bayesian estimator as

$$\hat{\theta}_{\mathrm{B}}(y) = \arg\min_{a \in \mathcal{A}} C(a|y), \quad y \in \mathcal{Y}, \tag{11.1}$$

where

$$C(a|y) = \mathbb{E}[C(a, \Theta)|Y = y] = \int_{\mathcal{X}} C(a, \theta)\pi(\theta|y)d\nu(\theta)$$

is the *posterior risk* of action a given observation y. Here $d\nu(\theta)$ typically denotes the Lebesgue measure $d\theta$ (for θ taking values in Euclidean spaces). The *posterior distribution* of Θ, given $Y = y$, is expressed as

$$\pi(\theta|y) = \frac{p_\theta(y)\pi(\theta)}{\int_{\mathcal{X}} p_\theta(y)\pi(\theta)d\nu(\theta)}.$$

We now specialize the cost function to three common choices, for which we can compute $\hat{\theta}_{\mathrm{B}}$ more explicitly in terms of the posterior distribution $\pi(\theta|y)$. In Sections 11.3–11.6 we restrict our attention to the scalar parameter case, where $\mathcal{X} = \mathcal{A} \subseteq \mathbb{R}$. The vector case is considered in Section 11.7.

11.3 MMSE Estimation

Mean squared-error is a popular choice for the cost function because it gives rise to relatively tractable estimators (in fact linear in the case of Gaussian models) and strongly penalizes large errors. Assume thus $C(a, \theta) = (a - \theta)^2$, and $\mathbb{E}[\Theta^2] < \infty$. In this case,

$$C(a|y) = \mathbb{E}[(a - \Theta)^2|Y = y].$$

Consequently, the Bayes estimator is the MMSE estimator

$$\hat{\theta}_{\mathrm{MMSE}}(y) = \arg\min_{a \in \mathbb{R}} \mathbb{E}[(a - \Theta)^2|Y = y],$$

where the subscript MMSE stands for *minimum mean squared-error*. The minimand is a convex quadratic function of a. Hence, the minimizer can be obtained by setting the derivative of $C(a|y)$ with respect to a equal to zero:

$$0 = \frac{dC(a|y)}{da} = 2\mathbb{E}[(a - \Theta)|Y = y] = 2a - 2\mathbb{E}[\Theta|Y = y].$$

Consequently,

$$\hat{\theta}_{\mathrm{MMSE}}(y) = \mathbb{E}[\Theta|Y = y] = \int_{\mathcal{X}} \theta \; \pi(\theta|y) \; d\nu(\theta), \tag{11.2}$$

which is the *conditional mean* of $\pi(\theta|y)$. Hence the MMSE estimator may also be called *conditional-mean estimator*.

The MMSE solution (11.2) can also be obtained without differentiation using the *orthogonality principle*:

$$\mathbb{E}[(a - \Theta)^2|Y = y] = \mathbb{E}[(\Theta - \mathbb{E}[\Theta|Y = y])^2|Y = y] + \mathbb{E}[(a - \mathbb{E}[\Theta|Y = y])^2|Y = y],$$

which is minimized by setting $a = \mathbb{E}[\Theta|Y = y]$.

Example 11.1 (See Figure 11.1) Assume the prior

$$\pi(\theta) = \theta e^{-\theta} \mathbb{1}\{\theta \geq 0\}$$

and assume Y is uniformly distributed over the interval $[0, \Theta]$:

$$p_\theta(y) = \frac{1}{\theta} \mathbb{1}\{0 \leq y \leq \theta\}.$$

The unconditional pdf of Y is equal to

$$p(y) = \int_{\mathbb{R}} p_\theta(y) \pi(\theta) d\theta = \int_y^\infty e^{-\theta} d\theta = e^{-y}$$

for $y \geq 0$, and is equal to 0 otherwise. Thus, the posterior pdf of Θ given $Y = y$ is

$$\pi(\theta|y) = \frac{p_\theta(y)\pi(\theta)}{e^{-y}} = e^{-(\theta-y)} \mathbb{1}\{0 \leq y \leq \theta\},$$

as depicted in Figure 11.1.

The MMSE estimator is obtained by evaluating the conditional mean of the posterior pdf:

$$\begin{aligned}\hat{\theta}_{\text{MMSE}}(y) &= \int_0^\infty \theta \; \pi(\theta|y) \; d\theta = \int_y^\infty \theta \; e^{-(\theta-y)} d\theta \\ &= \int_y^\infty (\theta - y) e^{-(\theta-y)} d\theta + y \int_y^\infty e^{-(\theta-y)} d\theta.\end{aligned}$$

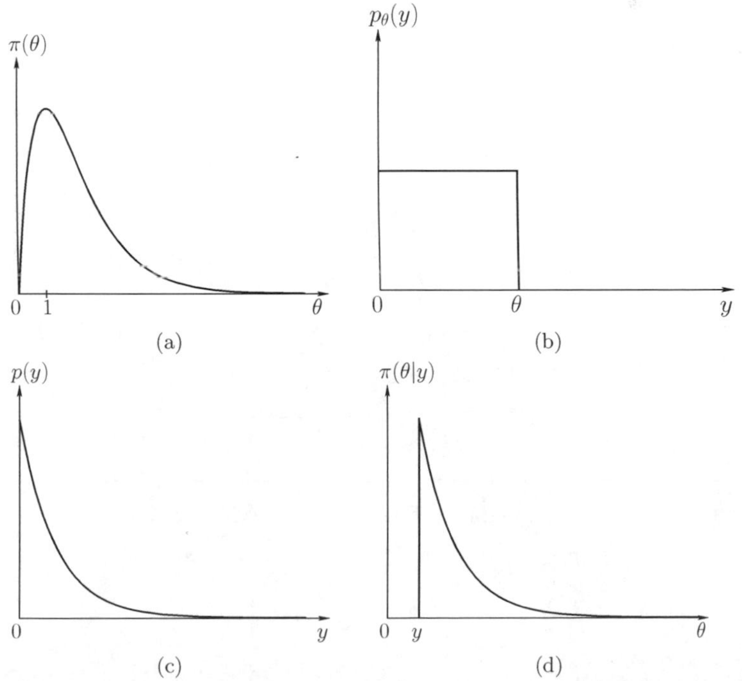

Figure 11.1 Pdfs for parameter estimation problem: (a) Prior pdf $\pi(\theta)$; (b) Conditional pdf $p_\theta(y)$; (c) Marginal pdf $p(y)$; (d) Posterior pdf $\pi(\theta|y)$

Make the change of variables $\theta' = \theta - y$. Then

$$\hat{\theta}_{\mathrm{MMSE}}(y) = \int_0^\infty \theta' e^{-\theta'} d\theta' + y \int_0^\infty e^{-\theta'} d\theta' = 1 + y.$$

11.4 MMAE Estimation

Another common cost function is the absolute-error cost function $C(a, \theta) = |a - \theta|$. This cost function has higher tolerance for large errors relative to the squared-error cost function. Assume that $\mathbb{E}[|\Theta|] < \infty$. Then the Bayes estimator is given by the MMAE estimator

$$\begin{aligned}\hat{\theta}_{\mathrm{MMAE}}(y) &= \arg\min_{a \in \mathbb{R}} \int_{\mathbb{R}} \pi(\theta|y)|a - \theta| d\theta \\ &= \arg\min_{a \in \mathbb{R}} \left[\int_{-\infty}^{a} \pi(\theta|y)(a - \theta) d\theta - \int_a^\infty \pi(\theta|y)(a - \theta) d\theta \right],\end{aligned}$$

where the subscript MMAE stands for *minimum mean absolute-error*. Since the minimand is a convex function of a, we may try to find the minimum by setting the derivative equal to zero,

$$\begin{aligned}0 &= \frac{d}{da} \int_{-\infty}^{a} \pi(\theta|y)(a - \theta) d\theta - \frac{d}{da} \int_a^\infty \pi(\theta|y)(a - \theta) d\theta \\ &= \int_{-\infty}^{a} \pi(\theta|y) d\theta - \int_a^\infty \pi(\theta|y) d\theta,\end{aligned}$$

where the last line holds by application of Leibnitz's rule for differentiation of integrals. Hence

$$\int_{-\infty}^{a} \pi(\theta|y) d\theta = \int_a^\infty \pi(\theta|y) d\theta = \frac{1}{2},$$

which means that a is the median of the posterior pdf. Consequently,

$$\hat{\theta}_{\mathrm{MMAE}}(y) = \mathrm{med}\left[\pi(\theta|y)\right] \tag{11.3}$$

is the *conditional-median* estimator. It is convenient to compute the MMAE estimate using the identity

$$\int_{\hat{\theta}_{\mathrm{MMAE}}}^\infty \pi(\theta|y) d\theta = \frac{1}{2}.$$

Example 11.2 For the estimation problem introduced in Example 11.1, we find that

$$\begin{aligned}\int_a^\infty e^{-(\theta - y)} d\theta = \frac{1}{2} &\Rightarrow e^{y-a} = \frac{1}{2} \\ &\Rightarrow y = -\ln 2 + a \\ &\Rightarrow \hat{\theta}_{\mathrm{MMAE}}(y) = y + \ln 2.\end{aligned}$$

11.5 MAP Estimation

Consider now the uniform cost, defined by

$$C(a,\theta) = \begin{cases} 0 & \text{if } |a-\theta| \le \epsilon \\ 1 & \text{if } |a-\theta| > \epsilon \end{cases}$$

for some small $\epsilon > 0$. The Bayes estimator under this cost function is given by

$$\begin{aligned} \hat{\theta}(y) = \arg\min_a \int_{\mathbb{R}} \pi(\theta|y)\mathbb{1}\{|a-\theta| > \epsilon\} d\theta &= \arg\min_a \int_{|a-\theta|>\epsilon} \pi(\theta|y) d\theta \\ &= \arg\min_a \left[1 - \int_{|a-\theta|\le\epsilon} \pi(\theta|y) d\theta\right] \\ &= \arg\max_a \int_{|a-\theta|\le\epsilon} \pi(\theta|y) d\theta. \end{aligned} \tag{11.4}$$

In order to solve this problem, we assume that the posterior pdf $\pi(\theta|y)$ is continuous at $\theta = a$ and approximate the integral in (11.4) with[1]

$$\int_{|a-\theta|\le\epsilon} \pi(\theta|y) d\theta \approx 2\epsilon\pi(a|y).$$

Therefore,

$$\hat{\theta}_{\text{MAP}}(y) = \arg\max_{\theta\in\mathbb{R}} \pi(\theta|y) = \arg\max_{\theta\in\mathbb{R}} p_\theta(y)\pi(\theta), \tag{11.5}$$

which is the *conditional-mode estimator*, more commonly called the *maximum a posteriori* (MAP) estimator.

Example 11.3 For the estimation problem introduced in Example 11.1, we find that

$$\hat{\theta}_{\text{MAP}}(y) = \arg\max_{\theta\ge 0} e^{-(\theta-y)} = y.$$

Example 11.4 Consider the estimation problem in which $\mathcal{X} = (0,\infty)$, $\mathcal{Y} = [0,\infty)$, and

$$p_\theta(y) = \theta\, e^{-\theta y}\, \mathbb{1}\{y > 0\}, \quad \pi(\theta) = \alpha e^{-\alpha\theta}\, \mathbb{1}\{\theta \ge 0\},$$

where $\alpha > 0$ is fixed.

Using the fact that $\int_0^\infty x^m e^{-\beta x} dx = m!/\beta^{m+1}$ for $\beta > 0$ and $m \in \mathbb{Z}^+$, we obtain

$$\pi(\theta|y) = (\alpha+y)^2\, \theta\, e^{-(\alpha+y)\theta}\, \mathbb{1}\{y \ge 0\}\, \mathbb{1}\{\theta \ge 0\}.$$

Therefore

$$\hat{\theta}_{\text{MMSE}}(y) = \int_0^\infty \theta\, \pi(\theta|y)\, d\theta = \frac{2}{\alpha+y}.$$

Also,

$$\int_{\hat{\theta}_{\text{MMAE}}(y)}^\infty \pi(\theta|y)\, d\theta = 0.5 \quad \Longrightarrow \quad \hat{\theta}_{\text{MMAE}}(y) \approx \frac{1.68}{\alpha+y}.$$

[1] If the posterior pdf is discontinuous at $\theta = a$, the integral can be approximated by $\epsilon[\pi(a^+|y) + \pi(a^-|y)]$ which results in the same MAP estimator as in (11.5).

Finally

$$\hat{\theta}_{\mathrm{MAP}}(y) = \arg\max_{\theta > 0} \theta e^{-(\alpha+y)\theta} = \frac{1}{\alpha + y}.$$

Note that for the MAP estimator, we did not use the conditional distribution of θ given y explicitly.

We see that $\hat{\theta}_{\mathrm{MAP}}(y) < \hat{\theta}_{\mathrm{MMAE}}(y) < \hat{\theta}_{\mathrm{MMSE}}(y)$ in this example. In general, since the median of a distribution usually lies between the mean and the mode, we expect such a relationship or the one with inequalities reversed to hold.

Example 11.5 Here we modify Example 11.3, assuming now a sequence of n observations $\boldsymbol{Y} = (Y_1, Y_2, \ldots, Y_n)$ drawn i.i.d. from the uniform distribution on the interval $[0, \theta]$. The parameter θ is also random with exponential prior distribution

$$\pi(\theta) = \theta e^{-\theta} \mathbb{1}\{\theta \geq 0\}.$$

We first derive the posterior pdf (see Figure 11.2)

$$\begin{aligned}
\pi(\theta | \boldsymbol{y}) = \frac{\pi(\theta) p_\theta^n(\boldsymbol{y})}{p(\boldsymbol{y})} &= \frac{\pi(\theta) \prod_{i=1}^n p_\theta(y_i)}{p(\boldsymbol{y})} \\
&= \frac{\theta e^{-\theta} \left(\frac{1}{\theta}\right)^n \prod_{i=1}^n \mathbb{1}\{0 \leq y_i \leq \theta\}}{p(\boldsymbol{y})} \mathbb{1}\{\theta \geq 0\} \\
&= \frac{1}{p(\boldsymbol{y})} \frac{e^{-\theta}}{\theta^{n-1}} \mathbb{1}\{\theta \geq \max_{1 \leq i \leq n} y_i\}.
\end{aligned}$$

Hence the MAP estimator is given by

$$\hat{\theta}_{\mathrm{MAP}}(\boldsymbol{y}) = \arg\max_{\theta \geq 0} \pi(\theta | \boldsymbol{y}) = \arg\max_{\theta \geq \max_{1 \leq i \leq n} y_i} \frac{e^{-\theta}}{\theta^{n-1}} = \max_{1 \leq i \leq n} y_i,$$

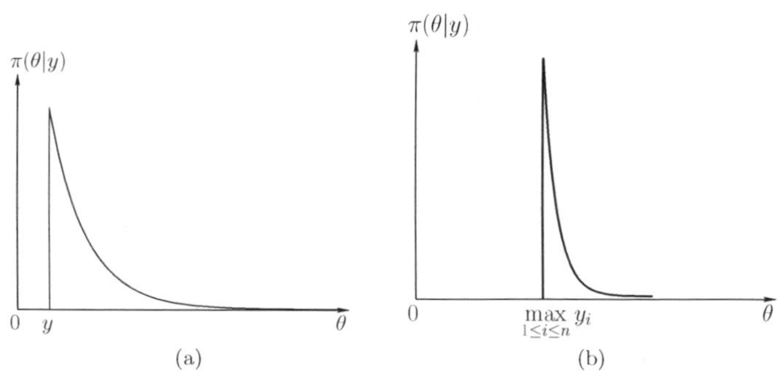

Figure 11.2 The posterior pdf, $\pi(\theta | \boldsymbol{y})$: (a) $n = 1$; (b) $n = 10$

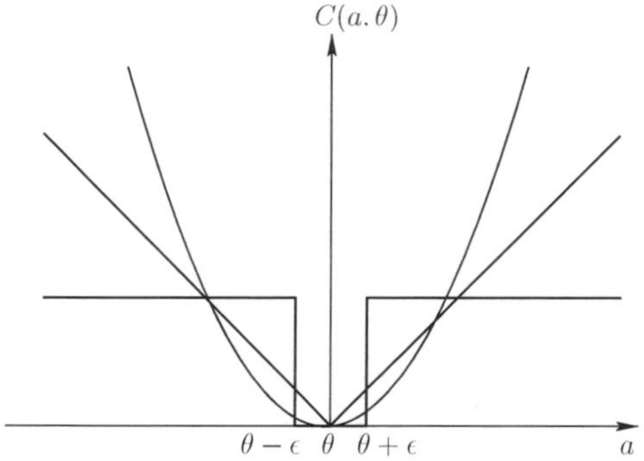

Figure 11.3 The three cost functions used in the scalar parameter estimation problems, resulting in MMSE, MMAE, and MAP estimators

where the last equality holds because the function $\theta^{1-n}e^{-\theta}$ is strictly decreasing over its domain, hence its maximum is achieved by the smallest feasible θ.

Note again that we avoided unnecessary evaluation of the marginal $p(\boldsymbol{y})$. Also note that the posterior pdf is sharply peaked for large n, indicating that the residual uncertainty upon seeing n observations is small. (Compare Figures 11.2 (a) and (b).) Finally, note that the computation of the MMSE and MMAE estimators would be more involved here.

Figure 11.3 compares the three different cost functions considered in MMSE, MMAE, and MAP estimation. In all cases, the estimator is derived from the posterior pdf – by finding its mean, its median, or its mode. The posterior pdf is fundamental in all Bayesian inference problems.

11.6 Parameter Estimation for Linear Gaussian Models

The case of linear Gaussian models is frequently encountered in practice and is insightful. Assume the parameter space $\mathcal{X} = \mathbb{R}$, and $\mathcal{Y} = \mathbb{R}^n$, and

$$\boldsymbol{Y} = \Theta\, \boldsymbol{s} + \boldsymbol{Z},$$

where $\boldsymbol{s}$ is a known (signal) vector, $\Theta \sim \mathcal{N}(\mu, \sigma^2)$, $\boldsymbol{Z} \sim \mathcal{N}(\boldsymbol{0}, \mathsf{C}_Z)$, and Θ and $\boldsymbol{Z}$ are independent.

Using the fact that Θ has a conditional distribution that is Gaussian, given $\boldsymbol{Y}$, and the fact that the MMSE estimate is the same as the linear MMSE estimate for this Gaussian problem (see Section 11.7.4), we get that Θ, given $\boldsymbol{Y} = \boldsymbol{y}$, has a Gaussian distribution with mean

$$\mathbb{E}[\Theta|\boldsymbol{Y} = \boldsymbol{y}] = \mathbb{E}[\Theta] + \mathrm{Cov}(\Theta, \boldsymbol{Y})\, \mathrm{Cov}(\boldsymbol{Y})^{-1}(\boldsymbol{y} - \mathbb{E}[\boldsymbol{Y}])$$

and variance

$$\mathrm{Var}(\Theta|\boldsymbol{Y} = \boldsymbol{y}) = \mathrm{Var}(\Theta) - \mathrm{Cov}(\Theta, \boldsymbol{Y})\, \mathrm{Cov}(\boldsymbol{Y})^{-1}\mathrm{Cov}(\boldsymbol{Y}, \Theta).$$

We can compute

$$\begin{aligned}\mathbb{E}[\Theta] &= \mu, \mathbb{E}[Y] = \mu s\\ \mathrm{Var}(\Theta) &= \sigma^2, \mathrm{Cov}(\Theta, \boldsymbol{Y}) = \sigma^2 \boldsymbol{s}^\top, \ \mathrm{Cov}(\boldsymbol{Y}, \Theta) = \sigma^2 \boldsymbol{s}, \ \mathrm{Cov}(\boldsymbol{Y}) = \sigma^2 \boldsymbol{s}\boldsymbol{s}^\top + \mathsf{C}_Z,\end{aligned}$$

and use these to conclude that

$$\mathbb{E}[\Theta|\boldsymbol{Y} = \boldsymbol{y}] = \mu + \sigma^2 \boldsymbol{s}^\top (\sigma^2 \boldsymbol{s}\boldsymbol{s}^\top + \mathsf{C}_Z)^{-1} (y - \mu \boldsymbol{s})$$

$$= \frac{\boldsymbol{s}^\top \mathsf{C}_Z^{-1} \boldsymbol{y} + \frac{\mu}{\sigma^2}}{d^2 + \frac{1}{\sigma^2}}, \tag{11.6}$$

$$\mathrm{Var}(\Theta|\boldsymbol{Y} = \boldsymbol{y}) = \sigma^2 - \sigma^4 \boldsymbol{s}^\top (\sigma^2 \boldsymbol{s}\boldsymbol{s}^\top + \mathsf{C}_Z)^{-1} \boldsymbol{s}$$

$$= \frac{1}{d^2 + \frac{1}{\sigma^2}}, \tag{11.7}$$

where $d^2 = \boldsymbol{s}^\top \mathsf{C}_Z^{-1} \boldsymbol{s}$ is the squared Mahalanobis distance (see Definition 5.1) of $\boldsymbol{s}$ from $\mathbf{0}$.

Note that due to the symmetry of the Gaussian distribution,

$$\hat{\theta}_{\mathrm{MMSE}}(\boldsymbol{y}) = \hat{\theta}_{\mathrm{MMAE}}(\boldsymbol{y}) = \hat{\theta}_{\mathrm{MAP}}(\boldsymbol{y}) = \mathbb{E}[\Theta|\boldsymbol{Y} = \boldsymbol{y}]. \tag{11.8}$$

If $d^2 \gg \frac{1}{\sigma^2}$, i.e., if the observations are more informative than the prior, then the estimated parameter is $\hat{\theta}(\boldsymbol{y}) \approx \boldsymbol{s}^\top \mathsf{C}_Z^{-1} \boldsymbol{y}/d^2$, and we essentially ignore the prior. On the other hand, if $d^2 \ll \frac{1}{\sigma^2}$, i.e., the prior is more informative than the observations, then $\hat{\theta}(\boldsymbol{y}) \approx \mu$ and we ignore the observations.

In the special case $\boldsymbol{s} = \mathbf{1}$ and $\mathsf{C}_Z = \sigma_Z^2 \mathsf{I}_n$ (the model reduces to $\{Y_i\}_{i=1}^n$ being conditionally i.i.d. $\mathcal{N}(\theta, \sigma_Z^2)$ given θ), we obtain

$$\hat{\theta}_{\mathrm{MMSE}}(\boldsymbol{y}) = \hat{\theta}_{\mathrm{MMAE}}(\boldsymbol{y}) = \hat{\theta}_{\mathrm{MAP}}(\boldsymbol{y}) = \frac{\frac{1}{n}\sum_{i=1}^n y_i + \mu \frac{\sigma_Z^2}{n\sigma^2}}{1 + \frac{\sigma_Z^2}{n\sigma^2}}.$$

11.7 Estimation of Vector Parameters

The cost functions introduced in Sections 11.3–11.5 for scalar parameters can be extended to the vector case. Here we assume that $\mathcal{X} = \mathcal{A} \subseteq \mathbb{R}^m$, and $\mathcal{Y} = \mathbb{R}^n$. Consider the special case of an additive cost function, where

$$C(\boldsymbol{a}, \boldsymbol{\theta}) = \sum_{i=1}^m C_i(a_i, \theta_i). \tag{11.9}$$

Then the posterior cost of action $\boldsymbol{a}$, given observation $\boldsymbol{y}$, can be written as

$$\begin{aligned}C(\boldsymbol{a}|\boldsymbol{y}) &= \mathbb{E}\left[C(\boldsymbol{a}, \boldsymbol{\Theta})|\boldsymbol{Y} = \boldsymbol{y}\right]\\ &= \sum_{i=1}^m \mathbb{E}\left[C(a_i, \Theta_i)|\boldsymbol{Y} = \boldsymbol{y}\right].\end{aligned}$$

Minimizing $C(\boldsymbol{a}|\boldsymbol{y})$ is equivalent to minimizing

$$C_i(a_i|\boldsymbol{y}) = \mathbb{E}\left[C_i(a_i, \Theta_i)|\boldsymbol{Y} = \boldsymbol{y}\right]$$

for each i, which implies that the Bayes solution $\hat{\boldsymbol{\theta}}_{\text{B}}(\boldsymbol{y})$ has components that are given by solving m independent minimization problems:

$$\hat{\boldsymbol{\theta}}_{\text{B,i}}(\boldsymbol{y}) = \arg\min_{a_i} C_i(a_i|\boldsymbol{y}), \quad i = 1, 2, \ldots, m.$$

11.7.1 Vector MMSE Estimation

Assume that $\mathbb{E}(\|\Theta\|^2) < \infty$. The squared-error cost function is additive:

$$C(\boldsymbol{a}, \boldsymbol{\theta}) = \|\boldsymbol{a} - \boldsymbol{\theta}\|^2 = \sum_{i=1}^{m} (a_i - \theta_i)^2.$$

Therefore the components of $\hat{\boldsymbol{\theta}}_{\text{MMSE}}(\boldsymbol{y})$ are given by

$$\hat{\theta}_{\text{MMSE,i}}(\boldsymbol{y}) = \mathbb{E}\left[\Theta_i|\boldsymbol{Y} = \boldsymbol{y}\right].$$

By the linearity of the expectation operator, we obtain

$$\hat{\boldsymbol{\theta}}_{\text{MMSE}}(\boldsymbol{y}) = \mathbb{E}\left[\boldsymbol{\Theta}|\boldsymbol{Y} = \boldsymbol{y}\right]. \tag{11.10}$$

11.7.2 Vector MMAE Estimation

The absolute-error cost function (ℓ_1 norm of the error vector) is also additive:

$$C(\boldsymbol{a}, \boldsymbol{\theta}) = \|\boldsymbol{a} - \boldsymbol{\theta}\|_1 = \sum_{i=1}^{m} |a_i - \theta_i|$$

and therefore the components of $\hat{\boldsymbol{\theta}}_{\text{MMAE}}(\boldsymbol{y})$ are given by

$$\hat{\theta}_{\text{MMAE,i}}(\boldsymbol{y}) = \text{med}\left[\pi(\theta_i|\boldsymbol{y})\right], \quad i = 1, 2, \ldots, m. \tag{11.11}$$

11.7.3 Vector MAP Estimation

There are two ways to define the uniform cost function that lead to MAP solutions. In the first, we consider an additive cost function with components:

$$C_i(a_i, \theta_i) = \begin{cases} 0 & \text{if } |a_i - \theta_i| \le \epsilon \\ 1 & \text{if } |a_i - \theta_i| > \epsilon. \end{cases}$$

This leads to a componentwise MAP estimate as $\epsilon \to 0$ with

$$\hat{\theta}_{\text{MAP,i}}(\boldsymbol{y}) = \arg\max_{\theta_i} \pi(\theta_i|\boldsymbol{y}).$$

A more natural way to define the uniform cost function is through a nonadditive cost function:

$$C(\boldsymbol{a}, \boldsymbol{\theta}) = \begin{cases} 0 & \text{if } \max_i |a_i - \theta_i| \le \epsilon \\ 1 & \text{otherwise.} \end{cases}$$

In this case, using the same argument as in the scalar case, we obtain as $\epsilon \to 0$, the optimum estimator converges to

$$\hat{\boldsymbol{\theta}}_{\text{MAP}}(\boldsymbol{y}) = \arg\max_{\boldsymbol{\theta}} \pi(\boldsymbol{\theta}|\boldsymbol{y}). \tag{11.12}$$

Note that in this case $\hat{\theta}_{\text{MAP,i}}(\boldsymbol{y})$ is not necessarily equal to $\arg\max_{\theta_i} \pi(\theta_i|\boldsymbol{y})$. However, the equality will hold if

$$\pi(\boldsymbol{\theta}|\boldsymbol{y}) = \prod_{i=1}^{m} \pi(\theta_i|\boldsymbol{y}).$$

Example 11.6 Suppose $\mathcal{Y} = (0, \infty)$ and $\mathcal{X} = [0, \infty)^2$, with

$$\pi(\boldsymbol{\theta}|y) = \begin{cases} y^2 e^{-y\theta_1} & \text{if } 0 \le \theta_2 \le \theta_1 \\ 0 & \text{otherwise.} \end{cases}$$

Then

$$\hat{\boldsymbol{\theta}}_{\text{MAP}}(y) = \mathbf{0}.$$

However, it is easy to see that

$$\pi(\theta_1|y) = y^2\theta_1 e^{-y\theta_1}\, \mathbb{1}\{\theta_1 \ge 0\}, \text{ and } \pi(\theta_2|y) = ye^{-y\theta_2}\, \mathbb{1}\{\theta_2 \ge 0\},$$

from which we get that

$$\arg\max_{\theta_1} \pi(\theta_1|y) = \frac{1}{y}, \text{ and } \arg\max_{\theta_2} \pi(\theta_2|y) = 0.$$

Thus $\hat{\theta}_{\text{MAP,i}}(y) \ne \arg\max_{\theta_i} \pi(\theta_i|y)$ for $i = 1$.

11.7.4 Linear MMSE Estimation

We saw in Section 11.7.2 that the (unconstrained) MMSE estimator of $\boldsymbol{\Theta}$, given $\boldsymbol{Y}$ is the conditional mean:

$$\hat{\boldsymbol{\theta}}_{\text{MMSE}}(\boldsymbol{y}) = \mathbb{E}[\boldsymbol{\Theta}|\boldsymbol{Y} = \boldsymbol{y}].$$

For general models, this conditional expectation is nonlinear in $\boldsymbol{y}$ and may be difficult to compute. To simplify the MMSE estimation problem, we add the constraint that the estimator should be an affine function of the observations of the form:

$$\hat{\boldsymbol{\theta}}(\boldsymbol{Y}) = \mathsf{A}\boldsymbol{Y} + \boldsymbol{b}.$$

As in the unconstrained MMSE estimation problem we ask to minimize the mean squared-error

$$\text{MSE} = \mathbb{E}[\|\hat{\boldsymbol{\theta}}(\boldsymbol{Y}) - \boldsymbol{\Theta}\|^2]$$

but with the restriction that the minimization is over the set of linear estimators. This problem can be solved using orthogonality principle (see, e.g., [1]). If $\boldsymbol{W} = \hat{\boldsymbol{\theta}}(\boldsymbol{Y}) - \boldsymbol{\Theta}$

denotes the estimation error, then the orthogonality principle states that the optimal $\hat{\boldsymbol{\theta}}$ satisfies

$$\mathbb{E}[\boldsymbol{W}] = 0 \tag{11.13}$$

$$\text{Cov}(\boldsymbol{W}, \boldsymbol{Y}) = 0. \tag{11.14}$$

These equations can solved to yield the linear MMSE (or LMMSE) estimate [1]:

$$\begin{aligned}\hat{\boldsymbol{\theta}}_{\text{LMMSE}}(\boldsymbol{Y}) &\triangleq \hat{\mathbb{E}}[\boldsymbol{\Theta}|\boldsymbol{Y}] \\ &= \mathbb{E}[\boldsymbol{\Theta}] + \text{Cov}(\boldsymbol{\Theta}, \boldsymbol{Y})\text{Cov}(\boldsymbol{Y})^{-1}(\boldsymbol{Y} - \mathbb{E}[\boldsymbol{Y}]),\end{aligned} \tag{11.15}$$

where the notation $\hat{\mathbb{E}}$ for the LMMSE estimate is used to denote the fact that the LMMSE estimate can be considered to be a linear approximation to the conditional expectation, which is the MMSE estimate. The covariance of the error can also be obtained:

$$\text{Cov}(\boldsymbol{W}) = \text{Cov}(\boldsymbol{\Theta}) - \text{Cov}(\boldsymbol{\Theta}, \boldsymbol{Y})\text{Cov}(\boldsymbol{Y})^{-1}\text{Cov}(\boldsymbol{Y}, \boldsymbol{\Theta}), \tag{11.16}$$

from which we can compute the MSE of the LMMSE estimator as

$$\begin{aligned}\text{MSE}(\hat{\boldsymbol{\theta}}_{\text{LMMSE}}) &= \mathbb{E}[\|\boldsymbol{W}\|^2] = \text{Tr}\left(\mathbb{E}[\|\boldsymbol{W}\|^2]\right) = \mathbb{E}\left[\text{Tr}\left(\|\boldsymbol{W}\|^2\right)\right] \\ &= \mathbb{E}\left[\text{Tr}\left(\boldsymbol{W}^\top \boldsymbol{W}\right)\right] = \text{Tr}\left(\mathbb{E}\left[\boldsymbol{W}\boldsymbol{W}^\top\right]\right) = \text{Tr}\left(\text{Cov}(\boldsymbol{W})\right),\end{aligned}$$

where, as in Section 5.10, we exploited the linearity of the expectation operation, the linearity of the trace operation, and invariance of the trace to circular rotations of matrix products (see also Appendix A).

Furthermore, for $(\boldsymbol{\Theta}, \boldsymbol{Y})$ jointly Gaussian it can be shown that

$$\hat{\boldsymbol{\theta}}_{\text{MMSE}}(\boldsymbol{y}) = \mathbb{E}[\boldsymbol{\Theta}|\boldsymbol{Y} = \boldsymbol{y}] = \hat{\mathbb{E}}[\boldsymbol{\Theta}|\boldsymbol{Y} = \boldsymbol{y}] = \hat{\boldsymbol{\theta}}_{\text{LMMSE}}(\boldsymbol{y}) \tag{11.17}$$

and that

$$\text{Cov}(\boldsymbol{W}) = \text{Cov}(\hat{\boldsymbol{\theta}}(\boldsymbol{Y}) - \boldsymbol{\Theta}) = \text{Cov}(\hat{\boldsymbol{\theta}}(\boldsymbol{Y}) - \boldsymbol{\Theta}|\boldsymbol{Y} = \boldsymbol{y}) = \text{Cov}(\boldsymbol{\Theta}|\boldsymbol{Y} = \boldsymbol{y}). \tag{11.18}$$

It is to be noted that, unlike the MMSE estimate, the computations of the LMMSE estimate and the covariance of the estimation error do not require the entire joint distribution of $\boldsymbol{\Theta}$ and $\boldsymbol{Y}$; moments up to second order are sufficient.

11.7.5 Vector Parameter Estimation in Linear Gaussian Models

This is the vector parameter version of the problem considered in Section 11.6. Assume $\mathcal{X} = \mathbb{R}^m$, $\mathcal{Y} = \mathbb{R}^n$, and

$$\boldsymbol{Y} = \mathsf{H}\,\boldsymbol{\Theta} + \boldsymbol{Z}, \tag{11.19}$$

where H is a known $n \times m$ matrix, $\boldsymbol{\Theta} \sim \mathcal{N}(\boldsymbol{\mu}, \mathsf{K})$, $\boldsymbol{Z} \sim \mathcal{N}(\boldsymbol{0}, \mathsf{C}_Z)$, and $\boldsymbol{\Theta}$ and $\boldsymbol{Z}$ are independent.

Using (11.17) and (11.18), we get that $\boldsymbol{\Theta}$, given $\boldsymbol{Y} = \boldsymbol{y}$, has a Gaussian distribution with mean and covariance matrix respectively given by

$$\mathbb{E}[\boldsymbol{\Theta}|\boldsymbol{Y} = \boldsymbol{y}] = \hat{\mathbb{E}}[\boldsymbol{\Theta}|\boldsymbol{Y} = \boldsymbol{y}] = \mathbb{E}[\boldsymbol{\Theta}] - \text{Cov}(\boldsymbol{\Theta}, \boldsymbol{Y})\,\text{Cov}(\boldsymbol{Y})^{-1}(\boldsymbol{y} - \mathbb{E}[\boldsymbol{Y}]) \tag{11.20}$$

and

$$\mathrm{Cov}(\boldsymbol{\Theta}|\boldsymbol{Y}=\boldsymbol{y}) = \mathrm{Cov}(\boldsymbol{\Theta}) - \mathrm{Cov}(\boldsymbol{\Theta},\boldsymbol{Y})\,\mathrm{Cov}(\boldsymbol{Y})^{-1}\,\mathrm{Cov}(\boldsymbol{Y},\boldsymbol{\Theta}). \tag{11.21}$$

We can compute

$$\mathbb{E}[\boldsymbol{\Theta}] = \boldsymbol{\mu}\mathbb{E}[Y] = \mathsf{H}\boldsymbol{\mu},$$
$$\mathrm{Cov}(\boldsymbol{\Theta}) = \mathsf{K}, \quad \mathrm{Cov}(\boldsymbol{\Theta},\boldsymbol{Y}) = \mathsf{KH}^\top,$$
$$\mathrm{Cov}(\boldsymbol{Y},\boldsymbol{\Theta}) = \mathsf{HK}, \ \mathrm{Cov}(\boldsymbol{Y}) = \mathsf{HKH}^\top + \mathsf{C}_Z,$$

and use these to conclude that

$$\mathbb{E}[\boldsymbol{\Theta}|\boldsymbol{Y}=\boldsymbol{y}] = \boldsymbol{\mu} + \mathsf{KH}^\top(\mathsf{HKH}^\top + \mathsf{K})^{-1}(\boldsymbol{y} - \mathsf{H}\boldsymbol{\mu})$$

and

$$\mathrm{Cov}(\boldsymbol{\Theta}|Y=y) = \mathsf{K} - \mathsf{KH}^\top(\mathsf{HKH}^\top + \mathsf{K})^{-1}\mathsf{HK}.$$

Note again that due to the symmetry of the Gaussian distribution,

$$\hat{\boldsymbol{\theta}}_{\mathrm{MMSE}}(\boldsymbol{y}) = \hat{\boldsymbol{\theta}}_{\mathrm{MMAE}}(\boldsymbol{y}) = \hat{\boldsymbol{\theta}}_{\mathrm{MAP}}(\boldsymbol{y}) = \mathbb{E}[\boldsymbol{\Theta}|\boldsymbol{Y}=\boldsymbol{y}]. \tag{11.22}$$

11.7.6 Other Cost Functions for Bayesian Estimation

Other cost functions are sometimes encountered. For instance, choose $q \geq 1$ and let

$$C(\boldsymbol{a},\boldsymbol{\theta}) = \sum_{i=1}^{m} |a_i - \theta_i|^q \tag{11.23}$$

be the ℓ_q norm $\|\boldsymbol{a}-\boldsymbol{\theta}\|_q$ of the estimation error, raised to the power q. This is an additive cost function, and the special cases $q = 1$ and $q = 2$ lead to the MMAE and MMSE estimators, respectively. For large q, large estimation errors are strongly penalized. Moreover the cost function is dominated by the largest component of the estimation error vector because

$$\lim_{q\to\infty} \|\boldsymbol{a}-\boldsymbol{\theta}\|_q = \max_{1\leq i\leq m} |a_i - \theta_i|.$$

Another possible cost function is the *Kullback–Leibler (KL) divergence loss* (see Section 6.1 for the definition of KL divergence),

$$C(\boldsymbol{a},\boldsymbol{\theta}) = D(p_{\boldsymbol{a}}\|p_{\boldsymbol{\theta}}), \tag{11.24}$$

which is encountered in some theoretical analyses. For the Gaussian family $p_{\boldsymbol{\theta}} = \mathcal{N}(\boldsymbol{\theta},\mathsf{I}_n)$, this reduces to the squared-error cost.

11.8 Exponential Families

DEFINITION 11.1 An m-dimensional exponential family $\{p_{\boldsymbol{\theta}},\ \boldsymbol{\theta}\in\mathcal{X}\}$ is a family of distributions with densities of the form

$$p_{\boldsymbol{\theta}}(y) = h(y)\,C(\boldsymbol{\theta})\,\exp\left\{\sum_{k=1}^{m}\eta_k(\boldsymbol{\theta})T_k(y)\right\}, \quad y\in\mathcal{Y}, \tag{11.25}$$

where $\{\eta_k(\boldsymbol{\theta}),\ 1 \le k \le m\}$ and $\{T_k(y),\ 1 \le k \le m\}$ are real-valued functions, and $C(\boldsymbol{\theta})$ is a normalization constant.

Note that m is not necessarily the dimension of $\boldsymbol{\theta}$ in Definition 11.1. Also, if we let $\boldsymbol{\eta}(\boldsymbol{\theta}) = [\eta_1(\boldsymbol{\theta}) \dots \eta_m(\boldsymbol{\theta})]$, then $\boldsymbol{\eta}$ maps the parameter space $\mathcal{X}$ to some subset of $\mathbb{R}^m$, which we denote by Γ.

Exponential families are widely used and encompass many common distributions such as Gaussian, Gamma, Poisson, and multinomial. However, the family of uniform distributions over $[0, \theta]$ is not in the exponential family, and neither is the family of Cauchy distributions with location parameter θ.

It is often simpler to view $\boldsymbol{\eta} = \{\eta_k\}_{k=1}^m$ as parameters of the distribution. The exponential family of (11.25) is then written in the following *canonical form*:

$$q_{\boldsymbol{\eta}}(y) = h(y) \exp\left\{\sum_{k=1}^m \eta_k T_k(y) - A(\boldsymbol{\eta})\right\}, \quad y \in \mathcal{Y}, \tag{11.26}$$

where $A(\boldsymbol{\eta}) = -\ln C(\boldsymbol{\theta})$. In order for the normalization constant to be well defined, we must have

$$\int_{\mathcal{Y}} h(y) \exp\left\{\sum_{k=1}^m \eta_k T_k(y)\right\} d\mu(y) < \infty. \tag{11.27}$$

Since the integral is a convex function of $\boldsymbol{\eta}$, the set of $\boldsymbol{\eta}$ satisfying the inequality (11.27) is a convex set, called the *natural parameter space*. Usually (but not always), this set is open.

11.8.1 Basic Properties

The following proposition gives some basic properties of a canonical exponential family. In what follows:

- the m-vector $\nabla A(\boldsymbol{\eta}) \triangleq \{\partial A(\boldsymbol{\eta})/\partial \eta_k,\ 1 \le k \le m\}$ is the gradient of the function $A(\boldsymbol{\eta})$, and
- the $m \times m$ matrix $\nabla^2 A(\boldsymbol{\eta}) \triangleq \{\partial^2 A(\boldsymbol{\eta})/\partial \theta_j \partial \eta_k,\ 1 \le j, k \le m\}$ is the Hessian of $A(\boldsymbol{\eta})$.

The identities given at the end of Appendix A are used in the derivations. The following technical result is also used (see e.g. [2, p. 27]): for any integrable function $f : \mathcal{Y} \to \mathbb{R}$ and $\boldsymbol{\eta}$ in the interior of $\mathcal{X}_{\text{nat}}$, the expectation $\mathbb{E}_{\boldsymbol{\eta}}[f(Y)] = \int_{\mathcal{Y}} q_{\boldsymbol{\eta}}(y) f(y)\, d\mu(y)$ is continuous and has derivatives of all orders with respect to $\boldsymbol{\eta}$, and these derivatives can be obtained by switching the order of integration and differentiation.

PROPOSITION 11.1 *The following statements hold:*

(i) *The mean of $\boldsymbol{T}(Y)$ is given by*

$$\mathbb{E}_{\boldsymbol{\eta}}[\boldsymbol{T}(Y)] = \nabla A(\boldsymbol{\eta}). \tag{11.28}$$

(ii) *The covariance of $\boldsymbol{T}(Y)$ is given by*

$$\mathrm{Cov}_{\boldsymbol{\eta}}[\boldsymbol{T}(Y)] = \nabla^2 A(\boldsymbol{\eta}). \tag{11.29}$$

(iii) *The cumulant-generating function of $q_{\boldsymbol{\eta}}$ is given by*

$$\kappa_{\boldsymbol{\eta}}(\boldsymbol{u}) = \ln \mathbb{E}_{\boldsymbol{\eta}}[e^{\boldsymbol{u}^\top \boldsymbol{T}}] = A(\boldsymbol{\eta} + \boldsymbol{u}) - A(\boldsymbol{\eta}), \quad \boldsymbol{u} \in \mathbb{R}^m. \tag{11.30}$$

(iv) *The KL divergence between $q_{\boldsymbol{\eta}}$ and $q_{\boldsymbol{\eta}'}$ is given by*

$$D(q_{\boldsymbol{\eta}} \| q_{\boldsymbol{\eta}'}) = (\boldsymbol{\eta} - \boldsymbol{\eta}')^\top \nabla A(\boldsymbol{\eta}) - A(\boldsymbol{\eta}) + A(\boldsymbol{\eta}'), \tag{11.31}$$

which is also a Bregman divergence with generator function $-A(\boldsymbol{\eta})$.

Proof

(i) Taking the gradient of the identity $1 = \int_{\mathcal{Y}} q_{\boldsymbol{\eta}} \, d\mu$ with respect to $\boldsymbol{\eta}$, we obtain

$$\begin{aligned}
0 &= \nabla \int_{\mathcal{Y}} h(y) \exp\left\{\sum_{k=1}^m \eta_k T_k(y) - A(\boldsymbol{\eta})\right\} d\mu(y) \\
&= \int_{\mathcal{Y}} h(y) \exp\left\{\sum_{k=1}^m \eta_k T_k(y) - A(\boldsymbol{\eta})\right\} [\boldsymbol{T}(y) - \nabla A(\boldsymbol{\eta})] \, d\mu(y) \\
&= \mathbb{E}_{\boldsymbol{\eta}}[\boldsymbol{T}(Y)] - \nabla A(\boldsymbol{\eta}).
\end{aligned}$$

(ii) Similarly, taking the Hessian of the identity $1 = \int_{\mathcal{Y}} q_{\boldsymbol{\eta}} \, d\mu$ with respect to $\boldsymbol{\eta}$, we obtain

$$\begin{aligned}
0 &= \nabla^2 \int_{\mathcal{Y}} h(y) \exp\left\{\sum_{k=1}^m \eta_k T_k(y) - A(\boldsymbol{\eta})\right\} d\mu(y) \\
&= \nabla \int_{\mathcal{Y}} h(y) \exp\left\{\sum_{k=1}^m \eta_k T_k(y) - A(\boldsymbol{\eta})\right\} [\boldsymbol{T}(y) - \nabla A(\boldsymbol{\eta})] \, d\mu(y) \\
&= \int_{\mathcal{Y}} h(y) \exp\left\{\sum_{k=1}^m \eta_k T_k(y) - A(\boldsymbol{\eta})\right\} \Big([\boldsymbol{T}(y) - \nabla A(\boldsymbol{\eta})][\boldsymbol{T}(y) - \nabla A(\boldsymbol{\eta})]^\top \\
&\qquad - \nabla^2 A(\boldsymbol{\eta}) \Big) \, d\mu(y) \\
&= \mathrm{Cov}_{\boldsymbol{\eta}}[\boldsymbol{T}(Y)] - \nabla^2 A(\boldsymbol{\eta}).
\end{aligned}$$

(iii) The moment-generating function of $q_{\boldsymbol{\eta}}$ is given by

$$\begin{aligned}
M_{\boldsymbol{\eta}}(\boldsymbol{u}) &= \mathbb{E}_{\boldsymbol{\eta}}[e^{\boldsymbol{u}^\top \boldsymbol{T}}] \\
&= \int_{\mathcal{Y}} h(y) \exp\left\{\sum_{k=1}^m \eta_k T_k(y) - A(\boldsymbol{\eta})\right\} e^{\sum_{k=1}^m u_k T_k(y)} \, d\mu(y) \\
&= e^{A(\boldsymbol{\eta}+\boldsymbol{u}) - A(\boldsymbol{\eta})} \int_{\mathcal{Y}} h(y) \exp\left\{\sum_{k=1}^m (\eta_k + u_k) T_k(y) - A(\boldsymbol{\eta} + \boldsymbol{u})\right\} d\mu(y) \\
&= e^{A(\boldsymbol{\eta}+\boldsymbol{u}) - A(\boldsymbol{\eta})}.
\end{aligned}$$

The claim follows by taking the logarithm of both sides.

(iv) The KL divergence between q_η and $q_{\eta'}$ is given by

$$D(q_\eta \| q_{\eta'}) = \mathbb{E}_\eta \left[\ln \frac{q_\eta(Y)}{q_{\eta'}(Y)} \right] = \mathbb{E}_\theta[(\boldsymbol{\eta} - \boldsymbol{\eta}')^\top \boldsymbol{T}(Y) - A(\boldsymbol{\eta}) + A(\boldsymbol{\eta}')]$$
$$= (\boldsymbol{\eta} - \boldsymbol{\eta}')^\top \nabla A(\boldsymbol{\eta}) - A(\boldsymbol{\eta}) + A(\boldsymbol{\eta}'). \qquad \square$$

Mean-Value Parameterization It will be convenient to view

$$\boldsymbol{\xi} = \boldsymbol{g}(\boldsymbol{\eta}) \triangleq \nabla_\eta A(\boldsymbol{\eta}) = \mathbb{E}_\eta[\boldsymbol{T}(Y)] \tag{11.32}$$

as a transformation of the vector $\boldsymbol{\eta}$. In view of the last equality in (11.32), $\boldsymbol{\xi}$ is also called the *mean-value parameterization* of the exponential family [2, p. 126].

i.i.d. Observations If $\{Y_i\}_{i=1}^n$ are drawn i.i.d. from p_θ of (11.25), then the pdf for the observation sequence $\boldsymbol{Y} = \{Y_i\}_{i=1}^n$ is given by

$$p_\theta^{(n)}(\boldsymbol{y}) = \prod_{i=1}^n p_\theta(y_i)$$
$$= \left(\prod_{i=1}^n h(y_i) \right) [C(\boldsymbol{\theta})]^n \exp \left\{ \sum_{k=1}^m \eta_k(\boldsymbol{\theta}) \sum_{i=1}^n T_k(y_i) \right\}, \tag{11.33}$$

which is again an exponential family.

11.8.2 Conjugate Priors

For a given distribution $p_\theta(y)$, there often exist several prior distributions $\pi(\boldsymbol{\theta})$ that are "well matched" to $p_\theta(y)$ in the sense that the posterior distribution $\pi(\boldsymbol{\theta}|y)$ is tractable. For instance, if both $p_\theta(y)$ and $\pi(\boldsymbol{\theta})$ are Gaussian, then so is $\pi(\boldsymbol{\theta}|y)$.

More generally, a similar concept holds when p_θ is taken from an arbitrary exponential family [3]. Consider the canonical family $\{q_\eta\}$ of (11.26) and the following *conjugate prior*, which belongs to a $(m + 1)$ dimensional exponential family parameterized by $c > 0$ and $\boldsymbol{\mu} \in \mathbb{R}^m$:

$$\tilde{\pi}(\boldsymbol{\eta}) = \tilde{h}(c, \boldsymbol{\mu}) \exp\{c\boldsymbol{\eta}^\top \boldsymbol{\mu} - cA(\boldsymbol{\eta})\}. \tag{11.34}$$

Assuming the mapping $\boldsymbol{\eta}(\boldsymbol{\theta})$ is invertible, the prior on $\boldsymbol{\eta}$ is related to the prior on π as follows:

$$\tilde{\pi}(\boldsymbol{\eta}) = \frac{\pi(\boldsymbol{\theta})}{|\mathsf{J}(\boldsymbol{\theta})|}, \tag{11.35}$$

where the $m \times m$ Jacobian matrix $\mathsf{J}(\boldsymbol{\theta})$ has components given by $\mathsf{J}_{kl} = \partial \eta_k(\boldsymbol{\theta})/\partial \theta_l$ for $1 \le k, l \le m$.

Consider again $\boldsymbol{Y} = (Y_1, \dots, Y_n)$, drawn i.i.d. from (11.26). Then

$$q_\eta(\boldsymbol{y})\tilde{\pi}(\boldsymbol{\eta}) = \left(\prod_{i=1}^n h(y_i) \right) \tilde{h}(c, \boldsymbol{\mu}) \exp \left\{ \boldsymbol{\eta}^\top \left(\sum_{i=1}^n \boldsymbol{T}(y_i) + c\boldsymbol{\mu} \right) - (n + c)A(\boldsymbol{\eta}) \right\}.$$

Therefore the posterior distribution of $\boldsymbol{\eta}$ given $Y = y$ is given by

$$\begin{aligned}\tilde{\pi}(\boldsymbol{\eta}|\boldsymbol{y}) &\propto \exp\left\{\boldsymbol{\eta}^\top\left(\sum_{i=1}^n \boldsymbol{T}(y_i) + c\boldsymbol{\mu}\right) - (n+c)A(\boldsymbol{\eta})\right\} \\ &= \exp\left\{c'\boldsymbol{\eta}^\top\boldsymbol{\mu}' - c'A(\boldsymbol{\eta})\right\}\end{aligned}$$

and belongs to the same exponential family as the prior (11.34), with parameters

$$c' = n + c \quad \text{and} \quad \boldsymbol{\mu}' = \frac{1}{n+c}\left(\sum_{i=1}^n \boldsymbol{T}(y_i) + c\boldsymbol{\mu}\right).$$

This greatly facilitates inference on $\boldsymbol{\eta}$ or some transformations of $\boldsymbol{\eta}$. In particular, the MAP estimator $\hat{\boldsymbol{\eta}}_{\text{MAP}}$ is the solution to

$$0 = \nabla \ln \tilde{\pi}(\boldsymbol{\eta}|\boldsymbol{y})\,|_{\boldsymbol{\eta}=\hat{\boldsymbol{\eta}}_{\text{MAP}}},$$

hence

$$\nabla A(\hat{\boldsymbol{\eta}}_{\text{MAP}}) = \frac{1}{n+c}\left(\sum_{i=1}^n \boldsymbol{T}(y_i) + c\boldsymbol{\mu}\right).$$

Inversion of the mapping ∇A yields $\hat{\boldsymbol{\eta}}_{\text{MAP}}$. Note that the ML estimator (see Chapter 14) is similarly obtained by maximizing the log-likelihood (11.33). Hence $\hat{\boldsymbol{\eta}}_{\text{ML}}$ is the solution to

$$0 = \nabla \ln q_{\boldsymbol{\eta}}^{(n)}(\boldsymbol{y})\Big|_{\boldsymbol{\eta}=\hat{\boldsymbol{\eta}}_{\text{ML}}},$$

which is[2]

$$\hat{\boldsymbol{\xi}}_{\text{ML}} = \nabla A(\hat{\boldsymbol{\eta}}_{\text{ML}}) = \frac{1}{n}\sum_{i=1}^n \boldsymbol{T}(y_i). \tag{11.36}$$

Observe the remarkable similarity between the ML and MAP estimators. It is convenient to think of c as a *number of pseudo-observations*, and of $\boldsymbol{\mu}$ as a prior mean for $\boldsymbol{T}(Y)$.

Conditional Expectation Similarly to Proposition 11.1(i), one can show

$$\mathbb{E}[\boldsymbol{\xi}] = \mathbb{E}[\nabla A(\boldsymbol{\eta})] = \boldsymbol{\mu}, \tag{11.37}$$

where the expectation is taken with respect to the prior $\tilde{\pi}(\boldsymbol{\eta})$. Similarly, for the problem with i.i.d. data $\boldsymbol{Y} = (Y_1, \dots, Y_n)$, we obtain

$$\hat{\boldsymbol{\xi}}_{\text{MMSE}}(\boldsymbol{y}) = \mathbb{E}[\boldsymbol{\xi}|\boldsymbol{y}] = \mathbb{E}[\nabla A(\boldsymbol{\eta})|\boldsymbol{y}] = \frac{1}{n+c}\left(\sum_{i=1}^n \boldsymbol{T}(y_i) + c\boldsymbol{\mu}\right) = \hat{\boldsymbol{\xi}}_{\text{MAP}}(\boldsymbol{y}). \tag{11.38}$$

Example 11.7 *Precision Estimation with Gamma Prior.* Consider $\boldsymbol{Y} = (Y_1, \dots, Y_n)$ drawn i.i.d. from the Gaussian distribution $\mathcal{N}(0, \sigma^2)$. Let $\eta = 1/\sigma^2$ (called precision parameter), then

[2] The first equality in (11.36) will be shown in Proposition 14.1.

$$q_\eta(y) = (2\pi)^{-1/2}\eta^{1/2}\exp\left\{-\frac{1}{2}\eta y^2\right\}$$

is in canonical form with $T(y) = -\frac{1}{2}y^2$ and $A(\eta) = -\frac{1}{2}\ln\eta$. The n-fold product distribution is given by

$$q_\eta(\mathbf{y}) = (2\pi)^{-n/2}\eta^{n/2}\exp\left\{-\frac{1}{2}\eta\sum_{i=1}^{n} y_i^2\right\}.$$

Consider the $\Gamma(a, b)$ prior for η:

$$\tilde{\pi}(\eta) = \frac{1}{\Gamma(a)b^a}\eta^{a-1}e^{-\eta/b}, \quad a, b, \eta > 0,$$

whose mode is at $b(a-1)$. Hence

$$\tilde{\pi}(\eta|\mathbf{y}) \propto \eta^{n/2+a-1}\exp\left\{-\eta\left(\frac{1}{b} + \frac{1}{2}\sum_{i=1}^{n} y_i^2\right)\right\}$$

is a $\Gamma(a', b')$ distribution with

$$a' = \frac{n}{2} + a \quad \text{and} \quad \frac{1}{b'} = \frac{1}{b} + \frac{1}{2}\sum_{i=1}^{n} y_i^2.$$

The MAP estimator of η given $\mathbf{y}$ is the mode of $\Gamma(a', b')$, i.e.,

$$\hat{\eta}_{\text{MAP}} = b'(a'-1) = \frac{2a - 2 + n}{\frac{2}{b} + \sum_{i=1}^{n} y_i^2}.$$

For this problem one can also derive (see Exercise 11.12)

$$\hat{\sigma}^2_{\text{MAP}} = \frac{\frac{1}{b} + \sum_{i=1}^{n} y_i^2}{2a + 2 + n}. \tag{11.39}$$

Observe that $\hat{\sigma}^2_{\text{MAP}} \neq 1/\hat{\eta}_{\text{MAP}}$, i.e., MAP estimation generally does not commute with nonlinear reparameterization.

Example 11.8 *Multinomial Distribution and Dirichlet Prior.* This model is frequently used in machine learning. Let $Y = (Y_1, \dots, Y_m)$ be a multinomial distribution with m categories, n trials, and probability vector $\boldsymbol{\theta}$; then

$$p_{\boldsymbol{\theta}}(y) = \frac{n!}{\prod_{i=1}^{m} y_i!}\prod_{i=1}^{m}\theta_i^{y_i}. \tag{11.40}$$

This is the pmf that results from drawing a category n times i.i.d. from the distribution $\boldsymbol{\theta}$, and recording the count for each category. Now assume that $\boldsymbol{\theta}$ was drawn from a *Dirichlet* probability distribution

$$\pi(\boldsymbol{\theta}) = \frac{\Gamma(\sum_{i=1}^{m}\alpha_i)}{\prod_{i=1}^{m}\Gamma(\alpha_i)}\prod_{i=1}^{m}\theta_i^{\alpha_i - 1}, \tag{11.41}$$

where $\Gamma(x) = \int_0^\infty e^{-t} t^{x-1}\, dt$, the vector $\boldsymbol{\alpha} = (\alpha_1, \ldots, \alpha_m)$ has positive components, $\theta_m = 1 - \sum_{i=1}^{m-1} \theta_i$, and ν is the Lebesgue measure on $\mathbb{R}^{m-1}$. For $m = 2$, (11.40) and (11.41) respectively reduce to the binomial and the beta distributions.

Note that π of (11.41) is the uniform distribution if $\boldsymbol{\alpha} = \mathbf{1}$ (all-ones vector), and π concentrates near the vertices of the probability simplex if α_i are large. The distribution of (11.41) is denoted by Dir$(m, \boldsymbol{\alpha})$. Its mean is equal to $(\sum_{i=1}^m \alpha_i)^{-1}\boldsymbol{\alpha}$ and its mode is at $(\sum_{i=1}^m \alpha_i - m)^{-1}(\boldsymbol{\alpha} - \mathbf{1})$ when $\alpha_i > 1$ for all i. The posterior distribution of $\boldsymbol{\theta}$ given Y is

$$\pi(\boldsymbol{\theta}|y) \propto p_{\boldsymbol{\theta}}(y)\pi(\boldsymbol{\theta}) \propto \prod_{i=1}^m \theta_i^{y_i+\alpha_i-1}$$

hence is Dir$(m, y + \boldsymbol{\alpha})$. We conclude that the Dirichlet distribution is the conjugate prior for the multinomial distribution.

The MMSE and MAP estimators of $\boldsymbol{\theta}$ are respectively the mean and the mode of the posterior distribution:

$$\hat{\theta}_{\text{MMSE,i}}(y) = \mathbb{E}[\Theta_i|Y = y] = \frac{y_i + \alpha_i}{n + \sum_{i=1}^m \alpha_i}$$

and

$$\hat{\theta}_{\text{MAP,i}}(y) = \frac{y_i + \alpha_i - 1}{n + \sum_{i=1}^m \alpha_i - m}, \qquad 1 \le i \le m.$$

If the prior is uniform ($\alpha_i \equiv 1$), the MAP estimator coincides with the ML estimator,

$$\hat{\theta}_{\text{ML,i}}(y) = \frac{y_i}{n}, \qquad 1 \le i \le m.$$

Otherwise $\{\alpha_i - 1\}_{i=1}^m$ may be viewed as *pseudocounts* that are added to the actual counts $\{y_i\}_{i=1}^m$ to obtain the MAP estimator of $\boldsymbol{\theta}$.

Exercises

11.1 *Gaussian Parameter Estimation.* Let $\theta \sim \mathcal{N}(0, 1)$, $Z_1 \sim \mathcal{N}(0, 1)$, and $Z_2 \sim \mathcal{N}(0, 2)$ be independent Gaussian variables. Find the MMSE estimator of θ given the observations

$$Y_1 = \theta + Z_1 + Z_2 \quad \text{and} \quad Y_2 = -\theta - Z_1 + Z_2.$$

11.2 *Bayesian Parameter Estimation.* Suppose Θ is a random parameter with exponential prior distribution

$$\pi(\theta) = e^{-\theta}\, \mathbb{1}\{\theta \ge 0\}$$

and suppose that given $\Theta = \theta$, the observation Y has a Rayleigh distribution with

$$p_\theta(y) = 2\theta y\, e^{-\theta y^2}\, \mathbb{1}\{y \ge 0\}.$$

(a) Find the MAP estimate of Θ given Y.
(b) Find the MMSE estimate of Θ given Y.

11.3 *Bayesian Parameter Estimation.* Suppose Λ is a random parameter with prior given by the Gamma density

$$\pi(\lambda) = \frac{e^{-\lambda}\,\lambda^{\alpha-1}}{\Gamma(\alpha)}\,\mathbb{1}\{\lambda \geq 0\},$$

where α is a known positive real number, and Γ is the Gamma function defined by the integral

$$\Gamma(x) = \int_0^\infty t^{x-1}\,e^{-t}\,dt\,, \quad \text{for } x > 0\,.$$

Our observation Y is Poisson with rate Λ, i.e.,

$$p_\lambda(y) = \mathrm{P}(\{Y = y\} \mid \{\Lambda = \lambda\}) = \frac{\lambda^y\,e^{-\lambda}}{y!}\,, \qquad y = 0, 1, 2, \ldots\,.$$

(a) Find the MAP estimate of Λ given Y.
(b) Now suppose we wish to estimate a parameter Θ that is related to Λ as

$$\Theta = e^{-\Lambda}.$$

Find the MMSE estimate of Θ given Y.

11.4 *MAP/MMSE/MMAE Estimation for Exponential Distribution and Uniform Prior.* Suppose Θ is uniformly distributed on $[0, 1]$ and that we observe

$$Y = \Theta + Z,$$

where Z is a random variable, independent of Θ, with exponential pdf

$$p_Z(z) = e^{-z}\mathbb{1}\{z \geq 0\}.$$

Find $\hat{\theta}_{\mathrm{MMSE}}(y)$, $\hat{\theta}_{\mathrm{MMAE}}(y)$, and $\hat{\theta}_{\mathrm{MAP}}(y)$.

11.5 *MMSE/MAP Estimation for Exponential Observations and Prior.* Let $Y_1, \ldots, Y_n$ be i.i.d. observations drawn from an exponential distribution with mean $1/\theta$, i.e.,

$$p_\theta(y_k) = \theta e^{-\theta y_k}\,\mathbb{1}\{y_k \geq 0\}.$$

The parameter θ is modeled as a unit exponential random variable, i.e.,

$$\pi(\theta) = e^{-\theta}\,\mathbb{1}\{\theta \geq 0\}.$$

Find the MMSE and MAP estimates of θ based on the n observations.

11.6 *MAP Estimator for Uniform Distribution and Uniform Prior.* Let $Y_1, \ldots, Y_n$ be i.i.d. observations drawn from a uniform distribution over $[1, \theta]$, i.e.,

$$p_\theta(y_k) = \frac{1}{\theta - 1}\mathbb{1}\{y_k \in [1, \theta]\}.$$

The prior distribution of θ is given by a uniform distribution over $[1, 10]$, i.e.,

$$\pi(\theta) = \frac{1}{9}\mathbb{1}\{\theta \in [1, 10]\}.$$

(a) Find $\hat{\theta}_{\text{MAP}}(\boldsymbol{y})$.

(b) Show that $\hat{\theta}_{\text{MAP}}(\boldsymbol{y})$ converges in distribution to the true value of θ as $n \to \infty$, i.e., show that

$$\lim_{n\to\infty} \text{P}\{\hat{\theta}_{\text{MAP}}(\boldsymbol{Y}) \le t\} = \text{P}\{\Theta \le t\} \quad \text{for all } t \in \mathbb{R}.$$

11.7 *Sunrise Probability.* In the 18th century, Laplace asked what is the probability that the Sun will rise tomorrow. He assumed that the Earth was formed n days ago, and that the Sun rises each day with probability θ. While the value of θ is unknown, we might model our uncertainty by viewing θ as a random parameter with uniform prior over $[0, 1]$. Under these assumptions, show that the estimate that the Sun will rise tomorrow is $\frac{n+1}{n+2}$.

11.8 *Distributions over Finite Sets.* Show that the set of probability distributions over a finite set $\mathcal{Y} = \{y_1, y_2, \dots, y_{m+1}\}$ forms an m-dimensional exponential family whose sufficient statistics are indicator functions:

$$T_k(y) = \mathbb{1}\{y = y_k\}, \, 1 \le k \le m.$$

Write this family in canonical form.

11.9 *Estimation for Mixed Discrete/Continuous Random Variables.* Let Y take value 0 with probability $1 - \epsilon$ and be $\mathcal{N}(0, \sigma^2)$ otherwise. Define $\boldsymbol{\theta} = (\epsilon, \sigma^2)$ where $0 < \epsilon < 1$ and $\sigma^2 > 0$.

(a) Show that the distribution $\text{P}_{\boldsymbol{\theta}}$ of Y belongs to a two-dimensional exponential family with respect to the measure μ which is the sum of the Lebesgue measure and the measure that puts a mass of 1 at $y = 0$; more explicitly, $\mu(\mathcal{A}) = |\mathcal{A}| + \mathbb{1}\{0 \in \mathcal{A}\}$ for any (open, closed, semiinfinite) interval $\mathcal{A}$ of the real line.

(b) Assume we are given observations $\{Y_i\}_{i=1}^n \sim$ i.i.d. $P_{\boldsymbol{\theta}}$ and assume a prior $\pi(\epsilon, \sigma^2) = e^{-\sigma^2}$ over $(0, 1) \times \mathbb{R}^+$. Derive the MAP estimator of ϵ and σ^2.

11.10 *Beta Distribution as Conjugate Prior.* Assume Y is a binomial random variable with n trials and parameter $\theta \in (0, 1)$:

$$p_\theta(y) = \frac{n!}{y! \, (n-y)!} \theta^y (1-\theta)^{n-y}, \quad y = 0, 1, \dots, n.$$

(a) Suppose $C(a, \theta) = (a - \theta)^2$. Let the prior distribution of Θ be the beta distribution $\beta(a, b)$ with density

$$\pi(\theta) = \frac{\Gamma(a+b)}{\Gamma(a)\,\Gamma(b)} \theta^{a-1} (1-\theta)^{b-1}, \quad 0 < \theta < 1,$$

where $a > 0$ and $b > 0$. Show that the posterior distribution of Θ given $Y = y$ is $\beta(a + y, b + n - y)$, and obtain the Bayes estimator.

(b) Repeat when the prior distribution of Θ is the uniform distribution on $[0, 1]$.

(c) Now, let $C(a, \theta) = \frac{(a-\theta)^2}{\theta(1-\theta)}$. Find the Bayes rule with respect to the uniform distribution on $[0, 1]$.

11.11 *Expectations in Conjugate Prior Family.* Prove (11.37) and (11.38).

11.12 *MAP Variance Estimation*. Derive the MAP estimator (11.39).
Hint: First express the prior for σ^2 in terms of the prior for $\eta = 1/(2\sigma^2)$, then evaluate the derivative of the log posterior density with respect to σ^2, and set it to zero.

References

[1] B. Hajek, *Random Processes for Engineers*, Cambridge University Press, 2015.
[2] E. L. Lehmann and G. Casella, *Theory of Point Estimation*, Springer, 1998.
[3] P. Diaconis and D. Ylvisaker, "Conjugate Priors for Exponential Families," *Ann. Stat.*, pp. 269–281, 1979.

12 Minimum Variance Unbiased Estimation

In this chapter we introduce several methods for constructing good estimators when prior probabilistic models are not available for the unknown parameter. First we define the notions of unbiasedness and minimum-variance unbiased estimators (MVUE). We then present the concept of sufficient statistics, followed by the factorization theorem and the Rao–Blackwell theorem. Parameter estimation for complete families of distributions and exponential families is studied in detail.

12.1 Nonrandom Parameter Estimation

In Chapter 11, we considered the Bayesian approach to the parameter estimation problem, where we assumed that the parameter $\theta \in \mathcal{X}$ is a realization of a random parameter Θ with prior $\pi(\theta)$. Now we consider the situation where θ is unknown but cannot be modeled as random.

The conditional risk of estimator $\hat{\theta}$ is given by

$$R_\theta(\hat{\theta}) = \mathbb{E}_\theta\left[C(\hat{\theta}(Y), \theta)\right], \quad \theta \in \mathcal{X}.$$

We could try the minimax approach (as we did in the case of hypothesis testing):

$$\hat{\theta}_{\mathrm{m}} \triangleq \arg\min_{\hat{\theta} \in \mathcal{D}} \max_{\theta \in \mathcal{X}} R_\theta(\hat{\theta}). \tag{12.1}$$

However, the minimax estimation rule is often hard to compute in practice. It is also often too conservative for many applications. We would like to explore estimators that are simple to compute and less conservative.

Consider first the case where θ is a scalar parameter, and we use the squared-error cost function $C(a, \theta) = (a - \theta)^2$. Then

$$R_\theta(\hat{\theta}) = \mathbb{E}_\theta\left[(\hat{\theta}(Y) - \theta)^2\right] \tag{12.2}$$

is the mean squared-error (MSE). Unfortunately it is generally difficult to construct an estimator that is good in terms of $R_\theta(\hat{\theta})$ *for all* $\theta \in \mathcal{X}$, let alone the "best" estimator. For instance the trivial estimator $\hat{\theta}$ that returns a fixed value θ_0 for all the observations has minimum squared-error risk (zero) if θ happens to be equal to θ_0. This is a silly estimator which performs poorly if the true θ is far from θ_0. However, this example shows one cannot expect to have a *uniformly best* estimator since the trivial estimator cannot be improved upon at $\theta = \theta_0$. This concept was already apparent in simple binary hypothesis testing, where a trivial detector that returns the same decision (say H_0) irrespective of the observations has zero conditional risk under H_0.

To avoid such trivialities, it is usually preferable to design an estimator that does not strongly favor one or more values of θ at the expense of others. We introduce two important properties that help us design good estimators.

DEFINITION 12.1 An estimator $\hat{\theta}$ is said to be unbiased if

$$\mathbb{E}_\theta(\hat{\theta}) = \theta, \quad \forall \theta \in \mathcal{X}. \tag{12.3}$$

Note that for unbiased estimators the squared-error risk $R_\theta(\hat{\theta})$ defined in (12.2) is the same as the variance $\mathrm{Var}_\theta[\hat{\theta}(Y)]$. This leads to the following definition.

DEFINITION 12.2 $\hat{\theta}$ is said to be a minimum-variance unbiased estimator (MVUE) if:

1. $\hat{\theta}$ is unbiased.
2. If $\tilde{\theta}$ is another unbiased estimator, then

$$\mathrm{Var}_\theta[\hat{\theta}(Y)] \le \mathrm{Var}_\theta[\tilde{\theta}(Y)], \quad \forall \theta \in \mathcal{X}.$$

An MVUE is a "good" estimator and is optimal in the sense that it has minimum variance among all unbiased estimators over all possible θ. Note that the existence of an MVUE is generally not guaranteed – in some problems, even an unbiased estimator does not exist.

Also note that an MVUE is not the same as an MMSE. There could be biased estimators that have lower MSE than MVUE estimators (see Section 13.7).

Thus far in this chapter we have assumed that θ is a scalar. The definition of an MVUE extends to the case where θ is a vector, i.e., $\mathcal{X} \subseteq \mathbb{R}^m$, for some $m > 1$, if instead of estimating θ, we estimate a scalar function $g(\theta)$. If we are interested in estimating the individual components of the vector θ, then we can allow the function to equal each of the components and estimate them separately. We have the following definition in this more general setting.

DEFINITION 12.3 $\hat{g}(Y)$ is said to be an MVUE of $g(\theta)$ if:

1. $\mathbb{E}_\theta[\hat{g}(Y)] = g(\theta)$ for all $\theta \in \mathcal{X}$ (unbiasedness).
2. If $\tilde{g}(Y)$ is another unbiased estimator, then

$$\mathrm{Var}_\theta[\hat{g}(Y)] \le \mathrm{Var}_\theta[\tilde{g}(Y)], \quad \forall \theta \in \mathcal{X}.$$

Our goal in the remainder of this chapter is to develop a procedure to find MVUE solutions, when they exist. The first step in this procedure requires the concept of a sufficient statistic.

12.2 Sufficient Statistics

DEFINITION 12.4 A sufficient statistic $T(Y)$ is any function of the data such that the conditional distribution of Y given $T(Y)$ is independent of θ.

Informally a sufficient statistic *compresses* the observations without losing any information that is relevant for estimating θ.

Example 12.1 *Empirical Frequency*. Let $\mathcal{Y} = \{0, 1\}^n$. We estimate the parameter $\theta \in [0, 1]$ of a Bernoulli distribution from n i.i.d. observations $Y_i \sim$ i.i.d. $\text{Be}(\theta)$, $1 \le i \le n$.

The pmf of $\boldsymbol{Y}$ given θ is

$$p_\theta(\boldsymbol{y}) = \prod_{i=1}^{n} p_\theta(y_i) = \theta^{n_1}(1-\theta)^{n-n_1}$$

where $n_1 = \sum_{i=1}^{n} \mathbb{1}\{y_i = 1\}$ is the number of 1's that occur in the sequence $\boldsymbol{y}$.
Claim: $T(\boldsymbol{y}) \triangleq n_1$ is a sufficient statistic for θ. Each sequence with n_1 1's and $n - n_1$ 0's has equal probability and the additional knowledge of θ does not change this probability. Specifically,

$$p_\theta(\boldsymbol{y}|n_1) = \text{P}_\theta\{\boldsymbol{Y} = \boldsymbol{y}|N_1 = n_1\} = 1/\binom{n}{n_1}$$

for all such sequences $\boldsymbol{y}$ and all $\theta \in [0, 1]$. Therefore $\boldsymbol{Y}$ is independent of θ conditioned on $T(\boldsymbol{Y})$.

Note that $T(\boldsymbol{Y})$ is not a unique sufficient statistic. Assume $n \ge 2$ and consider

$$n_{11} = \sum_{i=1}^{\lfloor n/2 \rfloor} \mathbb{1}\{Y_k = 1\} \quad \text{and} \quad n_{12} = \sum_{k=\lfloor n/2 \rfloor + 1}^{n} \mathbb{1}\{Y_k = 1\}.$$

Then $T'(\boldsymbol{y}) \triangleq \begin{pmatrix} n_{11} \\ n_{12} \end{pmatrix}$ is also a sufficient statistic for θ.

Example 12.2 *Trivial Sufficient Statistic*. Here the parameter is a vector $\boldsymbol{\theta} = [\theta_1, \dots, \theta_n] \in [0, 1]^n$, $Y_i \sim$ independent $\text{Be}(\theta_i)$, $1 \le i \le n$. In this example the Y_i's are independent but not identically distributed, and n_1, as defined in Example 12.1, is no longer a sufficient statistic. However, $T(\boldsymbol{y}) = \boldsymbol{y}$ is a *trivial sufficient statistic* since $\boldsymbol{Y}$ conditioned on $\boldsymbol{Y}$ is trivially independent of $\boldsymbol{\theta}$. A trivial sufficient statistic does not compress the information content available in the observations. This motivates Definition 12.5.

Example 12.3 *Largest Observation*. Here we revisit Example 11.5 but drop the assumption that θ is random. Let p_θ be the uniform distribution over $[0, \theta]$ where $\theta \in \mathcal{X} = \mathbb{R}^+$. Let Y_i, $1 \le i \le n$ be i.i.d. observations drawn from p_θ. Hence

$$p_\theta(\boldsymbol{y}) = \theta^{-n} \prod_{i=1}^{n} \mathbb{1}\{0 \le y_i \le \theta\} = \theta^{-n} \mathbb{1}\left\{\max_{1 \le i \le n} y_i \le \theta\right\}$$

hence $T(\boldsymbol{Y}) \triangleq \max_{1 \le i \le n} Y_i$ is a sufficient statistic for θ, as can be shown by verifying that the distribution of $\boldsymbol{Y}$ conditioned on $T(\boldsymbol{Y})$ is independent of θ, or by applying the factorization theorem in the next section.

DEFINITION 12.5 $T(Y)$ is a minimal sufficient statistic if $T(Y)$ is a function of any other sufficient statistic.

In Example 12.1, $T'(\boldsymbol{Y})$ is not a minimal sufficient statistic because it is not a function of $T(\boldsymbol{Y})$.

12.3 Factorization Theorem

THEOREM 12.1 *$T(Y)$ is a sufficient statistic if and only if there exist functions f_θ and h such that*

$$p_\theta(y) = f_\theta(T(y))\, h(y), \quad \forall \theta, y. \tag{12.4}$$

Return to Example 12.1, where $p_\theta(\boldsymbol{y}) = \theta^{n_1}(1-\theta)^{n-n_1}$. We can select $T(\boldsymbol{y}) = n_1$, choose the functions $h(\boldsymbol{y}) = 1$ and $f_\theta(t) = \theta^t(1-\theta)^{n-t}$, and apply the factorization theorem to verify that $T(\boldsymbol{y})$ is a sufficient statistic.

Example 12.4 Let Y_i be i.i.d. $\mathcal{N}(\theta, 1)$, $1 \le i \le n$, and $\theta \in \mathbb{R}$. Then

$$\begin{aligned} p_\theta(\boldsymbol{y}) = \prod_{i=1}^n p_\theta(y_i) &= (2\pi)^{-n/2} \exp\left\{-\frac{1}{2}\sum_{i=1}^n (y_i - \theta)^2\right\} \\ &= (2\pi)^{-n/2} \exp\left\{-\frac{1}{2}\sum_{i=1}^n y_i^2 + \theta \sum_{i=1}^n y_i - \frac{n\theta^2}{2}\right\}. \end{aligned}$$

Now let $T(\boldsymbol{y}) = \sum_{i=1}^n y_i$ and select the functions

$$h(\boldsymbol{y}) = (2\pi)^{-n/2} e^{-\frac{1}{2}\sum_{i=1}^n y_i^2}, \quad f_\theta(t) = e^{\theta t - n\theta^2/2}$$

and apply the factorization theorem to see that $T(\boldsymbol{y})$ is a sufficient statistic for θ.

Example 12.5 m-ary hypothesis testing. Let $\theta \in \mathcal{X} = \{0, 1, \dots, m-1\}$. Write

$$p_\theta(y) = \frac{p_\theta(y)}{p_0(y)}\, p_0(y), \quad \theta = 0, 1, \dots, m-1.$$

We obtain the desired factorization of (12.4) by choosing $T(y)$ as the vector of $m-1$ likelihood ratios $\{p_i(y)/p_0(y),\ 1 \le i \le m-1\}$ and letting

$$h(y) = p_0(y), \quad f_\theta(t) = \begin{cases} 1 & : \theta = 0 \\ t_\theta, & : \theta = 1, 2, \dots, m-1. \end{cases}$$

Proof of Theorem 12.1 We prove the result for discrete $\mathcal{Y}$ (see [1] for the general case).

(i) We first prove the "only if" condition. Assume $T(Y)$ is a sufficient statistic for θ. Then

$$\begin{aligned} p_\theta(y) = P_\theta\{Y = y\} &= P_\theta\{Y = y, T(Y) = T(y)\} \\ &= P_\theta\{Y = y|T(Y) = T(y)\}\ P_\theta\{T(Y) = T(y)\} \\ &= h(y)\ f_\theta(T(y)) \end{aligned}$$

with $h(y) \triangleq P_\theta\{Y = y|T(Y) = T(y)\}$ (by sufficiency) and $f_\theta(T(y)) \triangleq P_\theta\{T(Y) = T(y)\}$.

(ii) We now prove the "if" condition. Assume there exist functions f_θ, T, and h such that the factorization (12.4) holds. By Bayes' rule, we have

$$\begin{aligned} p_\theta(y|t) &\triangleq P_\theta\{Y = y|T(Y) = t\} \\ &= \frac{P_\theta\{T(Y) = t|Y = y\} p_\theta(y)}{P_\theta\{T(Y) = t\}}. \end{aligned} \tag{12.5}$$

Since $T(Y)$ is a function of Y, we have

$$P_\theta\{T(Y) = t|Y = y\} = \mathbb{1}\{T(y) = t\}. \tag{12.6}$$

Substituting (12.4) and (12.6) into (12.5), we obtain

$$p_\theta(y|t) = \frac{f_\theta(t)\, h(y)}{P_\theta\{T(Y) = t\}} \mathbb{1}\{T(y) = t\}. \tag{12.7}$$

Now

$$\begin{aligned} P_\theta\{T(Y) = t\} &= \sum_{y:T(y)=t} p_\theta(y) \\ &= \sum_{y:T(y)=t} f_\theta(T(y))h(y) \\ &= f_\theta(t) \sum_{y:T(y)=t} h(y). \end{aligned} \tag{12.8}$$

Substituting into (12.7), we obtain

$$p_\theta(y|t) = \frac{h(y)}{\sum_{y:T(y)=t} h(y)} \mathbb{1}\{T(y) = t\}, \tag{12.9}$$

which does not depend on θ, and therefore $T(Y)$ is a sufficient statistic. □

12.4 Rao–Blackwell Theorem

The Rao–Blackwell theorem [2, 3] provides a general framework for reducing the variance of unbiased estimators. Given a sufficient statistic $T(Y)$, any unbiased estimator can be potentially improved by averaging it with respect to the conditional distribution of Y given $T(Y)$. This procedure is known as *Rao–Blackwellization*.

THEOREM 12.2 *Assume $\hat{g}(Y)$ is an unbiased estimator of $g(\theta)$ and $T(y)$ is a sufficient statistic for θ, then the estimator*

$$\tilde{g}(T(Y)) \triangleq \mathbb{E}[\hat{g}(Y)|T(Y)] \tag{12.10}$$

satisfies:

(i) *$\tilde{g}(T(Y))$ is also an unbiased estimator of $g(\theta)$;*
(ii) *$\mathrm{Var}_\theta[\tilde{g}(T(Y))] \le \mathrm{Var}_\theta[\hat{g}(Y)]$ with equality if and only if $P_\theta\{\tilde{g}(T(Y)) = \hat{g}(Y)\} = 1$ for all $\theta \in \mathcal{X}$.*

COROLLARY 12.1 *If $T(Y)$ is a sufficient statistic for θ and $g^*(T(Y))$ is the only function of $T(Y)$ that is an unbiased estimator of $g(\theta)$, then g^* is an MVUE for $g(\theta)$.*

Example 12.6 We use the Rao–Blackwell theorem to design an estimator for thc sctup in Example 12.1 with $Y_i \sim$ i.i.d. Be(θ). Let $g(\theta) = \theta$. Consider $\hat{\theta}(\boldsymbol{Y}) = Y_1$, which is a fairly silly estimator (it uses only one out of n observations and has variance equal to $\theta(1-\theta)$). Nevertheless $\hat{\theta}$ is unbiased and can be improved dramatically using Rao–Blackwellization. Recall the sufficient statistic $T(\boldsymbol{Y}) = N_1 = \sum_{i=1}^n \mathbb{1}\{Y_i = 1\}$. Using Rao–Blackwellization we construct the estimator

$$\begin{aligned}\tilde{\theta}(N_1) &= \mathbb{E}[\hat{\theta}(Y)|N_1]\\ &= \mathbb{E}[Y_1|N_1]\\ &= \mathbb{E}[Y_i|N_1], \quad 1 \le i \le n,\end{aligned}$$

where the last line holds by symmetry. Averaging over all $1 \le i \le n$, we obtain

$$\begin{aligned}\tilde{\theta}(N_1) &= \frac{1}{n}\sum_{i=1}^n \mathbb{E}[Y_i|N_1]\\ &= \frac{1}{n}\mathbb{E}\left[\sum_{i=1}^n Y_i|N_1\right]\\ &= \frac{N_1}{n},\end{aligned}$$

which has variance equal to $\theta(1-\theta)/n$, i.e., n times lower than the original simple estimator. In fact, we shall soon see that $\tilde{\theta}$ is an MVUE.

Proof of Theorem 12.2

(i) For each $\theta \in \mathcal{X}$, we have

$$\begin{aligned}\mathbb{E}_\theta[\tilde{g}(T(Y))] &\overset{(a)}{=} \mathbb{E}_\theta\mathbb{E}[\hat{g}(Y)|T(Y)]\\ &\overset{(b)}{=} \mathbb{E}_\theta[\hat{g}(Y)]\\ &\overset{(c)}{=} g(\theta),\end{aligned}$$

where (a) holds by definition of $\tilde{g}$, (b) by the law of iterated expectations, and (c) by the unbiasedness of $\hat{g}$.

(ii) For each $\theta \in \mathcal{X}$, we have

$$\begin{aligned}\mathrm{Var}_\theta[\tilde{g}(T(Y))] &= \mathbb{E}_\theta\left[(\tilde{g}(T(Y)))^2\right] - \left[\mathbb{E}_\theta[\tilde{g}(T(Y))]\right]^2 \\ &= \mathbb{E}_\theta\left[(\tilde{g}(T(Y)))^2\right] - g^2(\theta).\end{aligned}$$

Similarly,

$$\mathrm{Var}_\theta[\hat{g}(Y)] = \mathbb{E}_\theta\left[(\hat{g}(Y))^2\right] - g^2(\theta).$$

Now

$$\begin{aligned}\mathbb{E}_\theta\left[(\tilde{g}(T(Y)))^2\right] &= \mathbb{E}_\theta\left[\left(\mathbb{E}[\hat{g}(Y)|T(Y)]\right)^2\right] \\ &\overset{(a)}{\le} \mathbb{E}_\theta\mathbb{E}\left[(\hat{g}(Y))^2|T(Y)\right] \\ &= \mathbb{E}_\theta\left[(\hat{g}(Y))^2\right],\end{aligned}$$

where (a) follows by Jensen's inequality: $\mathbb{E}^2[X] \le \mathbb{E}[X^2]$ for any random variable X, with equality if and only if X is equal to a constant with probability 1. Similarly $\mathbb{E}^2[X|T] \le \mathbb{E}[X^2|T]$, with equality if and only if X is a function of T with probability 1. This proves the claim. □

Proof of Corollary 12.1 Let $\hat{g}(Y)$ be any MVUE for $g(\theta)$. Then $\tilde{g}(T(Y)) = \mathbb{E}[\hat{g}(Y)|T(Y)]$ is unbiased and its variance cannot exceed that of $\hat{g}(Y)$, by the Rao–Blackwell theorem. But $\hat{g}(Y)$ is an MVUE, so equality is achieved in (ii). Since g^* is unique, we must have $g^* = \tilde{g}$, and g^* must be an MVUE. □

12.5 Complete Families of Distributions

DEFINITION 12.6 A family of distributions $\mathcal{P} = \{p_\theta,\ \theta \in \mathcal{X}\}$ is **complete** if for any function $f : \mathcal{Y} \to \mathbb{R}$, we have

$$\mathbb{E}_\theta[f(Y)] = 0 \quad \forall\theta \in \mathcal{X} \quad \Rightarrow \quad P_\theta\{f(Y) = 0\} = 1 \quad \forall\theta \in \mathcal{X}. \tag{12.11}$$

The latter statement may be written more concisely as $f(Y) = 0$ a.s. $\mathcal{P}$.

A geometric interpretation of completeness is the following. Consider finite $\mathcal{Y}$, with $|\mathcal{Y}| = k$, for simplicity and write $\mathbb{E}_\theta[f(Y)] = \sum_{y\in\mathcal{Y}} f(y)p_\theta(y)$ as the dot product $\boldsymbol{f}^\top \boldsymbol{p}_\theta$ of the two k-dimensional vectors $\boldsymbol{f} = \{f(y), y \in \mathcal{Y}\}$ and $\boldsymbol{p}_\theta = \{p_\theta(y), y \in \mathcal{Y}\}$. Condition (12.11) may be written as

$$\boldsymbol{f}^\top \boldsymbol{p}_\theta = 0 \quad \forall\theta \in \mathcal{X} \quad \Rightarrow \quad \boldsymbol{f} = \mathbf{0}.$$

In other words, the only vector that is orthogonal to all the vectors in the set $\{\boldsymbol{p}_\theta, \theta \in \mathcal{X}\}$ is the $\mathbf{0}$ vector.

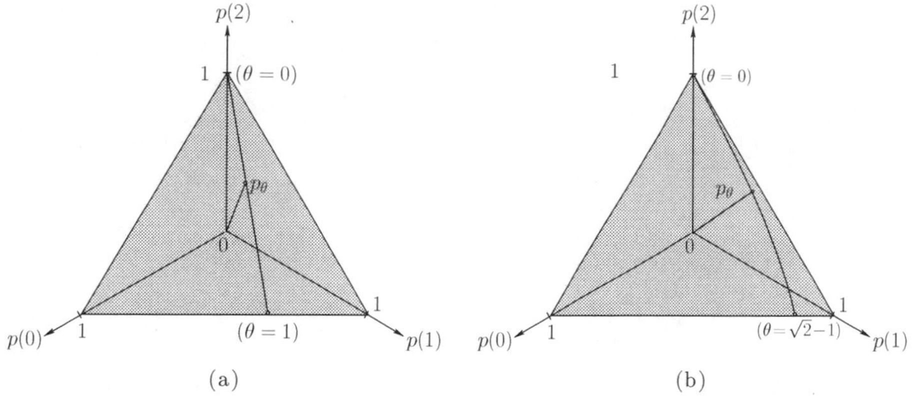

Figure 12.1 (a) Incomplete family of Example 12.7; (b) complete family of Example 12.8

Example 12.7 Let $\mathcal{Y} = \{0, 1, 2\}$ and consider the family

$$p_\theta(y) = \begin{cases} \theta/3 & : y = 0 \\ 2\theta/3 & : y = 1 \\ 1 - \theta & : y = 2, \end{cases} \tag{12.12}$$

where $\theta \in \mathcal{X} = [0, 1]$. The vector $\boldsymbol{f} = (2, -1, 0)$ is orthogonal to each vector $\boldsymbol{p}_\theta$, $\theta \in [0, 1]$. Since $\boldsymbol{f} \neq \mathbf{0}$, the family is not complete. A geometric representation of the problem is given in Figure 12.1, where the family is represented by a straight line in the probability simplex, and therefore there exists a nonzero vector $\boldsymbol{f}$ that is orthogonal to it.

Example 12.8 Let $\mathcal{Y} = \{0, 1, 2\}$ and consider the family

$$p_\theta(y) = \begin{cases} \theta^2 & : y = 0 \\ 2\theta & : y = 1 \\ 1 - \theta^2 - 2\theta & : y = 2, \end{cases} \tag{12.13}$$

where $\theta \in \mathcal{X} = [0, \sqrt{2} - 1]$. Here the family is represented by a curve in the simplex, and there exists no $\boldsymbol{f}$ such that $\boldsymbol{f} \cdot \boldsymbol{p}_\theta = 0, \forall \theta \in \mathcal{X}$, other than the trivial $\boldsymbol{f} = \mathbf{0}$.

The family of binomial distributions is another example of a complete family with discrete observations:

$$Y \sim \text{Bi}(n, \theta) \quad \Rightarrow \quad p_\theta(y) = \binom{n}{y} \theta^y (1 - \theta)^{n-y}, \quad y = 0, 1, \ldots, n, \quad \theta \in [0, 1].$$

We now provide an example of a complete family with continuous observations.

Example 12.9 Let $\mathcal{X} = (0, \infty)$, $\mathcal{Y} = \mathbb{R}$, and consider the family of distributions

$$p_\theta(y) = \theta\, e^{-\theta y} \mathbb{1}\{y \geq 0\}, \quad \theta \in \mathcal{X}.$$

Then

$$\mathbb{E}_\theta[\, f(Y)] = \int_0^\infty f(y)\, \theta\, e^{-\theta y}\, dy.$$

Therefore,

$$\mathbb{E}_\theta[\, f(Y)] = 0 \quad \forall \theta \in \mathcal{X} \quad \Longrightarrow \quad \int_0^\infty f(y)\, e^{-\theta y}\, dy = 0 \quad \forall \theta \in \mathcal{X}$$

which means that the Laplace transform of f equals 0 on the positive real line. Now by using the fact that the Laplace transform is an analytic function, and invoking analytic continuity, we can conclude that $f(y) = 0$ for almost all $y \in \mathcal{Y}$. This establishes the completeness of $\{p_\theta, \theta \in \mathcal{X}\}$.

More general results for exponential families are given in Section 12.5.3.

12.5.1 Link Between Completeness and Sufficiency

THEOREM 12.3 *If $\mathcal{P} = \{p_\theta,\ \theta \in \mathcal{X}\}$ is a complete family and $T(Y)$ is a sufficient statistic, then the mapping T is invertible, i.e., T is a trivial sufficient statistic.*

Proof Let $f(Y) = Y - \mathbb{E}[Y|T(Y)]$. Then

$$\begin{aligned}\mathbb{E}_\theta[\, f(Y)] &= \mathbb{E}_\theta[Y] - \mathbb{E}_\theta\, [\mathbb{E}[Y|T(Y)]] \\ &= \mathbb{E}_\theta[Y] - \mathbb{E}_\theta[Y] = 0, \quad \forall \theta \in \mathcal{X}.\end{aligned}$$

The completeness assumption (12.11) implies that $f(Y) = 0$ a.s. $\mathcal{P}$. By the definition of f, we conclude that

$$\begin{aligned} & Y = \mathbb{E}[Y|T(Y)] \quad \text{a.s. } \mathcal{P} \\ \Rightarrow \quad & Y = \text{a function of } T(Y) \quad \text{a.s. } \mathcal{P}.\end{aligned}$$

Since Y and $T(Y)$ are functions of each other, the mapping T is invertible. □

Theorem 12.3 implies that the completeness of $\mathcal{P}$ is not that useful since that would imply that all sufficient statistics are trivial. We will now see that it is the completeness of the family of distributions that $\mathcal{P}$ *induces* on a sufficient statistic that is more useful as it allows us to identify minimal sufficient statistics.

Denote by

$$q_\theta(t) = \int_{y\,:\,T(y)=t} p_\theta(y)\, d\mu(y) \tag{12.14}$$

the pdf of the sufficient statistic $T(Y)$ given θ.

DEFINITION 12.7 A sufficient statistic $T(Y)$ is complete if the family $\{q_\theta, \theta \in \mathcal{X}\}$ is complete.

THEOREM 12.4 *A complete sufficient statistic $T(Y)$ is always minimal.*

Proof Suppose $T(Y)$ is not minimal. Then there exists a non-invertible function h such that $h(T(Y))$ is a sufficient statistic. Let

$$f(T(Y)) = T(Y) - \mathbb{E}[T(Y)|h(T(Y))].$$

Then it is clear by law of iterated expectations that $\mathbb{E}_\theta[f(T(Y))] = 0$. Therefore, by completeness of $T(Y)$, we conclude that $f(T(Y)) = 0$, which implies that

$$T(Y) = \mathbb{E}[T(Y)|h(T(Y))].$$

Therefore $T(Y)$ is a function of $h(T(Y))$, i.e., h is invertible. This is a contradiction, and so $T(Y)$ must be minimal. □

12.5.2 Link Between Completeness and MVUE

THEOREM 12.5 *If $T(Y)$ is a complete sufficient statistic, and $\tilde{g}(T(Y))$ and $g^*(T(Y))$ are two unbiased estimators of $g(\theta)$, then*

$$\tilde{g}(T(Y)) = g^*(T(Y))$$

is the unique MVUE for $g(\theta)$.

Proof Let $f(T(y)) = \tilde{g}(T(y)) - g^*(T(y))$. Then

$$\begin{aligned}\mathbb{E}_\theta[f(T(Y))] &= \mathbb{E}_\theta[\tilde{g}(T(Y))] - \mathbb{E}_\theta[g^*(T(Y))]\\ &= g(\theta) - g(\theta)\\ &= 0, \quad \forall\theta \in \mathcal{X}.\end{aligned}$$

The completeness assumption implies that $f(T(Y)) = 0$ a.s. $\mathcal{P}$. By the definition of f, we conclude that $\tilde{g}(T(Y)) = g^*(T(Y))$ a.s. $\mathcal{P}$. Thus $g^*(T(Y))$ is the unique unbiased estimator of $g(\theta)$. By the corollary to the Rao–Blackwell theorem, this estimator is an MVUE. □

General Procedure For Finding MVUEs The following procedure can be applied to find an MVUE of $g(\theta)$ whenever an unbiased estimator exists and a complete sufficient statistic can be found:

1. Find some unbiased estimator $\hat{g}(Y)$ of $g(\theta)$.
2. Find a complete sufficient statistic $T(Y)$.
3. Apply Rao–Blackwellization and obtain $g^*(T(Y)) = \mathbb{E}\left[\hat{g}(Y)|T(Y)\right]$, which is an MVUE of $g(\theta)$.

12.5.3 Link Between Completeness and Exponential Families

The notions of sufficiency, completeness, and MVUEs are more straightforward if one deals with an exponential family. Recall the definition (11.26) of a family of m-dimensional exponential distributions in canonical form:

$$p_{\boldsymbol{\theta}}(y) = h(y)\exp\left\{\sum_{k=1}^{m}\theta_k T_k(y) - A(\boldsymbol{\theta})\right\}, \quad y \in \mathcal{Y}.$$

Denote by $q_{\boldsymbol{\theta}}$ the pdf of the m-vector $\boldsymbol{T}(Y) = [T_1(Y) \dots T_m(Y)]^\top$ under $p_{\boldsymbol{\theta}}$.

THEOREM 12.6 Sufficient Statistics for Exponential Family.

(i) *$\boldsymbol{T}(Y)$ is a sufficient statistic for $\boldsymbol{\theta}$.*
(ii) *The family $\{q_{\boldsymbol{\theta}}\}$ is an m-dimensional exponential family.*

Proof

(i) This follows directly from the factorization theorem applied to (11.26).
(ii) Integrating $p_{\boldsymbol{\theta}}$ over the set $\{y : \boldsymbol{T}(y) = \boldsymbol{t}\}$ yields

$$q_{\boldsymbol{\theta}}(\boldsymbol{t}) = h_T(\boldsymbol{t}) \exp\left\{\sum_{k=1}^{m} \theta_k t_k - A(\boldsymbol{\theta})\right\}$$

with

$$h_T(\boldsymbol{t}) = \int_{y:\boldsymbol{T}(y)=\boldsymbol{t}} h(y) d\mu(y).$$ □

THEOREM 12.7 Completeness Theorem for Exponential Families. *Consider the family of distributions defined in* (11.25), *with $\boldsymbol{\eta}$ mapping the parameter set $\mathcal{X}$ to the set $\Gamma \subset \mathbb{R}^m$. Then the vector $\boldsymbol{T}(Y) = [T_1(Y) \dots T_m(Y)]^\top$ is a complete sufficient (hence minimal) statistic for $\{p_{\boldsymbol{\theta}},\ \boldsymbol{\theta} \in \mathcal{X}\}$ if Γ contains an m-dimensional rectangle.*[1]

Proof The proof is a generalization of the argument given in Example 12.9. First of all, by the factorization theorem, $\boldsymbol{T}(Y)$ is a sufficient statistic for $\mathcal{P}$. Using an argument similar to that used in Part (ii) of Theorem 12.6, we see that the joint distribution of $\boldsymbol{T}(Y)$ is given by

$$q_{\boldsymbol{\theta}}(\boldsymbol{t}) = C(\boldsymbol{\theta}) \exp\left\{\sum_{k=1}^{m} \eta_k(\boldsymbol{\theta})\, t_k\right\} h_T(\boldsymbol{t})$$

for an appropriately defined $h_T(\boldsymbol{t})$. Therefore

$$\mathbb{E}_{\boldsymbol{\theta}}\left[f(\boldsymbol{T}(Y))\right] = C(\boldsymbol{\theta}) \int_{\mathbb{R}^m} f(\boldsymbol{t}) \exp\left\{\sum_{k=1}^{m} \eta_k(\boldsymbol{\theta})\, t_k\right\} h_T(\boldsymbol{t})\, d\boldsymbol{t}.$$

If Γ contains an m-dimensional rectangle, then $\mathbb{E}_{\boldsymbol{\theta}}\left[f(\boldsymbol{T}(Y))\right] = 0$, for all $\boldsymbol{\theta} \in \mathcal{X}$, which implies that the m-dimensional Laplace transform of $f(\boldsymbol{t})h_T(\boldsymbol{t})$ is equal to 0 on an m-dimensional rectangle of real parts of the complex variables defining the Laplace transform. By analytic continuity, we can conclude that $f(\boldsymbol{t})h_T(\boldsymbol{t}) = 0$ for almost all $\boldsymbol{t}$, which implies that $f(\boldsymbol{t}) = 0$ for almost all $\boldsymbol{t}$ [4].

Thus, $\{q_{\boldsymbol{\theta}},\ \boldsymbol{\theta} \in \mathcal{X}\}$ is a complete family, and $\boldsymbol{T}(y)$ is a complete sufficient statistic. □

COROLLARY 12.2 Completeness Theorem for Canonical Exponential Families. *For the canonical exponential family defined in* (11.26), *the vector $\boldsymbol{T}(Y) = [T_1(Y) \dots T_m(Y)]^\top$ is a complete sufficient (hence minimal) statistic for $\{p_{\boldsymbol{\theta}},\ \boldsymbol{\theta} \in \mathcal{X}\}$ if $\mathcal{X}$ contains an m-dimensional rectangle.*

[1] For $m = 1$ this is a one-dimensional interval.

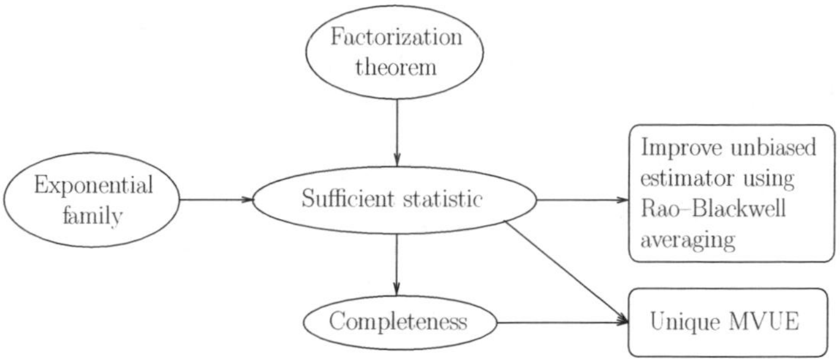

Figure 12.2 The Quest for MVUEs

12.6 Discussion

Figure 12.2 summarizes the key results in this chapter. The factorization theorem helps to identify sufficient statistics. The Rao–Blackwell theorem can be applied to reduce the variance of any unbiased estimator by averaging it over Y conditioned on the sufficient statistic $T(Y)$. If the sufficient statistic is complete, this produces the MVUE, which is unique. When the parametric family is exponential, it is easy to identify a sufficient statistic and to check whether it is complete.

12.7 Examples: Gaussian Families

Examples 12.10–12.12 are from Gaussian families.

Example 12.10 Consider the length-n sequence $\boldsymbol{Y} = \alpha \boldsymbol{s} + \boldsymbol{Z}$ where $\boldsymbol{Z} \sim \mathcal{N}(0, \sigma^2 \mathrm{I}_n) \in \mathbb{R}^n$. Thus $\boldsymbol{Y} \sim \mathcal{N}(\alpha \boldsymbol{s}, \sigma^2 \mathrm{I}_n)$. Here σ^2 and $\boldsymbol{s} \in \mathbb{R}^n$ are known, and the unknown parameter is $\alpha \in \mathbb{R}$. We identify θ with α and write

$$
\begin{aligned}
p_\theta(\boldsymbol{y}) &= (2\pi\sigma^2)^{-n/2} \exp\left\{-\frac{1}{2\sigma^2}\|\boldsymbol{y} - \alpha \boldsymbol{s}\|^2\right\} \\
&= (2\pi\sigma^2)^{-n/2} \exp\left\{-\frac{1}{2\sigma^2}(\|\boldsymbol{y}\|^2 + \alpha^2\|\boldsymbol{s}\|^2 - 2\alpha \boldsymbol{y}^\top \boldsymbol{s})\right\} \\
&= \underbrace{(2\pi\sigma^2)^{-n/2} \exp\left\{-\frac{1}{2\sigma^2}\alpha^2\|\boldsymbol{s}\|^2\right\}}_{=C(\theta)} \underbrace{\exp\{-\frac{1}{2\sigma^2}\|\boldsymbol{y}\|^2\}}_{=h(\boldsymbol{y})} \exp\left\{\frac{\alpha}{\sigma^2}\boldsymbol{y}^\top \boldsymbol{s}\right\}.
\end{aligned}
$$

We see that this is a one-dimensional (1-D) canonical exponential family where

$$
\begin{aligned}
C(\theta) &\triangleq (2\pi\sigma^2)^{-n/2} \exp\left\{-\frac{1}{2\sigma^2}\alpha^2\|\boldsymbol{s}\|^2\right\}, \\
h(\boldsymbol{y}) &\triangleq \exp\left\{-\frac{1}{2\sigma^2}\|\boldsymbol{y}\|^2\right\}, \\
\eta_1(\alpha) &\triangleq \frac{\alpha}{\sigma^2}, \\
T_1(\boldsymbol{y}) &\triangleq \boldsymbol{y}^\top \boldsymbol{s} \quad \text{(sufficient statistic).}
\end{aligned}
$$

Because $\{p_\theta,\ \theta \in \mathcal{X}\}$ is a 1-D exponential family and $\Gamma = \mathbb{R}$ includes a 1-D rectangle, $T_1(\boldsymbol{Y})$ is a complete sufficient statistic by Theorem 12.7.

Now we seek an MVUE for α. Observe that

$$\mathbb{E}_\theta[T_1(\boldsymbol{Y})] = \mathbb{E}_\theta[\boldsymbol{Y}^\top \boldsymbol{s}] = \alpha \boldsymbol{s}^\top \boldsymbol{s} = \alpha \|\boldsymbol{s}\|^2.$$

Hence

$$\hat{\alpha}(\boldsymbol{Y}) \triangleq \frac{T_1(\boldsymbol{Y})}{\|\boldsymbol{s}\|^2} = \frac{\boldsymbol{Y}^\top \boldsymbol{s}}{\|\boldsymbol{s}\|^2} \tag{12.15}$$

is an unbiased estimator of α. But $T_1(\boldsymbol{Y})$ is a complete sufficient statistic, so this estimator is also the unique MVUE. Note that the estimator $\hat{\alpha}(\boldsymbol{Y})$ does not depend on σ^2.

Example 12.11 Consider the same situation as in Example 12.10, except that here α is known and σ^2 is not. Thus we identify θ with σ^2 and write

$$p_\theta(\boldsymbol{y}) = (2\pi\sigma^2)^{-n/2} \exp\left\{-\frac{1}{2\sigma^2}\|\boldsymbol{y} - \alpha \boldsymbol{s}\|^2\right\},$$

which is again a 1-D canonical exponential family with

$$\begin{aligned} C(\theta) &= (2\pi\sigma^2)^{-n/2}, \\ \eta(\theta) &= -\frac{1}{2\sigma^2}, \\ T(\boldsymbol{y}) &= \|\boldsymbol{y} - \alpha \boldsymbol{s}\|^2, \\ h(\boldsymbol{y}) &= 1. \end{aligned}$$

We have

$$\mathbb{E}_\theta[T(\boldsymbol{Y})] = \mathbb{E}_\theta[\|\boldsymbol{Y} - \alpha \boldsymbol{s}\|^2] = n\sigma^2$$

hence

$$\hat{\sigma}^2(\boldsymbol{Y}) \triangleq \frac{T(\boldsymbol{Y})}{n} = \frac{1}{n}\|\boldsymbol{Y} - \alpha \boldsymbol{s}\|^2 \tag{12.16}$$

is an unbiased estimator of α. Since $\Gamma = \mathbb{R}^-$ includes a 1-D rectangle, $T(\boldsymbol{Y})$ is a complete sufficient statistic, and the estimator (12.16) is the unique MVUE.

Example 12.12 Consider the same example again, except that both α and σ^2 are unknown: $\boldsymbol{\theta} = (\alpha,\ \sigma^2) \in \mathcal{X} = \mathbb{R} \times \mathbb{R}^+$. We factor $p_{\boldsymbol{\theta}}$ as

$$p_{\boldsymbol{\theta}}(\boldsymbol{y}) = \underbrace{(2\pi\sigma^2)^{-n/2} \exp\left\{-\frac{\alpha^2}{2\sigma^2}\|\boldsymbol{s}\|^2\right\}}_{=C(\boldsymbol{\theta})} \exp\left\{\frac{\alpha}{\sigma^2}\boldsymbol{y}^\top \boldsymbol{s} - \frac{1}{2\sigma^2}\|\boldsymbol{y}\|^2\right\}.$$

This is a two-dimensional (2-D) exponential family where

$$\begin{aligned} C(\boldsymbol{\theta}) &\triangleq (2\pi\sigma^2)^{-n/2} \exp\left\{-\frac{\alpha^2}{2\sigma^2}\|\boldsymbol{s}\|^2\right\}, \\ h(\boldsymbol{y}) &\triangleq 1, \end{aligned}$$

$$\eta_1(\boldsymbol{\theta}) = \frac{\alpha}{\sigma^2},$$
$$T_1(\boldsymbol{y}) \triangleq \boldsymbol{y}^\top \boldsymbol{s},$$
$$\eta_2(\boldsymbol{\theta}) \triangleq -\frac{1}{2\sigma^2},$$
$$T_2(\boldsymbol{y}) = \|\boldsymbol{y}\|^2.$$

The sufficient statistic is the two-vector $T(\boldsymbol{Y}) = \begin{bmatrix} \boldsymbol{Y}^\top \boldsymbol{s} \\ \|\boldsymbol{Y}\|^2 \end{bmatrix}$. Moreover $T(\boldsymbol{Y})$ is complete because Γ contains a two-dimensional rectangle.

We already have an unbiased estimator of α in the form of (12.15). We now seek an unbiased estimator of σ^2. For notational simplicity, we shorten $\mathbb{E}_{\boldsymbol{\theta}}[\cdot]$ to $\mathbb{E}[\cdot]$. We have

$$\begin{aligned}\mathbb{E}[T_2(\boldsymbol{Y})] &= \mathbb{E}[\|\boldsymbol{Y}\|^2] = \mathbb{E}[\|\alpha\boldsymbol{s} + \boldsymbol{Z}\|^2] \\ &= \alpha^2\|\boldsymbol{s}\|^2 + \mathbb{E}[\|\boldsymbol{Z}\|^2] + 2\alpha\mathbb{E}[\boldsymbol{s}^\top \boldsymbol{Z}] \\ &= \alpha^2\|\boldsymbol{s}\|^2 + n\sigma^2 + 0 = \alpha^2\|\boldsymbol{s}\|^2 + n\sigma^2. \end{aligned} \tag{12.17}$$

To construct an unbiased estimator of σ^2, we need to combine $T_2(\boldsymbol{Y})$ with $T_1^2(\boldsymbol{Y})$.

$$\begin{aligned}\mathbb{E}[T_1^2(\boldsymbol{Y})] &= \mathbb{E}[(\boldsymbol{Y}^\top \boldsymbol{s})^2] = \mathbb{E}[((\alpha\boldsymbol{s} + \boldsymbol{Z})^\top \boldsymbol{s})^2] = \mathbb{E}[(\alpha\|\boldsymbol{s}\|^2 + \boldsymbol{Z}^\top \boldsymbol{s})^2] \\ &= \alpha^2\|\boldsymbol{s}\|^4 + \mathbb{E}[(\boldsymbol{Z}^\top \boldsymbol{s})^2] + 2\alpha\|\boldsymbol{s}\|^2\mathbb{E}[\boldsymbol{Z}^\top \boldsymbol{s}] \\ &= \alpha^2\|\boldsymbol{s}\|^4 + \sigma^2\|\boldsymbol{s}\|^2 + 0 = \alpha^2\|\boldsymbol{s}\|^4 + \sigma^2\|\boldsymbol{s}\|^2 \end{aligned}$$

and therefore

$$\frac{\mathbb{E}[T_1^2(\boldsymbol{Y})]}{\|\boldsymbol{s}\|^2} = \alpha^2\|\boldsymbol{s}\|^2 + \sigma^2. \tag{12.18}$$

Subtracting (12.18) from (12.17), we obtain

$$\mathbb{E}\left[T_2(\boldsymbol{Y}) - \frac{T_1^2(\boldsymbol{Y})}{\|\boldsymbol{s}\|^2}\right] = (n-1)\sigma^2.$$

Hence

$$\begin{aligned}\hat{\sigma}^2(\boldsymbol{Y}) &\triangleq \frac{1}{n-1}\left(T_2(\boldsymbol{Y}) - \frac{T_1^2(\boldsymbol{Y})}{\|\boldsymbol{s}\|^2}\right) \\ &= \frac{1}{n-1}\left(\|\boldsymbol{Y}\|^2 - \frac{(\boldsymbol{Y}^\top \boldsymbol{s})^2}{\|\boldsymbol{s}\|^2}\right) \\ &= \frac{1}{n-1}\left(\|\boldsymbol{Y}\|^2 - \hat{\alpha}^2\|\boldsymbol{s}\|^2\right) \\ &= \frac{1}{n-1}\|\boldsymbol{Y} - \hat{\alpha}\boldsymbol{s}\|^2 \end{aligned} \tag{12.19}$$

is an unbiased estimator of σ^2. Again, this is the unique MVUE estimator. Note the normalization factor is $\frac{1}{n-1}$ instead of $\frac{1}{n}$ as might have been expected. The reason is that α is unknown, and estimating it by the empirical mean $\hat{\alpha}$ minimizes the residual sum of squares $\|\boldsymbol{Y} - \hat{\alpha}\boldsymbol{s}\|^2$; the *expected* sum of squares is slightly larger.

Exercises

12.1 *Minimal Sufficient Statistics.* Consider the following families of distributions. In each case, find a minimal sufficient statistic, and explain why it must be minimal.

(a) $Y_i, 1 \le i \le n$ are i.i.d. samples from a Poisson distribution with parameter θ, where $\theta > 0$.
(b) $Y_i, 1 \le i \le n$ are i.i.d. samples from a uniform distribution over the interval $[-\theta, \theta]$, where $\theta > 0$.
(c) Y_i are independent $\mathcal{N}(\theta s_i, \sigma_i^2)$ random variables, for $1 \le i \le n$. Assume that $s_i, \sigma_i^2 \ne 0$ for all i.

12.2 *Minimal Sufficient Statistics.* Let $Y_i = X_i S_i, i \in \{1, 2, 3\}$ be three random variables generated as follows. X_1, X_2, X_3 are i.i.d random variables with distribution $\mathrm{P}\{X = 1\} = 1 - \mathrm{P}\{X = 2\} = \theta$, where $0 < \theta < 1$. S_1, S_2, S_3 are three independent (but not identically distributed) random variables with distributions $\mathrm{P}\{S_i = 1\} = 1 - \mathrm{P}\{S_i = -1\} = a_i$, where $a_1 = 0.5, a_2 = 1, a_3 = 0$.
Find minimal sufficient statistics for estimating θ from the data $\boldsymbol{Y} = [Y_1\ Y_2\ Y_3]^\top$.

12.3 *Factorization Theorem.* Let q_1 and q_2 be two pdfs with disjoint supports $\mathcal{Y}_1$ and $\mathcal{Y}_2$. Consider the mixture distribution

$$p_\theta(y) = \theta q_1(y) + (1 - \theta) q_2(y),$$

where $\theta \in [0, 1]$.
Use the factorization theorem to show that $T(y) = \mathbb{1}\{y \in \mathcal{Y}_1\}$ is a sufficient statistic for θ.

12.4 *Optimization of Unbiased Estimators.* Y_i are independent $\mathcal{N}(\theta, \sigma_i^2)$ random variables, for $1 \le i \le n$. Assume that $\sigma_i^2 > 0$ for all i. Show that

$$\hat{\theta}(\boldsymbol{Y}) = \sum_{i=1}^{n} a_i Y_i$$

is an unbiased estimator of θ for any sequence $\boldsymbol{a} = \{a_i\}_{i=1}^n$ summing to 1. Derive $\boldsymbol{a}$ that minimizes the variance of this estimator over all $\boldsymbol{a}$. Is this estimator an MVUE?

12.5 *Complete Sufficient Statistic.* Suppose that given a parameter $\theta > 1$, $Y_1, Y_2, \ldots, Y_n$ are i.i.d. observations each with pdf

$$p_\theta(y) = (\theta - 1) y^{-\theta} \mathbb{1}\{y \ge 1\}.$$

That is,

$$p_\theta(\boldsymbol{y}) = \prod_{k=1}^{n} (\theta - 1) y_k^{-\theta} \mathbb{1}\{y_k \ge 1\}.$$

Find a complete sufficient statistic for $\{p_\theta, \theta > 1\}$.

12.6 *MVUE for Exponential Family.* Show that the family $\{p_\theta\}$ of (12.12) is a one-dimensional exponential family, and derive the MVUE of θ given n i.i.d. random variables drawn from P_θ.

12.7 *MVUE for Uniform Family.* Suppose $\theta > 0$ is a parameter of interest and that given θ, $Y_1, \ldots, Y_n$ are i.i.d. observations with marginal densities

$$p_\theta(y) = \frac{1}{\theta}\mathbb{1}\{y \in [0, \theta]\}.$$

(a) Prove that $T(\boldsymbol{y}) = \max\{y_1, \ldots, y_n\}$ is a sufficient statistic for θ.
(b) Prove that $T(\boldsymbol{y})$ is also *complete*. Note that you cannot apply the Completeness Theorem for Exponential Families.

$$\mathit{Hint}: \text{ Use the fact that } \int_0^\theta f(t)\frac{t^{n-1}}{\theta^n}dt = 0,\ \forall\theta > 0 \ \Rightarrow\ f(t) = 0 \text{ on } [0, \infty).$$

(c) Find an MVUE of θ based on $\boldsymbol{y}$.

12.8 *Order Statistics.* Consider data $\boldsymbol{Y} = [Y_1\ Y_2 \ldots Y_n]^\top$, and the reordered data $\boldsymbol{T}(\boldsymbol{Y}) = [Y_{(1)}\ Y_{(2)} \ldots Y_{(n)}]^\top$, where $Y_{(1)} \geq Y_{(2)} \geq \ldots \geq Y_{(n)}$. Clearly $\boldsymbol{T}$ is a many-to-one map. Show that if $Y_1, Y_2, \ldots, Y_n$ are i.i.d. from a family of univariate distributions $\mathcal{P} = \{p_\theta, \theta \in \mathcal{X}\}$, then $\boldsymbol{T}(\boldsymbol{Y})$ are sufficient statistics for the i.i.d. extension of the family $\mathcal{P}$.

12.9 *Unbiased Estimators*. Consider the family of distributions $p_\theta(y) = 1+2\theta y$ where $|y| \leq 1/2$ and $|\theta| \leq 1/2$.

(a) Let k be an arbitrary odd number. Show that $\hat{\theta}_k(Y) \triangleq 2^k(k+2)Y^k$ is an unbiased estimator of θ, and derive its variance.
(b) Show that $v_k \triangleq \max_{|\theta|\leq 1} \mathrm{Var}_\theta[\hat{\theta}_k(Y)]$ is achieved at $\theta = 0$.
(c) Show that among all the unbiased estimators of Part (a), $\hat{\theta}_1(Y) = 6Y$ achieves $\inf\{v_1, v_3, \ldots\}$.
(d) Consider the convex combination $\hat{\theta}_\lambda(Y) \triangleq \lambda\hat{\theta}_1(Y)+(1-\lambda)\hat{\theta}_3(Y)$ where $\lambda \in [0, 1]$. Show that $\hat{\theta}_\lambda(Y)$ is an unbiased estimator of θ, and derive its variance. Show that the variance is maximized when $\theta = 0$, and derive λ that minimizes $\max_\theta \mathrm{Var}_\theta[\hat{\theta}_\lambda(Y)]$.

12.10 *MVUE for Poisson Family.* Suppose $\{Y_k\}_{k=1}^n$ (with $n \geq 2$) are i.i.d. Poisson random variables with parameter (mean) $\theta > 0$, i.e.,

$$p_\theta(y_k) = \frac{e^{-\theta}\theta^{y_k}}{y_k!} \qquad y_k = 0, 1, 2, \ldots$$

and $p_\theta(\boldsymbol{y}) = \prod_{k=1}^n p_\theta(y_k)$.

We wish to estimate the parameter λ that is given by

$$\lambda = e^{-\theta}.$$

(a) Show that $T(\boldsymbol{y}) = \sum_{k=1}^n y_k$ is a complete sufficient statistic for this estimation problem.

(b) Show that

$$\hat{\lambda}_1(\boldsymbol{y}) = \frac{1}{n}\sum_{k=1}^{n} \mathbb{1}\{y_k = 0\}$$

is an unbiased estimator of λ.

(c) Show that

$$\hat{\lambda}_2(\boldsymbol{y}) = \left[\frac{n-1}{n}\right]^{T(\boldsymbol{y})}$$

is an MVUE of λ.

Hint: You may use the fact that $T(\boldsymbol{y})$ is Poisson with parameter $n\theta$.

References

[1] E. L. Lehman and G. Casella, *Theory of Point Estimation*, 2nd edition, Springer-Verlag, 1998.

[2] C. R. Rao, "Information and the Accuracy Attainable in the Estimation of Statistical Parameters," *Bull. Calcutta Math. Soc.*, pp. 81–91, 1945.

[3] D. Blackwell, "Conditional Expectation and Unbiased Sequential Estimation," *Annal. Math. Stat.*, pp. 105–110, 1947.

[4] L. Hormander, *An Introduction to Complex Analysis in Several Variables*, Elsevier, 1973.

13 Information Inequality and Cramér–Rao Lower Bound

In this chapter, we introduce the information inequality, which provides a lower bound on the variance of any estimator of scalar parameters. When applied to unbiased estimators, this lower bound is known as the Cramér–Rao lower bound (CRLB). The bound is expressed in terms of the Fisher information, which quantifies the information present in the data to estimate the (nonrandom) parameter. We extend the bound to vector-valued parameters and to random parameters.

13.1 Fisher Information and the Information Inequality

In Chapter 12, we studied a systematic approach to finding estimators using the MVUE criterion. Unfortunately in many cases, with the exception of exponential families, it may be difficult to find an MVUE solution, or an MVUE solution may not even exist. One approach in these cases is to design good suboptimal detectors and compare their performance to pick the best one. The performance measure of interest is the conditional risk $R_\theta(\hat{\theta})$ for squared-error cost. If θ is real-valued then

$$\begin{aligned} R_\theta(\hat{\theta}) = \mathrm{MSE}_\theta(\hat{\theta}) &= \mathbb{E}_\theta\left[(\hat{\theta}(Y)-\theta)^2\right] \\ &= \mathrm{Var}_\theta(\hat{\theta}(Y)) + \left(b_\theta(\hat{\theta})\right)^2, \end{aligned} \tag{13.1}$$

where the quantity

$$b_\theta(\hat{\theta}) \triangleq \mathbb{E}_\theta[\hat{\theta}(Y)-\theta], \quad \theta \in \mathcal{X}, \tag{13.2}$$

is called the *bias* of the estimator $\hat{\theta}$.

For a vector parameter $\boldsymbol{\theta} \in \mathcal{X} \subseteq \mathbb{R}^m$, the bias of (13.2) is an m-vector $\boldsymbol{b}_{\boldsymbol{\theta}}(\hat{\boldsymbol{\theta}})$. We are interested in the correlation matrix of the estimation error,

$$\begin{aligned} \mathsf{R}_{\boldsymbol{\theta}}(\hat{\boldsymbol{\theta}}) &\triangleq \mathbb{E}_{\boldsymbol{\theta}}\left[\left(\hat{\boldsymbol{\theta}}(\boldsymbol{Y})-\boldsymbol{\theta}\right)\left(\hat{\boldsymbol{\theta}}(\boldsymbol{Y})-\boldsymbol{\theta}\right)^\top\right] \\ &= \mathrm{Cov}_{\boldsymbol{\theta}}(\hat{\boldsymbol{\theta}}) + \left(\mathbb{E}_{\boldsymbol{\theta}}\left[\hat{\boldsymbol{\theta}}(\boldsymbol{Y})-\boldsymbol{\theta}\right]\right)\left(\mathbb{E}_{\boldsymbol{\theta}}\left[\hat{\boldsymbol{\theta}}(\boldsymbol{Y})-\boldsymbol{\theta}\right]\right)^\top \\ &= \mathrm{Cov}_{\boldsymbol{\theta}}(\hat{\boldsymbol{\theta}}) + \boldsymbol{b}_{\boldsymbol{\theta}}(\hat{\boldsymbol{\theta}})^\top \boldsymbol{b}_{\boldsymbol{\theta}}(\hat{\boldsymbol{\theta}}), \end{aligned} \tag{13.3}$$

and especially in its diagonal elements which are the mean-squared values of the individual components of the estimator error vector. The total mean squared-error is given by the sum of these terms, i.e., the trace of the correlation matrix:

$$\text{MSE}(\hat{\boldsymbol{\theta}}) = \mathbb{E}_{\boldsymbol{\theta}} \|\hat{\boldsymbol{\theta}}(Y) - \boldsymbol{\theta}\|^2 = \text{Tr}[\text{R}_{\boldsymbol{\theta}}(\hat{\boldsymbol{\theta}})] = \text{Tr}[\text{Cov}_{\boldsymbol{\theta}}(\hat{\boldsymbol{\theta}})] + \|b_{\boldsymbol{\theta}}(\hat{\boldsymbol{\theta}})\|^2. \tag{13.4}$$

It is often difficult if not impossible to derive closed-form expressions for the MSE. One approach then is to perform Monte Carlo simulations, in which n i.i.d. observations $Y^{(i)}$, $i = 1, \ldots, n$, are generated from p_θ. The estimator $\hat{\theta}(Y^{(i)})$ can be computed for each of these n experiments, and the MSE can be estimated as

$$\widehat{\text{MSE}}(\hat{\theta}) = \frac{1}{n} \sum_{i=1}^{n} \left[\hat{\theta}(Y^{(i)}) - \theta\right]^2.$$

The main drawback of this approach is that it may not yield much insight, and that Monte Carlo simulations can be computationally expensive. This motivates a different approach, where a lower bound on the MSE is derived, which can be more easily evaluated. The main result is the *information inequality*, which is given in Theorem 13.1 for scalar parameters and is specialized to unbiased estimators in the next section. A key quantity that appears in the lower bound is Fisher information, which is defined next. We make the following assumptions, which are known as *regularity conditions*. In particular, Assumptions (iv) and (v) state that the order of differentiation with respect to θ and integration with respect to Y may be switched. These assumptions are generally satisfied when Assumption (ii) holds.

Regularity Assumptions

(i) $\mathcal{X}$ is an open interval of $\mathbb{R}$ (possibly semi-infinite or infinite).
(ii) The support set $\text{supp}\{p_\theta\} = \{y \in \mathcal{Y} : p_\theta(y) > 0\}$ is the same for all $\theta \in \mathcal{X}$.
(iii) $p'_\theta(y) \triangleq \frac{\partial p_\theta(y)}{\partial \theta}$ exists and is finite for all $\theta \in \mathcal{X}$ and $y \in \mathcal{Y}$.
(iv) $\int_{\mathcal{Y}} p'_\theta(y) d\mu(y) = \frac{\partial}{\partial \theta} \int_{\mathcal{Y}} p_\theta(y) d\mu(y) = 0, \forall \theta \in \mathcal{X}$.
(v) If $p''_\theta(y) \triangleq \frac{\partial^2 p_\theta(y)}{\partial \theta^2}$ exists and is finite $\forall \theta \in \mathcal{X}$, $\forall y \in \mathcal{Y}$ then $\int_{\mathcal{Y}} p''_\theta(y) d\mu(y) = \frac{\partial^2}{\partial \theta^2} \int_{\mathcal{Y}} p_\theta(y) d\mu(y) = 0$.

The notion of information contained in the observations about an unknown parameter was introduced by Edgeworth [1] and later fully developed by Fisher [2].

DEFINITION 13.1 The quantity

$$I_\theta \triangleq \mathbb{E}_\theta \left[\left(\frac{\partial \ln p_\theta(Y)}{\partial \theta} \right)^2 \right] \tag{13.5}$$

is called the *Fisher information* for θ.

LEMMA 13.1 *Under Assumptions (i)–(iv) we have*

$$\mathbb{E}_\theta \left[\frac{\partial \ln p_\theta(Y)}{\partial \theta} \right] = 0 \tag{13.6}$$

and

$$I_\theta = \text{Var}_\theta \left[\frac{\partial \ln p_\theta(Y)}{\partial \theta} \right]. \tag{13.7}$$

If in addition Assumption (v) holds, then

$$I_\theta = \mathbb{E}_\theta\left[-\frac{\partial^2 \ln p_\theta(Y)}{\partial\theta^2}\right]. \tag{13.8}$$

Proof The left hand side of (13.6) is equal to $\int_\mathcal{Y} p'_\theta \, d\mu$ which is zero by Assumption (iv). Then (13.7) follows directly from (13.5) and (13.6). Finally, if Assumption (v) holds, then $\int_\mathcal{Y} p''_\theta \, d\mu = 0$ and

$$\begin{aligned}\frac{\partial^2 \ln p_\theta(Y)}{\partial\theta^2} &= \frac{\partial}{\partial\theta}\left(\frac{p'_\theta(Y)}{p_\theta(Y)}\right)\\ &= \frac{p''_\theta(Y)p_\theta(Y) - [p'_\theta(Y)]^2}{[p_\theta(Y)]^2}\\ &= \frac{p''_\theta(Y)}{p_\theta(Y)} - \left(\frac{\partial \ln p_\theta(Y)}{\partial\theta}\right)^2.\end{aligned}$$

Taking the expectation of both sides with respect to p_θ establishes (13.8). □

THEOREM 13.1 Information Inequality. *Assume (i)–(iv) above hold. Let $\hat{\theta}(Y)$ be any estimator of $\theta \in \mathcal{X}$ satisfying the regularity condition*[1]

(vi) $\frac{d}{d\theta}\int_\mathcal{Y} \hat{\theta}(y)p_\theta(y)d\mu(y) = \int_\mathcal{Y} \hat{\theta}(y)p'_\theta(y)d\mu(y), \quad \forall\theta \in \mathcal{X}.$

Then

$$\mathrm{Var}_\theta\left[\hat{\theta}(Y)\right] \ge \frac{\left(\frac{d}{d\theta}\mathbb{E}_\theta\left[\hat{\theta}(Y)\right]\right)^2}{I_\theta} = \frac{\left(\frac{d}{d\theta}b_\theta(\hat{\theta}) + 1\right)^2}{I_\theta}. \tag{13.9}$$

Proof Without loss of generality, assume $\mathcal{Y} = \mathrm{supp}\,\{p_\theta\}$. The proof is an application of the Cauchy–Schwarz inequality. For any two random variables A and B with finite variances, we have

$$|\mathrm{Cov}(A, B)| \le \sqrt{\mathrm{Var}(A)\,\mathrm{Var}(B)}. \tag{13.10}$$

We apply this inequality with $A = \hat{\theta}(Y)$ and $B = \frac{\partial \ln p_\theta(Y)}{\partial\theta}$, both of which are functions of Y. For any $\theta \in \mathcal{X}$, we have

$$\mathrm{Var}_\theta(A) = \mathrm{Var}_\theta[\hat{\theta}(Y)], \quad \mathrm{Var}_\theta(B) = I_\theta,$$

where the second equality is the same as (13.7). By (13.6), $\mathbb{E}_\theta(B) = 0$, and thus

$$\mathrm{Cov}_\theta(A, B) = \mathbb{E}_\theta(AB) = \mathbb{E}_\theta\left[\hat{\theta}(Y)\frac{\partial \ln p_\theta(Y)}{\partial\theta}\right].$$

[1] Similarly to Assumption (iv), the order of differentiation with respect to θ and integration with respect to Y may be switched.

Therefore (13.10) yields

$$\begin{aligned}\mathrm{Var}_\theta[\hat{\theta}(Y)]\, I_\theta &\ge \left(\mathbb{E}_\theta\left[\hat{\theta}(Y)\frac{\partial \ln p_\theta(Y)}{\partial\theta}\right]\right)^2 \\ &= \left(\int_{\mathcal{Y}} \hat{\theta}(y) p'_\theta(y) d\mu(y)\right)^2 \\ &= \left(\frac{d}{d\theta}\mathbb{E}_\theta\left[\hat{\theta}(Y)\right]\right)^2,\end{aligned}$$

where the last equality follows from Assumption (vi). This proves the information inequality (13.9). □

13.2 Cramér–Rao Lower Bound

Observe that $\hat{\theta}$ appears on both sides of the information inequality (13.9). In particular, the right hand side might be difficult to evaluate if no closed-form expression for the bias of $\hat{\theta}$ is available. For unbiased estimators, however, the information inequality takes a simpler form. Then $b_\theta[\hat{\theta}] = 0$ for all $\theta \in \mathcal{X}$. The information inequality (13.9) reduces to the following lower bound on the variance, which is generally known as the Cramér–Rao lower bound [3, 4] although it was also discovered by Fréchet earlier [5]:

$$\mathrm{Var}_\theta[\hat{\theta}(Y)] \ge \frac{1}{I_\theta}. \tag{13.11}$$

DEFINITION 13.2 *Efficient Estimator.* An unbiased estimator whose variance achieves the CRLB (13.11) with equality is said to be efficient.

An efficient estimator is necessarily MVUE.

Example 13.1 *Known Signal with Unknown Scale Factor.* Let $\boldsymbol{Y} = \theta\boldsymbol{s} + \boldsymbol{Z}$ where $\boldsymbol{Z} \sim \mathcal{N}(0, \sigma^2 I_n)$, $\boldsymbol{s}$ is a known signal, and $\theta \in \mathbb{R}$. Thus

$$\ln p_\theta(\boldsymbol{y}) = -\frac{n}{2}\ln(2\pi\sigma^2) - \frac{1}{2\sigma^2}\|\boldsymbol{y} - \theta\boldsymbol{s}\|^2$$

is twice differentiable with respect to θ, and

$$\frac{\partial^2 \ln p_\theta(\boldsymbol{y})}{\partial\theta^2} = -\frac{\|\boldsymbol{s}\|^2}{\sigma^2}.$$

Thus we obtain Fisher information from (13.8) as

$$I_\theta = \frac{\|\boldsymbol{s}\|^2}{\sigma^2}, \tag{13.12}$$

which represents the SNR in the observations.

We have seen in (12.15), (12.18) that the variance of the MVUE

$$\hat{\theta}_{\mathrm{MVUE}}(\boldsymbol{Y}) = \frac{\boldsymbol{s}^\top \boldsymbol{Y}}{\|\boldsymbol{s}\|^2}$$

is equal to

$$\mathrm{Var}_\theta\left[\hat{\theta}_{\mathrm{MVUE}}\right] = \frac{\sigma^2}{\|s\|^2}.$$

In view of (13.12), the MVUE achieves the CRLB (13.11) and is thus efficient. In other problems, the MVUE need not achieve the CRLB.

Example 13.2 *Nonlinear Signal Model.* Let $s(\theta) \in \mathbb{R}^n$ be a signal which is a twice-differentiable function of a scalar parameter $\theta \in \mathbb{R}$. The observations are the sum of this signal and i.i.d. Gaussian noise: $Y = s(\theta) + Z$ where $Z \sim \mathcal{N}(0, \sigma^2 \mathsf{I}_n)$. Then

$$\ln p_\theta(y) = -\frac{n}{2}\ln(2\pi\sigma^2) - \frac{1}{2\sigma^2}\sum_{i=1}^n (y_i - s_i(\theta))^2$$

is twice differentiable with respect to θ, and

$$\frac{\partial^2 \ln p_\theta(Y)}{\partial\theta^2} = -\frac{1}{\sigma^2}\sum_{i=1}^n [(Y_i - s_i(\theta))s_i''(\theta) + (s_i'(\theta))^2]. \tag{13.13}$$

Since $\mathbb{E}[Y_i - s_i(\theta)] = \mathbb{E}[Z_i] = 0$ has zero mean, we obtain Fisher information from (13.8) and (13.13) as

$$I_\theta = \frac{1}{\sigma^2}\sum_{i=1}^n [s_i'(\theta)]^2, \tag{13.14}$$

which may be thought of as a SNR, where signal power is the energy in the derivative.

A particularly insightful application of this model is estimation of the angular frequency of a sinusoid: $s_i(\theta) = \cos(i\theta + \phi)$ where $\theta \in (0, \pi)$ is the angular frequency, and $\phi \in [0, 2\pi)$ is a known phase. Then $s_i'(\theta) = -i\sin(i\theta + \phi)$ and

$$I_\theta = \frac{1}{\sigma^2}\sum_{i=1}^n i^2 \sin^2(i\theta + \phi). \tag{13.15}$$

Note that I_θ is generally large for large n, but the case of $\theta \ll 1/n$ (in which case the observation window is much less than one cycle of the sinusoid) and $\phi = 0$ is rather different: then $\sin(i\theta + \phi) \approx i\theta$, and thus $I_\theta \approx \frac{\theta^2}{\sigma^2}\sum_{i=1}^n i^4$ could be very small. Intuitively, a good estimator of θ cannot exist because the signal $s_i(\theta) \approx 1 - i^2\theta^2/2$ depends very weakly on θ over the observation window.

Example 13.3 *Gaussian Observations with Parameterized Covariance Matrix.* Assume that the observations are an n-vector $Y \sim \mathcal{N}(0, \mathsf{C}(\theta))$ where the covariance matrix $\mathsf{C}(\theta)$ is twice differentiable in θ. Then Fisher information is given by (see derivation in Section 13.8)

$$I_\theta = \frac{1}{2}\mathrm{Tr}\left[\mathsf{C}^{-1}(\theta)\frac{d\mathsf{C}(\theta)}{d\theta}\mathsf{C}^{-1}(\theta)\frac{d\mathsf{C}(\theta)}{d\theta}\right]. \tag{13.16}$$

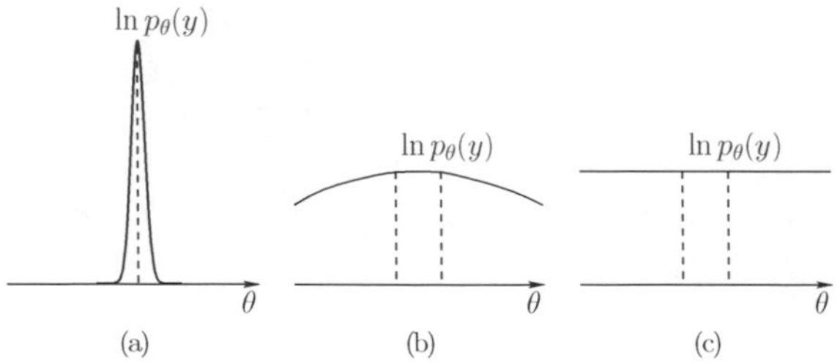

Figure 13.1 (a) A sharply peaked log-likelihood function; (b) a fairly flat log-likelihood function; (c) a completely flat log-likelihood function, for which the parameter θ is not identifiable

13.3 Properties of Fisher Information

Relation to Estimation Accuracy Both the information inequality and the Cramér–Rao bound suggest that the higher the Fisher information is, the higher estimation accuracy will *potentially* be. (The caveat does not apply to efficient estimators, for which the Cramér–Rao bound holds with equality.) In Example 13.1, Fisher information was directly related to SNR.

More generally, (13.8) suggests that Fisher information is the average curvature of the log-likelihood function. Figure 13.1 depicts different scenarios. In Figure 13.1(a), the log-likelihood function is sharply peaked, Fisher information is large, and high estimation accuracy can be expected. In Figure 13.1(b), the log-likelihood function is fairly flat, Fisher information is low, and poor estimation accuracy can be expected. In Figure 13.1(c), as an extreme case, the log-likelihood function is completely flat, Fisher information is zero, and θ is *not identifiable*.

Relation to KL Divergence Assume $p_\theta(Y)$ is twice differentiable with respect to θ, a.s. P_θ. Consider a small variation ϵ of the parameter θ. Assuming $\theta + \epsilon \in \mathcal{X}$ and taking the limit as ϵ tends to zero, we obtain

$$
\begin{aligned}
D(p_\theta \| p_{\theta+\epsilon}) &= \mathbb{E}_\theta\left[\ln \frac{p_\theta(Y)}{p_{\theta+\epsilon}(Y)}\right] \\
&= \mathbb{E}_\theta\left[-\epsilon \frac{\partial \ln p_\theta(Y)}{\partial\theta} + \frac{\epsilon^2}{2}\frac{\partial^2 \ln p_\theta(Y)}{\partial\theta^2} + o(\epsilon^2)\right] \\
&= \frac{\epsilon^2}{2} I_\theta + o(\epsilon^2) \quad \text{as } \epsilon \to 0, \qquad (13.17)
\end{aligned}
$$

where the last equality follows from (13.5) and (13.6). Hence Fisher information also represents the local curvature of the KL divergence function.

Bound on MSE Combining the expressions for bias and variance, we have

$$
\begin{aligned}
\mathrm{MSE}_\theta(\hat\theta) &= \mathrm{Var}_\theta(\hat\theta) + \left[b_\theta(\hat\theta)\right]^2 \\
&\ge \frac{\left(\frac{\partial}{\partial\theta} b_\theta(\hat\theta) + 1\right)^2}{I_\theta} + \left[b_\theta(\hat\theta)\right]^2.
\end{aligned}
$$

Bounds on Estimation of Functions of θ If we wish to estimate $g(\theta)$ using an estimator $\hat{g}(Y)$, then following the same steps that were used in proving Theorem 13.1, we get

$$\mathrm{Var}_\theta\left[\hat{g}(Y)\right] \geq \frac{\left(\frac{d}{d\theta}\mathbb{E}_\theta\left[\hat{g}(Y)\right]\right)^2}{I_\theta}. \tag{13.18}$$

If $\hat{g}(Y)$ is unbiased, i.e., if $\mathbb{E}_\theta\left[\hat{g}(Y)\right] = g(\theta)$, then

$$\mathrm{Var}_\theta\left[\hat{g}(Y)\right] \geq \frac{\left(\frac{d}{d\theta}g(\theta)\right)^2}{I_\theta} = \frac{\left(g'(\theta)\right)^2}{I_\theta}. \tag{13.19}$$

We can consider the ratio $\frac{I_\theta}{(g'(\theta))^2}$ to be the Fisher information for the estimation of $g(\theta)$. In fact, if g is a one-to-one (invertible) and differentiable function, then this notion can be made more precise, as shown next.

Reparameterization Let g be a one-to-one continuously differentiable function and suppose we wish to estimate $\eta = g(\theta)$. Then the family of distributions $\{p_\theta, \theta \in \mathcal{X}\}$ is the same as the family of distributions $\{q_\eta, \eta \in g(\mathcal{X})\}$, where $g(\mathcal{X})$ is the image of $\mathcal{X}$ under the mapping g. The family $\{q_\eta, \eta \in g(\mathcal{X})\}$ is simply a reparameterization of $\{p_\theta, \theta \in \mathcal{X}\}$.

Using the chain rule for differentiation we see that

$$\frac{\partial}{\partial\theta}\ln p_\theta(y) = \frac{\partial\eta}{\partial\theta}\frac{\partial}{\partial\eta}\ln q_\eta(y) = g'(\theta)\frac{\partial}{\partial\eta}\ln q_\eta(y). \tag{13.20}$$

Squaring and taking expectations on both sides we get

$$I_\theta = \left(g'(\theta)\right)^2 I_\eta. \tag{13.21}$$

Thus Fisher information is not invariant to reparameterization via invertible functions.[2]

Moreover, reparameterization does generally not preserve efficiency. For instance, as a special case of Example 13.1, we have that $\hat{\theta} = Y$ is an efficient estimator of θ when $Y \sim \mathcal{N}(\theta, 1)$. However, Y^2 is not an efficient estimator of θ^2 since it is biased ($\mathbb{E}_\theta(Y^2) = \theta^2 + 1$).

Exponential Families Consider the one-dimensional exponential family

$$p_\theta(y) = h(y)C(\theta)\, e^{g(\theta)T(y)}, \tag{13.22}$$

where the function $g(\cdot)$ is assumed to be *invertible* and continuously differentiable. To derive I_θ, it is convenient to parameterize the family in canonical form

$$q_\eta(y) = h(y)e^{\eta T(y) - A(\eta)}$$

[2] However, KL divergence is invariant to reparameterization: $D(p_{\theta'}\|p_\theta) = D(q_{\eta'}\|q_\eta)$ for $\eta = g(\theta)$, $\eta' = g(\theta')$.

with $\eta = g(\theta)$. The Fisher information for η is given by

$$I_\eta = \mathbb{E}_\eta \left(-\frac{\partial^2 \ln q_\eta(y)}{d\eta^2} \right) = A''(\eta) \overset{\text{(a)}}{=} \mathrm{Var}_\eta[T(Y)] = \mathrm{Var}_\theta[T(Y)], \tag{13.23}$$

where equality (a) follows from (11.29). Applying the reparameterization formula (13.21), we conclude that

$$I_\theta = (g'(\theta))^2 \, \mathrm{Var}_\theta[T(Y)]. \tag{13.24}$$

The Gaussian example of Section 13.2 is a special case of (13.22) with $T(\boldsymbol{Y}) = \boldsymbol{Y}^\top \boldsymbol{s}$ (hence $\mathrm{Var}_\theta[T(\boldsymbol{Y})] = \sigma^2 \|\boldsymbol{s}\|^2$) and $g(\theta) = \theta/\sigma^2$, thus $I_\theta = \|\boldsymbol{s}\|^2/\sigma^2$.

i.i.d. Observations Suppose the observations are a sequence $\boldsymbol{Y} = (Y_1, \ldots, Y_n)$ whose elements are i.i.d. p_θ. The Fisher information for the n-fold product pdf p_θ^n is n times the Fisher information for a single observation:

$$I_\theta^{(n)} = \mathbb{E}_\theta \left[-\frac{\partial^2 \ln p_\theta^n(\boldsymbol{Y})}{\partial \theta^2} \right] = n I_\theta. \tag{13.25}$$

We shall see in Chapter 14 (ML estimation) that the loss of unbiasedness and efficiency under nonlinear reparameterization is mitigated in the setting with many i.i.d. observations. This phenomenon is observed with other estimators too, as illustrated in Exercise 13.14.

Local Bound In some problems, the Cramér–Rao bound may be very loose. Consider the Gaussian family $\{p_\theta = \mathcal{N}(\sin\theta, \sigma^2),\ \theta \in \mathbb{R}\}$. Let $\eta = \sin\theta$. Using (13.12), we have $I_\eta = 1/\sigma^2$. From (13.21), we obtain $I_\theta = (\cos^2\theta)/\sigma^2$. As might be expected, I_θ is maximal for $\theta = 0$ and vanishes for $\theta = \pi/2$. However, note that accurate estimation performance cannot be expected for *any* value of θ, since the values $\theta + 2\pi k,\ k \in \mathbb{Z}$, are indistinguishable. Even if $\sigma^2 \ll 1$ and Fisher information is large, no estimator can perform well – unless it has some prior information about θ, which is not available here.

Example 13.4 *Variance Estimation.* Let $\boldsymbol{Y} \sim \mathcal{N}(\boldsymbol{0}, \theta \boldsymbol{I}_n)$ where $\theta > 0$. This is an exponential family where

$$\ln p_\theta(\boldsymbol{y}) = -\frac{n}{2} \ln(2\pi\theta) - \frac{1}{2\theta} \sum_{i=1}^n y_i^2. \tag{13.26}$$

Direct evaluation of (13.8) yields

$$I_\theta = \mathbb{E}_\theta \left[-\frac{n}{2\theta^2} + \frac{1}{\theta^3} \sum_{i=1}^n Y_i^2 \right] = \frac{n}{2\theta^2}. \tag{13.27}$$

Alternatively, the Fisher information could have been derived using the canonical representation

$$\ln q_\eta(\mathbf{y}) = \frac{n}{2} \ln \frac{\eta}{2\pi} + \eta T(\mathbf{y})$$

of this exponential family with $\eta = g(\theta) \triangleq \theta^{-1}$ and $T(\mathbf{Y}) = -\frac{1}{2}\sum_{i=1}^n Y_i^2$. Then $I_\eta = \mathrm{Var}[T(\mathbf{Y})] = \frac{n}{2\eta^2}$ by application of (13.23). Since $[g'(\theta)]^2 = \theta^{-4}$, application of (13.24) yields (13.27) again.

Another interesting parameterization of this family is given by $\xi = \ln\theta \in \mathbb{R}$, in which case $I_\xi = \frac{n}{2}$ independently of ξ (proof is left to the reader).

13.4 Conditions for Equality in Information Inequality

THEOREM 13.2 *Let $\mathcal{X}$ be an open interval (a, b) and assume the regularity conditions of Theorem 13.1 hold. The information inequality* (13.9) *holds with equality for some estimator $\hat{\theta}(Y)$ if and only if p_θ, $\theta \in \mathcal{X}$ is a one-dimensional exponential family and $\hat{\theta}(Y)$ is a sufficient statistic for θ. Moreover, if $\hat{\theta}(Y)$ is an unbiased estimator of θ, then $\hat{\theta}(Y)$ is an MVUE and*

$$p_\theta(y) = h(y) \exp\left\{\int_a^\theta I_{\theta'}(\hat{\theta}(y) - \theta')\, d\theta'\right\}. \tag{13.28}$$

Proof The Cauchy–Schwarz inequality (13.10) holds with equality if and only if there exists a constant c independent of (A, B) such that $B - \mathbb{E}(B) = c[A - \mathbb{E}(A)]$. Hence the information inequality (13.9) holds with equality if and only if there exists a function $c(\theta)$, $\theta \in \mathcal{X}$, such that

$$\frac{\partial \ln p_\theta(Y)}{\partial\theta} - \underbrace{\mathbb{E}_\theta\left[\frac{\partial \ln p_\theta(Y)}{\partial\theta}\right]}_{=0} = c(\theta)\left(\hat{\theta}(Y) - \mathbb{E}_\theta\left[\hat{\theta}(Y)\right]\right) \quad \text{a.s. } p_\theta. \tag{13.29}$$

The function $c(\theta)$ may be identified by taking the variance of both sides of this equation and using the expression (13.7) for Fisher information:

$$I_\theta = c^2(\theta)\mathrm{Var}_\theta[\hat{\theta}(Y)].$$

Substituting into (13.9), we obtain

$$c(\theta) = \frac{I_\theta}{\frac{d}{d\theta}\mathbb{E}_\theta[\hat{\theta}(Y)]}, \tag{13.30}$$

where the denominator is equal to 1 for unbiased estimators. Let $m(\theta) \triangleq \mathbb{E}_\theta[\hat{\theta}(Y)]$. Integrating (13.29) over θ, we obtain

$$\begin{aligned}
\ln p_\theta(y) &= \int_a^\theta c(\theta')\left[\hat{\theta}(y) - m(\theta')\right] d\theta' + b(y) \\
&= \int_a^\theta \frac{I_{\theta'}}{m'(\theta')}\left[\hat{\theta}(y) - m(\theta')\right] d\theta' + b(y) \qquad (13.31)\\
&= \hat{\theta}(y)\int_a^\theta \frac{I_{\theta'}}{m'(\theta')} d\theta' - \int_a^\theta \frac{I_{\theta'}\, m(\theta')}{m'(\theta')} d\theta' + b(y) \quad \text{a.s. } p_\theta,
\end{aligned}$$

where $b(y)$ is an arbitrary function of y but not of θ. Equivalently,

$$p_\theta(y) = C(\theta)h(y)e^{\hat{\theta}(y)\,\eta(\theta)}, \quad \theta \in \mathcal{X}$$

is an exponential family, where $\hat{\theta}(Y)$ is a sufficient statistic and

$$C(\theta) = \exp\left\{-\int_a^\theta \frac{I_{\theta'}\, m(\theta')}{m'(\theta')} d\theta'\right\},$$
$$h(y) = e^{b(y)},$$
$$\text{and } \eta(\theta) = \int_a^\theta \frac{I_{\theta'}}{m'(\theta')} d\theta'.$$

If $\hat{\theta}(Y)$ is an unbiased estimator, then $m(\theta) = \theta$, and therefore (13.28) follows from (13.31). □

13.5 Vector Parameters

The notions developed so far in this chapter can be extended to vector parameters. Consider a parametric family $p_{\boldsymbol{\theta}}(Y)$, $\boldsymbol{\theta} \in \mathcal{X} \subseteq \mathbb{R}^m$, where $\boldsymbol{\theta}$ is the unknown m-vector. Assume the pdf $p_{\boldsymbol{\theta}}(Y)$ is differentiable with respect to $\boldsymbol{\theta}$ and denote by $\nabla_{\boldsymbol{\theta}} f(\boldsymbol{\theta}, y) = [\frac{\partial f(\boldsymbol{\theta}, y)}{\partial \theta_i}]_{i=1}^m$ the gradient of a real-valued function $f(\boldsymbol{\theta}, y)$ with respect to $\boldsymbol{\theta}$ and by $\nabla_{\boldsymbol{\theta}}^2 f(\boldsymbol{\theta}, y) = [\frac{\partial^2 f(\boldsymbol{\theta}, y)}{\partial \theta_i \partial \theta_j}]_{i,j=1}^m$ the Hessian of f with respect to $\boldsymbol{\theta}$. The subscript $\boldsymbol{\theta}$ on the gradient and the Hessian will be omitted when the function f depends on $\boldsymbol{\theta}$ only.

DEFINITION 13.3 The $m \times m$ *Fisher information matrix* of $\boldsymbol{\theta}$ is

$$\mathsf{I}_{\boldsymbol{\theta}} \triangleq \mathbb{E}_{\boldsymbol{\theta}}\left[(\nabla_{\boldsymbol{\theta}} \ln p_{\boldsymbol{\theta}}(Y))\,(\nabla_{\boldsymbol{\theta}} \ln p_{\boldsymbol{\theta}}(Y))^\top\right]. \tag{13.32}$$

Note that $\mathsf{I}_{\boldsymbol{\theta}} \succeq 0$, i.e., it is a nonnegative definite (positive semi-definite) matrix. Its entries are given by

$$(\mathsf{I}_{\boldsymbol{\theta}})_{ij} = \mathbb{E}_{\boldsymbol{\theta}}\left[\frac{\partial \ln p_{\boldsymbol{\theta}}(Y)}{\partial \theta_i}\frac{\partial \ln p_{\boldsymbol{\theta}}(Y)}{\partial \theta_j}\right], \quad 1 \le i, j \le m. \tag{13.33}$$

Similarly to (i)–(v) in Section 13.1, we assume the following regularity conditions hold:

(i) $\mathcal{X}$ is an open hyperrectangle in $\mathbb{R}^m$.
(ii) The support set $\text{supp}\{p_{\boldsymbol{\theta}}\} = \{y \in \mathcal{Y} : p_{\boldsymbol{\theta}}(y) > 0\}$ is the same for all $\boldsymbol{\theta} \in \mathcal{X}$.
(iii) The partial derivatives $\frac{\partial p_{\boldsymbol{\theta}}(y)}{\partial \theta_i}$, $i = 1, 2, \ldots, m$, exist and are finite for all $\boldsymbol{\theta} \in \mathcal{X}$ and $y \in \mathcal{Y}$.
(iv) $\int_{\mathcal{Y}} \nabla_{\boldsymbol{\theta}} p_{\boldsymbol{\theta}}(y) d\mu(y) = \nabla_{\boldsymbol{\theta}} \int_{\mathcal{Y}} p_{\boldsymbol{\theta}}(y) d\mu(y) = 0, \forall \boldsymbol{\theta} \in \mathcal{X}$.
(v) If $\nabla_{\boldsymbol{\theta}}^2 p_{\boldsymbol{\theta}}(y)$ exists and is finite $\forall \boldsymbol{\theta} \in \mathcal{X}, \forall y \in \mathcal{Y}$ then

$$\int_{\mathcal{Y}} \nabla_{\boldsymbol{\theta}}^2 p_{\boldsymbol{\theta}}(y) d\mu(y) = \nabla_{\boldsymbol{\theta}}^2 \int_{\mathcal{Y}} p_{\boldsymbol{\theta}}(y) d\mu(y) = 0.$$

The following lemma is an extension of Lemma 13.1 to the vector case. The lemma is used to prove Theorem 13.4, which is an extension of the CRLB (13.11) to the vector case.

LEMMA 13.2 *Under Assumptions (i)–(iv) we have*

$$\mathbb{E}_{\boldsymbol{\theta}}\left[\nabla_{\boldsymbol{\theta}} \ln p_{\boldsymbol{\theta}}(Y)\right] = \mathbf{0} \tag{13.34}$$

and

$$\mathsf{I}_{\boldsymbol{\theta}} = \mathrm{Cov}_{\boldsymbol{\theta}}\left[\nabla_{\boldsymbol{\theta}} \ln p_{\boldsymbol{\theta}}(Y)\right]. \tag{13.35}$$

If in addition Assumption (v) holds, then

$$\mathsf{I}_{\boldsymbol{\theta}} = \mathbb{E}_{\boldsymbol{\theta}}\left[-\nabla^2_{\boldsymbol{\theta}} \ln p_{\boldsymbol{\theta}}(Y)\right]. \tag{13.36}$$

Analogously to (13.18) and (13.19), the following result holds by application of the Cauchy–Schwarz inequality.

THEOREM 13.3 *Let $g(\boldsymbol{\theta})$ be a real-valued function and $\hat{g}(Y)$ an estimator of $g(\boldsymbol{\theta})$. Assume the following regularity condition holds:*

(vi) $\nabla_{\boldsymbol{\theta}} \int_{\mathcal{Y}} \hat{g}(y)\, p_{\boldsymbol{\theta}}(y)\, d\mu(y) \overset{(a)}{=} \int_{\mathcal{Y}} \hat{g}(y)\, \nabla_{\boldsymbol{\theta}} p_{\boldsymbol{\theta}}(y)\, d\mu(y)$.

Define $\boldsymbol{d} \triangleq \nabla \mathbb{E}_{\boldsymbol{\theta}}[\hat{g}(Y)]$ (which generally depends on $\boldsymbol{\theta}$, but the dependency is not explicitly shown here, to simplify the notation). Then

$$\mathrm{Var}_{\boldsymbol{\theta}}[\hat{g}(Y)] \geq \boldsymbol{d}^\top \mathsf{I}_{\boldsymbol{\theta}}^{-1} \boldsymbol{d}. \tag{13.37}$$

If $\hat{g}(Y)$ is an unbiased estimator, then $\boldsymbol{d} = \nabla g(\boldsymbol{\theta})$ and the right hand side of (13.37) *does not depend on $\hat{g}$.*

Proof We have

$$\boldsymbol{d} = \nabla_{\boldsymbol{\theta}} \int_{\mathcal{Y}} \hat{g}(y)\, p_{\boldsymbol{\theta}}(y)\, d\mu(y) \overset{(a)}{=} \int_{\mathcal{Y}} \hat{g}(y)\, \nabla_{\boldsymbol{\theta}} p_{\boldsymbol{\theta}}(y)\, d\mu(y) = \mathbb{E}_{\boldsymbol{\theta}}[\hat{g}(Y)\, \nabla_{\boldsymbol{\theta}} \ln p_{\boldsymbol{\theta}}(Y)],$$

where equality (a) holds by the regularity assumption (vi). Fix any m-vector $\boldsymbol{a}$. By (13.34) and (13.35), the random variable $Z \triangleq \boldsymbol{a}^\top \nabla_{\boldsymbol{\theta}} \ln p_{\boldsymbol{\theta}}(Y)$ has mean $\mathbb{E}_{\boldsymbol{\theta}}(Z) = 0$ and variance $\mathrm{Var}_{\boldsymbol{\theta}}(Z) = \boldsymbol{a}^\top \mathsf{I}_{\boldsymbol{\theta}}\, \boldsymbol{a}$. By (13.10), we have

$$\mathrm{Var}_{\boldsymbol{\theta}}[\hat{g}(Y)]\, \mathrm{Var}_{\boldsymbol{\theta}}(Z) \geq \left(\mathrm{Cov}_{\boldsymbol{\theta}}(\hat{g}(Y), Z)\right)^2.$$

Moreover, $\mathrm{Cov}_{\boldsymbol{\theta}}(\hat{g}(Y), Z) = \mathbb{E}_{\boldsymbol{\theta}}[\hat{g}(Y)\, Z] = \boldsymbol{a}^\top \boldsymbol{d}$, and therefore

$$\mathrm{Var}_{\boldsymbol{\theta}}[\hat{g}(Y)] \geq \frac{(\boldsymbol{a}^\top \boldsymbol{d})^2}{\boldsymbol{a}^\top \mathsf{I}_{\boldsymbol{\theta}} \boldsymbol{a}}.$$

This inequality holds for all $\boldsymbol{a} \in \mathbb{R}^m$. We now show that the supremum of the right hand side over all $\boldsymbol{a} \in \mathbb{R}^m$ is equal to $\boldsymbol{d}^\top \mathsf{I}_{\boldsymbol{\theta}}^{-1} \boldsymbol{d}$, from which (13.37) follows.

Let $\boldsymbol{b} = \mathsf{I}_\theta^{1/2}\boldsymbol{a}$. By the Cauchy–Schwarz inequality applied to vectors in $\mathbb{R}^m$, we have

$$\frac{(\boldsymbol{a}^\top \boldsymbol{d})^2}{\boldsymbol{a}^\top \mathsf{I}_\theta \boldsymbol{a}} = \frac{(\boldsymbol{b}^\top \mathsf{I}_\theta^{-1/2} \boldsymbol{d})^2}{\boldsymbol{b}^\top \boldsymbol{b}} \le \|\mathsf{I}_\theta^{-1/2}\boldsymbol{d}\|^2 = \boldsymbol{d}^\top \mathsf{I}_\theta^{-1}\boldsymbol{d}$$

with equality if $\boldsymbol{b} = c\mathsf{I}_\theta^{-1/2}\boldsymbol{d}$ (hence $\boldsymbol{a} = c\mathsf{I}_\theta^{-1}\boldsymbol{d}$) for some constant $c \neq 0$. □

THEOREM 13.4 CRLB for Vector Parameters. *For any unbiased estimator* $\hat{\boldsymbol{\theta}}(Y)$,

$$\mathrm{Cov}_{\boldsymbol{\theta}}(\hat{\boldsymbol{\theta}}(Y)) \succeq \mathsf{I}_\theta^{-1}, \tag{13.38}$$

i.e., $\mathrm{Cov}_{\boldsymbol{\theta}}(\hat{\boldsymbol{\theta}}(Y)) - \mathsf{I}_\theta^{-1}$ *is positive semi-definite.*

Proof Fix any $\boldsymbol{a} \in \mathbb{R}^m$ and let $\hat{g}(Y) = \boldsymbol{a}^\top \hat{\boldsymbol{\theta}}(Y)$. Then

$$\boldsymbol{d} = \nabla_{\boldsymbol{\theta}} \mathbb{E}_{\boldsymbol{\theta}}[\hat{g}(Y)] = \nabla_{\boldsymbol{\theta}}(\boldsymbol{a}^\top \boldsymbol{\theta}) = \boldsymbol{a}$$

and (13.37) yields

$$\boldsymbol{a}^\top \mathrm{Cov}_{\boldsymbol{\theta}}(\hat{\boldsymbol{\theta}}(Y))\boldsymbol{a} \ge \boldsymbol{a}^\top \mathsf{I}_\theta^{-1}\boldsymbol{a},$$

which implies that

$$\boldsymbol{a}^\top \left(\mathrm{Cov}_{\boldsymbol{\theta}}(\hat{\boldsymbol{\theta}}(Y)) - \mathsf{I}_\theta^{-1}\right)\boldsymbol{a} \ge 0.$$

Since this inequality holds for any $\boldsymbol{a} \in \mathbb{R}^m$, the claim is proved. □

By carefully following the proof of this theorem, one can also obtain the following result (see Exercise 13.5).

COROLLARY 13.1 Attainment of CRLB in Exponential Families. *Equality holds in* (13.38) *if* $p_{\boldsymbol{\theta}}$ *satisfies the vector identity*

$$\nabla_{\boldsymbol{\theta}} \ln p_{\boldsymbol{\theta}}(y) = \mathsf{I}_{\boldsymbol{\theta}}(\hat{\boldsymbol{\theta}}(y) - \boldsymbol{\theta}) \quad \forall y \in \mathcal{Y}.$$

DEFINITION 13.4 *Efficient Estimator.* An unbiased estimator whose covariance matrix achieves the CRLB (13.38) with equality is said to be efficient.

Of special interest are the variances of the individual components of $\hat{\boldsymbol{\theta}}$, which are also the diagonal terms of the covariance matrix. Since the diagonal entries of a nonnegative definite matrix are nonnegative, we have

$$\mathrm{Var}_{\boldsymbol{\theta}}[\hat{\theta}_i] \ge \left[\mathsf{I}_\theta^{-1}\right]_{i,i} \qquad i = 1, \dots, m. \tag{13.39}$$

On the other hand, the CRLB for scalar parameters gives

$$\mathrm{Var}_{\boldsymbol{\theta}}[\hat{\theta}_i] \ge \frac{1}{\left[\mathsf{I}_{\boldsymbol{\theta}}\right]_{i,i}}, \tag{13.40}$$

where

$$\left[\mathsf{I}_{\boldsymbol{\theta}}\right]_{i,i} = \mathbb{E}_\theta\left[-\frac{\partial^2 \ln p_{\boldsymbol{\theta}}(Y)}{\partial \theta_i^2}\right].$$

Since

$$\left[\mathsf{I}_{\boldsymbol{\theta}}^{-1}\right]_{i,i} \geq \frac{1}{\left[\mathsf{I}_{\boldsymbol{\theta}}\right]_{i,i}} \tag{13.41}$$

for a positive definite matrix, the CRLB for the vector parameter gives a tighter bound (see also Exercise 13.7). If $\mathsf{I}_{\boldsymbol{\theta}}$ is a diagonal matrix, the bounds given in (13.39 and 13.40) are equal. In this case, the components of $\boldsymbol{\theta}$ are said to be orthogonal parameters.

Rank The Fisher information matrix need not have full rank. For instance, let $m = 2$ and consider a variation of Example 13.1 where θ is replaced by $\theta_1 + \theta_2$, and $\mathcal{X} = \mathbb{R}^2$. Then the 2×2 Fisher information matrix

$$\mathsf{I}_{\boldsymbol{\theta}} = \frac{\|s\|^2}{\sigma^2} \begin{pmatrix} 1 & 1 \\ 1 & 1 \end{pmatrix}$$

has rank one. In this problem the parameter $\boldsymbol{\theta}$ appears only through the sum $\theta_1 + \theta_2$ of its components, and the difference $\theta_1 - \theta_2$ is not identifiable.

Another way one can have a rank-deficient Fisher information matrix is via redundancy in the parameterization. Consider for instance $\mathcal{X} = \{\boldsymbol{\theta} \in \mathbb{R}^2 \, : \, \theta_1 = \theta_2\}$, which describes a line in $\mathbb{R}^2$. Then all partial derivatives in (13.33) are equal, and $\mathsf{I}_{\boldsymbol{\theta}}$ is rank-deficient.

Relation to KL Divergence Assume $\nabla_{\boldsymbol{\theta}}^2 \ln p_{\boldsymbol{\theta}}(Y)$ exists so that the expression (13.36) for the Fisher information matrix holds. Consider a small variation $\epsilon\, \boldsymbol{a}$ of the parameter $\boldsymbol{\theta}$ in direction $\boldsymbol{a}$. Assuming $\boldsymbol{\theta} + \epsilon\, \boldsymbol{a} \in \mathcal{X}$, we obtain, similarly to (13.17),

$$\begin{aligned} D(p_{\boldsymbol{\theta}} \| p_{\boldsymbol{\theta}+\epsilon \boldsymbol{a}}) &= \mathbb{E}_{\boldsymbol{\theta}}\left[\ln \frac{p_{\boldsymbol{\theta}}(Y)}{p_{\boldsymbol{\theta}+\epsilon \boldsymbol{a}}(Y)}\right] \\ &= \mathbb{E}_{\boldsymbol{\theta}}\left[-\epsilon \boldsymbol{a}^\top \nabla_{\boldsymbol{\theta}} \ln p_{\boldsymbol{\theta}}(Y) + \frac{\epsilon^2}{2} \boldsymbol{a}^\top (\nabla_{\boldsymbol{\theta}}^2 \ln p_{\boldsymbol{\theta}}(Y)) \boldsymbol{a} + o(\epsilon^2)\right] \\ &= \frac{\epsilon^2}{2} \boldsymbol{a}^\top \mathsf{I}_{\boldsymbol{\theta}}\, \boldsymbol{a} + o(\epsilon^2) \quad \text{as } \epsilon \to 0. \end{aligned} \tag{13.42}$$

The quadratic form $\boldsymbol{a}^\top \mathsf{I}_{\boldsymbol{\theta}}\, \boldsymbol{a}$ is positive definite and therefore induces a metric

$$d_{\mathrm{F}}(\boldsymbol{\theta}, \boldsymbol{\theta}') \triangleq \sqrt{(\boldsymbol{\theta} - \boldsymbol{\theta}')^\top \mathsf{I}_{\boldsymbol{\theta}} (\boldsymbol{\theta} - \boldsymbol{\theta}')} \tag{13.43}$$

on the parameter space $\mathcal{X}$; this distance is called the *Fisher information metric.*

Reparameterization Let $\boldsymbol{\eta} = g(\boldsymbol{\theta})$ where the mapping g is assumed to be continuously differentiable with Jacobian $\mathsf{J} = [\partial \eta_j / \partial \theta_i]_{1 \le i, j \le m}$. By the chain rule for differentiation, we have

$$\frac{\partial \ln p_{\boldsymbol{\theta}}(Y)}{\partial \theta_i} = \sum_{j=1}^{m} \frac{\partial \eta_j}{\partial \theta_i} \frac{\partial \ln q_{\boldsymbol{\eta}}(Y)}{\partial \eta_i}, \quad i = 1, 2, \ldots, m,$$

or equivalently, in vector form,

$$\nabla_{\theta} \ln p_{\theta}(Y) = \mathsf{J}\, \nabla_{\eta} \ln q_{\eta}(Y). \tag{13.44}$$

Taking the outer product of this vector with itself and taking expectations, we see that the Fisher information matrix for η satisfies

$$\mathsf{I}_{\theta} = \mathsf{J}\, \mathsf{I}_{\eta}\, \mathsf{J}^{\top}. \tag{13.45}$$

Exponential Family For an m-dimensional exponential family in canonical form, combining (11.26), (11.29), and (13.36) we obtain

$$\mathsf{I}_{\eta} = \nabla_{\eta}^{2} A(\eta) = \mathrm{Cov}_{\eta}[\boldsymbol{T}(Y)] = \mathrm{Cov}_{\theta}[\boldsymbol{T}(Y)]. \tag{13.46}$$

Example 13.5 *Fisher Information Matrix for Covariance Estimation in Gaussian Model.* Let $\boldsymbol{Y}_i \sim$ i.i.d. $\mathcal{N}(\mathbf{0}, \mathsf{C})$ for $1 \le i \le n$ where C is an $r \times r$ covariance matrix. Viewing C as the parameter of the family, we have

$$\begin{aligned} \ln p_{\mathsf{C}}(\boldsymbol{Y}) &= -\frac{rn}{2} \ln(2\pi) - \frac{n}{2} \ln|\mathsf{C}| - \frac{1}{2} \sum_{i=1}^{n} \boldsymbol{Y}_i^{\top} \mathsf{C}^{-1} \boldsymbol{Y}_i \\ &= -\frac{rn}{2} \ln(2\pi) - \frac{n}{2} \ln|\mathsf{C}| - \frac{n}{2} \mathrm{Tr}[\boldsymbol{S}\mathsf{C}^{-1}], \end{aligned}$$

where $\boldsymbol{S} = \frac{1}{n} \sum_{i=1}^{n} \boldsymbol{Y}_i^{\top} \boldsymbol{Y}_i$ is the $r \times r$ sample covariance matrix. Again it is convenient to use the canonical representation of the family, where $\boldsymbol{T}(\boldsymbol{Y}) = -\frac{n}{2}\boldsymbol{S}$ and $\boldsymbol{\eta} = \mathsf{C}^{-1}$ is the inverse covariance matrix, or precision matrix. Then

$$\ln q_{\eta}(\boldsymbol{S}) = -\frac{rn}{2} \ln(2\pi) + \frac{n}{2} \ln|\boldsymbol{\eta}| - \frac{n}{2} \sum_{i,j=1}^{r} S_{ji} \eta_{ij}.$$

By symmetry of the precision matrix, this is an exponential family of order $\frac{r(r+1)}{2}$. We have

$$\mathbb{E}_{\mathsf{C}}[\boldsymbol{T}(\boldsymbol{Y})] = -\frac{n}{2}\mathsf{C}$$

and the Fisher information matrix for $\boldsymbol{\eta}$ is given by (proof left to the reader)

$$\begin{aligned} (\mathsf{I}_{\eta})_{ij,kl} &= \mathrm{Cov}_{\mathsf{C}}[T_{ij}(\boldsymbol{Y}) T_{kl}(\boldsymbol{Y})] \\ &= \begin{cases} \frac{n}{2}\mathsf{C}_{ij}^{2} & : (ij) = (kl) \\ \frac{n}{2}\mathsf{C}_{ik}\mathsf{C}_{jl} & : i = k,\ j \ne l \text{ or } i \ne k,\ j = l \le i, j, k, l \le r \\ \frac{n}{4}[\mathsf{C}_{ik}\mathsf{C}_{jl} + \mathsf{C}_{il}\mathsf{C}_{jk}] & : i \ne k,\ j \ne l. \end{cases} \end{aligned}$$

Alternatively, this result could have been derived using the formulas for first and second partial derivatives of the log-determinant function $\ln|\boldsymbol{\eta}|$.

Finally, note that since $\boldsymbol{\eta}$ has r^2 entries, the Fisher information matrix I_{η} has dimensions $r^2 \times r^2$. However, $\boldsymbol{\eta}$ is symmetric and has only $\frac{r(r+1)}{2}$ degrees of freedom, hence the rank of I_{η} is at most $\frac{r(r+1)}{2}$.

13.6 Information Inequality for Random Parameters

If $\boldsymbol{\theta}$ is a random vector, a lower bound on the correlation matrix of the estimation error vector can also be derived. Denote by $\pi(\boldsymbol{\theta})$ the prior density and by $p(\boldsymbol{\theta}, y) = p_{\boldsymbol{\theta}}(y)\pi(\boldsymbol{\theta})$ the joint density, assumed to be twice differentiable in $\boldsymbol{\theta}$, for each $y \in \mathcal{Y}$. Define the *total Fisher information matrix*

$$\mathsf{I}^{\text{tot}} \triangleq \mathbb{E}\left\{\left[\nabla_{\boldsymbol{\Theta}} \ln p(\boldsymbol{\Theta}, Y)\right]\left[\nabla_{\boldsymbol{\Theta}} \ln p(\boldsymbol{\Theta}, Y)\right]^{\top}\right\}$$
$$= \mathbb{E}\left[-\nabla^2_{\boldsymbol{\Theta}} \ln p(\boldsymbol{\Theta}, Y)\right], \tag{13.47}$$

where the expectation is with respect to both Y and $\boldsymbol{\Theta}$. Since $\ln p(\boldsymbol{\theta}, Y) = \ln p_{\boldsymbol{\theta}}(Y) + \ln \pi(\boldsymbol{\theta})$, the total Fisher information matrix can be split into two parts

$$\mathsf{I}^{\text{tot}} = \mathsf{I}^{\text{D}} + \mathsf{I}^{\text{P}}, \tag{13.48}$$

where

$$\mathsf{I}^{\text{D}} = \mathbb{E}\left[-\nabla^2_{\boldsymbol{\Theta}} \ln p_{\boldsymbol{\Theta}}(Y)\right] \succeq 0 \tag{13.49}$$

is the average Fisher information matrix associated with the observations, and

$$\mathsf{I}^{\text{P}} = \mathbb{E}\left[-\nabla^2_{\boldsymbol{\Theta}} \ln \pi(\boldsymbol{\Theta})\right] \succeq 0 \tag{13.50}$$

is the Fisher information matrix associated with the prior. Using a derivation analogous to the information inequality, one may show that the correlation matrix for the estimation error vector satisfies the following inequality [6, Ch. 2.4.3]:

$$\mathbb{E}\left[(\hat{\boldsymbol{\theta}} - \boldsymbol{\Theta})(\hat{\boldsymbol{\theta}} - \boldsymbol{\Theta})^{\top}\right] \succeq \left(I^{\text{total}}\right)^{-1} \tag{13.51}$$

with equality if and only if

$$\nabla^2_{\boldsymbol{\Theta}} \ln p(\boldsymbol{\Theta}, Y) = -\mathsf{Q} \quad \text{a.s.}$$

for some constant nonnegative definite matrix Q. Since $\ln p(\boldsymbol{\theta}, y) = \ln \pi(\boldsymbol{\theta}|y) + \ln p(y)$, this implies

$$\nabla^2_{\boldsymbol{\theta}} \ln \pi(\boldsymbol{\theta}|y) = -\mathsf{Q}.$$

Integrating twice over $\boldsymbol{\theta}$, we obtain

$$\ln \pi(\boldsymbol{\theta}|y) = -\frac{1}{2}\boldsymbol{\theta}^{\top}\mathsf{Q}\boldsymbol{\theta} + \boldsymbol{a}(y)^{\top}\boldsymbol{\theta} + b(y)$$

for some vector $\boldsymbol{a}(y)$ and scalar $b(y)$. Hence the posterior distribution $\pi(\boldsymbol{\theta}|y)$ must be Gaussian for equality to hold in (13.51) and therefore for the estimator to be efficient.

Example 13.6 Let $\boldsymbol{Y} \sim \mathcal{N}(\boldsymbol{\theta}, \mathsf{C})$ be a Gaussian vector with unknown mean $\boldsymbol{\theta} \in \mathbb{R}^m$ and known covariance C. Assume $\boldsymbol{\theta}$ is also random with Gaussian prior $\mathcal{N}(\mathbf{0}, \mathsf{C}_{\boldsymbol{\Theta}})$. The marginal distribution of $\boldsymbol{Y}$ is $\mathcal{N}(\mathbf{0}, \mathsf{C}_{\boldsymbol{\Theta}} + \mathsf{C})$, and the posterior distribution of $\boldsymbol{Y}$ given $\boldsymbol{\theta}$ is $\mathcal{N}(\hat{\boldsymbol{\theta}}_{\text{MMSE}}(Y), \mathsf{C}_E)$, where

$$\hat{\boldsymbol{\theta}}_{\text{MMSE}}(\boldsymbol{Y}) = \mathsf{C}_{\boldsymbol{\Theta}}(\mathsf{C}_{\boldsymbol{\Theta}} + \mathsf{C})^{-1}\boldsymbol{Y} \tag{13.52}$$

is the (linear) MMSE estimator, and

$$\begin{aligned}\mathsf{C}_E &= \mathsf{C}_\Theta - \mathsf{C}_\Theta(\mathsf{C}_\Theta + \mathsf{C})^{-1}\mathsf{C}_\Theta \\ &= (\mathsf{C}^{-1} + \mathsf{C}_\Theta^{-1})^{-1}\end{aligned}$$

is its covariance matrix. We have

$$\ln p_{\boldsymbol{\theta}}(\boldsymbol{Y}) = -\frac{m}{2}\ln(2\pi) - \frac{1}{2}\ln|\mathsf{C}| - \frac{1}{2}(\boldsymbol{Y}-\boldsymbol{\theta})^\top \mathsf{C}^{-1}(\boldsymbol{Y}-\boldsymbol{\theta}),$$

and thus (13.49) yields

$$\mathsf{I}^{\mathrm{D}} = \mathsf{C}^{-1}.$$

Likewise

$$\ln \pi(\boldsymbol{\theta}) = -\frac{m}{2}\ln(2\pi) - \frac{1}{2}\ln|\mathsf{C}_\Theta| - \frac{1}{2}\boldsymbol{\theta}^\top \mathsf{C}_\Theta^{-1}\boldsymbol{\theta},$$

and thus (13.50) yields

$$\mathsf{I}^{\mathrm{P}} = \mathsf{C}_\Theta^{-1}.$$

The total Fisher information (13.47) is the sum of the inverse covariance matrices:

$$\mathsf{I}^{\mathrm{tot}} = \mathsf{C}^{-1} + \mathsf{C}_\Theta^{-1}.$$

Then

$$\mathbb{E}\left[(\hat{\boldsymbol{\theta}}-\boldsymbol{\theta})(\hat{\boldsymbol{\theta}}-\boldsymbol{\theta})^\top\right] \succeq (\mathsf{C}^{-1} + \mathsf{C}_\Theta^{-1})^{-1}$$

for any estimator $\hat{\boldsymbol{\theta}}$. Equality is achieved by the MMSE estimator (13.52).

13.7 Biased Estimators

Recall that an MVUE is said to be *efficient* if its variance achieves the CRLB. However, some biased estimator might achieve lower mean squared-error (MSE) than does the MVUE, as illustrated in Example 13.7.

Example 13.7 Given $Y_i \overset{\text{i.i.d.}}{\sim} \mathcal{N}(0, \theta)$, $1 \le i \le n$, estimate the unknown variance θ. The likelihood function is given by

$$p_\theta(\boldsymbol{y}) = (2\pi\theta)^{-n/2} \exp\left\{-\frac{1}{2\theta}\sum_{i=1}^n y_i^2\right\}. \tag{13.53}$$

We have seen in Example 12.11 that the MVUE of θ is

$$\hat{\theta}_{\mathrm{MVUE}}(\boldsymbol{Y}) = \frac{1}{n}\sum_{i=1}^n Y_i^2. \tag{13.54}$$

Now, consider a biased estimator of the form

$$\hat{\theta}_a(\boldsymbol{Y}) = a\,\hat{\theta}_{\mathrm{MVUE}}(\boldsymbol{Y}) = \frac{a}{n}\sum_{i=1}^n Y_i^2, \tag{13.55}$$

where $a \in \mathbb{R}$ is a scaling factor to be optimized. The MSE of the estimator $\hat{\theta}_a$ is given by

$$\begin{aligned}
\mathrm{MSE}_\theta(\hat{\theta}_a) &= \mathbb{E}_\theta\left[\hat{\theta}_a(\boldsymbol{Y}) - \theta\right]^2 \\
&= \mathbb{E}_\theta\left(\frac{a}{n}\sum_{i=1}^{n} Y_i^2 - \theta\right)^2 \\
&= \mathbb{E}_\theta\left(\frac{a^2}{n^2}\left(\sum_{i=1}^{n} Y_i^2\right)^2 - \frac{2\theta a}{n}\sum_{i=1}^{n} Y_i^2 + \theta^2\right) \\
&= \left[a^2\left(1+\frac{2}{n}\right) - 2a + 1\right]\theta^2,
\end{aligned} \tag{13.56}$$

where the last line follows from

$$\begin{aligned}
\mathbb{E}_\theta\left[\left(\sum_{i=1}^{n} Y_i^2\right)^2\right] &= \mathbb{E}_\theta\left[\sum_{i,j=1}^{n} Y_i^2 Y_j^2\right] \\
&= \sum_{i=1}^{n} \mathbb{E}_\theta\left[Y_i^4\right] + \sum_{i\neq j=1}^{n} \mathbb{E}_\theta\left[Y_i^2\right]\mathbb{E}_\theta\left[Y_j^2\right] \\
&= 3n\theta^2 + n(n-1)\theta^2 \\
&= (n^2+2n)\theta^2.
\end{aligned}$$

Using (13.56) with $a = 1$, we obtain $\mathrm{MSE}_\theta(\hat{\theta}_{\mathrm{MVUE}}) = 2\theta^2/n$.

However, the value of a that minimizes (13.56) is

$$a_{\mathrm{opt}} = \frac{1}{1+2/n} < 1 \tag{13.57}$$

and the resulting MSE is obtained by substituting (13.57) into (13.56):

$$\mathrm{MSE}_\theta(\hat{\theta}_{a_{\mathrm{opt}}}) = \left(\frac{1}{1+2/n} - \frac{2}{1+2/n} + 1\right)\theta^2 = \frac{2}{n+2}\theta^2 = \frac{n}{n+2}\mathrm{MSE}_\theta(\hat{\theta}_{\mathrm{MVUE}}).$$

It is remarkable that, for $n = 1$, we have $a_{\mathrm{opt}} = 3$, and $\hat{\theta}_{\mathrm{MMSE}}(Y) = Y^2/3$. The MSE of this biased estimator is *three times smaller* than that of the MVUE.

More generally, owing to the bias–variance decomposition (13.1), we can consider the two extreme cases of the bias–variance tradeoff:

- For any unbiased estimator, $\mathrm{MSE}_\theta(\hat{\theta}) = \mathrm{Var}_\theta(\hat{\theta})$.
- Any estimator returning a constant c has zero variance, and

$$\mathrm{MSE}_\theta(\hat{\theta}) = \left[b_\theta(\hat{\theta})\right]^2 = (c-\theta)^2.$$

13.8 Appendix: Derivation of (13.16)

To derive (13.16) we make use of the following differentiation formulas:

$$\frac{d \ln |\mathsf{C}(\theta)|}{d\theta} = \mathrm{Tr}\left[\mathsf{C}^{-1}(\theta)\frac{d\mathsf{C}(\theta)}{d\theta}\right], \tag{13.58}$$

$$\frac{d\mathsf{C}^{-1}(\theta)}{d\theta} = -\mathsf{C}^{-1}(\theta)\frac{d\mathsf{C}(\theta)}{d\theta}\mathsf{C}^{-1}(\theta), \tag{13.59}$$

and the expectation formulas

$$\mathbb{E}\left[\boldsymbol{Y}^\top \mathsf{A}\boldsymbol{Y}\right] = \mathrm{Tr}[\mathsf{CA}], \tag{13.60}$$

which holds for any matrix A and any random vector $\boldsymbol{Y}$ with mean zero and covariance matrix C, and

$$\mathbb{E}\mathrm{Tr}[\boldsymbol{Y}\boldsymbol{Y}^\top \mathsf{A}\boldsymbol{Y}\boldsymbol{Y}^\top \mathsf{B}] = \mathrm{Tr}[\mathsf{AC}]\mathrm{Tr}[\mathsf{BC}] + 2\mathrm{Tr}[\mathsf{ACBC}], \tag{13.61}$$

which holds for any symmetric matrices A and B and any random vector $\boldsymbol{Y}$ with mean zero and covariance matrix C.

Differentiating the log-likelihood

$$\ln p_\theta(\boldsymbol{Y}) = -\frac{n}{2}\ln(2\pi) - \frac{1}{2}\ln|\mathsf{C}(\theta)| - \frac{1}{2}\boldsymbol{Y}^\top \mathsf{C}^{-1}(\theta)\boldsymbol{Y}$$

with respect to θ, we obtain

$$\begin{aligned}\frac{\partial \ln p_\theta(\boldsymbol{Y})}{\partial\theta} &= -\frac{1}{2}\frac{d\ln|\mathsf{C}(\theta)|}{d\theta} - \frac{1}{2}\boldsymbol{Y}^\top\frac{d\mathsf{C}^{-1}(\theta)}{d\theta}\boldsymbol{Y}\\ &= -\frac{1}{2}\mathrm{Tr}\left[\mathsf{C}^{-1}(\theta)\frac{d\mathsf{C}(\theta)}{d\theta}\right] + \frac{1}{2}\boldsymbol{Y}^\top\mathsf{C}^{-1}(\theta)\frac{d\mathsf{C}(\theta)}{d\theta}\mathsf{C}^{-1}(\theta)\boldsymbol{Y},\end{aligned} \tag{13.62}$$

where the last equality results from the differentiation formulas (13.58) and (13.59). Squaring both sides of (13.62), we obtain

$$\begin{aligned}&\left(\frac{\partial \ln p_\theta(\boldsymbol{Y})}{\partial\theta}\right)^2\\ &= \frac{1}{4}\mathrm{Tr}^2\left[\mathsf{C}^{-1}(\theta)\frac{d\mathsf{C}(\theta)}{d\theta}\right] - \frac{1}{2}\mathrm{Tr}\left[\mathsf{C}^{-1}(\theta)\frac{d\mathsf{C}(\theta)}{d\theta}\right]\boldsymbol{Y}^\top\underbrace{\mathsf{C}^{-1}(\theta)\frac{d\mathsf{C}(\theta)}{d\theta}\mathsf{C}^{-1}(\theta)}_{=\mathsf{A}}\boldsymbol{Y}\\ &\quad + \frac{1}{4}\boldsymbol{Y}^\top\underbrace{\mathsf{C}^{-1}(\theta)\frac{d\mathsf{C}(\theta)}{d\theta}\mathsf{C}^{-1}(\theta)}_{=\mathsf{A}}\boldsymbol{Y}\boldsymbol{Y}^\top\underbrace{\mathsf{C}^{-1}(\theta)\frac{d\mathsf{C}(\theta)}{d\theta}\mathsf{C}^{-1}(\theta)}_{=\mathsf{A}}\boldsymbol{Y}.\end{aligned}$$

Taking expectations and applying the formulas (13.60) and (13.61) with $\mathsf{A} = \mathsf{B} = \mathsf{C}^{-1}(\theta)\frac{d\mathsf{C}(\theta)}{d\theta}\mathsf{C}^{-1}(\theta)$, we obtain Fisher information as

$$
\begin{aligned}
I_\theta &= \mathbb{E}_\theta \left(\frac{\partial \ln p_\theta(\mathbf{Y})}{\partial \theta} \right)^2 \\
&= \frac{1}{4}\mathrm{Tr}^2\left[\mathsf{C}^{-1}(\theta)\frac{d\mathsf{C}(\theta)}{d\theta}\right] - \frac{1}{2}\mathrm{Tr}^2\left[\mathsf{C}^{-1}(\theta)\frac{d\mathsf{C}(\theta)}{d\theta}\right] \\
&\quad + \frac{1}{4}\left(\mathrm{Tr}^2\left[\mathsf{C}^{-1}(\theta)\frac{d\mathsf{C}(\theta)}{d\theta}\right] + 2\mathrm{Tr}\left[\mathsf{C}^{-1}(\theta)\frac{d\mathsf{C}(\theta)}{d\theta}\mathsf{C}^{-1}(\theta)\frac{d\mathsf{C}(\theta)}{d\theta}\right]\right).
\end{aligned}
$$

The three terms involving squares of traces cancel out, and (13.16) follows.

Exercises

13.1 *Fisher information and CRLB for Poisson Family.* Let Y be Poisson with parameter $\theta > 0$.

Find the Fisher information and CRLB for estimation of θ and $\sqrt{\theta}$.

13.2 *Fisher Information and CRLB for i.i.d. Observations.* Let $\mathcal{X} \subset \mathbb{R}$ and i_θ be the Fisher information for $p_\theta, \theta \in \mathcal{X}$.

Derive an expression for the Fisher information and the CRLB for the i.i.d., n-variate extension of that family.

13.3 *Fisher Information for Independent Observations.* Let $\boldsymbol{\theta}$ be a vector parameter, and $\{Y_i\}_{i=1}^n$ be independent random variables with respective pdfs $p_{\boldsymbol{\theta},i}$ with respect to a common measure μ.

Show that the Fisher information matrix for $\boldsymbol{\theta}$ is the sum of the Fisher information matrices for the individual components: $\mathsf{I}_{\boldsymbol{\theta}}^{(n)} = \sum_{i=1}^n \mathsf{I}_{\boldsymbol{\theta},i}$.

13.4 *Fisher Information for Location-Scale Family.* Let q be a pdf with support set equal to $\mathbb{R}$. Show that for the so-called location-scale family

$$
\left\{ p_\theta(y) = \frac{1}{\theta_2} q\left(\frac{y-\theta_1}{\theta_2}\right), \quad y, \theta_1 \in \mathbb{R},\ \theta_2 > 0 \right\},
$$

the elements of the Fisher information matrix are

$$
\begin{aligned}
I_{11} &= \frac{1}{\theta_2^2} \int_{\mathbb{R}} \left(\frac{q'(y)}{q(y)}\right)^2 q(y)\, dy, \\
I_{22} &= \frac{1}{\theta_2^2} \int_{\mathbb{R}} \left(\frac{yq'(y)}{q(y)} + 1\right)^2 q(y)\, dy, \\
I_{12} &= \frac{1}{\theta_2^2} \int_{\mathbb{R}} y \left(\frac{q'(y)}{q(y)}\right)^2 q(y)\, dy.
\end{aligned}
$$

If the prototype distribution $q(y)$ is symmetric about zero, the crossterm $I_{12} = 0$.

13.5 *Attainment of CRLB in Exponential Families.* Prove Corollary 13.1.

13.6 *Efficient Estimator.* Let X_1, X_2, X_3 be i.i.d. $\mathcal{N}(\theta, 1)$ where $\theta \in \mathbb{R}$. You are given the partial sums $Y_1 = X_1 + X_2$ and $Y_2 = X_2 + X_3$.

Derive the MVUE of θ and show it is an efficient estimator.

13.7 *Vector CRLB.* Suppose Y_1 and Y_2 are independent random variables, with $Y_1 \sim \mathcal{N}(\theta_1, 1)$ and $Y_2 \sim \mathcal{N}(\theta_1 + \theta_2, 1)$.

(a) Find the Fisher information matrix $\mathsf{I}_{\boldsymbol{\theta}}$ and the vector CRLB for the estimation of $\boldsymbol{\theta} = [\theta_1\ \theta_2]^\top$ from $\boldsymbol{Y} = [Y_1\ Y_2]^\top$.
(b) Now compute the scalar CRLB individually for each of θ_1 and θ_2 (see (13.40)). Compare these bounds with those obtained from part (a).

13.8 *Linear Regression.*

$$Y_k = \theta_1 + \theta_2 k + Z_k,\ k = 1, 2, \ldots, n,$$

where $\{Z_k\}$ are i.i.d. $\mathcal{N}(0, 1)$ random variables.

(a) Find the Fisher information matrix for the vector parameter $\boldsymbol{\theta} = [\theta_1\ \theta_2]$ and its inverse. You may leave your answer in terms of the sums:

$$s_1(n) = \sum_{k=1}^{n} k = \frac{n(n+1)}{2}, \qquad s_2(n) = \sum_{k=1}^{n} k^2 = \frac{n(n+1)(2n+1)}{6}.$$

(b) Now suppose $n = 2$. Explain why unique MVUE estimators of θ_1 and θ_2 exist, and give their variance.

13.9 *Estimating a 2-D Sinusoid in Gaussian Noise.* Consider an image patch consisting of a 2-D sine pattern

$$s_{\boldsymbol{i}}(\boldsymbol{\theta}) = \cos(\boldsymbol{i}^\top \boldsymbol{\theta}),\ \boldsymbol{i} = (i_1, i_2) \in \{0, 1, \ldots, 7\}^2$$

at angular frequency

$$\boldsymbol{\theta} = (\theta_1, \theta_2) \in [0, \pi)^2$$

corrupted by i.i.d. $\mathcal{N}(0, 1)$ noise.

(a) Derive and plot the CRLB for an unbiased estimator of $\boldsymbol{\theta}$.
(b) Comment on the dependency of the CRLB on $\boldsymbol{\theta}$, and identify the frequencies for which the Fisher information matrix is singular.

13.10 *Estimating the Parameters of a General Gaussian Model.* In (13.14) and (13.16), we derived the Fisher information for a scalar θ that parameterizes the mean and the covariance of a Gaussian observation vector $\boldsymbol{Y}$, respectively. In this problem, we generalize those results.

(a) Show that the Fisher information matrix for an m-vector $\boldsymbol{\theta}$ in the model $\boldsymbol{Y} \sim \mathcal{N}(\boldsymbol{s}(\boldsymbol{\theta}), \mathsf{C})$ (for a fixed $n \times n$ covariance matrix C) is given by

$$(\mathsf{I}_{\boldsymbol{\theta}})_{jk} = \frac{\partial \boldsymbol{s}(\boldsymbol{\theta})^\top}{\partial \theta_j} \mathsf{C}^{-1} \frac{\partial \boldsymbol{s}(\boldsymbol{\theta})}{\partial \theta_k}, \quad 1 \le j, k \le m.$$

(b) Show that the Fisher information matrix for an m-vector $\boldsymbol{\theta}$ in the model $\boldsymbol{Y} \sim \mathcal{N}(\boldsymbol{s}, \mathsf{C}(\boldsymbol{\theta}))$ (for a fixed mean $\boldsymbol{s}$) is given by

$$(\mathsf{I}_{\boldsymbol{\theta}})_{jk} = \frac{1}{2}\mathrm{Tr}\left[\mathsf{C}^{-1}(\theta)\frac{\partial\mathsf{C}(\theta)}{\partial\theta_j}\mathsf{C}^{-1}(\theta)\frac{\partial\mathsf{C}(\theta)}{\partial\theta_k}\right], \quad 1 \le j, k \le m,$$

and is therefore independent of $\boldsymbol{s}$.

(c) Show that the Fisher information matrix for an m-vector $\boldsymbol{\theta}$ in the general model $\boldsymbol{Y} \sim \mathcal{N}(\boldsymbol{s}(\boldsymbol{\theta}), \mathsf{C}(\boldsymbol{\theta}))$ is given by the sum of the two Fisher information matrices in parts (a) and (b).

13.11 *Estimating Linear Filter Parameters.* You observe a random signal whose samples $\{Y_i\}_{i=1}^n$ satisfy the equation $Y_i = \theta_0 Z_i + \theta_1 Z_{i-1}$ where $Z_0 = 0$ and $\{Z_i\}_{i=1}^n$ are i.i.d. $\mathcal{N}(0, 1)$. Hence the observations are filtered white Gaussian noise.

(a) Derive the mean and covariance matrix for the observations.
(b) Using a result of Exercise 13.10, derive the Fisher information matrix for (θ_0, θ_1).
(c) Give the CRLB on the variance of any unbiased estimator of θ_0 and θ_1.

13.12 *Estimation of Log-Variance.* Derive the Fisher information of the log-variance $\xi = \ln\theta$ for the estimation problem of (13.26).

13.13 *Efficiency and Linear Reparameterization.* Show that if $\hat{\boldsymbol{\theta}}$ is an efficient estimator of $\boldsymbol{\theta} \in \mathbb{R}^m$, then $\mathsf{M}\hat{\boldsymbol{\theta}} + \boldsymbol{b}$ is an efficient estimator of $\mathsf{M}\boldsymbol{\theta} + \boldsymbol{b}$, for any $m \times m$ matrix M and m-vector $\boldsymbol{b}$.

13.14 *Efficiency and Nonlinear Reparameterization.* Assume that the observations $\{Y_i\}_{i=1}^n$ are drawn i.i.d. $\mathcal{N}(\theta, 1)$. Recall that the sample mean

$$\hat{\theta} = \frac{1}{n}\sum_{i=1}^{n} Y_i$$

is an efficient estimator of θ. Now consider estimating θ^2.

(a) Compute the CRLB for any unbiased estimator of θ^2.
(b) Derive the bias and the variance of $\hat{\theta}^2$ as an estimator of θ^2. Evaluate them for $n = 1$.
(c) Now consider large n. Show that $\hat{\theta}^2$ is asymptotically unbiased (its bias tends to 0 as $n \to \infty$) and asymptotically efficient (the ratio of its variance to the CRLB tends to 1 as $n \to \infty$).

13.15 *Averaging Unbiased Estimators.* Assume $\hat{\theta}_1$ and $\hat{\theta}_2$ are unbiased estimators of a scalar parameter θ. The estimators have respective variances σ_1^2 and σ_2^2 and normalized correlation $\rho = \mathrm{Cov}(\hat{\theta}_1, \hat{\theta}_2)/(\sigma_1\sigma_2)$, all independent of θ.

(a) Show that the averaged estimator

$$\hat{\theta}_\lambda \triangleq \lambda\theta_1 + (1 - \lambda)\theta_2$$

is unbiased for all $\lambda \in \mathbb{R}$.

(b) Find λ that minimizes the variance of $\hat{\theta}_\lambda$ in part (a).
(c) Specialize your answer in the case of uncorrelated estimators ($\rho = 0$).

References

[1] F. Y. Edgeworth, "On the Probable Errors of Frequency Constraints," *J. Royal Stat. Soc. Series B*, Vol. 71, pp. 381–397, 499–512, 651–678; Vol. 72, pp. 81–90, 1908 and 1909.

[2] R. A. Fisher, "On the Mathematical Foundations of Theoretical Statistics," *Philos. Trans. Roy. Stat. Soc. London*, Series A, Vol. 222, pp. 309–368, 1922.

[3] H. Cramér, "A Contribution to the Theory of Statistical Estimation," *Skand. Akt. Tidskr.*, Vol. 29, pp. 85–94, 1946.

[4] C. R. Rao, "Information and the Accuracy Attainable in the Estimation of Statistical Parameters," *Bull. Calc. Math. Soc.*, Vol. 37, pp. 81–91, 1945.

[5] M. Fréchet, "Sur l'Extension de Certaines Évaluations Statistiques au cas de Petits Échantillons," *Rev. Int. Stat.*, Vol. 11, No. 3/4, pp. 182–205, 1943.

[6] H. L. Van Trees, *Detection, Estimation and Modulation Theory, Part I*, Wiley, 1968.

14 Maximum Likelihood Estimation

Maximum likelihood (ML) is one of the most popular approaches to parameter estimation. We formulate the problem as a maximization over the parameter set $\mathcal{X}$ and identify closed-form solutions when the parametric family is an exponential family. We identify conditions under which the ML estimator is also an MVUE. For the problem of estimation given n conditionally i.i.d. observations, we show the ML estimator to be asymptotically consistent, efficient, and Gaussian as n tends to infinity, under certain regularity conditions. We then study recursive ways to compute (approximations to) ML estimators, along with the practically useful expectation-maximization (EM) algorithm.

14.1 Introduction

Denote by $\mathcal{P} = \{\mathrm{P}_\theta\}_{\theta \in \mathcal{X}}$ the parametric family of distributions on the observations. Given the observation $Y = y$, one may view $\{p_\theta(y),\ \theta \in \mathcal{X}\}$ as a function of θ, which we defined earlier as the likelihood function. The maximum-likelihood (ML) estimator seeks $\hat{\theta}_{\mathrm{ML}}(y)$ that achieves the supremum of this function over $\mathcal{X}$, i.e.,

$$p_{\hat{\theta}_{\mathrm{ML}}(y)}(y) = \sup_{\theta \in \mathcal{X}} p_\theta(y). \tag{14.1}$$

(Note that this is not a probability distribution over $\mathcal{Y}$.) The likelihood function may have multiple local maxima over the feasible set $\mathcal{X}$. In case several global maxima exist, all of them are ML estimates. The definition (14.1) is often casually replaced by

$$\hat{\theta}_{\mathrm{ML}}(y) = \arg\max_{\theta \in \mathcal{X}} p_\theta(y). \tag{14.2}$$

However, there exist problems where the supremum of (14.1) is not achieved, in which case the ML estimator does not exist; see Section 14.8.

The ML approach can be motivated in several ways:

- If $\mathcal{X}$ is a bounded subset of $\mathbb{R}^m$, ML estimation is equivalent to MAP estimation with a uniform prior. Indeed

  $$\hat{\theta}_{\mathrm{MAP}}(y) = \arg\max_{\theta \in \mathcal{X}} [p_\theta(y)\pi(\theta)],$$

  where $\pi(\theta) = 1/|\mathcal{X}|$ for all $\theta \in \mathcal{X}$. Still, justifying ML on those grounds may be dubious because the assumptions of a uniform prior and uniform cost may be questionable. It should also be noted that the equivalence to MAP estimation with uniform prior does not formally hold if $\mathcal{X}$ is an unbounded set such as the real line because the uniform distribution does not exist on such sets. One may then think of ML estimation as a limiting case of MAP estimation with a prior whose flatness increases – for instance, a Gaussian prior with variance increasing to infinity.

- One could think of $p_\theta(y)$ as a probability that should be made as large as possible so that the observations are deemed more "likely." It is unclear whether such a heuristic approach should work well.
- More fundamentally, we shall see that strong performance guarantees can be given for the ML estimator in asymptotic settings where the number of observations tends to infinity.

14.2 Computation of ML Estimates

Maximizing the likelihood function or any monotonic function thereof are equivalent operations. It is often convenient to maximize $\ln p_\theta(y)$, which is called the log-likelihood function. For simplicity we consider scalar θ first. The roots of the so-called *likelihood equation*

$$0 = \frac{d}{d\theta} \ln p_\theta(y) \tag{14.3}$$

give the local extremas of $p_\theta(y)$ *in the interior of* $\mathcal{X}$. These local extrema could be maxima, minima, or saddlepoints. The type of extremum can be identified by examining the second derivative of the log-likelihood function at the root (respectively negative, positive, and zero). In many problems, the ML estimator is a root of the likelihood equation. In other problems, the maximum of the likelihood function is achieved on the boundary of $\mathcal{X}$ (assuming such boundary exists); see Section 14.5.

Example 14.1 Let $Y \sim \mathcal{N}(0, \theta)$ where $\theta > 0$. The log-likelihood function is given by

$$\ln p_\theta(y) = -\frac{1}{2}\ln(2\pi\theta) - \frac{y^2}{2\theta}, \qquad \theta > 0,$$

as illustrated in Figure 14.1. Its derivative is given by

$$\frac{d}{d\theta} \ln p_\theta(y) = -\frac{1}{2\theta} + \frac{y^2}{2\theta^2}.$$

The root of the likelihood equation (14.3) is unique and is equal to y^2. The second derivative of the log-likelihood function is given by

$$\frac{d^2}{d\theta^2} \ln p_\theta(y) = \frac{1}{2\theta^2} - \frac{y^2}{\theta^3}.$$

Evaluating it at the root, we obtain

$$\left.\frac{d^2}{d\theta^2} \ln p_\theta(y)\right|_{\theta=y^2} = -\frac{1}{2y^4},$$

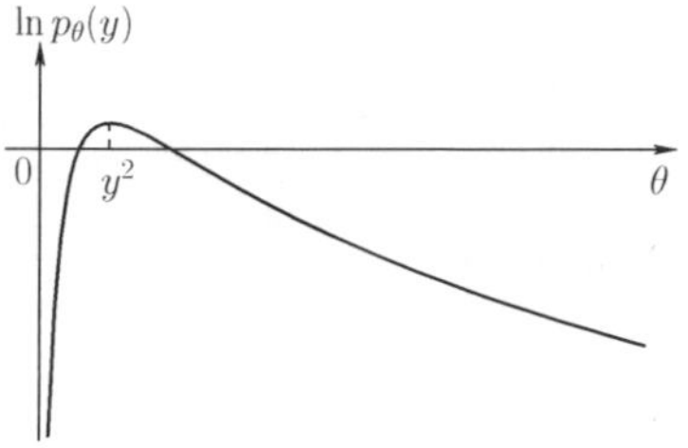

Figure 14.1 Log-likelihood function for Example 14.1

which is a negative number. Hence the root is a maximizer of the likelihood function, and

$$\hat{\theta}_{\text{ML}}(y) = y^2.$$

Example 14.2 Assume as in Example 14.1 that $Y \sim \mathcal{N}(0, \theta)$, but now $\theta \geq 1$. If the root y^2 of the likelihood equation satisfies $y^2 \geq 1$, we again have $\hat{\theta}_{\text{ML}}(y) = y^2$. However, if $y^2 < 1$, the likelihood function is monotonically decreasing over the feasible set $[1, \infty)$, and the maximum is achieved on the boundary: $\hat{\theta}_{\text{ML}}(y) = 1$. Combining both cases, we conclude that

$$\hat{\theta}_{\text{ML}}(y) = \max\{1,\ y^2\}.$$

Consider now vector parameters. If $\boldsymbol{\theta} \in \mathcal{X} \subset \mathbb{R}^m$ is an m-vector, the likelihood equation is given in vector form by

$$\mathbf{0} = \nabla_{\boldsymbol{\theta}} \ln p_{\boldsymbol{\theta}}(y) = \begin{bmatrix} \frac{\partial}{\partial \theta_1} \ln p_{\boldsymbol{\theta}}(y) \\ \vdots \\ \frac{\partial}{\partial \theta_m} \ln p_{\boldsymbol{\theta}}(y) \end{bmatrix} \in \mathbb{R}^m. \tag{14.4}$$

To verify that a solution to (14.4) is indeed a local maximum of $\ln p_{\boldsymbol{\theta}}(y)$, we would need to verify that the Hessian $\nabla^2_{\boldsymbol{\theta}} \ln p_{\boldsymbol{\theta}}(y)$ is nonpositive definite at the solution.

Example 14.3 $\boldsymbol{Y} = \alpha \boldsymbol{s} + \boldsymbol{Z}$, where $\boldsymbol{Z} \sim \mathcal{N}(0, \sigma^2 \mathsf{I}_n)$, $n \geq 2$, $\boldsymbol{s} \in \mathbb{R}^n$ is a known signal, $\alpha \in \mathbb{R}$, and $\sigma^2 > 0$. The unknown parameters are α and σ^2. Let $\boldsymbol{\theta} = [\alpha,\ \sigma^2] \in \mathcal{X} = \mathbb{R} \times \mathbb{R}^+$.

The MVUEs of α and σ^2 were obtained in (12.15) and (12.19):

$$\hat{\alpha}_{\text{MVUE}}(\boldsymbol{y}) = \frac{\boldsymbol{s}^\top \boldsymbol{y}}{\|\boldsymbol{s}\|^2},$$

$$\hat{\sigma}^2_{\text{MVUE}}(\boldsymbol{y}) = \frac{1}{n-1} \|\boldsymbol{y} - \hat{\alpha}_{\text{MVUE}}(\boldsymbol{y})\, \boldsymbol{s}\|^2.$$

The log-likelihood function is

$$\ln p_{\boldsymbol{\theta}}(\boldsymbol{y}) = -\frac{n}{2}\ln(2\pi\sigma^2) - \frac{1}{2\sigma^2}\|\boldsymbol{y} - \alpha\boldsymbol{s}\|^2, \quad \boldsymbol{\theta} = [\alpha, \sigma^2]^\top. \tag{14.5}$$

The roots of the likelihood equation are obtained from (14.4) as follows. First,

$$0 = \frac{\partial}{\partial\alpha}\ln p_{\boldsymbol{\theta}}(\boldsymbol{y}) = \frac{1}{\sigma^2}(\boldsymbol{y} - \alpha\boldsymbol{s})^\top\boldsymbol{s}$$
$$\Rightarrow \hat{\alpha}_{\text{ML}} = \frac{\boldsymbol{y}^\top\boldsymbol{s}}{\|\boldsymbol{s}\|^2} = \hat{\alpha}_{\text{MVUE}}, \tag{14.6}$$

and so the ML estimator of α is an MVUE. Second,

$$0 = \frac{\partial}{\partial(\sigma^2)}\ln p_{\boldsymbol{\theta}}(\boldsymbol{y}) = -\frac{n}{2\sigma^2} + \frac{1}{(\sigma^2)^2}\|\boldsymbol{y} - \hat{\alpha}_{\text{ML}}\boldsymbol{s}\|^2$$
$$\Rightarrow \hat{\sigma}^2_{\text{ML}} = \frac{1}{n}\|\boldsymbol{y} - \hat{\alpha}_{\text{ML}}\boldsymbol{s}\|^2 = \frac{n-1}{n}\hat{\sigma}^2_{\text{MVUE}}, \tag{14.7}$$

and so the ML estimator of σ^2 is biased. Note that it is easy to verify that the $\nabla^2_{\boldsymbol{\theta}}\ln p_{\boldsymbol{\theta}}(y)$ is negative definite at the solutions to the likelihood equation for α and σ^2 given in (14.6) and (14.7) respectively.

Let us compare the MSEs of $\hat{\sigma}^2_{\text{MVUE}}$ and $\hat{\sigma}^2_{\text{ML}}$:

$$\text{MSE}_{\boldsymbol{\theta}}(\hat{\sigma}^2_{\text{MVUE}}) = \text{Var}_{\boldsymbol{\theta}}(\hat{\sigma}^2_{\text{MVUE}}) = \frac{2\sigma^4}{n-1},$$
$$b_{\boldsymbol{\theta}}(\hat{\sigma}^2_{\text{ML}}) = \mathbb{E}_{\boldsymbol{\theta}}\left(\frac{n-1}{n}\hat{\sigma}^2_{\text{MVUE}} - \sigma^2\right) = -\frac{\sigma^2}{n},$$
$$\text{Var}_{\boldsymbol{\theta}}(\hat{\sigma}^2_{\text{ML}}) = \sigma^4\frac{2(n-1)}{n^2},$$
$$\text{MSE}_{\boldsymbol{\theta}}(\hat{\sigma}^2_{\text{ML}}) = \text{Var}_{\boldsymbol{\theta}}(\hat{\sigma}^2_{\text{ML}}) + b^2_{\boldsymbol{\theta}}(\hat{\sigma}^2_{\text{ML}}) = \sigma^4\frac{2n-1}{n^2}.$$

Since $\frac{2n-1}{n^2} < \frac{2}{n-1}$ for all $n \geq 2$, we conclude that $\text{MSE}_{\boldsymbol{\theta}}(\hat{\sigma}^2_{\text{ML}}) < \text{MSE}_{\boldsymbol{\theta}}(\hat{\sigma}^2_{\text{MVUE}})$, showing again the potential benefits of a biased estimator.

14.3 Invariance to Reparameterization

Consider a reparameterization $\eta = g(\theta)$ of the family $\{p_\theta,\ \theta \in \mathcal{X}\}$ where g is invertible. Denote by $g(\mathcal{X}) = \{g(\theta),\ \theta \in \mathcal{X}\}$ the image of the original parameter set $\mathcal{X}$ under g, and by $\{q_\eta(y) = p_{g^{-1}(\eta)}(y),\ \eta \in g(\mathcal{X})\}$ the reparameterized family. Given the MLE $\hat{\theta}_{\text{ML}}$, one would like to obtain the MLE $\hat{\eta}_{\text{ML}}$ for the reparameterized family.

PROPOSITION 14.1 *The MLE for the family $\{q_\eta,\ \eta \in g(\mathcal{X})\}$ satisfies*

$$\hat{\eta}_{ML} = g(\hat{\theta}_{\text{ML}}). \tag{14.8}$$

Proof We have

$$q_{\hat{\eta}_{\text{ML}}}(y) \overset{(a)}{=} \sup_{\eta \in g(\mathcal{X})} q_\eta(y)$$

$$\stackrel{(b)}{=} \sup_{\eta \in g(\mathcal{X})} p_{g^{-1}(\eta)}(y)$$
$$\stackrel{(c)}{=} \sup_{\theta \in \mathcal{X}} p_\theta(y)$$
$$\stackrel{(d)}{=} p_{\hat{\theta}_{\mathrm{ML}}}(y),$$

where (a) and (d) follow from the definition (14.1) of the MLE, (b) from the definition of q_η, and (c) from the change in variables $\theta = g^{-1}(\eta)$. □

Hence ML estimation commutes with nonlinear mappings.[1] We have encountered an example of reparameterization in Example 12.12. For the $\mathcal{N}(\mu, \sigma^2)$ family, the transformation from $\boldsymbol{\theta} = (\mu, \sigma^2)$ to

$$\boldsymbol{\eta} = g(\boldsymbol{\theta}) = \left(\frac{\mu}{\sigma^2}, -\frac{1}{2\sigma^2}\right)$$

yields a representation of the exponential family in canonical form.

If θ is a scalar and g is invertible and differentiable over $\mathcal{X}$, its derivative $g'(\theta)$ is nonzero over all $\theta \in \mathcal{X}$. If $\boldsymbol{\theta}$ is an m-vector and g is invertible and differentiable over $\mathcal{X}$, the Jacobian $\mathsf{J} = \left(\frac{\partial g_j(\boldsymbol{\theta})}{\partial \theta_k}\right)_{j,k=1}^n$ has full rank.

Some problems require estimating a noninvertible function $\eta = g(\theta)$ of the parameter – for instance, estimate θ^2 when $\theta \in \mathbb{R}$. Then one can view (14.8) as a *definition* of MLE for η, as discussed by Zehna [1]. Alternatively, one can use asymptotic consistency of the MLE (see Section 14.6) to justify the use of $g(\hat{\theta}_{\mathrm{ML}})$ as the MLE for $g(\theta)$.

14.4 MLE in Exponential Families

MLEs in exponential families are quite easy to derive and analyze, exploiting the results of Section 11.8 where the first- and second-order moments and Fisher information were derived in closed form. The general expression for an m-dimensional exponential family in canonical form is

$$p_{\boldsymbol{\theta}}(y) = h(y) e^{\boldsymbol{\theta}^\top \boldsymbol{T}(y) - A(\boldsymbol{\theta})}, \tag{14.9}$$

where $\boldsymbol{T}(Y) \in \mathbb{R}^m$ is a sufficient statistic for $\boldsymbol{\theta} \in \mathcal{X}$. We assume that $\mathcal{X}$ is the *natural parameter set*

$$\mathcal{X} = \left\{\boldsymbol{\theta} \in \mathbb{R}^m : \int_{\mathcal{Y}} h(y) e^{\boldsymbol{\theta}^\top \boldsymbol{T}(y)} d\mu(y) < \infty\right\},$$

which is convex and assumed to be open. In other words, there is no constraint on $\boldsymbol{\theta}$ other than having finite $A(\boldsymbol{\theta})$. From (11.28), (11.29), and (13.46) we obtain the expectation and the covariance of $\boldsymbol{T}(Y)$:

$$\mathbb{E}_{\boldsymbol{\theta}}[\boldsymbol{T}(Y)] = \nabla A(\boldsymbol{\theta}), \tag{14.10}$$
$$\mathsf{I}_{\boldsymbol{\theta}} = \mathrm{Cov}_{\boldsymbol{\theta}}[\boldsymbol{T}(Y)] = \nabla^2 A(\boldsymbol{\theta}). \tag{14.11}$$

[1] Recall from Example 11.7 that the same does generally not hold for MAP estimators.

The log-likelihood function is concave because $\nabla_{\boldsymbol{\theta}}^2 \ln p_{\boldsymbol{\theta}}(y) = -\mathsf{I}_{\boldsymbol{\theta}}$ is nonpositive definite. Since $\mathcal{X}$ is an open set, the maximum of the likelihood function over $\mathcal{X}$, if it exists, satisfies the likelihood equation. Moreover the concavity of the log-likelihood function implies that any root of the likelihood equation is a global maximum, hence a ML solution. Therefore

$$\begin{aligned}\mathbf{0} &= \nabla_{\boldsymbol{\theta}} \ln p_{\boldsymbol{\theta}}(y)|_{\boldsymbol{\theta}=\hat{\boldsymbol{\theta}}_{\text{ML}}} \\ &= \boldsymbol{T}(y) - \nabla A(\hat{\boldsymbol{\theta}}_{\text{ML}}).\end{aligned} \tag{14.12}$$

14.4.1 Mean-Value Parameterization

As in (11.32), it will be convenient to use the mean-value parameterization of the exponential family:

$$\boldsymbol{\xi} = \boldsymbol{g}(\boldsymbol{\theta}) \triangleq \nabla A(\boldsymbol{\theta}) = \mathbb{E}_{\boldsymbol{\theta}}[\boldsymbol{T}(Y)], \tag{14.13}$$

which defines a transformation of the vector $\boldsymbol{\theta}$. The Jacobian of $\boldsymbol{g}$, which is given by

$$\mathsf{J} = \left\{ \frac{\partial g_j(\boldsymbol{\theta})}{\partial \theta_k} \right\}_{j,k=1}^{m} = \nabla^2 A(\boldsymbol{\theta}) = \mathsf{I}_{\boldsymbol{\theta}} = \text{Cov}_{\boldsymbol{\theta}}[\boldsymbol{T}(Y)], \tag{14.14}$$

is positive definite and invertible if $\mathsf{I}_{\boldsymbol{\theta}}$ is (strictly) positive definite. The positive definiteness of J implies that the mapping $\boldsymbol{g}$ is invertible (see [2]), and therefore that $\boldsymbol{\theta} = \boldsymbol{g}^{-1}(\boldsymbol{\xi})$.

The Fisher information matrix for $\boldsymbol{\xi}$ is obtained from (13.45) as

$$\mathsf{I}_{\boldsymbol{\xi}} = \mathsf{J}^{-1} \mathsf{I}_{\boldsymbol{\theta}} \mathsf{J}^{-1} = \mathsf{I}_{\boldsymbol{\theta}}^{-1}. \tag{14.15}$$

Since ML estimators commute with nonlinear invertible operations, the MLEs of $\boldsymbol{\xi}$ and $\boldsymbol{\theta}$ are related via

$$\hat{\boldsymbol{\xi}}_{\text{ML}} = \boldsymbol{g}(\hat{\boldsymbol{\theta}}_{\text{ML}}) = \nabla A(\hat{\boldsymbol{\theta}}_{\text{ML}}) = \boldsymbol{T}(y).$$

From (14.13) we conclude that $\mathbb{E}_{\boldsymbol{\theta}}[\hat{\boldsymbol{\xi}}_{\text{ML}}] = \boldsymbol{\xi}$, and hence that $\hat{\boldsymbol{\xi}}_{\text{ML}}$ is *unbiased*. From (14.14) and (14.15), we conclude that $\text{Cov}_{\boldsymbol{\theta}}[\hat{\boldsymbol{\xi}}_{\text{ML}}] = \mathsf{I}_{\boldsymbol{\theta}} = \mathsf{I}_{\boldsymbol{\xi}}^{-1}$, and hence that $\hat{\boldsymbol{\xi}}_{\text{ML}}$ is *efficient*.

14.4.2 Relation to MVUEs

For a one-dimensional exponential family $\{p_\theta\}_{\theta \in \mathcal{X}}$ with parameter set $\mathcal{X} = (a, b)$, we have seen in (13.28) that subject to regularity conditions, an efficient estimator $\hat{\theta}_{\text{MVUE}}$ exists if and only if $\ln p_\theta(y)$ can be written as

$$\ln p_\theta(y) = \int_a^\theta \mathsf{I}_{\theta'} \left[\hat{\theta}_{\text{MVUE}}(y) - \theta' \right] d\theta' + \ln h(y). \tag{14.16}$$

Differentiating (14.16) with respect to θ, we obtain

$$\frac{d}{d\theta} \ln p_{\boldsymbol{\theta}}(y) = \mathsf{I}_\theta \left[\hat{\theta}_{\text{MVUE}}(y) - \theta \right]. \tag{14.17}$$

Hence $\hat{\theta}_{\text{MVUE}}(y)$ is the unique root of the likelihood equation (14.3):

$$0 = \frac{d}{d\theta} \ln p_\theta(y)\bigg|_{\theta=\hat{\theta}_{\text{MVUE}}(y)}.$$

If the ML estimator satisfies the likelihood equation, we must therefore have

$$\hat{\theta}_{\text{ML}}(y) = \hat{\theta}_{\text{MVUE}}(y). \tag{14.18}$$

14.4.3 Asymptotics

While the MLE $\hat{\boldsymbol{\xi}}_{\text{ML}}$ for the mean-value parameter $\boldsymbol{\xi} = \boldsymbol{g}(\boldsymbol{\theta})$ is unbiased and efficient, the same can generally not be said about the ML estimator $\hat{\boldsymbol{\theta}}_{\text{ML}}$ of $\boldsymbol{\theta}$ due to the nonlinearity of the mapping $\boldsymbol{g}$. However, these properties hold *asymptotically* in a model with many independent observations. Two limit theorems of probability theory will be useful in this context: the *Continuous Mapping Theorem* and the *Delta Theorem*, both of which are given in Appendix D.

The asymptotic setup is as follows. Let $\boldsymbol{Y} = (Y_1, \ldots, Y_n)$ be an i.i.d. sequence drawn from the exponential family (14.9). Then

$$p_{\boldsymbol{\theta}}(\boldsymbol{y}) = \left(\prod_{i=1}^{n} h(y_i)\right) \exp\left\{\boldsymbol{\theta}^\top \sum_{i=1}^{n} \boldsymbol{T}(y_i) - nA(\boldsymbol{\theta})\right\}, \quad \boldsymbol{\theta} \in \mathcal{X},$$

is still an m-dimensional exponential family in canonical form.

Since the random variables $\{Y_i\}_{i=1}^n$ are i.i.d., so are $\{\boldsymbol{T}(Y_i)\}_{i=1}^n$. We can see from (14.10) and (14.11) that the following relationships hold:

$$\mathbb{E}_{\boldsymbol{\theta}}\left[\frac{1}{n}\sum_{i=1}^{n} \boldsymbol{T}(Y_i)\right] = \mathbb{E}_{\boldsymbol{\theta}}[\boldsymbol{T}(Y)] = \nabla A(\boldsymbol{\theta}) = \boldsymbol{\xi},$$

$$\text{Cov}_{\boldsymbol{\theta}}\left[\frac{1}{n}\sum_{i=1}^{n} \boldsymbol{T}(Y_i)\right] = \frac{1}{n}\text{Cov}_{\boldsymbol{\theta}}[\boldsymbol{T}(Y)] = \frac{1}{n}\nabla^2 A(\boldsymbol{\theta}) = \frac{1}{n}\mathsf{I}_{\boldsymbol{\theta}}. \tag{14.19}$$

The likelihood equation is given by

$$\nabla A(\hat{\boldsymbol{\theta}}_{\text{ML}}) = \frac{1}{n}\sum_{i=1}^{n} \boldsymbol{T}(Y_i). \tag{14.20}$$

Therefore, if we consider the mean-value reparameterization as in (14.13) and assume that $\boldsymbol{\theta}$ is in the interior of $\mathcal{X}$, then the MLE of $\boldsymbol{\theta}$ given $\boldsymbol{Y}$ satisfies

$$\hat{\boldsymbol{\xi}}_{\text{ML}} = \boldsymbol{g}(\hat{\boldsymbol{\theta}}_{\text{ML}}) = \nabla A(\hat{\boldsymbol{\theta}}_{\text{ML}}) = \frac{1}{n}\sum_{i=1}^{n} \boldsymbol{T}(Y_i). \tag{14.21}$$

From (14.19)–(14.21), we can conclude that

$$\mathbb{E}_{\boldsymbol{\theta}}[\hat{\boldsymbol{\xi}}_{\text{ML}}] = \boldsymbol{\xi}, \quad \text{Cov}_{\boldsymbol{\theta}}[\hat{\boldsymbol{\xi}}_{\text{ML}}] = \frac{1}{n}\mathsf{I}_{\boldsymbol{\theta}} = \frac{1}{n}\mathsf{I}_{\boldsymbol{\xi}}^{-1}. \tag{14.22}$$

The convergence of $\hat{\boldsymbol{\theta}}_{\text{ML}}$ to the true $\boldsymbol{\theta}$ can be studied through the corresponding properties of $\hat{\boldsymbol{\xi}}_{\text{ML}}$.

Consistency By the Strong Law of Large Numbers we have

$$\hat{\boldsymbol{\xi}}_{\text{ML}}(\boldsymbol{Y}) \stackrel{\text{a.s.}}{\to} \boldsymbol{\xi} \tag{14.23}$$

and $\hat{\boldsymbol{\xi}}_{\text{ML}}$ is said to be *strongly consistent*. Since a.s. convergence implies i.p. convergence, we also have $\hat{\boldsymbol{\xi}}_{\text{ML}}(\boldsymbol{Y}) \stackrel{i.p.}{\to} \boldsymbol{\xi}$, a property known as *weak consistency*. Also by (14.22), the MSE $\mathbb{E}_{\boldsymbol{\theta}}(\|\hat{\boldsymbol{\xi}}_{\text{ML}} - \boldsymbol{\xi}\|^2) = \frac{1}{n}\text{Tr}(\mathsf{I}_{\boldsymbol{\xi}})$ tends to zero, hence $\hat{\boldsymbol{\xi}}_{\text{ML}}$ is said to be *m.s. consistent*. By the continuous mapping theorem (see Appendix D), $\hat{\boldsymbol{\theta}}_{\text{ML}} = \boldsymbol{g}^{-1}(\hat{\boldsymbol{\xi}}_{\text{ML}})$ inherits the strong and weak consistency properties of $\hat{\boldsymbol{\xi}}_{\text{ML}}$. However, the mean squares (m.s.) consistency property is not necessarily inherited by $\hat{\boldsymbol{\theta}}_{\text{ML}}$.

Asymptotic Unbiasedness The asymptotic unbiasedness of $\hat{\boldsymbol{\theta}}_{\text{ML}}$ will follow from strong consistency, only if some additional conditions are satisfied. In particular, if there exists a random variable X such that $\mathbb{E}_{\boldsymbol{\theta}}[X] < \infty$ for all $\boldsymbol{\theta} \in \mathcal{X}$ and $\|\hat{\boldsymbol{\theta}}_{\text{ML}}(\boldsymbol{Y})\| \leq X$ a.s. $\mathcal{P}$ for all n, then by the dominated convergence theorem, we have $\lim_{n\to\infty} \mathbb{E}_{\boldsymbol{\theta}}[\hat{\boldsymbol{\theta}}_{\text{ML}}] = \boldsymbol{\theta}$, i.e., $\hat{\boldsymbol{\theta}}_{\text{ML}}$ is *asymptotically unbiased*.

Asymptotic Efficiency and Gaussianity By the Central Limit Theorem, the distribution of $\frac{1}{n}\sum_{i=1}^N \boldsymbol{T}(Y_i)$ is asymptotically Gaussian, more precisely,

$$\sqrt{n}\left(\frac{1}{n}\sum_{i=1}^N \boldsymbol{T}(Y_i) - \mathbb{E}_{\boldsymbol{\theta}}[\boldsymbol{T}(Y)]\right) \stackrel{\text{d}}{\to} \mathcal{N}(0, \text{Cov}_{\boldsymbol{\theta}}[\boldsymbol{T}(Y)]). \tag{14.24}$$

Substituting (14.11), (14.13), and (14.21) into (14.24), we obtain

$$\sqrt{n}\left(\hat{\boldsymbol{\xi}}_{\text{ML}} - \boldsymbol{\xi}\right) \stackrel{\text{d}}{\to} \mathcal{N}(0, \mathsf{I}_{\boldsymbol{\theta}}).$$

Applying the Delta theorem (see Appendix D) and using the fact that $\mathsf{J}^{-1}\mathsf{I}_{\boldsymbol{\theta}}\mathsf{J} = \mathsf{I}_{\boldsymbol{\theta}}^{-1}$, we conclude that

$$\sqrt{n}\left(\hat{\boldsymbol{\theta}}_{\text{ML}} - \boldsymbol{\theta}\right) \stackrel{\text{d}}{\to} \mathcal{N}(0, \mathsf{I}_{\boldsymbol{\theta}}^{-1}). \tag{14.25}$$

Hence $\hat{\boldsymbol{\theta}}_{\text{ML}}$ is *asymptotically efficient*, i.e., it asymptotically satisfies the Cramér–Rao lower bound for unbiased estimators.

We have shown that when the observation sequence $\boldsymbol{Y}$ is drawn i.i.d. from the m-dimensional canonical exponential family (14.9) with $\boldsymbol{\theta} \in \text{interior}(\mathcal{X})$, the likelihood equation takes the form (14.21). Then $\hat{\boldsymbol{\theta}}_{\text{ML}}$ is asymptotically unbiased (under uniform integrability conditions), efficient, strongly consistent, and Gaussian.

Example 14.4 *Covariance Matrix Estimation in Gaussian Model.* This is a variation on Example 13.5. Let $\boldsymbol{Y}_i \sim$ i.i.d. $\mathcal{N}(\mathbf{0}, \boldsymbol{C}(\boldsymbol{\theta}))$ for $1 \leq i \leq n$ where $\boldsymbol{C}(\boldsymbol{\theta})$ is an $r \times r$ covariance matrix parameterized by $\boldsymbol{\theta} \in \mathbb{R}^m$, where $m \leq \frac{1}{2}r(r+1)$. For simplicity we will assume that the inverse covariance matrix $\mathsf{M}(\boldsymbol{\theta}) \triangleq \boldsymbol{C}(\boldsymbol{\theta})^{-1}$ depends linearly on $\boldsymbol{\theta}$. For instance, consider the following model with $r = 3$ and $m = 4$:

$$\mathsf{M}(\boldsymbol{\theta}) = \begin{bmatrix} \theta_1 & 0 & \theta_4 \\ 0 & \theta_2 & 0 \\ \theta_4 & 0 & \theta_3 \end{bmatrix}.$$

We have

$$\begin{aligned} \ln p_{\boldsymbol{\theta}}(\boldsymbol{Y}) &= -\frac{rn}{2}\ln(2\pi) - \frac{n}{2}\ln|\mathsf{C}(\boldsymbol{\theta})| + \frac{1}{2}\sum_{i=1}^{n} \boldsymbol{Y}_i^\top \mathsf{C}(\boldsymbol{\theta})^{-1}\boldsymbol{Y}_i \\ &= -\frac{rn}{2}\ln(2\pi) + \frac{n}{2}\ln|\mathsf{M}(\boldsymbol{\theta})| - \frac{n}{2}\mathrm{Tr}[\mathsf{S}\mathsf{M}(\boldsymbol{\theta})], \end{aligned}$$

where $\mathsf{S} = \frac{1}{n}\sum_{i=1}^{n} \boldsymbol{Y}_i^\top \boldsymbol{Y}_i$ is the $r \times r$ sample covariance matrix. The likelihood equation is given by

$$0 = \frac{\partial \ln p_{\boldsymbol{\theta}}(\boldsymbol{Y})}{\partial \theta_i} = \frac{n}{2}\mathrm{Tr}\left[\mathsf{M}(\boldsymbol{\theta})^{-1}\frac{\partial \mathsf{M}(\boldsymbol{\theta})}{\partial \theta_i}\right] - \frac{n}{2}\mathrm{Tr}\left[\mathsf{S}\frac{\partial \mathsf{M}(\boldsymbol{\theta})}{\partial \theta_i}\right], \quad 1 \le i \le m,$$

where we have used the differentiation formulas (13.58) and (13.59). Since the matrix $\frac{\partial \mathsf{M}(\boldsymbol{\theta})}{\partial \theta_i}$ contains only 0's and 1's, the second partial derivatives of $\mathsf{M}(\boldsymbol{\theta})$ are all zero. Hence the Hessian of $\ln p_{\boldsymbol{\theta}}(\boldsymbol{Y})$ is given by

$$\frac{\partial^2 \ln p_{\boldsymbol{\theta}}(\boldsymbol{Y})}{\partial \theta_i \partial \theta_j} = -\frac{n}{2}\mathrm{Tr}\left[\mathsf{M}(\boldsymbol{\theta})^{-1}\frac{\partial \mathsf{M}(\boldsymbol{\theta})}{\partial \theta_j}\mathsf{M}(\boldsymbol{\theta})^{-1}\frac{\partial \mathsf{M}(\boldsymbol{\theta})}{\partial \theta_i}\right], \quad 1 \le i, j \le m$$

(independent of $\boldsymbol{Y}$), and its negative is the Fisher information matrix $\mathsf{I}_{\boldsymbol{\theta}}$.

14.5 Estimation of Parameters on Boundary

We have seen in Example 14.2 that when θ lies on the boundary of $\mathcal{X}$, the MLE does not necessarily satisfy the likelihood equation. To illustrate the properties of MLEs in such cases, we consider a simple one-dimensional Gaussian family and investigate whether the MLE is still unbiased, consistent, efficient, and asymptotically Gaussian.

Example 14.5 Derive the MLE of θ when $p_\theta = \mathcal{N}(\theta, 1)$ and $\mathcal{X} = [0, \infty)$.
Recall that the empirical mean estimator

$$\overline{Y} \triangleq \frac{1}{n}\sum_{i=1}^{n} Y_i$$

is the MVUE, as well as the MLE when $\mathcal{X} = \mathbb{R}$. However, $\overline{Y} \sim \mathcal{N}(0, 1/n)$ and thus $\overline{Y}$ is negative (is outside $\mathcal{X}$, therefore not feasible) with probability 1/2. The MLE for this problem is given by

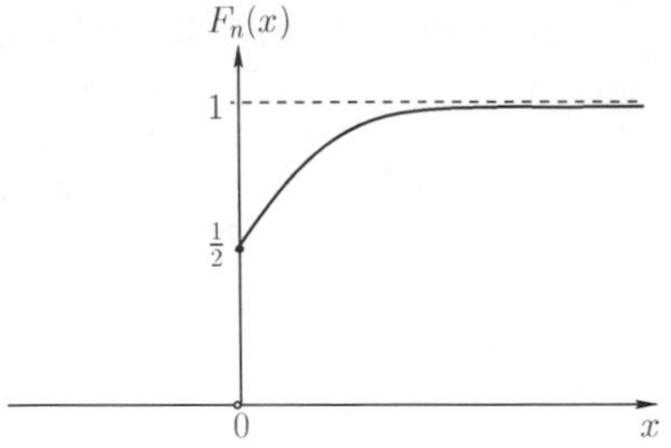

Figure 14.2 Cumulative distribution function of $\sqrt{n}\,\hat{\theta}_{\text{ML}}$

$$\begin{aligned}\hat{\theta}_{\text{ML}}(\boldsymbol{y}) &= \arg\max_{\theta\ge 0} \ln p_\theta(y_i)\\ &= \arg\max_{\theta\ge 0} \sum_{i=1}^{n}\left[-\frac{1}{2}\ln 2\pi - \frac{1}{2}(y_i-\theta)^2\right]\\ &= \arg\max_{\theta\ge 0}[2\theta\overline{y} - \theta^2].\end{aligned}$$

The expression in brackets is quadratic in θ and decreasing for $\theta \ge \overline{y}$. Hence it is minimized by $\hat{\theta}_{\text{ML}} = \overline{y}$ when $\overline{y} \ge 0$ and by $\hat{\theta}_{\text{ML}} = 0$ when $\overline{y} \le 0$. The MLE is given by the compact formula

$$\hat{\theta}_{\text{ML}}(\boldsymbol{y}) = \max\{\overline{y}, 0\} = \max\left\{\frac{1}{n}\sum_{i=1}^{n} y_i, 0\right\}.$$

We now derive the statistics of $\hat{\theta}_{\text{ML}}$ for the case when $\theta = 0$, i.e., θ is on the boundary of $\mathcal{X}$. The cdf of the normalized error $\sqrt{n}\hat{\theta}_{\text{ML}}$ is given by

$$F_n(x) = \mathrm{P}_0\{\sqrt{n}\hat{\theta}_{\text{ML}} \le x\} = \begin{cases} 0 & : x < 0 \\ \Phi(x) & : x \ge 0 \end{cases} \tag{14.26}$$

(as shown in Figure 14.2) and is strongly non-Gaussian. The first two moments of $\hat{\theta}_{\text{ML}}$ are

$$\begin{aligned}\mathbb{E}_0[\hat{\theta}_{\text{ML}}(\boldsymbol{Y})] &= \int_{-\infty}^{\infty} \max\{0, x\}\frac{e^{-nx^2/2}}{\sqrt{2\pi/n}}dx\\ &= \int_0^{\infty} x\frac{e^{-nx^2/2}}{\sqrt{2\pi/n}}dx = \int_0^{\infty}\frac{e^{-t}}{\sqrt{2\pi n}}dt = \frac{1}{\sqrt{2\pi n}}\end{aligned}$$

and, similarly,

$$\mathbb{E}[\hat{\theta}_{\text{ML}}^2] = \int_0^{\infty} x^2\frac{e^{-nx^2/2}}{\sqrt{2\pi/n}}dx = \frac{1}{2n}.$$

Note that the MSE is *half* of that when the point is in the interior of $\mathcal{X}$. While $\hat{\theta}_{\text{ML}}$ is biased, it is asymptotically unbiased and consistent. Since (14.26) holds for each n, it is also the limiting (non-Gaussian) distribution of the normalized error.

A general discussion of MLEs with parameter on the boundary may be found in [3, pp. 517, 518].

14.6 Asymptotic Properties for General Families

We now ask whether the MLE is asymptotically unbiased, consistent, and efficient for general parametric families. To emphasize its dependency on n, the MLE is denoted by $\hat{\boldsymbol{\theta}}_n$. The analysis is more involved than in the case of exponential families because the MLE can generally not be expressed in terms of easily analyzed sufficient statistics. Still, under some regularity conditions, we show that the MLE is:

1. asymptotically unbiased: $\lim_{n\to\infty} \mathbb{E}_{\boldsymbol{\theta}}[\hat{\boldsymbol{\theta}}_n] = \boldsymbol{\theta}$;
2. strongly consistent: $\hat{\boldsymbol{\theta}}_n \xrightarrow{a.s.} \boldsymbol{\theta}$;
3. asymptotically efficient: $\lim_{n\to\infty} n\mathrm{Cov}_{\boldsymbol{\theta}}[\hat{\boldsymbol{\theta}}_n] = \mathsf{I}_{\boldsymbol{\theta}}^{-1}$;
4. asymptotically Gaussian: $\sqrt{n}(\hat{\boldsymbol{\theta}}_n - \boldsymbol{\theta}) \xrightarrow{d} \mathcal{N}(0, I_{\boldsymbol{\theta}}^{-1})$.

As will be discussed in Section 14.7, examples abound where some of these regularity conditions are not satisfied.

The regularity conditions are sufficient but can be relaxed to some extent. The conditions under which the MLE is consistent are weaker than the conditions under which the estimator is asymptotically Gaussian.

14.6.1 Consistency

Let us start with Lemma 14.1, which implies that for any fixed $\theta \neq \theta'$, the likelihood ratio $\frac{p_{\theta'}(\boldsymbol{Y})}{p_\theta(\boldsymbol{Y})}$ vanishes exponentially with n, at the rate $D(p_\theta \| p_{\theta'})$, when $\boldsymbol{Y} \sim$ i.i.d. p_θ.

LEMMA 14.1 *Fix any $\theta' \neq \theta$ both in $\mathcal{X}$ and assume $D(p_\theta \| p_{\theta'}) < \infty$. Let $\{Y_i\}_{i=1}^n$ be i.i.d. P_θ. Then*

$$\frac{1}{n} \ln \frac{p_\theta(\boldsymbol{Y})}{p_{\theta'}(\boldsymbol{Y})} \xrightarrow{a.s.} D(p_\theta \| p_{\theta'}). \tag{14.27}$$

Proof Define the random variable $L = \ln \frac{p_\theta(Y)}{p_{\theta'}(Y)}$ where $Y \sim p_\theta$. The expected value of L under P_θ is

$$\mathbb{E}_\theta[L] = \mathbb{E}_\theta\left[\ln \frac{p_\theta(Y)}{p_{\theta'}(Y)}\right] = \int_{\mathcal{Y}} p_\theta(y) \ln \frac{p_\theta(y)}{p_{\theta'}(y)} d\mu(y) = D(p_\theta \| p_{\theta'}).$$

Since $\{Y_i\}_{i=1}^n$ are i.i.d., so are the random variables $L_i = \ln \frac{p_\theta(Y_i)}{p_{\theta'}(Y_i)}$, $1 \le i \le n$. It follows from the Strong Law of Large Numbers that the random variable

$$\frac{1}{n} \ln \frac{p_\theta(\boldsymbol{Y})}{p_{\theta'}(\boldsymbol{Y})} = \frac{1}{n} \sum_{i=1}^n L_i$$

converges a.s. to its expectation $D(p_\theta \| p_{\theta'})$. □

If $\mathcal{X}$ were a finite set (m-ary hypothesis testing problem with $m = |\mathcal{X}|$), and if $p_\theta \neq p_{\theta'}$ for each $\theta \neq \theta'$ (identifiability condition), then Lemma 14.1 and the union bound could be applied directly to show strong consistency of the MLE: $\hat{\theta}_n \xrightarrow{\text{a.s.}} \theta$ (proof is left to the reader). Since $\mathcal{X}$ is not a finite set, a more powerful version of Lemma 14.1 is needed to ensure that convergence holds uniformly over θ' outside some small ball around θ; this is the key tool needed to prove strong consistency of the MLE. The following lemma is a consequence of the *Uniform Strong Law of Large Numbers* [4, Ch. 16] [5, p. 36].

LEMMA 14.2 [4, Theorem 16(b)] *Assume that:*

(A1) *$p_\theta \neq p_{\theta'}$ for all $\theta \neq \theta'$ (identifiability).*
(A2) *$\theta \in \mathcal{X}$, a compact subset of $\mathbb{R}^m$.*
(A3) *$p_\theta(y)$ is upper semicontinuous in θ for all y.*
(A4) *There exists a function $M(Y)$ such that $\mathbb{E}_\theta|M(Y)| < \infty$ and*

$$\ln \frac{p_{\theta'}(y)}{p_\theta(y)} \leq M(y) \quad \forall y \text{ and } \theta, \theta' \in \mathcal{X}.$$

Let $\mathcal{X}'$ be a compact subset of $\mathcal{X}$. Then

$$\sup_{\theta' \in \mathcal{X}'} \left| \frac{1}{n} \ln \frac{p_\theta(\mathbf{Y})}{p_{\theta'}(\mathbf{Y})} - D(p_\theta \| p_{\theta'}) \right| \xrightarrow{\text{a.s.}} 0. \tag{14.28}$$

REMARK 14.1 The upper semicontinuity assumption (A3) means that

$$\limsup_{\theta' \to \theta} p_{\theta'(y)} \leq p_\theta(y) \quad \forall \theta, y \tag{14.29}$$

which is weaker than continuity and applies among others to families of uniform distributions

$$p_\theta(y) = \theta^{-1} \mathbb{1}\{0 \leq y \leq \theta\}, \quad \theta > 0. \tag{14.30}$$

THEOREM 14.1 Strong Consistency of MLE. *Under the assumptions of Lemma 14.2, any sequence of ML estimates $\hat{\theta}_n$ converges a.s. to θ.*

Proof Fix an arbitrarily small $\epsilon > 0$ and define the subset $\mathcal{X}_\epsilon \triangleq \{\theta' \in \mathcal{X} : |\theta' - \theta| > \epsilon\}$. Let $\delta_\epsilon \triangleq \inf_{\theta' \in \mathcal{X}_\epsilon} D(p_\theta \| p_{\theta'})$ which is positive by the identifiability assumption **(A1)**. By Lemma 14.2, we have

$$P_\theta \left\{ \lim_{n \to \infty} \sup_{\theta' \in \mathcal{X}_\epsilon} \left[\ln \frac{p_{\theta'}(\mathbf{Y})}{p_\theta(\mathbf{Y})} - nD(p_\theta \| p_{\theta'}) \right] = 0 \right\} = 1,$$

hence

$$P_\theta \left\{ \lim_{n \to \infty} \sup_{\theta' \in \mathcal{X}_\epsilon} \ln \frac{p_{\theta'}(\mathbf{Y})}{p_\theta(\mathbf{Y})} \leq -n\delta_\epsilon \right\} = 1. \tag{14.31}$$

Hence there exists some $n_0(\epsilon)$ such that for all $n > n_0(\epsilon)$,

$$\sup_{\theta' \in \mathcal{X}_\epsilon} \ln \frac{p_{\theta'}(\mathbf{Y})}{p_\theta(\mathbf{Y})} \leq -\frac{n}{2}\delta_\epsilon. \tag{14.32}$$

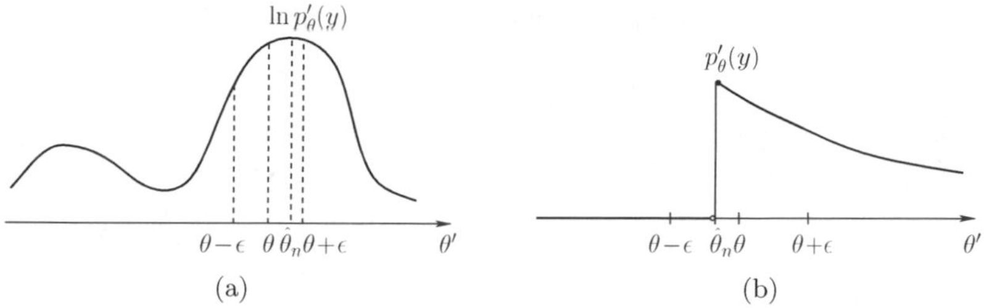

Figure 14.3 Consistency of MLEs. (a) Continuous likelihood function. (b) Upper semicontinuous likelihood function

Since $\hat{\theta}_n$ satisfies

$$\ln \frac{p_{\hat{\theta}_n}(\boldsymbol{Y})}{p_\theta(\boldsymbol{Y})} = \sup_{\theta' \in \mathcal{X}} \ln \frac{p_{\theta'}(\boldsymbol{Y})}{p_\theta(\boldsymbol{Y})} = 0, \tag{14.33}$$

it follows from (14.32) that $\hat{\theta}_n \notin \mathcal{X}_\epsilon$ or equivalently $|\hat{\theta}_n - \theta| \le \epsilon$, for all $n > n_0(\epsilon)$. Since this statement holds for arbitrarily small $\epsilon > 0$, the claim follows. □

Figure 14.3 depicts the likelihood function for some large n and some typical observation sequence $\boldsymbol{y}$. The event that $|\hat{\theta}_n - \theta| < \epsilon$ for all $n \ge n_0(\epsilon)$ has probability 1. This holds whether the likelihood function is continuous (a) or upper semicontinuous (b). The scenario depicted in Figure 14.3(b) corresponds to the family of Uniform$[0, \theta]$ distributions which is treated in detail in Section 14.7.

Finally, note that almost-sure convergence of $\hat{\theta}_n$ does not necessarily imply asymptotic unbiasedness ($\lim_{n\to\infty} \mathbb{E}_\theta[\hat{\theta}_n] = \theta$). However, by the dominated convergence theorem, existence of a random variable X such that

$$|\hat{\theta}_n(\boldsymbol{Y})| \le X \quad \text{a.s } \mathcal{P}, \quad \mathbb{E}_\theta[X] < \infty \quad \forall \theta \in \mathcal{X},$$

ensures asymptotic unbiasedness.

14.6.2 Asymptotic Efficiency and Normality

Next we study the asymptotic distribution of the MLE. We first consider the case of scalar θ. Define the *score function*

$$\psi(y, \theta) = \frac{\partial}{\partial \theta} \ln p_\theta(y) \tag{14.34}$$

and denote by $\psi'(y, \theta) = \frac{\partial^2}{\partial \theta^2} \ln p_\theta(y)$ and $\psi''(y, \theta) = \frac{\partial^3}{\partial \theta^3} \ln p_\theta(y)$, its second and third derivatives with respect to θ, which are all well defined provided regularity conditions apply. The likelihood equation (14.3) may be written as

$$0 = \sum_{i=1}^n \psi(Y_i, \theta). \tag{14.35}$$

The main idea in the following theorem is to apply a Taylor series expansion of the score function around θ and to show that $\hat{\theta}_n$ (properly scaled) converges to an average of i.i.d. scores. The proof is given in Section 14.13.

THEOREM 14.2 [Cramér] Asymptotic Normality of MLE. *Assume that:*

(A1) $p_\theta \neq p'_\theta$ *for all* $\theta \neq \theta'$ *(identifiability).*
(A2) $\theta \in \mathcal{X}$, *an open subset of* $\mathbb{R}$.
(A3) $\{p_\theta, \theta \in \mathcal{X}\}$ *have the same support*

$$\mathcal{Y} = \text{supp}\{p_\theta\} = \{y : p_\theta(y) > 0\} \quad \forall \theta \in \mathcal{X}.$$

(A4) $\psi(Y, \theta)$ *is twice differentiable with respect to* θ *a.s.*
(A5) *There exists a function* $M(Y)$ *such that* $|\psi''(Y, \theta)| \leq M(Y)$ *a.s.* $\mathcal{P}$ *and* $\mathbb{E}_\theta|M(Y)| < \infty \; \forall \theta \in \mathcal{X}$.
(A6) *The Fisher information* $I_\theta = \mathbb{E}_\theta[\psi^2(Y, \theta)] = \mathbb{E}_\theta[-\psi'(Y, \theta)]$ *satisfies* $0 < I_\theta < \infty$.
(A7) $\frac{\partial}{\partial\theta} \int_\mathcal{Y} p_\theta(y) d\mu(y) = \int_\mathcal{Y} \frac{\partial p_\theta(y)}{\partial\theta} d\mu(y)$ *and* $\frac{\partial^2}{\partial\theta^2} \int_\mathcal{Y} p_\theta(y) d\mu(y) = \int_\mathcal{Y} \frac{\partial^2 p_\theta(y)}{\partial\theta^2} d\mu(y)$.

Then there exists a strongly consistent sequence $\hat{\theta}_n$ *of roots of the likelihood equation such that*

$$\sqrt{n}(\hat{\theta}_n - \theta) \xrightarrow{d.} \mathcal{N}(0, I_\theta^{-1}). \tag{14.36}$$

REMARK 14.2 Equation (14.36) does not necessarily imply that

$$\lim_{n\to\infty} n\text{Var}_\theta(\hat{\theta}_n) = I_\theta^{-1},$$

which is the property required for asymptotic efficiency. Some mild additional technical assumptions are needed to show that the MLE is asymptotically efficient and asymptotically unbiased.

REMARK 14.3 For any estimator $\hat{\theta}_n$, the property

$$\lim_{n\to\infty} n\text{Var}_\theta(\hat{\theta}_n) = O(1)$$

implies that the estimation errors are $O_P(n^{-1/2})$ and is known as $\sqrt{n}$ consistency.

REMARK 14.4 Theorem 14.2 does not guarantee that $\hat{\theta}_n$ is MLE.

REMARK 14.5 Assumption (A5) on ψ'' can be replaced by a weaker condition on the modulus of continuity of ψ' [4].

Example 14.6 Let $p_\theta(y) = \frac{1}{2}e^{-|y-\theta|}$, $y \in \mathbb{R}$ be the Laplacian distribution with mean θ and variance equal to 2. The log-likelihood function is given by

$$\ln p_\theta(\mathbf{y}) = -n \ln 2 - \frac{1}{2}\sum_{i=1}^{n} |\theta - y_i|$$

or equivalently by

$$\ln p_\theta(\mathbf{y}) = -n \ln 2 - \frac{1}{2}\sum_{i=1}^{n} |\theta - y_{(i)}|$$

where $\{y_{(i)}\}_{i=1}^n$ denote the *order statistics*, i.e., the observations reordered such as $y_{(1)} \le y_{(2)} \le \cdots \le y_{(n)}$. The log-likelihood function is maximized by the median estimator

$$\hat{\theta}_{\text{ML}}(\boldsymbol{y}) = \text{med}\{y_i\}_{i=1}^n.$$

For odd n, the solution $\hat{\theta}_{\text{ML}}(\boldsymbol{y}) = y_{(\frac{n+1}{2})}$ is the unique MLE. For even n any point in the interval $[y_{(\frac{n}{2})},\ y_{(\frac{n}{2}+1)}]$ is a MLE.

Consistency, efficiency, and asymptotic Gaussianity of the median estimator can be inferred from Theorem 14.2, but also from asymptotic properties of median estimator in a much broader family of distributions [5, p. 92]. Consider a cdf F whose median shall be assumed to be 0 without loss of generality. Assume the associated pdf is continuous at 0 where it takes the value $p(0)$. Denote by $Y_{(k)}$ the median of $n = 2k-1$ i.i.d. random variables with common cdf F. Then by application of [5, Theorem 7.1] (also see [6]), the normalized median estimator

$$\sqrt{n}(Y_{(k)} - \text{med}(F)) \xrightarrow{d} \mathcal{N}\left(0, \frac{1}{4p(0)^2}\right).$$

Moreover, the median estimator is unbiased for all n and has asymptotic variance $\lim_n n\text{Var}[Y_{(k)}] = \frac{1}{4p(0)^2}$. For the Laplace distribution, $p(0) = \frac{1}{2}$ and thus $\lim_n n\text{Var}[Y_{(k)}] = 1$. This coincides with the CRLB because Fisher information for θ in the Laplace family $\{p_\theta\}$ is $I_\theta = \mathbb{E}_\theta(1) = 1$.

In contrast, the sample mean estimator $\overline{Y} = \frac{1}{n}\sum_{i=1}^n Y_i$, while also unbiased and $\sqrt{n}$-consistent, is not efficient because $n\text{Var}_\theta[\overline{Y}] = \text{Var}[Y] = 2$ for all $n \ge 1$.

Vector Parameters Consider a parametric family $\{p_{\boldsymbol{\theta}}\}_{\boldsymbol{\theta}\in\mathcal{X}}$ where $\mathcal{X} \subseteq \mathbb{R}^m$. We define the score function as the gradient vector $\psi(y, \boldsymbol{\theta}) \triangleq \nabla_{\boldsymbol{\theta}} \ln p_{\boldsymbol{\theta}}(y) \in \mathbb{R}^m$ and the $m \times m$ Fisher information matrix as

$$\mathsf{I}_{\boldsymbol{\theta}} = \mathbb{E}_{\boldsymbol{\theta}}[\psi(Y, \boldsymbol{\theta})\psi^\top(Y, \boldsymbol{\theta})] = \mathbb{E}_{\boldsymbol{\theta}}[-\nabla_{\boldsymbol{\theta}}\psi(Y, \boldsymbol{\theta})],$$

assuming the derivatives exist. The counterpart of Theorem 14.2 in the vector case is given in Theorem 14.3.

THEOREM 14.3 Asymptotic Normality of MLE [4, p. 121]. *Assume:*

(B1) $p_{\boldsymbol{\theta}} \ne p'_{\boldsymbol{\theta}}$ *for all* $\boldsymbol{\theta} \ne \boldsymbol{\theta}'$ *(identifiability).*
(B2) $\mathcal{X}$ *is an open subset of* $\mathbb{R}^m$.
(B3) $\{p_{\boldsymbol{\theta}}\}_{\boldsymbol{\theta}\in\mathcal{X}}$ *have the same support*

$$\mathcal{Y} = \text{supp}\{p_{\boldsymbol{\theta}}\} = \{y : p_{\boldsymbol{\theta}}(y) > 0\} \quad \forall \boldsymbol{\theta} \in \mathcal{X}.$$

(B4) $p_{\boldsymbol{\theta}}(y)$ *is twice continuously differentiable with respect to* $\boldsymbol{\theta}$.
(B5) *There exists a function* $M(Y)$ *such that* $\mathbb{E}_{\boldsymbol{\theta}}[M(Y)] < \infty \forall \boldsymbol{\theta} \in \mathcal{X}$ *and all third-order partial derivatives of* $p_{\boldsymbol{\theta}}(Y)$ *are bounded in magnitude by* $M(Y)$.
(B6) *The Fisher information matrix* $\mathsf{I}_{\boldsymbol{\theta}}$ *is finite and has full rank.*
(B7) $\nabla^2_{\boldsymbol{\theta}} \int_{\mathcal{Y}} p_{\boldsymbol{\theta}}(y) d\mu(y) = \int_{\mathcal{Y}} \nabla^2 p_{\boldsymbol{\theta}}(y)\, d\mu(y)$.

Then there exists a strongly consistent sequence $\hat{\boldsymbol{\theta}}_n$ of roots of the likelihood equation such that

$$\sqrt{n}(\hat{\boldsymbol{\theta}}_n - \boldsymbol{\theta}) \xrightarrow{d} \mathcal{N}(0, \mathsf{I}_{\boldsymbol{\theta}}^{-1}). \tag{14.37}$$

14.7 Nonregular ML Estimation Problems

Consider the case where the distributions p_θ do not have common support, so Assumption (B3) of Theorem 14.3 does not hold. Theorem 14.1 still applies but Theorem 14.3 does not. As an example, consider

$$Y_i \sim \text{i.i.d. Uniform}[0, \theta], \quad \theta \in \mathcal{X} = (0, \infty).$$

Then

$$\begin{aligned}
p_\theta(\boldsymbol{Y}) &= \prod_{i=1}^n p_\theta(Y_i) \\
&= \theta^{-n} \prod_{i=1}^n \mathbb{1}\{0 \le Y_i \le \theta\} \\
&= \theta^{-n} \mathbb{1}\{Z_n \le \theta\},
\end{aligned}$$

where

$$Z_n \triangleq \max_{1 \le i \le n} Y_i$$

is a complete sufficient statistic for θ. The MLE is given by

$$\hat{\theta}_n = \arg\max_{\theta > 0} p_\theta(\boldsymbol{Y}) = \arg\max_{\theta \ge Z_n} \theta^{-n} = Z_n.$$

The cdf of $\hat{\theta}_n = Z_n$ is given by

$$\begin{aligned}
P_\theta\{\hat{\theta}_n \le z\} &= P_\theta\{Z_n \le z\} \\
&= P_\theta\{\forall i : Y_i \le z\} \\
&= (P_\theta\{Y \le z\})^n \\
&= \left(\frac{z}{\theta}\right)^n, \quad 0 \le z \le \theta.
\end{aligned}$$

The estimation error is $O_P(n^{-1})$ in this problem, as can be seen from the following analysis. Define the normalized error

$$X_n = n(\theta - \hat{\theta}_n) \ge 0$$

and denote by F_n its cdf. We have

$$\begin{aligned}
1 - F_n(x) &= P_\theta\{X_n > x\} \\
&= P_\theta\left\{Z_n < \theta - \frac{x}{n}\right\} \\
&= \left(1 - \frac{x/\theta}{n}\right)^n, \quad x \ge 0.
\end{aligned}$$

Hence

$$\lim_{n\to\infty} F_n[x] = 1 - e^{-x/\theta}, \quad x \geq 0,$$

i.e., X_n converges in distribution to an $\mathrm{Exp}(\theta)$ random variable.

The asymptotics of the estimation error are thus completely different from those in the regular case, both in the scaling factor ($1/n$ instead of $1/\sqrt{n}$) and in the asymptotic distribution (exponential instead of Gaussian).

14.8 Nonexistence of MLE

Since the MLE is defined as any $\hat{\theta}$ that *achieves the supremum* of the likelihood function over $\mathcal{X}$, one must be concerned about problems in which the supremum is not achieved, hence a MLE *does not exist*. This is typically the case when the likelihood function is unbounded and may happen even when *all the regularity conditions are satisfied.* While the ML approach technically fails, it can often be salvaged in the sense that the likelihood equation admits a "good root," as illustrated in Example 14.7.

Example 14.7 *Gaussian Observations.* Let $Y_1 \sim \mathcal{N}(\mu, \sigma^2)$ and $Y_i \overset{i.i.d.}{\sim} \mathcal{N}(\mu, 1)$ for $i = 2, \ldots, n$. The unknown parameter is $\theta = (\mu, \sigma^2) \in \mathbb{R}\times\mathbb{R}^+$. Note that σ^2 influences Y_1 but not $Y_2, \ldots, Y_n$.

Before attempting to derive the MLE, note there exists a simple and reasonable estimator for this problem, i.e.,

$$\tilde{\mu} = \frac{1}{n-1}\sum_{i=2}^{n} Y_i, \qquad \tilde{\sigma}^2 = (Y_1 - \tilde{\mu})^2, \tag{14.38}$$

where $\tilde{\mu}$ ignores Y_1 but is asymptotically unbiased, consistent, and efficient as $n \to \infty$, and $\tilde{\sigma}^2$ is asymptotically MVUE (albeit not consistent) since $\tilde{\mu} \overset{m.s.}{\to} \mu$ as $n \to \infty$.

The log-likelihood function for (μ, σ^2) may be written as

$$\begin{aligned} l(\mu, \sigma^2) &= -\frac{1}{2}\ln(2\pi\sigma^2) - \frac{(Y_1-\mu)^2}{2\sigma^2} - \frac{(n-1)}{2}\ln(2\pi) - \sum_{i=2}^{n}\frac{(Y_i-\mu)^2}{2} \\ &= -\frac{n}{2}\ln(2\pi) - \frac{1}{2}\ln\sigma^2 - \frac{(Y_1-\mu)^2}{2\sigma^2} - \sum_{i=2}^{n}\frac{(Y_i-\mu)^2}{2} \end{aligned} \tag{14.39}$$

to be maximized over μ and σ^2. Unfortunately $l(\mu, \sigma^2)$ is unbounded. To see this, consider $\mu_\epsilon = Y_1 - \epsilon$ and $\sigma_\epsilon^2 = \epsilon^2$. We have

$$\begin{aligned} l(\mu_\epsilon, \sigma_\epsilon^2) &= \ln\frac{1}{\sqrt{2\pi}} - \frac{1}{2}\ln\epsilon^2 - \frac{1}{2} - \frac{1}{2}\sum_{i=2}^{n}(Y_i - Y_1 + \epsilon)^2, \\ &= C_\epsilon - \ln\epsilon \end{aligned}$$

where C_ϵ tends to a finite limit as $\epsilon \downarrow 0$, but $-\ln\epsilon$ tends to $+\infty$. Hence

$$\sup_{\mu,\sigma^2} l(\mu, \sigma^2) \geq \sup_{\epsilon>0} l(\mu_\epsilon, \sigma_\epsilon^2) = \infty.$$

An iterative optimization algorithm will often produce a sequence with unbounded likelihood that converges to $(Y_1, 0)$ which is infeasible, hence is not a MLE. For instance a gradient-descent algorithm will converge to a root of the likelihood equation

$$0 = \frac{\partial l(\mu, \sigma^2)}{\partial \mu} = \frac{Y_1 - \mu}{\sigma^2} + \sum_{i=2}^{n}(Y_i - \mu),$$

$$0 = \frac{\partial l(\mu, \sigma^2)}{\partial \sigma^2} = -\frac{1}{2\sigma^2} + \frac{(Y_1 - \mu)^2}{2\sigma^4}.$$

Hence any solution $(\hat{\mu}, \hat{\sigma}^2)$ to the likelihood equation satisfies

$$\hat{\mu} = \frac{Y_1/\hat{\sigma}^2 + \sum_{i=2}^{n} Y_i}{1/\hat{\sigma}^2 + n - 1},$$
$$\hat{\sigma}^2 = (Y_1 - \hat{\mu})^2. \tag{14.40}$$

This nonlinear system admits a solution that converges a.s. to the infeasible point $(Y_1, 0)$ as $n \to \infty$.

However, the system (14.40) also admits a feasible solution that converges a.s. to the "simple" estimator $(\tilde{\mu}, \tilde{\sigma}^2)$ of (14.38) as $n \to \infty$ (see Exercise 14.8). It may also be verified[2] that the Hessian of the log-likelihood at that point is negative definite, hence the root is a local maximum of the log-likelihood.

Example 14.8 *Mixture of Gaussians.* Here we consider i.i.d. observations $\{Y_i\}_{i=1}^n \sim i.i.d.\ p_{\boldsymbol{\theta}}$ where $\boldsymbol{\theta} = (\mu, \sigma^2)$ and $p_{\boldsymbol{\theta}}$ is a mixture of two Gaussians with common mean μ:

$$P_{\boldsymbol{\theta}} = \frac{1}{2}\mathcal{N}(\mu, 1) + \frac{1}{2}\mathcal{N}(\mu, \sigma^2).$$

The log-likelihood function for (μ, σ^2) is given by

[2] These derivations are simplified if one works with the reparameterization $\xi = 1/\sigma^2$. Then maximization of (14.39) is equivalent to maximization of

$$l(\mu, \xi) = \ln \frac{1}{\sqrt{2\pi}} + \frac{1}{2}\ln \xi - \frac{1}{2}(Y_1 - \mu)^2 \xi - \frac{1}{2}\sum_{i=2}^{n}(Y_i - \mu)^2,$$

$$\Rightarrow \quad \nabla l(\mu, \xi) = \begin{bmatrix} (Y_1 - \mu)\xi + \sum_{i=2}^{n}(Y_i - \mu) \\ \frac{1}{2\xi} - \frac{1}{2}(Y_1 - \mu)^2 \end{bmatrix}$$

$$\nabla^2 l(\mu, \xi) = \begin{bmatrix} -\xi - n + 1 & Y_1 - \mu \\ Y_1 - \mu & -1/2\xi^2 \end{bmatrix},$$

which is negative definite in a neighborhood of $(\tilde{\mu}, \tilde{\xi})$, for n sufficiently large.

$$
\begin{aligned}
l(\mu, \sigma^2) &= \sum_{i=1}^{n} \ln p_{\boldsymbol{\theta}}(Y_i) \\
&= \sum_{i=1}^{n} \ln \left[\frac{1}{2\sqrt{2\pi}} \exp\left\{ -\frac{(Y_i-\mu)^2}{2} \right\} + \frac{1}{2\sqrt{2\pi\sigma^2}} \exp\left\{ -\frac{(Y_i-\mu)^2}{2\sigma^2} \right\} \right]
\end{aligned}
$$

and is again unbounded. To see this, pick any $i \in \{1, 2, \ldots, n\}$ and consider $\mu_\epsilon = Y_i - \epsilon$ and $\sigma_\epsilon^2 = \epsilon^2$. Then

$$
\begin{aligned}
l(\mu_\epsilon, \sigma_\epsilon^2) &= \sum_{j=1}^{n} \ln \left[\frac{1}{2\sqrt{2\pi}} \exp\left\{ -\frac{(Y_j - Y_i + \epsilon)^2}{2} \right\} \right. \\
&\qquad \left. + \frac{1}{2\sqrt{2\pi\epsilon^2}} \exp\left\{ -\frac{(Y_j - Y_i + \epsilon)^2}{2\epsilon^2} \right\} \right] \\
&= C_\epsilon + \ln \left[\frac{1}{2\sqrt{2\pi}} e^{-\epsilon^2/2} + \frac{1}{2\sqrt{2\pi\epsilon^2}} e^{-1/2} \right],
\end{aligned}
$$

where C_ϵ is the sum of the terms indexed by $j \neq i$ and tends to a finite limit as $\epsilon \downarrow 0$. Hence $l(\mu_\epsilon, \sigma_\epsilon^2) \sim -\ln \epsilon$ as $\epsilon \downarrow 0$, and

$$
\sup_{\mu, \sigma^2} l(\mu, \sigma^2) \geq \sup_{\epsilon > 0} l(\mu_\epsilon, \sigma_\epsilon^2) = \infty.
$$

Thus the likelihood function is unbounded as (μ, σ^2) approaches $(Y_i, 0)$ for each $1 \leq i \leq n$, and the MLE does not exist.

Example 14.9 *General Mixtures.* The same problem of unbounded likelihood function arises with more general mixture models

$$
\mathrm{P}_{\boldsymbol{\theta}} = \sum_{j=1}^{k} \pi_j \mathcal{N}(\mu_j, \sigma_j^2), \tag{14.41}
$$

where k is the number of mixture components, π is a pmf over $\{1, 2, \ldots, k\}$, and the parameter $\boldsymbol{\theta} = \{\pi_j, \mu_j, \sigma_j^2\}_{j=1}^k$. Yet it may be shown that if the Fisher information matrix $\mathsf{I}_{\boldsymbol{\theta}}$ has full rank the following result holds: for all large n with probability 1, there exists a unique root $\hat{\boldsymbol{\theta}}_n$ of the likelihood equation such that $\sqrt{n}(\hat{\boldsymbol{\theta}}_n - \boldsymbol{\theta}) \xrightarrow{d} \mathcal{N}(0, \mathsf{I}_{\boldsymbol{\theta}}^{-1})$ [5, Ch. 33]. The same result applies to mixtures of Gaussian vectors

$$
P_{\boldsymbol{\theta}} = \sum_{j=1}^{k} \pi_j \mathcal{N}(\mu_j, \mathsf{C}_j), \tag{14.42}
$$

where μ_j and C_j are respectively the mean vector and covariance matrix for mixture component j, and $\boldsymbol{\theta} = \{\pi_j, \mu_j, \mathsf{C}_j\}_{j=1}^k$. The same result applies to even more general mixture models of the form

$$
p_{\boldsymbol{\theta}} = \sum_{j=1}^{k} \pi_j \, p_{\xi_j},
$$

where $\{p_\xi\}_{\xi\in\Xi}$ is a parametric family that satisfies some regularity conditions, and $\boldsymbol{\theta} = \{\pi_j, \xi_j\}_{j=1}^k$.

14.9 Non-i.i.d. Observations

The methods of Section 14.6 can be used to study asymptotics in problems where the observations are not i.i.d. Consider for instance the problem of estimating the scale parameter $\theta \in \mathbb{R}$ for a signal received in white Gaussian noise:

$$\boldsymbol{Y} \sim \mathcal{N}(\theta \boldsymbol{s}, \mathsf{I}_n),$$

where $\boldsymbol{s} \in \mathbb{R}^n$ is a known sequence. As we saw in Example 14.3, the MLE is an MVUE for this problem:

$$\hat{\theta}_{\text{ML}} = \hat{\theta}_{\text{MVUE}} = \frac{\boldsymbol{Y}^\top \boldsymbol{s}}{\|\boldsymbol{s}\|^2}.$$

The estimator is Gaussian with mean θ and variance $1/\|\boldsymbol{s}\|^2$ equal to the reciprocal of Fisher information. Hence the estimator is unbiased and efficient. The estimator is consistent if and only if $\lim_{n\to\infty} \|\boldsymbol{s}\|^2 = \infty$. Note that the estimation error need not scale as $1/\sqrt{n}$. For instance, if $s_i = i^{-a}$ with $-\infty < a \le \frac{1}{2}$, the estimation error variance is $O(n^{2a-1})$. For $a \ge \frac{1}{2}$, the estimator is not consistent.

14.10 M-Estimators and Least-Squares Estimators

Often the pdf p_θ is not known precisely, or the likelihood maximization problem is difficult. It is common in such problems to use an *M-estimator* [5]

$$\hat{\theta} \triangleq \arg\min_{\theta\in\mathcal{X}} \sum_{i=1}^n \varphi(Y_i, \theta),$$

where φ is a function that possesses nice properties, such as convexity. This approach reduces to ML if $\varphi(Y, \theta) = -\ln p_\theta(Y)$ is the negative log-likelihood; one could also let $\varphi(Y, \theta) = -\ln q_\theta(Y)$ for some other pdf q_θ. If q_θ is Gaussian, the method is known as *least-squares* estimation. The asymptotic properties of M-estimators can be derived using the same approach as for MLEs and often result in estimators that are asymptotically unbiased and consistent, but not efficient: while the estimation error scales as $1/\sqrt{n}$ under regularity conditions, the asymptotic variance is larger than the reciprocal of Fisher information. This is the penalty for using a *mismatched model* (q_θ instead of the actual p_θ).

We explore the special case of least-squares estimation in more detail now. Consider the estimation of a parameter $\boldsymbol{\theta}$ from a sequence of observations, where each observation is a linear transformation of $\boldsymbol{\theta}$ in zero-mean additive noise, i.e., the observation sequence given by

$$\boldsymbol{Y}_k = \mathsf{H}_k \boldsymbol{\theta} + \boldsymbol{V}_k, \quad k = 1, 2, \ldots,$$

where H_k is known, and $\boldsymbol{V}_k$ is assumed to be zero mean. If we obtain a block of n observations, then by concatenation

$$\boldsymbol{Y} = [\boldsymbol{Y}_1^\top \dots \boldsymbol{Y}_n^\top]^\top, \quad \mathsf{H} = [\mathsf{H}_1^\top \dots \mathsf{H}_n^\top]^\top, \quad \boldsymbol{V} = [\boldsymbol{V}_1^\top \dots \boldsymbol{V}_n^\top]^\top$$

we obtain the block model

$$\boldsymbol{Y} = \mathsf{H}\boldsymbol{\theta} + \boldsymbol{V}.$$

The least-squares estimate for $\boldsymbol{\theta}$ is defined as

$$\hat{\boldsymbol{\theta}}_{\mathrm{LS}}(\boldsymbol{y}) \triangleq \arg\min_{\boldsymbol{\theta}} \|\boldsymbol{y} - \mathsf{H}\boldsymbol{\theta}\|^2, \tag{14.43}$$

which equals $\hat{\boldsymbol{\theta}}_{\mathrm{ML}}(\boldsymbol{y}) = \hat{\boldsymbol{\theta}}_{\mathrm{MVUE}}(\boldsymbol{y})$ if $\boldsymbol{V}$ has i.i.d. zero-mean Gaussian components. The optimization problem in (14.43) is a convex optimization problem in $\boldsymbol{\theta}$, the solution to which is easily seen to be (see Exercise 14.9):

$$\hat{\boldsymbol{\theta}}_{\mathrm{LS}}(\boldsymbol{y}) = (\mathsf{H}^\top\mathsf{H})^{-1}\mathsf{H}^\top \boldsymbol{y}. \tag{14.44}$$

The matrix $(\mathsf{H}^\top\mathsf{H})^{-1}\mathsf{H}^\top$ is called the pseudo-inverse of H.

In applications where the components of the noise vector are known to be correlated with covariance matrix C, the following modification to the least-squares problem called the weighted least-squares problem is of interest:

$$\hat{\boldsymbol{\theta}}_{\mathrm{WLS}}(\boldsymbol{y}) \triangleq \arg\min_{\boldsymbol{\theta}} (\boldsymbol{y} - \mathsf{H}\boldsymbol{\theta})^\top\mathsf{C}^{-1}(\boldsymbol{y} - \mathsf{H}\boldsymbol{\theta}), \tag{14.45}$$

which equals $\hat{\boldsymbol{\theta}}_{\mathrm{ML}}(\boldsymbol{y}) = \hat{\boldsymbol{\theta}}_{\mathrm{MVUE}}(\boldsymbol{y})$ if $\boldsymbol{V}$ is zero-mean and Gaussian, with covariance matrix C.

The solution to the problem in (14.45), which is also a convex optimization problem, is easily seen to be (see Exercise 14.9):

$$\hat{\boldsymbol{\theta}}_{\mathrm{WLS}}(\boldsymbol{y}) = (\mathsf{H}^\top\mathsf{C}^{-1}\mathsf{H})^{-1}\mathsf{H}^\top\mathsf{C}^{-1}\boldsymbol{y}. \tag{14.46}$$

While the solution to least-squares and weighted least-squares problems can be written in closed-form, computing the required pseudo-inverses can be computationally expensive especially when the number of observations is large. For this reason, in practice recursive solutions such as the ones described in Section 14.12.2 are preferable.

14.11 Expectation-Maximization (EM) Algorithm

Numerical evaluation of maximum-likelihood (ML) estimates is often difficult. The likelihood function may have multiple extrema and the parameter θ may be multidimensional, all of which are problematic for any numerical algorithm. In this section we introduce the expectation-maximization (EM) algorithm, a powerful optimization method that has been used with great success in many applications. The idea can be traced back to Hartley [7] and was fully developed in a landmark paper by Dempster, Laird, and Rubin [8]. The EM algorithm relies on the concept of a *complete data space*, to be selected in some convenient way. The EM algorithm is iterative and alternates between conditional expectation and maximization steps. If the complete data space is properly chosen, this method can be very effective as it exploits the inherent statistical structure of the parameter estimation problem. The EM algorithm is particularly adept at dealing with problems involving *missing data* which arise in many applications in the areas of data communications, signal processing, imaging, and biology. Comprehensive surveys include the tutorial paper by Meng and van Dijk [9] and the book by McLachlan and Krishnan [10].

14.11.1 General Structure of the EM Algorithm

The EM algorithm is an iterative algorithm that starts from an initial estimate of the unknown parameter θ and produces a sequence of estimates $\hat{\theta}^{(k)}$, $k = 1, 2, \ldots$. The EM algorithm uses the following ingredients:

- an incomplete-data space $\mathcal{Y}$ (measurement space);
- a complete-data space $\mathcal{Z}$ (many choices are possible);
- a many-to-one mapping $h : \mathcal{Z} \to \mathcal{Y}$.

With some abuse of notation, we denote the pdf for the complete data by $p_\theta(z)$. The pdf for the incomplete data is given by

$$p_\theta(y) = \int_{h^{-1}(y)} p_\theta(z)\, dz,$$

where $h^{-1}(y) \subset \mathcal{Z}$ denotes the set of z such that $h(z) = y$.

The incomplete-data and complete-data log-likelihood functions are respectively denoted by

$$\ell_{\text{id}}(\theta, y) \triangleq \ln p_\theta(y), \quad \ell_{\text{cd}}(\theta, z) \triangleq \ln p_\theta(z).$$

The complete data space should be chosen in such a way that maximization of $\ell_{\text{cd}}(\theta, z)$ is easy. However, the complete data are not available, so one uses a *surrogate cost function*: the expectation of $\ell_{\text{cd}}(\theta, z)$ conditioned on Y and the current estimate of θ. Maximizing this surrogate function results in an updated estimate of θ, and these steps are repeated till convergence.

Upon selection of an initial estimate $\hat{\theta}^{(0)}$, the EM algorithm alternates between expectation (E) and maximization (M) steps for updating the estimate $\hat{\theta}^{(k)}$ of the unknown parameter θ at iterations $k = 1, 2, 3, \ldots$.

E-Step: Compute

$$Q(\theta|\hat{\theta}^{(k)}) \triangleq \mathbb{E}_{\hat{\theta}^{(k)}}[\ell_{\text{cd}}(\theta, Z)|Y = y], \tag{14.47}$$

where the conditional expectation is evaluated assuming that the true parameter is equal to $\hat{\theta}^{(k)}$.

M-step: Compute

$$\hat{\theta}^{(k+1)} = \arg\max_{\theta \in \mathcal{X}} Q(\theta|\hat{\theta}^{(k)}). \tag{14.48}$$

If the EM sequence $\hat{\theta}^{(k)}$, $k \geq 0$ converges, the limit θ^* is called a stable point of the algorithm. In practice, the iterations are terminated when the estimates are stable enough (using some stopping criterion).

14.11.2 Convergence of EM Algorithm

The convergence properties of the EM algorithm were first studied by Dempster, Laird, and Rubin [8] and by Wu [11]. The main result is as follows. The proof is given in Section 14.14.

THEOREM 14.4 *The sequence of log-likelihoods of the incomplete data $\ell_{id}(\hat{\theta}^{(k)}, y)$, $k = 0, 1, 2, \ldots$ is nondecreasing.*

While this theorem guarantees monotonicity of the log-likelihood sequence $\ell_{\text{id}}(\hat{\theta}^{(k)})$, convergence to an extremum of $\ell_{\text{id}}(\theta, y)$, i.e., a solution to the likelihood equation, requires additional differentiability assumptions (see, e.g., [11]).[3]

REMARK 14.6 Some remarks about the EM algorithm are in order.

1. Even though the sequence $\ell_{\text{id}}(\hat{\theta}^{(k)})$ may converge, there is in general no guarantee that $\hat{\theta}^{(k)}$ converges, e.g., the EM algorithm may jump back and forth between two attractors.
2. Even if the algorithm has a stable point, there is in general no guarantee that this stable point is a global maximum of the likelihood function, or even a local maximum. For many problems, however, convergence to a local maximum has been demonstrated.
3. The solution in general depends on the initialization.
4. Several variants of the EM algorithm have been developed. One of the most promising ones is Fessler and Hero's SAGE algorithm (space-alternating generalized EM) whose convergence rate is often far superior to that of the EM algorithm [12].
5. There is a close relationship between convergence rate of the EM algorithm and Fisher information, as described in [8, 12].
6. Often the choice for Z is a pair (Y, U) where U are thought of as missing data, or *latent data*. Such is the case with the estimation of mixtures in Example 14.11, where the missing data were the labels associated with each observation.
7. The E-step is relatively simple to implement when Z conditioned on Y is from an exponential family. Then the E-step is reduced to evaluating the conditional expectation of sufficient statistics, and the M-step is often tractable.

14.11.3 Examples

We now provide some examples of applications of the EM algorithm.

[3] To illustrate why such differentiability assumptions are necessary, consider the function $f(u, v) = \phi(u - v)\cos(\pi(u + v)/2)$ where $\phi(u) = \max\{0, 1 - |u|\}$ and $-1 \leq u, v \leq 1$. This function admits a unique local maximum (which is therefore also a global maximum) at $(u, v) = (0, 0)$. However, the graph of the function has a ridge along the direction $v = u$. A coordinate-ascent algorithm (that would successively maximize f along the u and v directions) would reach the ridge after just one step of the algorithm and remain stuck there.

Example 14.10 *Convergence of EM Algorithm* Suppose the observation $Y = X + W$, where $X \sim \mathcal{N}(\theta, 1)$ and $W \sim \mathcal{N}(0, 1)$ are independent, with θ being an unknown parameter to be estimated from Y. It is easy to see that

$$p_\theta(y) = \frac{1}{\sqrt{4\pi}} e^{-(y-\theta)^2/4}$$

and that the log-likelihood

$$\ell_{\text{id}}(\theta, y) = \ln p_\theta(y) = -\frac{(y-\theta)^2}{4} - \frac{1}{2}\ln(4\pi).$$

Maximizing over θ, we obtain

$$\hat{\theta}_{\text{ML}}(y) = y.$$

While the MLE is easy to obtain in this example, it is nevertheless instructive to study the solution that is produced by the EM algorithm applied to this problem with complete data given by $\mathbf{Z} = (X, W)$. Note that the mapping that takes us from $\mathbf{Z}$ to Y is many-to-one (non-invertible). By the independence of X and W, we obtain

$$p_\theta(z) = p_\theta(x, w) = \frac{1}{2\pi} e^{-(x-\theta)^2/2} e^{-w^2/2}$$

and the log-likelihood of the complete data is given by

$$\ell_{\text{cd}}(\theta, z) = -\frac{(x-\theta)^2}{2} - \frac{w^2}{2} - \ln(2\pi).$$

E-Step

$$Q(\theta|\hat{\theta}^{(k)}) = \mathbb{E}_{\hat{\theta}^{(k)}}\left[\ell_{\text{cd}}(\theta, \mathbf{Z})|Y = y\right].$$

In computing $Q(\theta|\hat{\theta}^{(k)})$, we can ignore additive terms in $\ell_{\text{cd}}(\theta, \mathbf{Z})$ that are not functions of θ, since they are irrelevant in the M-step. In particular,

$$\ell_{\text{cd}}(\theta, \mathbf{Z}) = -\frac{\theta^2}{2} + \theta X + \text{terms that not functions of } \theta.$$

Therefore

$$Q(\theta|\hat{\theta}^{(k)}) = -\frac{\theta^2}{2} + \theta\, \mathbb{E}_{\hat{\theta}^{(k)}}\left[X|Y = y\right].$$

This is a (concave) quadratic function in θ, which is maximized by setting the derivative with respect to θ to 0.

M-Step

$$\hat{\theta}^{(k+1)} = \arg\max_{\theta \in \mathcal{X}} Q(\theta|\hat{\theta}^{(k)}) = \mathbb{E}_{\hat{\theta}^{(k)}}\left[X|Y = y\right].$$

By the LMMSE formula (11.15), we obtain the conditional expectation as

$$\mathbb{E}_{\hat{\theta}^{(k)}}\left[X|Y = y\right] = \hat{\theta}^{(k)} + \frac{1}{2}(y - \hat{\theta}^{(k)}) = \frac{1}{2}(y + \hat{\theta}^{(k)})$$

hence

$$\hat{\theta}^{(k+1)} = \frac{1}{2}(y + \hat{\theta}^{(k)}).$$

As $k \to \infty$, $\hat{\theta}^{(k)}$ converges to θ^{∞} that satisfies the equation

$$\theta^{\infty} = \frac{1}{2}\left(y + \theta^{\infty}\right),$$

which can be solved to yield

$$\theta^{\infty}(y) = y = \hat{\theta}_{\mathrm{ML}}(y).$$

That is, the EM algorithm indeed converges to the ML solution.

Example 14.11 *Estimation in Gaussian Mixture Models.* This is one of the most famous problems to which the EM algorithm has been (successfully) applied. Assume the data $\boldsymbol{Y} = \{Y_i,\ 1 \le i \le n\} \in \mathbb{R}^n$, are drawn i.i.d. from a pdf $p_\theta(y)$ which is the mixture of two Gaussian distributions with means μ and ν and variances equal to 1, with probabilities ρ and $(1-\rho)$, respectively:

$$p_{\boldsymbol{\theta}}(y) = \rho\phi(\mu, y) + (1-\rho)\phi(\nu, y),$$

where

$$\phi(m, y) \triangleq \frac{1}{\sqrt{2\pi}} e^{-(y-m)^2/2}$$

is the Gaussian pdf with mean m and variance 1.

Let $\boldsymbol{\theta} = (\mu, \nu, \rho)$. The log-likelihood function is given by

$$\ln p_{\boldsymbol{\theta}}(\boldsymbol{y}) = \sum_{i=1}^{n} \ln\left(\rho\phi(\mu, y_i) + (1-\rho)\phi(\nu, y_i)\right),$$

which is non-convex in $\boldsymbol{\theta}$. The likelihood equation $\nabla_{\boldsymbol{\theta}} \ln p_{\boldsymbol{\theta}}(\boldsymbol{y}) = \boldsymbol{0}$ takes the form

$$\begin{aligned}
0 &= \frac{\partial}{\partial\rho} \ln p_{\boldsymbol{\theta}}(\boldsymbol{y}) = \sum_{i=1}^{n} \frac{\phi(\mu, y_i) - \phi(\nu, y_i)}{\rho\phi(\mu, y_i) + (1-\rho)\phi(\nu, y_i)},\\
0 &= \frac{\partial}{\partial\mu} \ln p_{\boldsymbol{\theta}}(\boldsymbol{y}) = \sum_{i=1}^{n} \frac{\rho(y_i - \mu)\phi(\mu, y_i)}{\rho\phi(\mu, y_i) + (1-\rho)\phi(\nu, y_i)},\\
0 &= \frac{\partial}{\partial\nu} \ln p_{\boldsymbol{\theta}}(\boldsymbol{y}) = \sum_{i=1}^{n} \frac{(1-\rho)(y_i - \nu)\phi(\nu, y_i)}{\rho\phi(\mu, y_i) + (1-\rho)\phi(\nu, y_i)}.
\end{aligned} \tag{14.49}$$

This nonlinear system of three equations with three unknowns cannot be solved in closed form and may have multiple local maxima, or even local minima and stationary points.

The EM approach to this problem adds auxiliary variables to $\boldsymbol{y}$ in such a way as to make $Q(\boldsymbol{\theta}|\hat{\boldsymbol{\theta}}^{(k)})$ easy to maximize. There is some guesswork involved here, but a good choice for the auxiliary variables are the binary random variables that determine which of $\mathcal{N}(\mu, 1)$ or $\mathcal{N}(\nu, 1)$ is chosen as the distribution for the observations. In particular, let $\{X_i\}_{i=1}^n$ be i.i.d. Bernouilli (ρ) random variables such that

$$Y_i \sim \begin{cases} \mathcal{N}(\mu, 1) & \text{if } X_i = 1 \\ \mathcal{N}(\nu, 1) & \text{if } X_i = 0. \end{cases}$$

The complete data is set to be $\boldsymbol{Z} = (\boldsymbol{Y}, \boldsymbol{X})$. Then

$$\ell_{\text{cd}}(\boldsymbol{\theta}, z) = \ln p_{\boldsymbol{\theta}}(z) - \sum_{i=1}^{n} \ln p_{\boldsymbol{\theta}}(z_i),$$

where

$$p_{\boldsymbol{\theta}}(z_i) = p_{\boldsymbol{\theta}}(y_i, x_i) = p_{\boldsymbol{\theta}}(y_i|x_i)p_{\boldsymbol{\theta}}(x_i) = \begin{cases} \phi(\mu, y_i)\ \rho & \text{if } x_i = 1 \\ \phi(\nu, y_i)\ (1-\rho) & \text{if } x_i = 0, \end{cases}$$

and so

$$\begin{aligned} \ln p_{\boldsymbol{\theta}}(z_i) &= \begin{cases} \ln(\phi(\mu, y_i)\ \rho) & \text{if } x_i = 1 \\ \ln(\phi(\nu, y_i)\ (1-\rho)) & \text{if } x_i = 0 \end{cases} \\ &= x_i \ln(\phi(\mu, y_i)\ \rho) + (1-x_i)\ln(\phi(\nu, y_i)\ (1-\rho)). \end{aligned}$$

E-Step

$$\begin{aligned} Q(\boldsymbol{\theta}|\hat{\boldsymbol{\theta}}^{(k)}) &= \mathbb{E}_{\hat{\boldsymbol{\theta}}^{(k)}}\left[\ell_{\text{cd}}(\boldsymbol{\theta}, \boldsymbol{Z})|\boldsymbol{Y} = \boldsymbol{y}\right] \\ &= \sum_{i=1}^{n} x_i^{(k)} \ln(\phi(\mu, y_i)\ \rho) + (1 - x_i^{(k)}) \ln(\phi(\nu, y_i)\ (1-\rho)), \end{aligned} \tag{14.50}$$

where

$$x_i^{(k)} \triangleq \mathbb{E}_{\hat{\boldsymbol{\theta}}^{(k)}}\left[X_i|\boldsymbol{Y} = \boldsymbol{y}\right].$$

Since X_i is independent of Y_k, for $k \neq i$, we have

$$\begin{aligned} x_i^{(k)} &= \mathbb{E}_{\hat{\boldsymbol{\theta}}^{(k)}}\left[X_i|Y_i = y_i\right] \\ &= \mathrm{P}_{\hat{\boldsymbol{\theta}}^{(k)}}(X_i = 1|Y_i = y_i) \\ &= \frac{\hat{\rho}^{(k)}\phi(\hat{\mu}^{(k)}, y_i)}{\hat{\rho}^{(k)}\phi(\hat{\mu}^{(k)}, y_i) + (1-\hat{\rho}^{(k)})\phi(\hat{\nu}^{(k)}, y_i)}. \end{aligned}$$

M-Step Note that

$$\ln(\phi(\mu, y_i)\ \rho) = \ln\rho - \frac{(y_i - \mu)^2}{2} + (\text{constant in } \boldsymbol{\theta})$$

and

$$\ln(\phi(\nu, y_i)\ (1-\rho)) = \ln(1-\rho) - \frac{(y_i - \nu)^2}{2} + (\text{constant in } \boldsymbol{\theta}).$$

Therefore $Q(\boldsymbol{\theta}|\hat{\boldsymbol{\theta}}^{(k)})$ in (14.50) is a concave function of $\boldsymbol{\theta} = (\mu, \nu, \rho)$. Taking derivatives and setting them equal to 0 we obtain $\hat{\boldsymbol{\theta}}^{(k+1)}$ as follows:

$$\frac{\partial}{\partial \rho} Q(\boldsymbol{\theta}|\hat{\boldsymbol{\theta}}^{(k)}) = 0 \implies \frac{1}{\rho}\sum_{i=1}^{n} x_i^{(k)} - \frac{1}{1-\rho}\sum_{i=1}^{n}(1 - x_i^{(k)}) = 0$$

$$\implies \hat{\rho}^{(k+1)} = \frac{1}{n}\sum_{i=1}^{n} x_i^{(k)},$$

$$\frac{\partial}{\partial \mu} Q(\boldsymbol{\theta}|\hat{\boldsymbol{\theta}}^{(k)}) = 0 \implies \sum_{i=1}^{n} x_i^{(k)}(y_i - \mu) = 0$$

$$\implies \hat{\mu}^{(k+1)} = \frac{\sum_{i=1}^{n} x_i^{(k)} y_i}{\sum_{i=1}^{n} x_i^{(k)}},$$

$$\frac{\partial}{\partial \nu} Q(\boldsymbol{\theta}|\hat{\boldsymbol{\theta}}^{(k)}) = 0 \implies \sum_{i=1}^{n}(1 - x_i^{(k)})(y_i - \nu) = 0$$

$$\implies \hat{\nu}^{(k+1)} = \frac{\sum_{i=1}^{n}(1 - x_i^{(k)})y_i}{\sum_{i=1}^{n}(1 - x_i^{(k)})}.$$

The solutions for $\hat{\rho}^{(k+1)}$, $\hat{\mu}^{(k+1)}$, and $\hat{\nu}^{(k+1)}$ have the following intuitive interpretation. The quantity $x_i^{(k)}$ is an estimate of the probability that Y_i has mean μ at iteration k. Then the average of the $\{x_i^{(k)}\}_{i=1}^{n}$ is a reasonable candidate for $\hat{\rho}^{(k+1)}$. The weighted average of the observations corresponding to those estimated to have mean μ and those estimated to have mean ν are reasonable candidates for $\hat{\mu}^{(k+1)}$ and $\hat{\nu}^{(k+1)}$, respectively.

Example 14.12 *Joint Estimation of Signal Amplitude and Noise Variance.*

$$\boldsymbol{Y} = a\boldsymbol{X} + \boldsymbol{W},$$

where $\boldsymbol{X} \sim \mathcal{N}(0, \mathsf{C})$ and $\boldsymbol{W} \sim \mathcal{N}(0, v\mathsf{I})$ are independent Gaussian vectors, and a, v are unknown parameters. With $\boldsymbol{\theta} = (a, v)$, the log-likelihood function of the observations is given by:

$$\ln p_{\boldsymbol{\theta}}(\boldsymbol{y}) = -\frac{1}{2}\boldsymbol{y}^\top(a^2\mathsf{C} + v\,\mathsf{I})^{-1}\boldsymbol{y} - \frac{1}{2}\ln\left|a^2\mathsf{C} + v\,\mathsf{I}\right| - \frac{n}{2}\ln(2\pi)$$

and the corresponding likelihood equation is given by

$$0 = \frac{\partial}{\partial a}\ln p_{\boldsymbol{\theta}}(\boldsymbol{y}) = a\boldsymbol{y}^\top(a^2\mathsf{C} + v\,\mathsf{I})^{-1}\mathsf{C}(a^2\mathsf{C} + v\,\mathsf{I})^{-1}\boldsymbol{y} - a\mathrm{Tr}\left((a^2\mathsf{C} + v\,\mathsf{I})^{-1}\mathsf{C}\right)$$

$$0 = \frac{\partial}{\partial v}\ln p_{\boldsymbol{\theta}}(\boldsymbol{y}) = \frac{1}{2}\boldsymbol{y}^\top(a^2\mathsf{C} + v\,\mathsf{I})^{-1}(a^2\mathsf{C} + v\,\mathsf{I})^{-1}\boldsymbol{y} - \frac{1}{2}\mathrm{Tr}\left((a^2\mathsf{C} + v\,\mathsf{I})^{-1}\right)$$

where we have used the results at the end of Appendix A. While the log-likelihood function is easily seen to be concave in $\boldsymbol{\theta}$, the likelihood equation does not have a closed-form solution. Therefore one has to resort to iterative gradient ascent techniques to obtain $\hat{a}_{\mathrm{ML}}$ and $\hat{v}_{\mathrm{ML}}$. The EM algorithm, which we discuss next, provides an iterative solution for $\hat{a}_{\mathrm{ML}}$ and $\hat{v}_{\mathrm{ML}}$, with an intuitive interpretation.

For the EM approach, we set the complete data to be $\boldsymbol{Z} = (\boldsymbol{Y}, \boldsymbol{X})$. Then

$$\begin{aligned}\ell_{\text{cd}}(\boldsymbol{\theta}, z) &= \ln p_{\boldsymbol{\theta}}(z) = \ln p_{\boldsymbol{\theta}}(\boldsymbol{y}|\boldsymbol{x}) + \ln p_{\boldsymbol{\theta}}(\boldsymbol{x}) \\ &= -\frac{\|\boldsymbol{y} - a\boldsymbol{x}\|^2}{2v} - \frac{n}{2}\ln v + (\text{constant in } \boldsymbol{\theta}).\end{aligned}$$

E-Step

$$Q(\boldsymbol{\theta}|\hat{\boldsymbol{\theta}}^{(k)}) = -\frac{1}{2v}\mathbb{E}_{\hat{\boldsymbol{\theta}}^{(k)}}\left[\|\boldsymbol{Y} - a\boldsymbol{X}\|^2 | \boldsymbol{Y} = \boldsymbol{y}\right] - \frac{n}{2}\ln v.$$

To compute the expectation we need the pdf of $\boldsymbol{X}$, given $\boldsymbol{Y}$, under $\hat{\boldsymbol{\theta}}^{(k)}$. Since $\boldsymbol{X}$ and $\boldsymbol{Y}$ are jointly Gaussian,

$$p_{\hat{\boldsymbol{\theta}}^{(k)}}(\boldsymbol{x}|\boldsymbol{y}) \sim \mathcal{N}(\hat{\boldsymbol{x}}^{(k)}, \hat{\mathsf{C}}^{(k)}),$$

where $\hat{\boldsymbol{x}}^{(k)}$ and $\hat{\mathsf{C}}^{(k)}$ are computed using the LMMSE formulas (see Section 11.7.4) as follows:

$$\begin{aligned}\hat{\boldsymbol{x}}^{(k)} &= \text{Cov}_{\hat{\boldsymbol{\theta}}^{(k)}}(\boldsymbol{X}, \boldsymbol{Y})\text{Cov}_{\hat{\boldsymbol{\theta}}^{(k)}}(\boldsymbol{Y})^{-1}\boldsymbol{y} \\ &= \hat{a}^{(k)}\mathsf{C}\left((\hat{a}^{(k)})^2\mathsf{C} + \hat{v}^{(k)}\mathsf{I}\right)^{-1}\boldsymbol{y} \\ &= \frac{\hat{a}^{(k)}}{\hat{v}^{(k)}}\left(\frac{(\hat{a}^{(k)})^2}{\hat{v}^{(k)}}\mathsf{I} + \mathsf{C}^{-1}\right)^{-1}\boldsymbol{y}\end{aligned}$$

and

$$\begin{aligned}\hat{\mathsf{C}}^{(k)} &= \mathsf{C} - \text{Cov}_{\hat{\boldsymbol{\theta}}^{(k)}}(\boldsymbol{X}, \boldsymbol{Y})\text{Cov}_{\hat{\boldsymbol{\theta}}^{(k)}}(\boldsymbol{Y})^{-1}\text{Cov}_{\hat{\boldsymbol{\theta}}^{(k)}}(\boldsymbol{Y}, \boldsymbol{X}) \\ &= \mathsf{C} - (\hat{a}^{(k)})^2\mathsf{C}\left((\hat{a}^{(k)})^2\mathsf{C} + \hat{v}^{(k)}\mathsf{I}\right)^{-1}\mathsf{C} \\ &= \left(\frac{(\hat{a}^{(k)})^2}{\hat{v}^{(k)}}\mathsf{I} + \mathsf{C}^{-1}\right)^{-1},\end{aligned}$$

where the last line follows from the Matrix Inversion Lemma (see Lemma A.1 in Appendix A). Note that we have

$$\hat{\boldsymbol{x}}^{(k)} = \frac{\hat{a}^{(k)}}{\hat{v}^{(k)}}\hat{\mathsf{C}}^{(k)}\boldsymbol{y}.$$

We are now ready to compute $Q(\boldsymbol{\theta}|\hat{\boldsymbol{\theta}}^{(k)})$ as follows:

$$\begin{aligned}\mathbb{E}_{\hat{\boldsymbol{\theta}}^{(k)}}\left[\|\boldsymbol{Y} - a\boldsymbol{X}\|^2|\boldsymbol{Y} = \boldsymbol{y}\right] &= \mathbb{E}_{\hat{\boldsymbol{\theta}}^{(k)}}\left[\|(\boldsymbol{Y} - a\hat{\boldsymbol{x}}^{(k)}) - a(\boldsymbol{X} - \hat{\boldsymbol{x}}^{(k)})\|^2|\boldsymbol{Y} = \boldsymbol{y}\right] \\ &= \|\boldsymbol{y} - a\hat{\boldsymbol{x}}^{(k)}\|^2 + a^2\mathbb{E}_{\hat{\boldsymbol{\theta}}^{(k)}}\left[\|\boldsymbol{X} - \hat{\boldsymbol{x}}^{(k)}\|^2|\boldsymbol{Y} = \boldsymbol{y}\right] \\ &= \|\boldsymbol{y} - a\hat{\boldsymbol{x}}^{(k)}\|^2 + a^2\,\text{Tr}\left(\hat{\mathsf{C}}^{(k)}\right),\end{aligned}$$

where the second line follows from the orthogonality principle (see Section 11.7.4). Thus

$$Q(\boldsymbol{\theta}|\hat{\boldsymbol{\theta}}^{(k)}) = -\frac{1}{2v}\left(\|\boldsymbol{y} - a\hat{\boldsymbol{x}}^{(k)}\|^2 + a^2\,\mathrm{Tr}\left(\hat{\mathsf{C}}^{(k)}\right)\right) - \frac{n}{2}\ln v.$$

M-Step It is easy to see that $Q(\boldsymbol{\theta}|\hat{\boldsymbol{\theta}}^{(k)})$ is concave in (a, v). Taking derivatives and setting them equal to 0 we obtain the iterates $\hat{a}^{(k+1)}$ and $\hat{v}^{(k+1)}$ as follows:

$$\begin{aligned}\frac{\partial}{\partial a}Q(\boldsymbol{\theta}|\hat{\boldsymbol{\theta}}^{(k)}) = 0 &\implies 2a\mathrm{Tr}\left(\hat{\mathsf{C}}^{(k)}\right) + 2a\|\hat{\boldsymbol{x}}^{(k)}\|^2 - 2\boldsymbol{y}^\top|\hat{\boldsymbol{x}}^{(k)} = 0\\ &\implies \hat{a}^{(k+1)} = \frac{\boldsymbol{y}^\top|\hat{\boldsymbol{x}}^{(k)}}{\mathrm{Tr}\left(\hat{\mathsf{C}}^{(k)}\right) + \|\hat{\boldsymbol{x}}^{(k)}\|^2},\end{aligned}$$

$$\begin{aligned}\frac{\partial}{\partial v}Q(\boldsymbol{\theta}|\hat{\boldsymbol{\theta}}^{(k)}) = 0 &\implies \frac{1}{v^2}\left[\|\boldsymbol{y} - a\hat{\boldsymbol{x}}^{(k)}\|^2 + a^2\,\mathrm{Tr}\left(\hat{\mathsf{C}}^{(k)}\right)\right] - \frac{n}{2v} = 0\\ &\implies \hat{v}^{(k+1)} = \frac{1}{n}\left[\|\boldsymbol{y} - \hat{a}^{(k+1)}\hat{\boldsymbol{x}}^{(k)}\|^2 + (\hat{a}^{(k+1)})^2\,\mathrm{Tr}\left(\hat{\mathsf{C}}^{(k)}\right)\right].\end{aligned}$$

14.12 Recursive Estimation

In many applications, the observations used for estimation are obtained sequentially in time, and it is desirable to produce a parameter estimate at each time k based on only the observations obtained up to time k, which can easily be updated to produce a parameter estimate at time $k+1$ based on the estimate at time k and the observation at time $k+1$. Such estimation procedures are called *recursive* estimation procedures. Note that recursive estimation procedures should not be confused with iterative estimation procedures such as the EM algorithm, where an estimate based on one scalar or vector observation is refined over multiple steps to approach an optimal estimate.

14.12.1 Recursive MLE

In this subsection we use $\hat{\theta}_k$ to refer to the MLE of θ obtained using k i.i.d. observations $Y_1, Y_2, \ldots, Y_k$ that each has distribution p_θ. We start with an example that illustrates that, in some cases, the exact MLE solution can be computed recursively, without the need for any approximation.

Example 14.13 For each i, let

$$p_\theta(y_i) = \frac{1}{\theta}e^{-\frac{y_i}{\theta}}\mathbb{1}\{y_i \geq 0\}.$$

Then

$$\ln p_\theta(y_1, \ldots, y_k) = \sum_{i=1}^{k} \ln p_\theta(y_i)$$
$$= -k \ln \theta - \frac{1}{\theta} \sum_{i=1}^{k} y_i,$$

from which we can easily conclude that the MLE with k observations is given by

$$\hat{\theta}_k = \frac{1}{k} \sum_{i=1}^{k} y_i.$$

Extending this solution to $k + 1$ observations, we see that

$$\hat{\theta}_{k+1} = \frac{1}{k+1} \sum_{i=1}^{k+1} y_i$$
$$= \frac{n\hat{\theta}_k + y_{k+1}}{k+1},$$

which implies that the MLE can be computed recursively.

Such a recursion does not always hold for the MLE, but fortunately a good approximation to the MLE always has a recursion. For simplicity we assume here that θ is a scalar, but the development can easily be extended to the case where θ is a vector.

Define

$$\psi(y_i, \theta) \triangleq \frac{\partial}{\partial \theta} \ln p_\theta(y_i).$$

Then, $\hat{\theta}_k$ satisfies the likelihood equation:

$$\sum_{i=1}^{k} \psi(y_i, \hat{\theta}_k) = 0.$$

Now, assuming that $\hat{\theta}_k$ is consistent (see Section 14.6), we have that

$$\hat{\theta}_{k+1} - \hat{\theta}_k \to 0 \text{ as } k \to \infty \quad \text{i.p. or a.s.}$$

This justifies the Taylor series-based approximation:

$$\psi(y_i, \hat{\theta}_{k+1}) \approx \psi(y_i, \hat{\theta}_k) + (\hat{\theta}_{k+1} - \hat{\theta}_k)\psi'(y_i, \hat{\theta}_k). \tag{14.51}$$

Note that

$$\psi'(y_i, \theta) = \frac{\partial^2}{\partial \theta^2} \ln p_\theta(y_i).$$

Since, $\hat{\theta}_{k+1}$ is the MLE obtained using $(k + 1)$ i.i.d. observations, it must satisfy the likelihood equation:

$$\sum_{i=1}^{k+1} \psi(y_i, \hat{\theta}_{k+1}) = 0.$$

Based on the approximation obtained for $\psi(y_i, \hat{\theta}_{k+1})$ in (14.51), we have

$$\sum_{i=1}^{k+1} \left(\psi(y_i, \hat{\theta}_k) + (\hat{\theta}_{k+1} - \hat{\theta}_k)\psi'(y_i, \hat{\theta}_k) \right) \approx 0,$$

which can rewritten as

$$\hat{\theta}_{k+1} \approx \hat{\theta}_k + \frac{\sum_{i=1}^{k+1} \psi(y_i, \hat{\theta}_k)}{-\sum_{i=1}^{k+1} \psi'(y_i, \hat{\theta}_k)}. \tag{14.52}$$

This approximation does not yield a recursive estimator since the right hand side depends explicitly on all of the observations up to time $(k+1)$, but we can further simplify it to obtain a recursive estimator as follows. First, since $\sum_{i=1}^{k} \psi(y_i, \hat{\theta}_k) = 0$, the numerator on the right hand side of (14.52) reduces to $\psi(y_{k+1}, \hat{\theta}_k)$. For the denominator, note that by the Strong Law of Large Numbers:

$$-\frac{1}{k+1} \sum_{i=1}^{k+1} \psi'(y_i, \theta) = -\frac{1}{k+1} \sum_{i=1}^{k+1} \frac{\partial^2}{\partial \theta^2} \ln p_\theta(y_i) \xrightarrow{\text{a.s.}} -\mathbb{E}\left[\frac{\partial^2}{\partial \theta^2} \ln p_\theta(Y_1) \right] = I_\theta.$$

This allows us to make the approximation

$$-\left(\sum_{i=1}^{k+1} \psi'(y_i, \hat{\theta}_k) \right) \approx (k+1) I_{\hat{\theta}_k}. \tag{14.53}$$

Plugging (14.53) into (14.52) yields the following recursive approximation to the MLE:

$$\hat{\theta}_{k+1} \approx \hat{\theta}_k + \frac{\psi(y_{k+1}, \hat{\theta}_k)}{(k+1) I_{\hat{\theta}_k}}. \tag{14.54}$$

The estimator obtained by treating this approximation as an equality is called the *recursive MLE*. It can be shown that the recursion in (14.54) is special case of a stochastic approximation procedure [13], and yields an estimate with the same asymptotic properties as the MLE (see, e.g., [14]).

14.12.2 Recursive Approximations to Least-Squares Solution

We now develop a recursive approximation to the least-squares solution $\hat{\boldsymbol{\theta}}_{\text{LS}}(\boldsymbol{y})$ given in (14.44). First note that the function being optimized in least-squares can be rewritten as

$$\|\boldsymbol{y} - \mathsf{H}\boldsymbol{\theta}\|^2 = \sum_{k=1}^{n} \|\boldsymbol{y}_k - \mathsf{H}_k \boldsymbol{\theta}\|^2$$

and its gradient with respect to $\boldsymbol{\theta}$ is given by

$$\nabla_{\boldsymbol{\theta}} \|\boldsymbol{y} - \mathsf{H}\boldsymbol{\theta}\|^2 = \sum_{k=1}^{n} 2\mathsf{H}_k^\top (\mathsf{H}_k \boldsymbol{\theta} - \boldsymbol{y}_k).$$

The terms in the summation on the right hand side can be used in a *stochastic gradient descent* procedure to produce a recursive (online) estimator for $\boldsymbol{\theta}$, which is called the least mean squares (LMS) algorithm:

$$\hat{\boldsymbol{\theta}}_{k+1} = \hat{\boldsymbol{\theta}}_k - \mu_k \mathsf{H}_{k+1}^{\top}(\mathsf{H}_{k+1}\hat{\boldsymbol{\theta}}_k - \boldsymbol{y}_{k+1}), \tag{14.55}$$

where $\mu_k > 0$ is called the step-size at the stage k.

Under appropriate conditions on the step-sizes $\{\mu_k\}$ and the distribution of the observations, it can be shown that the LMS update has the desirable property [15]:

$$\hat{\boldsymbol{\theta}}_k \xrightarrow{\text{i.p.}} \hat{\boldsymbol{\theta}}_{\text{LS}}. \tag{14.56}$$

The rate of convergence in (14.56) can be improved by using a modification to LMS akin to Newton's method [16] for gradient descent. The resulting algorithm is called the *recursive least squares* (RLS) algorithm, which has the following update equation:

$$\hat{\boldsymbol{\theta}}_{k+1} = \hat{\boldsymbol{\theta}}_k - \mathsf{C}_k \mathsf{H}_{k+1}^{\top}(\mathsf{H}_{k+1}\hat{\boldsymbol{\theta}}_k - \boldsymbol{y}_{k+1}), \tag{14.57}$$

where C_k can be interpreted as the covariance of the estimation error at time k and is updated using the Riccati equation (which we will study in more detail in Chapter 15, (15.28)):

$$\mathsf{C}_{k+1} = \mathsf{C}_k - \mathsf{C}_k \mathsf{H}_{k+1}^{\top}(\mathsf{I} + \mathsf{H}_{k+1}\mathsf{C}_k\mathsf{H}_{k+1}^{\top})^{-1}\mathsf{H}_{k+1}\mathsf{C}_k.$$

14.13 Appendix: Proof of Theorem 14.2

By (14.31) and (14.33), there exists a strongly consistent sequence $\hat{\theta}_n$ of local maxima of $p_{\theta'}(\boldsymbol{Y})$ in the set $\{\theta' \in \mathcal{X} : |\theta - \theta'| \le \epsilon\}$. Such a sequence satisfies the likelihood equation (14.35). By the differentiability assumption (A4), Taylor's remainder theorem may be applied to the likelihood equation around θ. We obtain

$$\begin{aligned}
0 &= \frac{1}{\sqrt{n}} \sum_{i=1}^{n} \psi(Y_i, \hat{\theta}_n) \\
&= \frac{1}{\sqrt{n}} \sum_{i=1}^{n} \left[\psi(Y_i, \theta) + (\hat{\theta}_n - \theta)\psi'(Y_i, \theta) + \frac{1}{2}(\hat{\theta}_n - \theta)^2 \psi''(Y_i, \overline{\theta}_n) \right] \\
&= \underbrace{\frac{1}{\sqrt{n}} \sum_{i=1}^{n} \psi(Y_i, \theta)}_{=U_n} + \sqrt{n}(\hat{\theta}_n - \theta) \left[\underbrace{\frac{1}{n} \sum_{i=1}^{n} \psi'(Y_i, \theta)}_{=V_n} + \underbrace{(\hat{\theta}_n - \theta)}_{=W_n} \underbrace{\frac{1}{2n} \sum_{i=1}^{n} \psi''(Y_i, \overline{\theta}_n)}_{=X_n} \right]
\end{aligned} \tag{14.58}$$

for some $\overline{\theta}_n \in [\theta, \hat{\theta}_n]$. By Assumptions (A3), (A4), and (A7), (13.6) holds, and

$$\mathbb{E}_\theta[\psi(Y, \theta)] = \mathbb{E}_\theta \left[\frac{\partial \ln p_\theta(Y)}{\partial \theta} \right] = 0.$$

By Assumption (A6) and the Central Limit Theorem,

$$U_n \triangleq \frac{1}{\sqrt{n}} \sum_{i=1}^{n} \psi(Y_i, \theta) \xrightarrow{d.} \mathcal{N}(0, I_\theta).$$

By the WLLN,

$$V_n \triangleq \frac{1}{n} \sum_{i=1}^{n} \psi'(Y_i, \theta) \xrightarrow{i.p.} \mathbb{E}_\theta[\psi'(Y, \theta)] = -I_\theta$$

and

$$X_n \triangleq \frac{1}{n} \sum_{i=1}^{n} \psi''(Y_i, \theta) \xrightarrow{i.p.} \mathbb{E}_\theta[\psi''(Y, \theta)],$$

which is finite by Assumption (A5) since

$$|\mathbb{E}_\theta[\psi''(Y, \theta)]| \le \mathbb{E}_\theta[M(Y)] < \infty.$$

Since $\hat{\theta}_n \xrightarrow{a.s.} \theta$, this implies that $W_n \xrightarrow{i.p.} 0$. Since X_n converges i.p. to a constant, it follows from Slutsky's theorem (see Appendix D) that the product $W_n X_n$ converges i.p. to 0. Since $V_n \xrightarrow{i.p} -I_\theta$, it also follows from Slutsky's theorem that the sum $V_n + W_n X_n$ converges i.p. to $-I_\theta$. Then (14.58) yields

$$\sqrt{n}(\hat{\theta}_n - \theta) = \frac{U_n}{-V_n - W_n X_n} \xrightarrow{d} \frac{U_n}{I_\theta} \sim \mathcal{N}\left(0, \frac{1}{I_\theta}\right).$$

This completes the proof.

14.14 Appendix: Proof of Theorem 14.4

Since $p_\theta(z) = p_\theta(z|y)p_\theta(y)$, we have that

$$\ell_{\text{cd}}(\theta, z) = \ell_{\text{id}}(\theta, y) + \ln p_\theta(z|y). \tag{14.59}$$

Now using the fact that

$$\int p_{\hat{\theta}^{(k)}}(z|y) d\mu(z) = 1$$

we get that

$$\ell_{\text{id}}(\theta, y) = \int \ell_{\text{id}}(\theta, y) p_{\hat{\theta}^{(k)}}(z|y) d\mu(z). \tag{14.60}$$

Plugging (14.59) into (14.60), we get

$$\ell_{\text{id}}(\theta, y) = \int \ell_{\text{cd}}(\theta, z) p_{\hat{\theta}^{(k)}}(z|y) d\mu(z) - \int \ln p_\theta(z|y) p_{\hat{\theta}^{(k)}}(z|y) d\mu(z). \tag{14.61}$$

The KL divergence between distributions $p_{\hat{\theta}^{(k)}}(z|y)$ and $p_\theta(z|y)$ is given by

$$\int \ln p_{\hat{\theta}^{(k)}}(z|y) p_{\hat{\theta}^{(k)}}(z|y) d\mu(z) - \int \ln p_\theta(z|y) p_{\hat{\theta}^{(k)}}(z|y) d\mu(z).$$

Since this KL divergence is nonnegative, we get from (14.61) that

$$\ell_{\text{id}}(\theta, y) \geq \int \ell_{\text{cd}}(\theta, z) p_{\hat{\theta}^{(k)}}(z|y) d\mu(z) - \int \ln p_{\hat{\theta}^{(k)}}(z|y) p_{\hat{\theta}^{(k)}}(z|y) d\mu(z). \quad (14.62)$$

Therefore if

$$\int \ell_{\text{cd}}(\hat{\theta}^{(k+1)}, z) p_{\hat{\theta}^{(k)}}(z|y) d\mu(z) \geq \int \ell_{\text{cd}}(\hat{\theta}^{(k)}, z) p_{\hat{\theta}^{(k)}}(z|y) d\mu(z) \quad (14.63)$$

then, by (14.62), we obtain

$$\begin{aligned}
\ell_{\text{id}}(\hat{\theta}^{(k+1)}, y) &\geq \int \ell_{\text{cd}}(\hat{\theta}^{(k+1)}, z) p_{\hat{\theta}^{(k)}}(z|y) d\mu(z) - \int \ln p_{\hat{\theta}^{(k)}}(z|y) p_{\hat{\theta}^{(k)}}(z|y) d\mu(z) \\
&\geq \int \ell_{\text{cd}}(\hat{\theta}^{(k)}, z) p_{\hat{\theta}^{(k)}}(z|y) d\mu(z) - \int \ln p_{\hat{\theta}^{(k)}}(z|y) p_{\hat{\theta}^{(k)}}(z|y) d\mu(z) \\
&= \int \ell_{\text{id}}(\hat{\theta}^{(k)}, y) p_{\hat{\theta}^{(k)}}(z|y) d\mu(z) \\
&= \ell_{\text{id}}(\hat{\theta}^{(k)}, y),
\end{aligned}$$

where the last two lines follow from (14.59) and (14.60).

Therefore the statement of the theorem holds if (14.63) can be met. One way to meet the condition in (14.63) is by choosing

$$\begin{aligned}
\hat{\theta}^{(k+1)} &= \arg\max_{\theta \in \mathcal{X}} \int \ell_{\text{cd}}(\theta, z) p_{\hat{\theta}^{(k)}}(z|y) d\mu(z) \\
&= \arg\max_{\theta \in \mathcal{X}} \mathbb{E}_{\hat{\theta}^{(k)}}[\ell_{\text{cd}}(\theta, Z)|Y = y],
\end{aligned}$$

which can be broken up into the two steps of the EM algorithm given in (14.47) and (14.48).

Exercises

14.1 *Point on a Unit Circle.* Consider a point $\boldsymbol{s}$ on the unit circle in $\mathbb{R}^2$. You are given n i.i.d. points

$$\boldsymbol{Y}_i = \boldsymbol{s} + \boldsymbol{Z}_i, \quad 1 \leq i \leq n,$$

where $\{\boldsymbol{Z}_i\}_{i=1}^n$ are i.i.d. and $\mathcal{N}(\mathbf{0}, \mathsf{C})$ where the covariance matrix $\mathsf{C} = \begin{bmatrix} 1 & \rho \\ \rho & 1 \end{bmatrix}$, and $|\rho| < 1$.

(a) Determine the ML estimator of $\boldsymbol{s}$.
(b) Is the ML estimator unbiased?
(c) Is the ML estimator consistent (in the a.s and m.s. senses) as $n \to \infty$?

14.2 *MLE in Gaussian Noise.*

(a) Derive the ML estimator of $\theta \geq 0$ given the observations

$$Y_i = \frac{\theta}{i^2} + Z_i, \quad 1 \leq i \leq n,$$

where $\{Z_i\}_{i=1}^n$ are i.i.d. $\mathcal{N}(0, 1)$.

(b) Is the ML estimator unbiased?
(c) Is the ML estimator consistent (in the a.s and m.s. senses) as $n \to \infty$?

14.3 *MLE in Poisson Family.* Suppose $\{Y_k\}_{k=1}^n$ are i.i.d. Poisson random variables with parameter $\theta > 0$, i.e.,

$$p_\theta(y_k) = \frac{e^{-\theta}\theta^{y_k}}{y_k!} \qquad y_k = 0, 1, 2, \ldots$$

and $p_\theta(\mathbf{y}) = \prod_{k=1}^n p_\theta(y_k)$.

Hint: You may use the fact that under P_θ, both the mean and the variance of Y_k equal θ.

(a) Find $\hat{\theta}_{\mathrm{ML}}(\mathbf{y})$.
(b) Is $\hat{\theta}_{\mathrm{ML}}$ unbiased? If not, find its bias.
(c) Find the variance of $\hat{\theta}_{\mathrm{ML}}$.
(d) Find the CRLB on the variance of unbiased estimators of θ.
(e) Find an MVUE for θ based on $\mathbf{y}$.

14.4 *MLE in Rayleigh Family.* Suppose $\{Y_k\}_{k=1}^n$ are i.i.d. Rayleigh random variables with parameter $\theta > 0$, i.e.,

$$p_\theta(y_k) = \frac{2y_k}{\theta} e^{-\frac{y_k^2}{\theta}} \mathbb{1}_{\{y_k \geq 0\}}$$

and $p_\theta(\mathbf{y}) = \prod_{k=1}^n p_\theta(y_k)$.

We wish to estimate $\phi = \theta^2$.

Hint: You may use the fact that, under P_θ, $X = \sum_{k=1}^n Y_k^2$ has a Gamma pdf given by

$$p_\theta(x) = \frac{x^{n-1}\, e^{-x/\theta}}{(n-1)!\, \theta^n} \mathbb{1}\{x \geq 0\}.$$

(a) Find the MLE of ϕ and compute its bias.
(b) Find the MVUE of ϕ.
(c) Which of these two estimates has a smaller MSE?
(d) Compute the CRLB on the variance of unbiased estimators of ϕ.
(e) Is the MLE asymptotically unbiased as $n \to \infty$?
(f) Is the MLE asymptotically efficient as $n \to \infty$?

14.5 *Continuation of Exercise 13.8.*

(a) Suppose $n = 2$. Find the MLE of $\boldsymbol{\theta}$.
(b) Show that the MLE of part (a) is unbiased.
(c) Show that the MLE of part (a) is efficient, i.e., that the covariance of the MLE equals the inverse of the Fisher information matrix.

14.6 *End Point Estimation.* Denote by p_θ the uniform pdf over the union of two disjoint intervals $\mathcal{Y} = [0, \theta_1] \cup [\theta_2, 1]$, where $0 \le \theta_1 \le \theta_2 < 1$. The parameters θ_1 and θ_2 are unknown and are to be determined from n i.i.d. samples $Y_1, \dots, Y_n$ drawn from the distribution p_θ.

Determine the ML estimator of θ_1 and θ_2.

14.7 *Randomized Estimator.* Let Y be a Bernoulli random variable with $\mathrm{P}\{Y = 1\} = \theta = 1 - \mathrm{P}\{Y = 0\}$. Consider the randomized estimator

$$\hat{\theta} = \begin{cases} 1/2 & \text{with prob. } \gamma \\ Y & \text{with prob. } 1 - \gamma \end{cases}$$

whose performance will be assessed using the squared-error cost function $C(\theta, \hat{\theta}) = (\theta - \hat{\theta})^2$. Note that the MLE is obtained as a special case when $\gamma = 0$ (no randomization).

(a) Give an expression for the conditional risk of this estimator, in terms of θ and γ.
(b) Find the value of γ that minimizes the worst-case risk over $\theta \in [0, 1]$.
(c) Can you argue that the estimator $\hat{\theta}$ for the value of γ calculated in part (b) is minimax?

14.8 *Solutions to Likelihood Equation.* Prove that the nonlinear system of (14.40) also admits a feasible solution that is a perturbation of (14.38) and converges a.s. to the "simple" estimator $(\tilde{\mu}, \tilde{\sigma}^2)$ of (14.38) as $n \to \infty$.

14.9 *Least-Squares Estimation.*

(a) Prove the result in (14.44).
(b) Prove the result in (14.46).

14.10 *EM Algorithm for Estimating the Parameter of an Exponential Distribution.* Suppose that given a parameter $\theta > 0$, the random variable X is exponential with parameter θ, i.e.,

$$p_\theta(x) = \theta e^{-\theta x} \, \mathbb{1}\{x \ge 0\}.$$

The random variable X is not observed directly, rather a noisy version Y is observed, with

$$Y = X + W,$$

where W is exponential with parameter β, and independent of X.

Our goal is to estimate θ from Y.

(a) Find the likelihood function $p_\theta(y)$ and write down the corresponding likelihood equation.
(b) Can you find $\hat{\theta}_{\mathrm{ML}}(y)$ in closed-form? If your answer is yes, provide the solution.
(c) Now consider computing $\hat{\theta}_{\mathrm{ML}}(y)$ iteratively using the E-M algorithm based on $Z = (Y, X)$. Compute the E-step of the algorithm, i.e., find $Q(\theta|\hat{\theta}^{(k)})$.
(d) Compute the M-step in closed form, i.e., find $\hat{\theta}^{(k+1)}$ in terms of $\hat{\theta}^{(k)}$ and y.

14.11 *EM Algorithm for Interference Cancellation.* Consider the observation model:

$$\boldsymbol{Y}_1 = \boldsymbol{X}_1 + \theta \boldsymbol{X}_2 + \boldsymbol{W}_1,$$

where $\boldsymbol{X}_1 \sim \mathcal{N}(\mathbf{0}, \mathsf{C}_1)$ is the desired signal, $\boldsymbol{X}_2 \sim \mathcal{N}(\mathbf{0}, \mathsf{C}_2)$ is additive interference, and $\boldsymbol{W}_1 \sim \mathcal{N}(\mathbf{0}, \sigma^2 I)$ is white noise. We wish to cancel $\boldsymbol{X}_2$ by placing a sensor close to the interferer that produces observation:

$$\boldsymbol{Y}_2 = \boldsymbol{X}_2 + \boldsymbol{W}_2,$$

where $\boldsymbol{W}_2 \sim \mathcal{N}(\mathbf{0}, \sigma^2 I)$ is white noise.

Assume that $\boldsymbol{X}_1$, $\boldsymbol{X}_2$, $\boldsymbol{W}_1$, and $\boldsymbol{W}_2$ are independent, and that C_1, C_2, and σ^2 are known.

To cancel $\boldsymbol{X}_2$ in $\boldsymbol{Y}_1$, we need to estimate θ. Develop an EM algorithm to estimate θ from $\boldsymbol{Y}_1, \boldsymbol{Y}_2$.

14.12 *EM Algorithm Convergence.* Consider a random variable W that has the Gamma density that is parameterized by θ:

$$p_\theta(w) = \frac{e^{-w\theta}\ \theta^\alpha\ w^{\alpha-1}}{\Gamma(\alpha)}\ \mathbb{1}_{\{w \geq 0\}},$$

where $\alpha > 0$ is known and Γ is the Gamma function defined by the integral

$$\Gamma(x) = \int_0^\infty t^{x-1}\ e^{-t}\ dt\ , \quad \text{for } x > 0\ .$$

Now, given $W = w$, suppose $Y_1, Y_2, \ldots, Y_n$ are i.i.d. $\text{Exp}(1/w)$ random variables, i.e.,

$$p_\theta(\boldsymbol{y}|w) = \prod_{k=1}^n w\ e^{-y_k w}\ \mathbb{1}_{\{y_k \geq 0\}}.$$

(a) Find $p_\theta(\boldsymbol{y})$ and use it to compute $\hat{\theta}_{\text{ML}}(\boldsymbol{y})$.
(b) Now even though we can compute $\hat{\theta}_{\text{ML}}(\boldsymbol{y})$ directly, develop an EM algorithm for computing $\hat{\theta}_{\text{ML}}(\boldsymbol{y})$ using as complete data $(\boldsymbol{Y}, W)$.
(c) Show that $\hat{\theta}_{\text{ML}}(\boldsymbol{y})$ that you found in part (a) is indeed a stationary point (fixed point) of the EM iteration you developed in part (b).

14.13 *James–Stein Estimator.* Consider the estimation of a vector parameter $\boldsymbol{\theta} \in \mathbb{R}^m$, with $m \geq 3$, from observations $\boldsymbol{Y} \in \mathbb{R}^m$ with conditional distribution

$$p_{\boldsymbol{\theta}}(\boldsymbol{y}) \sim \mathcal{N}(\boldsymbol{\theta}, \mathsf{I}).$$

(a) Find $\hat{\boldsymbol{\theta}}_{\text{ML}}(\boldsymbol{y})$ and $\text{MSE}_{\boldsymbol{\theta}}(\hat{\boldsymbol{\theta}}_{\text{ML}}(\boldsymbol{Y}))$.
(b) Show that $\hat{\boldsymbol{\theta}}_{\text{MVUE}}(\boldsymbol{y}) = \hat{\boldsymbol{\theta}}_{\text{ML}}(\boldsymbol{y})$.
(c) Now consider the following *James–Stein Estimator* [17]:

$$\hat{\boldsymbol{\theta}}_{\text{S}}(\boldsymbol{y}) = \left(1 - \frac{m-2}{\|\boldsymbol{y}\|^2}\right)\boldsymbol{y}.$$

Show that

$$\text{MSE}_{\boldsymbol{\theta}}(\hat{\boldsymbol{\theta}}_{\text{S}}) = \text{MSE}_{\boldsymbol{\theta}}(\hat{\boldsymbol{\theta}}_{\text{ML}}) - (m-2)^2 \mathbb{E}_{\boldsymbol{\theta}}\left[\frac{1}{\|\boldsymbol{Y}\|^2}\right],$$

i.e., $\hat{\boldsymbol{\theta}}_{\text{S}}$ is strictly better than $\hat{\boldsymbol{\theta}}_{\text{ML}}$ in terms of MSE.

Hint: First show using integration by parts that

$$\mathbb{E}_{\boldsymbol{\theta}}\left[\frac{Y_i(Y_i-\theta_i)}{\|\boldsymbol{Y}\|^2}\right] = \mathbb{E}_{\boldsymbol{\theta}}\left[\frac{1}{\|\boldsymbol{Y}\|^2} - \frac{2Y_i^2}{\|\boldsymbol{Y}\|^4}\right], \text{ for } i = 1, 2, \ldots, m.$$

(d) Now consider the special case where $\boldsymbol{\theta} = \mathbf{0}$, and find the value of $\mathrm{MSE}_{\mathbf{0}}(\hat{\boldsymbol{\theta}}_{\mathrm{S}})$. Compare this value with $\mathrm{MSE}_{\mathbf{0}}(\hat{\boldsymbol{\theta}}_{\mathrm{ML}})$.

14.14 *Method of Moments.* Suppose $\{Y_k\}_{k=1}^n$ are i.i.d. random variables with the Gamma pdf

$$p_{\boldsymbol{\theta}}(y) = \frac{1}{\Gamma(\theta_1)\,{\theta_2}^{\theta_1}}\, y^{\theta_1-1}\, e^{-\frac{y}{\theta_2}}.$$

where $\boldsymbol{\theta} = [\theta_1 \ \theta_2]^\top$ is an unknown parameter vector. It is easy to see that the MLE for $\boldsymbol{\theta}$ cannot be computed in closed form for this example.

An alternative to the MLE, which is often more computationally feasible, is the so called method of moments (MOM) estimator, in which we compute empirical approximations to a set of moments of the random variable Y_1 and use them to get estimates of the unknown parameter $\boldsymbol{\theta}$. In particular, for the problem at hand, we set

$$\overline{Y} \triangleq \frac{1}{n}\sum_{k=1}^{n} Y_k \approx \mathbb{E}_{\boldsymbol{\theta}}[Y_1] = g_1(\theta_1, \theta_2) \tag{14.64}$$

$$\overline{Y^2} \triangleq \frac{1}{n}\sum_{k=1}^{n} Y_k^2 \approx \mathbb{E}_{\boldsymbol{\theta}}[Y_1^2] = g_2(\theta_1, \theta_2). \tag{14.65}$$

We treat the approximations in the above equations as equalities, and solve them jointly to obtain MOM estimates for θ_1 and θ_2 in terms of $\overline{Y}$ and $\overline{Y^2}$.

(a) Find g_1 and g_2 in (14.64) and (14.65), and solve the equations to find the MOM estimates $\hat{\theta}_1^{\mathrm{MOM}}$ and $\hat{\theta}_2^{\mathrm{MOM}}$ in terms of $\overline{Y}$ and $\overline{Y^2}$.

(b) Prove that the MOM estimates are strongly consistent, i.e., show that for $i = 1, 2$:

$$\hat{\theta}_i^{\mathrm{MOM}} \xrightarrow{\text{a.s.}} \theta_i, \text{ as } n \to \infty.$$

References

[1] P. W. Zehna, "Invariance of Maximum Likelihood Estimators," *Ann. Math. Stat.*, Vol. 37, No. 3, p. 744, 1966.

[2] B. A. Coomes, "On Conditions Sufficient for Injectivity of Maps," Institute for Mathematics and its Applications (USA), Preprint #544, 1989.

[3] E. L. Lehmann and G. Casella, *Theory of Point Estimation*, Springer, 1998.

[4] T. S. Ferguson, *A Course in Large Sample Theory*, Chapman and Hall, 1996.

[5] A. DasGupta, *Asymptotic Theory of Statistics and Probability*, Springer, 2008.

[6] R. Serfling, *Approximation Theorems of Mathematical Statistics*, Wiley, 1980.

[7] H. Hartley, "Maximum Likelihood Estimation from Incomplete Data," *Biometrics*, Vol. 14, pp. 174–194, 1958.

[8] A. P. Dempster, N. M. Laird, and D. B. Rubin, "Maximum Likelihood from Incomplete Data via the EM Algorithm," (with discussion) *J. Roy. Stat. Soc. B.*, Vol. 39, No. 1, pp. 1–38, 1977.

[9] X.-L. Meng and D. van Dijk, "The EM algorithm: An Old Folk Song Sung to a Fast New Tune," (with discussion) *J. Roy. Stat. Soc. B.*, Vol. 59, No. 3, pp. 511–567, 1997.

[10] G. McLachlan and T. Krishnan, *The EM Algorithm and Extensions*, 2nd edition, Wiley, 2008.

[11] C. F. J. Wu, "On the Convergence Properties of the EM Algorithm," *Ann. Stat.*, Vol. 11, No. 1, pp. 95–103, 1983.

[12] J. A. Fessler and A. O. Hero, "Space-Alternating Generalized Expectation-Maximization Algorithm," *IEEE Trans. Sig. Proc.*, Vol. 42, No. 10, pp. 2664–2677, 1994.

[13] H. Robbins and S. Monro, "A Stochastic Approximation Method," *Ann. Math. Stat.*, pp. 400–407, 1951.

[14] D. Sakrison, "Efficient Recursive Estimation of the Parameters of a Radar or Radio Astronomy Target," *IEEE Trans. Inf. Theory,* Vol. 12, No. 1, pp. 35–41, 1966.

[15] A. H. Sayed, *Adaptive Filters*, Wiley, 2011.

[16] D. P. Bertsekas, *Nonlinear Programming*, Athena Scientific, 1999.

[17] B. Efron and T. Hastie, *Computer Age Statistical Inference*, Cambridge University Press, 2016.

15 Signal Estimation

In this chapter, we consider the problem of estimating a discrete-time random signal from noisy observations of the signal. We will begin by adopting the MMSE estimation setting that was introduced in Chapter 11, and more specifically focusing on estimators with linear constraints (see Section 11.7.4) that allow for solutions to the estimation problem with fewer assumptions on the joint distributions of the observation and the state. In this context, we discuss the important notion of *linear innovations*, which is a key tool in the development of the discrete-time Kalman filter, where the signal is a sequence of states that evolves as a linear dynamical system driven by process noise, and the observation sequence is a linear, causal, and noisy version of the state. We then consider the more general problem of nonlinear filtering, where we drop the linearity assumptions on the state evolution and observation equations, and discuss some approaches to solving this problem. We end the chapter with a discussion of estimation in finite alphabet hidden Markov models (HMMs).

15.1 Linear Innovations

We first consider the computation of the LMMSE estimate described in Section 11.7.4 for the case where we wish to estimate the state $\boldsymbol{X} \in \mathbb{R}^p$, from a sequence of observations $\boldsymbol{Y}_1, \ldots, \boldsymbol{Y}_n \in \mathbb{R}^q$, where p and q are positive integers. If the observations are uncorrelated and all the random vectors are zero mean, then the LMMSE estimate of $\boldsymbol{X}$ can be obtained as a sum of the LMMSE estimates of the $\boldsymbol{X}$ from the individual observations as shown in the following result.

PROPOSITION 15.1 *Let $\boldsymbol{X}, \boldsymbol{Y}_1, \ldots, \boldsymbol{Y}_n$ be zero-mean random vectors with finite second moments.*

If $Cov(\boldsymbol{Y}_k, \boldsymbol{Y}_\ell) = 0$ for $k \neq \ell$, then the linear MMSE estimate of $\boldsymbol{X}$, given $\boldsymbol{Y}_1, \ldots, \boldsymbol{Y}_n$, satisfies

$$\hat{\mathbb{E}}[\boldsymbol{X}|\boldsymbol{Y}_1, \ldots, \boldsymbol{Y}_n] = \sum_{k=1}^{n} \hat{\mathbb{E}}[\boldsymbol{X}|\boldsymbol{Y}_k]. \tag{15.1}$$

Proof To prove this result it is enough to verify that the right hand side of (15.1) satisfies the orthogonality principle condition for LMMSE estimation of $\boldsymbol{X}$ from $\boldsymbol{Y}_1, \ldots, \boldsymbol{Y}_n$, i.e., if

$$\boldsymbol{Z} = \boldsymbol{X} - \sum_{k=1}^{n} \hat{\mathbb{E}}[\boldsymbol{X}|\boldsymbol{Y}_k]$$

then we need to show that

$$\mathbb{E}[\boldsymbol{Z}] = \mathbf{0}, \tag{15.2}$$

$$\mathrm{Cov}(\boldsymbol{Z}, \boldsymbol{Y}_k) = 0, \text{ for } k = 1, \dots, n. \tag{15.3}$$

We first note that the right hand side of (15.1) is linear in the observations. Next, because the observations are zero mean, we have that (see (11.15))

$$\hat{\mathbb{E}}[\boldsymbol{X}|\boldsymbol{Y}_k] = \mathsf{B}_k \boldsymbol{Y}_k \quad \text{for some deterministic matrix } \mathsf{B}_k.$$

Now, exploiting the fact that $\boldsymbol{X}$ is also zero mean,

$$\mathbb{E}[\boldsymbol{Z}] = \mathbb{E}\left[\boldsymbol{X} - \sum_{k=1}^{n} \mathsf{B}_k \boldsymbol{Y}_k\right] = \mathbf{0},$$

and hence (15.2) holds. Furthermore, for each $k \in \{1, 2, \dots, n\}$,

$$\begin{aligned}
\mathrm{Cov}(\boldsymbol{Z}, \boldsymbol{Y}_k) = \mathbb{E}[\boldsymbol{Z}\boldsymbol{Y}_k^\top] &= \mathbb{E}\left[\left(\boldsymbol{X} - \sum_{j=1}^{n} \mathsf{B}_j \boldsymbol{Y}_j\right)\boldsymbol{Y}_k^\top\right] \\
&\overset{(a)}{=} \mathbb{E}[\boldsymbol{X}\boldsymbol{Y}_k^\top] - \mathsf{B}_k \mathbb{E}[\boldsymbol{Y}_k \boldsymbol{Y}_k^\top] \\
&= \mathbb{E}[(\boldsymbol{X} - \mathsf{B}_k \boldsymbol{Y}_k)\boldsymbol{Y}_k^\top] \\
&\overset{(b)}{=} 0.
\end{aligned}$$

where (a) holds because the n vectors $\{\boldsymbol{Y}_j\}_{j=1}^n$ are uncorrelated, and (b) holds due to the orthogonality principle applied to the estimation of $\boldsymbol{X}$ using the vector $\boldsymbol{Y}_k$. This verifies (15.3), thus proving the result. □

If the observations $\boldsymbol{Y}_1, \boldsymbol{Y}_2, \dots, \boldsymbol{Y}_n$ are not zero mean and pairwise uncorrelated as in the statement of Proposition 15.1, we can create an *innovation sequence* that has the properties of being zero mean and uncorrelated as follows:

$$\begin{aligned}
\tilde{\boldsymbol{Y}}_1 &= \boldsymbol{Y}_1 - \mathbb{E}[\boldsymbol{Y}_1], \\
\tilde{\boldsymbol{Y}}_2 &= \boldsymbol{Y}_2 - \hat{\mathbb{E}}[\boldsymbol{Y}_2|\boldsymbol{Y}_1], \\
&\vdots \\
\tilde{\boldsymbol{Y}}_k &= \boldsymbol{Y}_k - \hat{\mathbb{E}}[\boldsymbol{Y}_k|\boldsymbol{Y}_1, \dots, \boldsymbol{Y}_{k-1}].
\end{aligned}$$

There is a one-to-one mapping between $\boldsymbol{Y}_1, \dots, \boldsymbol{Y}_k$ and $\tilde{\boldsymbol{Y}}_1, \dots, \tilde{\boldsymbol{Y}}_k$, for each k:

$$\begin{aligned}
\boldsymbol{Y}_1 &= \tilde{\boldsymbol{Y}}_1 + \mathbb{E}[\boldsymbol{Y}_1], \\
\boldsymbol{Y}_2 &= \tilde{\boldsymbol{Y}}_2 + \hat{\mathbb{E}}[\boldsymbol{Y}_2|\boldsymbol{Y}_1], \\
&\vdots \\
\boldsymbol{Y}_k &= \tilde{\boldsymbol{Y}}_k + \hat{\mathbb{E}}[\boldsymbol{Y}_k|\boldsymbol{Y}_1, \dots, \boldsymbol{Y}_{k-1}].
\end{aligned} \tag{15.4}$$

Note that in (15.4) $\hat{\mathbb{E}}[\boldsymbol{Y}_k|\boldsymbol{Y}_1, \ldots, \boldsymbol{Y}_{k-1}]$ is an affine function of $\boldsymbol{Y}_1, \ldots, \boldsymbol{Y}_k$. Also any affine combination of $\boldsymbol{Y}_1, \ldots, \boldsymbol{Y}_k$ is also an affine combination of $\tilde{\boldsymbol{Y}}_1, \ldots, \tilde{\boldsymbol{Y}}_k$, for each k. Hence, using Proposition 15.1 we have

$$\hat{\mathbb{E}}[\boldsymbol{X}|\boldsymbol{Y}_1, \ldots, \boldsymbol{Y}_n] = \hat{\mathbb{E}}[\boldsymbol{X}|\tilde{\boldsymbol{Y}}_1, \ldots, \tilde{\boldsymbol{Y}}_n] = \sum_{k=1}^{n} \hat{\mathbb{E}}[\boldsymbol{X}|\tilde{\boldsymbol{Y}}_k]. \tag{15.5}$$

Using the same arguments, the innovation sequence can also be computed recursively:

$$\begin{aligned}
\tilde{\boldsymbol{Y}}_1 &= \boldsymbol{Y}_1 - \mathbb{E}[\boldsymbol{Y}_1], \\
&\vdots \\
\tilde{\boldsymbol{Y}}_k &= \boldsymbol{Y}_k - \hat{\mathbb{E}}[\boldsymbol{Y}_k|\boldsymbol{Y}_1, \ldots, \boldsymbol{Y}_{k-1}] \\
&= \boldsymbol{Y}_k - \hat{\mathbb{E}}[\boldsymbol{Y}_k|\tilde{\boldsymbol{Y}}_1, \ldots, \tilde{\boldsymbol{Y}}_{k-1}] \\
&= \boldsymbol{Y}_k - \sum_{i=1}^{k-1} \hat{\mathbb{E}}[\boldsymbol{Y}_k|\tilde{\boldsymbol{Y}}_i].
\end{aligned}$$

We note that if $\boldsymbol{Y}_1, \ldots, \boldsymbol{Y}_k$ are jointly Gaussian, then $\tilde{\boldsymbol{Y}}_1, \ldots, \tilde{\boldsymbol{Y}}_n$ are independent.

Finally, if both the observations $\boldsymbol{Y}_1, \boldsymbol{Y}_2, \ldots, \boldsymbol{Y}_n$ and $\boldsymbol{X}$ are not zero mean, we have the following corollary to Proposition 15.1, which follows easily from (15.5) (see Exercise 15.1).

COROLLARY 15.1 *Let $\boldsymbol{X}, \boldsymbol{Y}_1, \ldots, \boldsymbol{Y}_n$ be random vectors with possibly nonzero means and finite second moments. Then*

$$\hat{\mathbb{E}}[\boldsymbol{X}|\boldsymbol{Y}_1, \ldots, \boldsymbol{Y}_n] = \mathbb{E}[\boldsymbol{X}] + \sum_{k=1}^{n} \hat{\mathbb{E}}\left[(\boldsymbol{X} - \mathbb{E}[\boldsymbol{X}])|\tilde{\boldsymbol{Y}}_k\right] \tag{15.6}$$

$$= \sum_{k=1}^{n} \hat{\mathbb{E}}\left[\boldsymbol{X}|\tilde{\boldsymbol{Y}}_k\right] - (n-1)\mathbb{E}[\boldsymbol{X}]. \tag{15.7}$$

15.2 Discrete-Time Kalman Filter

Consider a sequence of states $\{\boldsymbol{X}_k\}$ evolving as a linear dynamical system, where the observation sequence $\{\boldsymbol{Y}_k\}$ is a linear, causal, and noisy version of the state, i.e.,

$$\text{State:} \quad \boldsymbol{X}_{k+1} = \mathsf{F}_k \boldsymbol{X}_k + \boldsymbol{U}_k, \tag{15.8}$$

$$\text{Observations:} \quad \boldsymbol{Y}_k = \mathsf{H}_k \boldsymbol{X}_k + \boldsymbol{V}_k, \quad k = 0, 1, 2, \ldots, \tag{15.9}$$

where $\boldsymbol{U}_k$ and $\boldsymbol{V}_k$ are, respectively, the *state noise* (or *process noise*) and the *measurement noise*. Assume that for each k, $\boldsymbol{X}_k \in \mathbb{R}^p$ and $\boldsymbol{Y}_k \in \mathbb{R}^q$, where p and q are positive integers. Linear dynamical systems arise in a large variety of applications, including object tracking, navigation, robotics, etc.

The general problem is to produce an optimal linear estimator of the state sequence given the measurements. This problem comes in two main flavors: (1) estimation of

X_k given observations up to time k, and (2) one-step prediction of X_{k+1} given the same observations. These problems are closely related and can be solved efficiently and recursively, in such a way that the computational and storage complexity of the algorithm is independent of k. As a integral part of this computation, one obtains covariance matrices for the estimation error and the prediction error vectors at each time k. If the joint process $(X_k, Y_k)_{k\geq 0}$ is Gaussian, the linear MMSE estimators are also (unconstrained) MMSE estimators.

We use the compact notation $Y_0^k \triangleq (Y_0, Y_1, \ldots, Y_k)$ to denote all observations until time k; the sequences U_0^k and V_0^k are defined similarly. We denote by $\overline{X}_k$ the mean of X_k. The assumptions about the dynamical system are as follows:

1. $X_0, U_0, U_1, \ldots, V_0, V_1, \ldots$ are uncorrelated.
2. The following moments are known:

$$\mathbb{E}[X_0] = \overline{X}_0, \quad \mathrm{Cov}(X_0) = \Sigma_0,$$
$$\mathbb{E}[U_k] = 0, \quad \mathbb{E}[V_k] = 0, \quad \mathrm{Cov}(U_k) = \mathsf{Q}_k, \quad \mathrm{Cov}(V_k) = \mathsf{R}_k, \quad k \geq 0.$$

3. The sequence of matrices $\mathsf{F}_k, \mathsf{H}_k,\ k \geq 0$ is known.

The objective is to solve for the following two quantities recursively in time.
Estimator:

$$\hat{X}_{k|k} \triangleq \hat{\mathbb{E}}[X_k | Y_0^k], \tag{15.10}$$

Predictor:

$$\hat{X}_{k+1|k} \triangleq \hat{\mathbb{E}}[X_{k+1} | Y_0^k]. \tag{15.11}$$

The covariance matrix of the estimation error is given by

$$\Sigma_{k|k} \triangleq \mathrm{Cov}(X_k - \hat{X}_{k|k}) \tag{15.12}$$

and the covariance matrix of the prediction error by

$$\Sigma_{k+1|k} \triangleq \mathrm{Cov}(X_{k+1} - \hat{X}_{k+1|k}). \tag{15.13}$$

The recursive algorithm is comprised of a *measurement update*, where the new measurement at time k is used to update the state estimate, and a *time update*, where $\hat{X}_{k+1|k}$ is obtained in terms of $\hat{X}_{k|k}$.

Measurement Update We express $\hat{X}_{k|k}$ in terms of $\hat{X}_{k|k-1}$ and the new measurement Y_k. We denote by

$$\tilde{Y}_k = Y_k - \hat{Y}_{k|k-1} \tag{15.14}$$

the innovation at time k, which by construction is orthogonal to $\tilde{Y}_0^{k-1}$. Hence

$$\begin{aligned}\hat{X}_{k|k} = \hat{\mathbb{E}}[X_k | Y_0^k] &= \hat{\mathbb{E}}[X_k | \tilde{Y}_0^k] \\ &= \hat{\mathbb{E}}[X_k | \tilde{Y}_0^{k-1}, \tilde{Y}_k] \\ &\overset{(a)}{=} \overline{X}_k + \hat{\mathbb{E}}[X_k - \overline{X}_k | \tilde{Y}_0^{k-1}] + \hat{\mathbb{E}}[X_k - \overline{X}_k | \tilde{Y}_k]\end{aligned}$$

$$= \hat{\boldsymbol{X}}_{k|k-1} + \hat{\mathbb{E}}[\boldsymbol{X}_k - \overline{\boldsymbol{X}}_k | \tilde{\boldsymbol{Y}}_k]$$
$$\stackrel{(b)}{=} \hat{\boldsymbol{X}}_{k|k-1} + \mathrm{Cov}(\boldsymbol{X}_k, \tilde{\boldsymbol{Y}}_k)\mathrm{Cov}(\tilde{\boldsymbol{Y}}_k)^{-1}\tilde{\boldsymbol{Y}}_k, \tag{15.15}$$

where equality (a) follows from Corollary 15.1, and (b) from (11.15).

To evaluate (15.15), we derive expressions for $\tilde{\boldsymbol{Y}}_k$, $\mathrm{Cov}(\boldsymbol{X}_k, \tilde{\boldsymbol{Y}}_k)$, and $\mathrm{Cov}(\tilde{\boldsymbol{Y}}_k)$ and substitute them into (15.15). We first apply the linear operator $\hat{\mathbb{E}}[.|\boldsymbol{Y}_0^{k-1}]$ on the measurement equation (15.9) to obtain

$$\hat{\boldsymbol{Y}}_{k|k-1} = \hat{\mathbb{E}}[Y_k|Y_0^{k-1}] = \hat{\mathbb{E}}[\mathsf{H}_k \boldsymbol{X}_k + \boldsymbol{V}_k | \boldsymbol{Y}_0^{k-1}] = \mathsf{H}_k \hat{\boldsymbol{X}}_{k|k-1} = \mathsf{H}_k \hat{\boldsymbol{X}}_{k|k-1}.$$

Hence the innovation (15.14) can be expressed as

$$\tilde{\boldsymbol{Y}}_k = \boldsymbol{Y}_k - \hat{\boldsymbol{Y}}_{k|k-1} = \boldsymbol{Y}_k - \mathsf{H}_k \hat{\boldsymbol{X}}_{k|k-1}$$
$$= \mathsf{H}_k(\boldsymbol{X}_k - \hat{\boldsymbol{X}}_{k|k-1}) + \boldsymbol{V}_k. \tag{15.16}$$

From (15.16) we obtain $\mathrm{Cov}(\boldsymbol{X}_k, \tilde{\boldsymbol{Y}}_k)$ and $\mathrm{Cov}(\tilde{\boldsymbol{Y}}_k)$ as follows. First,

$$\begin{aligned}
\mathrm{Cov}(\boldsymbol{X}_k, \tilde{\boldsymbol{Y}}_k) &\stackrel{(a)}{=} \mathrm{Cov}(\boldsymbol{X}_k, \mathsf{H}_k(\boldsymbol{X}_k - \hat{\boldsymbol{X}}_{k|k-1}) + \boldsymbol{V}_k) \\
&\stackrel{(b)}{=} \mathrm{Cov}(\boldsymbol{X}_k, \boldsymbol{X}_k - \hat{\boldsymbol{X}}_{k|k-1})\,\mathsf{H}_k^\top \\
&\stackrel{(c)}{=} \mathrm{Cov}(\boldsymbol{X}_k - \hat{\boldsymbol{X}}_{k|k-1}, \boldsymbol{X}_k - \hat{\boldsymbol{X}}_{k|k-1})\,\mathsf{H}_k^\top \\
&= \Sigma_{k|k-1}\mathsf{H}_k^\top,
\end{aligned} \tag{15.17}$$

where (a) follows from (15.16), (b) holds because $\mathrm{Cov}(\boldsymbol{X}_k, \boldsymbol{V}_k) = 0$, and likewise (c) holds because $\hat{\boldsymbol{X}}_{k|k-1}$ is orthogonal to $\boldsymbol{X}_k - \hat{\boldsymbol{X}}_{k|k-1}$ by the orthogonality principle. Next,

$$\begin{aligned}
\mathrm{Cov}(\tilde{\boldsymbol{Y}}_k) &= \mathrm{Cov}(\mathsf{H}_k(\boldsymbol{X}_k - \hat{\boldsymbol{X}}_{k|k-1}) + \boldsymbol{V}_k) \\
&= \mathsf{H}_k \Sigma_{k|k-1}\mathsf{H}_k^\top + \mathsf{R}_k,
\end{aligned} \tag{15.18}$$

where the last equality holds because $\boldsymbol{V}_k$ is orthogonal to $\boldsymbol{X}_k - \hat{\boldsymbol{X}}_{k|k-1}$, which is a linear function of $(\boldsymbol{X}_0, \boldsymbol{U}_0^{k-1}, \boldsymbol{V}_0^{k-1})$. Substituting the covariance matrices (15.17) and (15.18) into (15.15), we obtain

$$\begin{aligned}
\hat{\boldsymbol{X}}_{k|k} &= \hat{\boldsymbol{X}}_{k|k-1} + \mathsf{K}_k \tilde{\boldsymbol{Y}}_k \\
&= \hat{\boldsymbol{X}}_{k|k-1} + \mathsf{K}_k(\boldsymbol{Y}_k - \mathsf{H}_k \hat{\boldsymbol{X}}_{k|k-1}),
\end{aligned}$$

where we have defined the *Kalman gain* matrix

$$\begin{aligned}
\mathsf{K}_k &\triangleq \mathrm{Cov}(\boldsymbol{X}_k, \tilde{\boldsymbol{Y}}_k)\mathrm{Cov}(\tilde{\boldsymbol{Y}}_k)^{-1} \\
&= \Sigma_{k|k-1}\mathsf{H}_k^\top(\mathsf{H}_k \Sigma_{k|k-1}\mathsf{H}_k^\top + \mathsf{R}_k)^{-1}.
\end{aligned} \tag{15.19}$$

Time Update We apply the operator $\hat{\mathbb{E}}[.|\boldsymbol{Y}_0^k]$ to the state equation (15.8). Since $\boldsymbol{Y}_0^k$ is a function of $(\boldsymbol{X}_0, \boldsymbol{U}_0^{k-1}, \boldsymbol{V}_0^k)$, which are uncorrelated with $\boldsymbol{U}_k$, we have $\hat{\mathbb{E}}[\boldsymbol{U}_k|\boldsymbol{Y}_0^k] = 0$, hence the predicted state estimate is

$$\begin{aligned}\hat{X}_{k+1|k} &= \hat{\mathbb{E}}[X_{k+1}|Y_0^k] = \hat{\mathbb{E}}[\mathsf{F}_k X_k + U_k|Y_0^k] \\ &= \mathsf{F}_k\hat{\mathbb{E}}[X_k|Y_0^k] + \hat{\mathbb{E}}[U_k|Y_0^k] \\ &= \mathsf{F}_k\hat{X}_{k|k}. \end{aligned} \tag{15.20}$$

Thus, the Kalman filter updates are given by

$$\hat{X}_{k|k} = \hat{X}_{k|k-1} + \mathsf{K}_k(Y_k - \mathsf{H}_k\hat{X}_{k|k-1}), \tag{15.21}$$

$$\hat{X}_{k+1|k} = \mathsf{F}_k\hat{X}_{k|k}, \tag{15.22}$$

and the filter is initialized with $\hat{X}_{0|-1} = \overline{X}_0$. These equations can be combined to form a single equation for the predictor update,

$$\hat{X}_{k+1|k} = \mathsf{F}_k\hat{X}_{k|k-1} + \mathsf{F}_k\mathsf{K}_k(Y_k - \mathsf{H}_k\hat{X}_{k|k-1}), \quad \hat{X}_{0|-1} = \overline{X}_0, \tag{15.23}$$

and a single equation for the estimator update,

$$\hat{X}_{k|k} = \mathsf{F}_{k-1}\hat{X}_{k-1|k-1} + \mathsf{K}_k(Y_k - \mathsf{H}_k\mathsf{F}_{k-1}\hat{X}_{k-1|k-1}), \tag{15.24}$$

with $\hat{X}_{0|0}$ computed from (15.21) with $\hat{X}_{0|-1} = \overline{X}_0$.

Kalman Gain Calculation To evaluate these recursions efficiently we need an efficient way to update the Kalman gain K_k in (15.19). This is achieved by obtaining a recursion for $\Sigma_{k|k-1}$ as follows. The measurement update for the error covariance matrix is obtained as follows. From (15.15), the estimation and prediction errors are related by

$$X_k - \hat{X}_{k|k} = X_k - \hat{X}_{k|k-1} - \mathrm{Cov}(X_k, \tilde{Y}_k)\mathrm{Cov}(\tilde{Y}_k)^{-1}\tilde{Y}_k.$$

Taking covariances, we obtain the following relation between estimation and prediction covariance error matrices:

$$\Sigma_{k|k} = \Sigma_{k|k-1} - \mathrm{Cov}(X_k, \tilde{Y}_k)\mathrm{Cov}(\tilde{Y}_k)^{-1}\mathrm{Cov}(\tilde{Y}_k, X_k). \tag{15.25}$$

Substituting (15.17) and (15.18) into (15.25), we obtain

$$\Sigma_{k|k} = \Sigma_{k|k-1} - \Sigma_{k|k-1}\mathsf{H}_k^\top(\mathsf{H}_k\Sigma_{k|k-1}\mathsf{H}_k^\top + \mathsf{R}_k)^{-1}\mathsf{H}_k\Sigma_{k|k-1}. \tag{15.26}$$

The time update for the error covariance matrix is given by

$$\begin{aligned}\Sigma_{k+1|k} &= \mathrm{Cov}(X_{k+1} - \hat{X}_{k+1|k}) \\ &\overset{(a)}{=} \mathrm{Cov}(\mathsf{F}_k(X_k - \hat{X}_{k|k}) + U_k) \\ &\overset{(b)}{=} \mathrm{Cov}(\mathsf{F}_k(X_k - \hat{X}_{k|k})) + \mathrm{Cov}(U_k) \\ &= \mathsf{F}_k\Sigma_{k|k}\mathsf{F}_k^\top + \mathsf{Q}_k, \end{aligned} \tag{15.27}$$

where (a) follows from (15.8) and (15.20), and (b) holds because U_k is orthogonal to $X_k - \hat{X}_{k|k}$, which is a linear function of (X_0, U_0^{k-1}, V_0^k).

Substituting (15.26) into (15.27) yields the desired covariance update

$$\Sigma_{k+1|k} = \mathsf{F}_k(\Sigma_{k|k-1} - \Sigma_{k|k-1}\mathsf{H}_k^\top(\mathsf{H}_k\Sigma_{k|k-1}\mathsf{H}_k^\top + \mathsf{R}_k)^{-1}\mathsf{H}_k\Sigma_{k|k-1})\mathsf{F}_k^\top + \mathsf{Q}_k, \tag{15.28}$$

which is called the *Riccati equation*, and is updated with $\Sigma_{0|-1} = \Sigma_0$.

The Kalman filter equations can be summarized as follows:

Estimator Update:

$$\boxed{\hat{X}_{k|k} = \mathsf{F}_{k-1}\hat{X}_{k-1|k-1} + \mathsf{K}_k(Y_k - \mathsf{H}_k\mathsf{F}_{k-1}\hat{X}_{k-1|k-1})}$$

Predictor Update:

$$\boxed{\hat{X}_{k+1|k} = \mathsf{F}_k\hat{X}_{k|k-1} + \mathsf{F}_k\mathsf{K}_k(Y_k - \mathsf{H}_k\hat{X}_{k|k-1})}$$

Kalman Gain:

$$\boxed{\mathsf{K}_k = \Sigma_{k|k-1}\mathsf{H}_k^\top(\mathsf{H}_k\Sigma_{k|k-1}\mathsf{H}_k^\top + \mathsf{R}_k)^{-1}}$$

Covariance Update:

$$\boxed{\Sigma_{k+1|k} = \mathsf{F}_k(\Sigma_{k|k-1} - \Sigma_{k|k-1}\mathsf{H}_k^\top(\mathsf{H}_k\Sigma_{k|k-1}\mathsf{H}_k^\top + \mathsf{R}_k)^{-1}\mathsf{H}_k\Sigma_{k|k-1})\mathsf{F}_k^\top + \mathsf{Q}_k}$$

REMARK 15.1 We make the following observations regarding the Kalman filtering equations:

1. The covariance updates are data-independent and can be performed offline to produce the Kalman gain matrices K_k for all k, which can then be used in the filtering equations.
2. The Kalman filter updates have a recursive structure which means that there is no need to store all the observations.
3. The Kalman gain matrix optimally weighs the innovation to update the state estimate.
4. A variant to the estimation and prediction problems considered here is the *fixed-lag smoothing* problem, where an estimate $\hat{X}_{k-\ell|k}$ of the state at time $k - \ell$ is desired given observations till time k. Here $\ell \geq 1$ is the lag.
5. The performance of the Kalman filter may be unsatisfactory for heavily non-Gaussian noise, in which case the optimal MMSE estimator could be heavily nonlinear.

Example 15.1 A vehicle is moving along a straight road. Samples of the trajectory $g(t)$ at multiples of a sampling period τ are denoted by $G_k = g(k\tau)$, $k = 0, 1, 2, \ldots$. The velocity and acceleration of the vehicle are the first and second derivatives of $g(t)$, and their samples are denoted by $S_k = g'(k\tau)$ and $g''(k\tau)$, $k = 0, 1, 2, \ldots$. Define the state X_k as a two-dimensional vector $[G_k, S_k]^\top$. Newtonian dynamics suggest the following state model:

$$X_{k+1} = \begin{bmatrix} 1 & \tau \\ 0 & 1 \end{bmatrix} X_k + \begin{bmatrix} \tau^2/2 \\ \tau \end{bmatrix} A_k,$$

where, by Taylor's remainder theorem, A_k is an approximation to $g''(k\tau)$. We model A_k as zero-mean uncorrelated random variables with common variance σ_A^2. We further assume noisy measurements of the vehicle's position (obtained, e.g., using a radar). This motivates the observation model

$$Y_k = [1, 0]\boldsymbol{X}_k + V_k, \quad k = 0, 1, 2, \ldots,$$

where V_k are zero-mean uncorrelated random variables with common variance σ_V^2.

15.2.1 Time-Invariant Case

The case of a system whose dynamics do not change over time, and in which the second-order statistics of the state and measurement noises do not change over time, is of interest in many applications. Specifically,

$$\begin{aligned} \text{State:} \quad \boldsymbol{X}_{k+1} &= \mathsf{F}\boldsymbol{X}_k + \boldsymbol{U}_k, \\ \text{Observations:} \quad \boldsymbol{Y}_k &= \mathsf{H}\boldsymbol{X}_k + \boldsymbol{V}_k, \quad k = 0, 1, 2, \ldots, \end{aligned} \tag{15.29}$$

where $\text{Cov}(\boldsymbol{U}_k) = \mathsf{Q}$ and $\text{Cov}(\boldsymbol{V}_k) = \mathsf{R}$ for all k.

In this case, the estimator and predictor equations simplify to

$$\hat{\boldsymbol{X}}_{k|k} = \mathsf{F}\hat{\boldsymbol{X}}_{k-1|k-1} + \mathsf{K}_k(\boldsymbol{Y}_k - \mathsf{HF}\hat{\boldsymbol{X}}_{k-1|k-1}) \tag{15.30}$$

and

$$\hat{\boldsymbol{X}}_{k+1|k} = \mathsf{F}\hat{\boldsymbol{X}}_{k|k-1} + \mathsf{FK}_k(\boldsymbol{Y}_k - \mathsf{H}\hat{\boldsymbol{X}}_{k|k-1}). \tag{15.31}$$

Furthermore, the various covariance matrices converge to a limit value, assuming the system is stable. For instance, taking the covariance matrix of both sides of (15.8), we have

$$\text{Cov}(\boldsymbol{X}_{k+1}) = \mathsf{F}\text{Cov}(\boldsymbol{X}_k)\mathsf{F}^\top + \mathsf{Q}, \quad \text{for } k \geq 0.$$

If this recursion converges, the limit $\mathsf{C} \triangleq \lim_{k\to\infty} \text{Cov}(\boldsymbol{X}_k)$ satisfies the so-called discrete Lyapunov equation

$$\mathsf{C} = \mathsf{FCF}^\top + \mathsf{Q},$$

which can be expressed as a linear system. Convergence is guaranteed if the singular values of F are all less than 1 in magnitude (see, e.g., [1]). In this case, the prediction error covariance matrix $\Sigma_{k+1|k}$ converges to a limit Σ_∞ as $k \to \infty$, where Σ_∞ satisfies the *steady-state Riccati equation* (see (15.28)):

$$\Sigma_\infty = \mathsf{F}(\Sigma_\infty - \Sigma_\infty\mathsf{H}^\top(\mathsf{H}\Sigma_\infty\mathsf{H}^\top + \mathsf{R})^{-1}\mathsf{H}\Sigma_\infty)\mathsf{F}^\top + \mathsf{Q} \tag{15.32}$$

and the Kalman gain K_k (see (15.19)) converges to a limit K_∞ as $k \to \infty$, where

$$\mathsf{K}_\infty = \Sigma_\infty\mathsf{H}^\top(\mathsf{H}\Sigma_\infty\mathsf{H}^\top + \mathsf{R})^{-1}. \tag{15.33}$$

Furthermore, it can be shown that the singular values of $\mathsf{F}(\mathsf{I} - \mathsf{KH})$ are all less than 1 in magnitude (see, e.g., [1]), so that the linear systems defining the estimator and predictor updates in (15.30) and (15.31) are asymptotically stable.

Example 15.2 *Steady-State Kalman Filter – Scalar Case.* The equations for the steady-state Kalman filter are easiest to describe and solve in the scalar case, for which we replace matrices F, H, Q, and R by the corresponding scalar variables f, h, q, and r, and the system model becomes

$$\begin{aligned} \text{State:} \quad & X_{k+1} = f X_k + U_k, \\ \text{Observations:} \quad & Y_k = h X_k + V_k, \quad k = 0, 1, 2, \ldots, \end{aligned}$$

with U_k and V_k being zero-mean random variables with variances q and r, respectively, for all k.

The estimator and predictor updates are scalar versions of (15.30) and (15.31). In particular, the estimator update can be written as

$$\hat{X}_{k|k} = f(1 - h\kappa_k)\hat{X}_{k-1|k-1} + \kappa_k Y_k, \tag{15.34}$$

where κ_k is the Kalman gain (scalar) given by

$$\kappa_k = \sigma^2_{k|k-1} h (h^2 \sigma^2_{k|k-1} + r)^{-1} = \frac{\sigma^2_{k|k-1} h}{h^2 \sigma^2_{k|k-1} + r}. \tag{15.35}$$

The update of the prediction-error variance (which is the scalar version of the Riccati equation in (15.28)) is given by

$$\begin{aligned} \sigma^2_{k+1|k} &= f^2 \sigma^2_{k|k-1} \left(1 - \frac{\sigma^2_{k|k-1} h^2}{h^2 \sigma^2_{k|k-1} + r} \right) + q \\ &= \frac{f^2 r \sigma^2_{k|k-1}}{h^2 \sigma^2_{k|k-1} + r} + q. \end{aligned} \tag{15.36}$$

Suppose this sequence of variances converges to a limit σ^2 as $k \to \infty$, then σ^2_∞ satisfies the equation

$$\sigma^2 = \frac{f^2 r \sigma^2}{h^2 \sigma^2 + r} + q, \tag{15.37}$$

which can be written as the following quadratic equation in σ^2:

$$h^2 \sigma^4 + r(1 - f^2)\sigma^2 - q = 0. \tag{15.38}$$

There is only one positive solution to this quadratic equation, which is given by

$$\sigma^2_\infty = \frac{-r(1 - f^2) + \sqrt{r^2(1 - f^2)^2 + 4h^2 q^2}}{2h^2}. \tag{15.39}$$

In order to prove that the sequence of variances $\sigma^2_{k+1|k}$ indeed converges to a limit as $k \to \infty$ under some conditions, we first see from (15.36) and (15.37) that

$$\begin{aligned} \left| \sigma^2_{k+1|k} - \sigma^2_\infty \right| &= f^2 \left| \frac{r \sigma^2_{k|k-1}}{h^2 \sigma^2_{k|k-1} + r} - \frac{r \sigma^2_\infty}{h^2 \sigma^2_\infty + r} \right| \\ &= f^2 \left| \frac{r^2 \sigma^2_{k|k-1} - r^2 \sigma^2_\infty}{(h^2 \sigma^2_{k|k-1} + r)(h^2 \sigma^2_\infty + r)} \right| \\ &\le f^2 \left| \sigma^2_{k|k-1} - \sigma^2_\infty \right|. \end{aligned}$$

Therefore, if $|f| < 1$, $\sigma^2_{k+1|k}$ converges to σ^2_∞ given in (15.39).

With $\sigma^2_{k+1|k}$ converging to σ^2_∞, the Kalman gain of (15.35) converges to

$$\mathsf{K} = \frac{\sigma^2 h}{h^2\sigma^2 + r}.$$

For very noisy observations ($r \gg h^2\sigma^2$), we have $\mathsf{K}_k \approx 0$, in which case the measurement update is $\hat{X}_{k+1|k} \approx f\hat{X}_{k|k-1}$, and the current observation receives very little weight. For very clean observations ($r \ll h^2\sigma^2$), we have $\mathsf{K}_k \approx 1/h$, in which case the measurement update is $\hat{X}_{k+1|k} \approx \frac{f}{h}Y_k$: the past is ignored, and only the current observation matters.

15.3 Extended Kalman Filter

In developing the Kalman filter we assumed that the sequence of states $\{\boldsymbol{X}_k\}$ evolved as a *linear* dynamical system with additive noise, and the observation sequence $\{\boldsymbol{Y}_k\}$ is a *linear*, causal version of the state with additive noise. Interestingly the Kalman filtering approach can be used to derive a good approximation to the optimal estimator and predictor even when the state evolution and the observation equations are nonlinear. In particular, consider the following system model:

$$\text{State:} \quad \boldsymbol{X}_{k+1} = \boldsymbol{f}_k(\boldsymbol{X}_k) + \boldsymbol{U}_k, \tag{15.40}$$

$$\text{Observations:} \quad \boldsymbol{Y}_k = \boldsymbol{h}_k(\boldsymbol{X}_k) + \boldsymbol{V}_k, \quad k = 0, 1, 2, \ldots, \tag{15.41}$$

where $\boldsymbol{f}_k : \mathbb{R}^p \mapsto \mathbb{R}^p$ and $\boldsymbol{h}_k : \mathbb{R}^p \mapsto \mathbb{R}^q$ are nonlinear mappings, for some positive integers p, q, and the state noise $\{\boldsymbol{U}_k\}$ and the measurement noise $\{\boldsymbol{V}_k\}$ have the same second-order statistics as in the Kalman filter described in Section 15.2.

The key idea behind the extended Kalman filter (EKF) is to linearize $\boldsymbol{f}_k$ and $\boldsymbol{h}_k$ around the current estimate $\hat{\boldsymbol{X}}_{k|k}$ and predictor $\hat{\boldsymbol{X}}_{k|k-1}$of the state at time k, respectively, assuming that the estimation errors $(\boldsymbol{X}_k - \hat{\boldsymbol{X}}_{k|k})$ and $(\boldsymbol{X}_k - \hat{\boldsymbol{X}}_{k|k-1})$ are small. In particular, using a Taylor series approximation, we can write

$$\boldsymbol{f}_k(X_k) \approx \boldsymbol{f}_k(\hat{\boldsymbol{X}}_{k|k}) + \mathsf{F}_k(\boldsymbol{X}_k - \hat{\boldsymbol{X}}_{k|k}),$$
$$\boldsymbol{h}_k(X_k) \approx \boldsymbol{h}_k(\hat{\boldsymbol{X}}_{k|k-1}) + \mathsf{H}_k(\boldsymbol{X}_k - \hat{\boldsymbol{X}}_{k|k-1}),$$

where

$$\mathsf{F}_k = \left.\frac{\partial \boldsymbol{f}(\boldsymbol{x})}{\partial \boldsymbol{x}}\right|_{\boldsymbol{x}=\hat{\boldsymbol{X}}_{k|k}} \quad \text{and} \quad \mathsf{H}_k = \left.\frac{\partial \boldsymbol{h}(\boldsymbol{x})}{\partial \boldsymbol{x}}\right|_{\boldsymbol{x}=\hat{\boldsymbol{X}}_{k|k-1}}.$$

A linear approximation to the system model given in (15.40) and (15.41) can then be written as

$$\boldsymbol{X}_{k+1} = \mathsf{F}_k\boldsymbol{X}_k + \boldsymbol{U}_k + \boldsymbol{f}_k(\hat{\boldsymbol{X}}_{k|k}) - \mathsf{F}_k\hat{\boldsymbol{X}}_{k|k}, \tag{15.42}$$

$$\boldsymbol{Y}_k = \mathsf{H}_k\boldsymbol{X}_k + \boldsymbol{V}_k + \boldsymbol{h}_k(\hat{\boldsymbol{X}}_{k|k-1}) - \mathsf{H}_k\hat{\boldsymbol{X}}_{k|k-1}, \quad k = 0, 1, 2, \ldots. \tag{15.43}$$

The EKF equations can then be derived by mimicking the corresponding (linear) Kalman filter updates in (15.44) and (15.45) to get

$$\hat{\boldsymbol{X}}_{k|k} = \hat{\boldsymbol{X}}_{k|k-1} + \mathsf{K}_k(\boldsymbol{Y}_k - \boldsymbol{h}_k(\hat{\boldsymbol{X}}_{k|k-1})), \tag{15.44}$$

$$\hat{\boldsymbol{X}}_{k+1|k} = \boldsymbol{f}_k(\hat{\boldsymbol{X}}_{k|k}). \tag{15.45}$$

The Kalman gain and covariance updates for the EKF are identical to those for the Kalman filter, i.e., the terms that depend on the estimates $\hat{\boldsymbol{X}}_{k|k}$ and $\hat{\boldsymbol{X}}_{k|k-1}$ in (15.42) and (15.43) are ignored in developing these updates:

$$\mathsf{K}_k = \Sigma_{k|k-1}\mathsf{H}_k^\top(\mathsf{H}_k\Sigma_{k|k-1}\mathsf{H}_k^\top + \mathsf{R}_k)^{-1}, \tag{15.46}$$

$$\Sigma_{k+1|k} = \mathsf{F}_k(\Sigma_{k|k-1} - \Sigma_{k|k-1}\mathsf{H}_k^\top(\mathsf{H}_k\Sigma_{k|k-1}\mathsf{H}_k^\top + \mathsf{R}_k)^{-1}\mathsf{H}_k\Sigma_{k|k-1})\mathsf{F}_k^\top + \mathsf{Q}_k. \tag{15.47}$$

While the covariance and Kalman gain update equations for the Kalman filter and the EKF are identical, one major difference between the updates for the two filters is that in the Kalman filter the updates can be performed offline before collecting the observations, whereas in the EKF these updates have to be performed online since F_k and H_k are functions of the estimates $\hat{\boldsymbol{X}}_{k|k}$ and $\hat{\boldsymbol{X}}_{k|k-1}$.

Also, unlike the Kalman filter, which is optimal for the linear dynamical system model, the EKF is in general not optimal for the nonlinear system model described in (15.40) and (15.41). Nevertheless, the EKF does provide reasonable performance in many applications as has become the *de facto* standard in navigation systems and global positioning systems [2].

Example 15.3 *Tracking the Phase of a Sinusoid.* The state in this example is the (random) phase of a sinusoid, and the observation is a noisy version of the sinusoid. The system model is as follows:

$$\begin{aligned} \text{State:} \quad & X_{k+1} = X_k + U_k, \\ \text{Observations:} \quad & Y_k = \sin(\omega_0 k + X_k) + V_k, \quad k = 0, 1, 2, \ldots, \end{aligned}$$

where ω_0 is the angular frequency of the sinusoid, $\text{Var}(U_k) = q$, and $\text{Var}(V_k) = r$ for all k.

Referring to (15.40) and (15.41), we see that $f(x) = x$ and $h(x) = \sin(\omega_0 k + x)$ for this example, with

$$\mathsf{F}_k = \left.\frac{\partial f(x)}{\partial x}\right|_{x=\hat{X}_{k|k}} = 1, \qquad \mathsf{H}_k = \left.\frac{\partial h(x)}{\partial x}\right|_{x=\hat{X}_{k|k-1}} = \cos\left(\omega_0 k + \hat{X}_{k|k-1}\right).$$

This leads to the following equations for the EKF:

$$\hat{X}_{k+1|k} = \hat{X}_{k|k} = \hat{X}_{k|k-1} + \kappa_k\left(Y_k - \sin\left(\omega_0 k + \hat{X}_{k|k-1}\right)\right),$$

$$\kappa_k = \frac{\sigma_{k|k-1}^2 \cos\left(\omega_0 k + \hat{X}_{k|k-1}\right)}{\cos^2\left(\omega_0 k + \hat{X}_{k|k-1}\right)\sigma_{k|k-1}^2 + r},$$

$$\sigma_{k+1|k}^2 = \frac{r\sigma_{k|k-1}^2}{\cos^2\left(\omega_0 k + \hat{X}_{k|k-1}\right)\sigma_{k|k-1}^2 + r} + q.$$

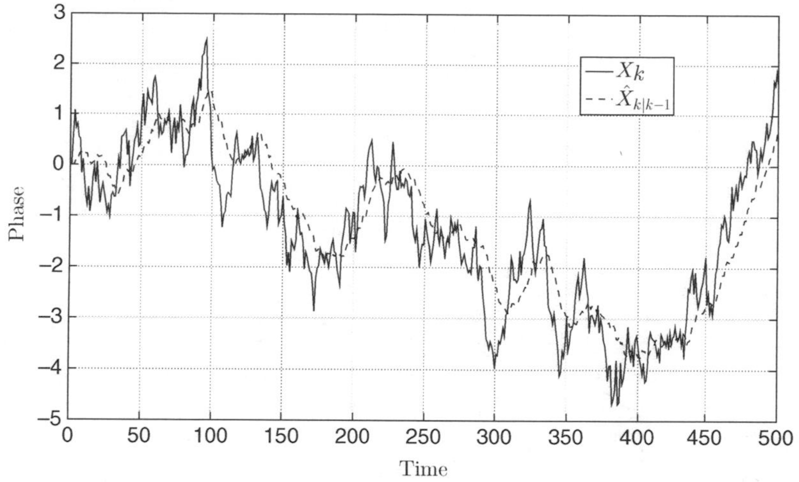

Figure 15.1 Simulation of the EKF for tracking the phase of a sinusoid, with parameters $r = q = 0.1$ and $\omega_0 = 1$

In Figure 15.1, we simulate the performance of the EKF for specific values of the parameters r, q, and ω_0, with the state and measurement noise being Gaussian. We can see that the EKF tracks the true phase reasonably well, but the finer variations in the phase are "smoothed out." See also Exercise 15.5, where some variants of this simulation example are explored.

15.4 Nonlinear Filtering for General Hidden Markov Models

The extended Kalman filtering problem discussed in the previous section is a special case of the more general nonlinear filtering problem, which can be described as follows:

$$\text{State:} \quad \boldsymbol{X}_{k+1} = \boldsymbol{f}_k(\boldsymbol{X}_k, \boldsymbol{U}_k), \tag{15.48}$$

$$\text{Observations:} \quad \boldsymbol{Y}_k = \boldsymbol{h}_k(\boldsymbol{X}_k, \boldsymbol{V}_k), \quad k = 0, 1, 2, \ldots, \tag{15.49}$$

where the noise sequences $\{\boldsymbol{U}_k, k = 0, 1, \ldots\}$ and $\{\boldsymbol{V}_k, k = 0, 1, \ldots\}$ satisfy the following Markov property with respect to the state sequence: $\boldsymbol{U}_k$, conditioned on $\boldsymbol{X}_k$ is independent of $\boldsymbol{U}_0^{k-1}$ and $\boldsymbol{X}_0^{k-1}$; likewise, $\boldsymbol{V}_k$, conditioned on $\boldsymbol{X}_k$, is independent of $\boldsymbol{V}_0^{k-1}$ and $\boldsymbol{X}_0^{k-1}$. A consequence of this Markov property is that the state evolves as a Markov process, with $\boldsymbol{X}_{k+1}$, conditioned on $\boldsymbol{X}_k$, being independent of $\boldsymbol{X}_0^{k-1}$. Furthermore, the distribution of $\boldsymbol{X}_{k+1}$, conditioned on $\boldsymbol{X}_k$, denoted by $p(\boldsymbol{x}_{k+1}|\boldsymbol{x}_k)$, is determined by the distribution of $\boldsymbol{U}_k$, conditioned on $\boldsymbol{X}_k$, and the distribution of $\boldsymbol{Y}_k$, conditioned on $\boldsymbol{X}_k$, denoted by $p(\boldsymbol{y}_k|\boldsymbol{x}_k)$, is determined by the distribution of $\boldsymbol{V}_k$, conditioned on $\boldsymbol{X}_k$. This model is also referred to as a *hidden Markov model* (HMM), and is illustrated in Figure 15.2.

The general goal in nonlinear filtering is to find at each time k the posterior distribution of the state $\boldsymbol{X}_k$ conditioned on the observations $\boldsymbol{Y}_0^k$ (estimation), which we denote

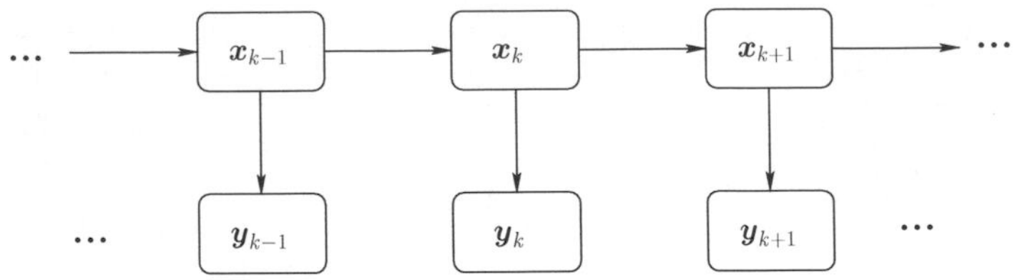

Figure 15.2 Hidden Markov model (HMM)

by $p(\boldsymbol{x}_k|\boldsymbol{y}_0^k)$, or the posterior distribution of the state $\boldsymbol{X}_{k+1}$ conditioned on the observations $\boldsymbol{Y}_0^k$ (prediction), which we denote by $p(\boldsymbol{x}_{k+1}|\boldsymbol{y}_0^k)$. In practice, we may be interested in the lesser goal of determining the posterior means $\mathbb{E}[X_k|\boldsymbol{Y}_0^k]$ and $\mathbb{E}[X_{k+1}|\boldsymbol{Y}_0^k]$ of these distributions, which would correspond to MMSE estimation, or determining their maximizers, which would correspond to MAP estimation.

Similar to the Kalman filter, the construction of the nonlinear filter (NLF) update consists of two steps. The first is a *measurement* update that takes us from $p(\boldsymbol{x}_k|\boldsymbol{y}_0^{k-1})$ to $p(\boldsymbol{x}_k|\boldsymbol{y}_0^k)$, and the second is a *time* update that takes us from $p(\boldsymbol{x}_k|\boldsymbol{y}_0^k)$ to $p(\boldsymbol{x}_{k+1}|\boldsymbol{y}_0^k)$.

The measurement update essentially follows from Bayes' rule:

$$
\begin{aligned}
p\left(\boldsymbol{x}_k|\boldsymbol{y}_0^k\right) &= \frac{p\left(\boldsymbol{y}_0^k|\boldsymbol{x}_k\right)p\left(\boldsymbol{x}_k\right)}{p\left(\boldsymbol{y}_0^k\right)} \\
&= \frac{p\left(\boldsymbol{y}_k,\boldsymbol{y}_0^{k-1}|\boldsymbol{x}_k\right)p\left(\boldsymbol{x}_k\right)}{p\left(\boldsymbol{y}_k,\boldsymbol{y}_0^{k-1}\right)} \\
&= \frac{p\left(\boldsymbol{y}_k|\boldsymbol{y}_0^{k-1},\boldsymbol{x}_k\right)p\left(\boldsymbol{y}_0^{k-1}|\boldsymbol{x}_k\right)p\left(\boldsymbol{x}_k\right)}{p\left(\boldsymbol{y}_k|\boldsymbol{y}_0^{k-1}\right)p(\boldsymbol{y}_0^{k-1})} \\
&= \frac{p\left(\boldsymbol{y}_k|\boldsymbol{y}_0^{k-1},\boldsymbol{x}_k\right)p\left(\boldsymbol{x}_k|\boldsymbol{y}_0^{k-1}\right)p\left(\boldsymbol{y}_0^{k-1}\right)p\left(\boldsymbol{x}_k\right)}{p\left(\boldsymbol{y}_k|\boldsymbol{y}_0^{k-1}\right)p(\boldsymbol{y}_0^{k-1})p\left(\boldsymbol{x}_k\right)} \\
&= \frac{p\left(\boldsymbol{y}_k|\boldsymbol{x}_k\right)p\left(\boldsymbol{x}_k|\boldsymbol{y}_0^{k-1}\right)}{p\left(\boldsymbol{y}_k|\boldsymbol{y}_0^{k-1}\right)}.
\end{aligned}
\tag{15.50}
$$

From the HMM property of the system, the denominator in (15.50) can be written as

$$
\begin{aligned}
p\left(\boldsymbol{y}_k|\boldsymbol{y}_0^{k-1}\right) &= \int p\left(\boldsymbol{y}_k,\boldsymbol{x}_k|\boldsymbol{y}_0^{k-1}\right)d\nu(\boldsymbol{x}_k) \\
&= \int p\left(\boldsymbol{y}_k|\boldsymbol{x}_k,\boldsymbol{y}_0^{k-1}\right)p\left(\boldsymbol{x}_k|\boldsymbol{y}_0^{k-1}\right)d\nu(\boldsymbol{x}_k) \\
&= \int p\left(\boldsymbol{y}_k|\boldsymbol{x}_k\right)p\left(\boldsymbol{x}_k|\boldsymbol{y}_0^{k-1}\right)d\nu(\boldsymbol{x}_k).
\end{aligned}
\tag{15.51}
$$

The time update follows from Markov property of the state evolution. In particular,

$$\begin{aligned} p(\boldsymbol{x}_{k+1}|\boldsymbol{y}_0^k) &= \int p\left(\boldsymbol{x}_{k+1}, \boldsymbol{x}_k|\boldsymbol{y}_0^k\right) d\nu(\boldsymbol{x}_k) \\ &= \int p\left(\boldsymbol{x}_{k+1}|\boldsymbol{x}_k, \boldsymbol{y}_0^k\right) p\left(\boldsymbol{x}_k|\boldsymbol{y}_0^k\right) d\nu(\boldsymbol{x}_k) \\ &= \int p\left(\boldsymbol{x}_{k+1}|\boldsymbol{x}_k\right) p\left(\boldsymbol{x}_k|\boldsymbol{y}_0^k\right) d\nu(\boldsymbol{x}_k). \end{aligned} \tag{15.52}$$

In general the NLF update equations (15.50)–(15.52) are difficult to calculate in closed form, and finding good approximations to these updates has been the subject of intense research over the last few decades – a popular approach is the *particle filter* [3]. An exception is the (linear) Kalman filtering problem with the random variables being jointly Gaussian. In the following we provide an example of a nonlinear filtering problem for which it is easy to compute the update equations. In Section 15.5, we study the special case of finite-state HMMs in more detail, for which the NLF update equations (15.50)–(15.52) can be computed in closed form.

Example 15.4 *Bayesian Quickest Change Detection (QCD).* The Bayesian QCD problem introduced in Section 9.2.2 can be considered to be a nonlinear filtering problem with a binary valued state, and 0 denoting the pre-change state and 1 denoting the post-change state. The state (evolution) is completely determined by the changepoint:

$$X_{k+1} = X_k \mathbb{1}\{\Lambda > k+1\} + \mathbb{1}\{\Lambda \le k+1\}, \quad X_0 = 0, \tag{15.53}$$

where the change point Λ is a *geometric* random variable with parameter ρ, i.e., for $0 < \rho < 1$,

$$\pi_\lambda = \mathrm{P}\{\Lambda = \lambda\} = \rho(1-\rho)^{\lambda-1}, \quad \lambda \ge 1.$$

The observation equation can be described as

$$Y_k = V_k^{(0)} \mathbb{1}\{X_k = 0\} + V_k^{(1)} \mathbb{1}\{X_k = 1\}, \tag{15.54}$$

where $\{V_k^{(j)}, k = 1, 2, \ldots\}$ are i.i.d. with pdf p_j, $j = 0, 1$. Note that there are no observations at time 0.

The NLF equations involve computing recursions for

$$G_k \triangleq \mathrm{P}(X_k = 1|Y_1^k) \quad \text{and} \quad \tilde{G}_{k+1} = \mathrm{P}(X_{k+1} = 1|Y_1^k)$$

since the posterior distributions are completely characterized by these two probabilities with the state being binary. In this case the NLF update equations (15.50)–(15.52) can be computed straightforwardly as

$$G_k = \frac{\tilde{G}_k L(Y_k)}{\tilde{G}_k L(Y_k) + (1 - \tilde{G}_k)}$$

and

$$\tilde{G}_{k+1} = G_k + (1 - G_k)\rho$$

with $G_0 = 0$ and $L(y) = \frac{p_1(y)}{p_0(y)}$ being the likelihood ratio. See also Exercise 9.8 in Chapter 9.

15.5 Estimation in Finite Alphabet Hidden Markov Models

Consider the HMM described in Figure 15.2, where both the state and the observation variables take on values in finite sets (alphabets) $\mathcal{X}$ and $\mathcal{Y}$, respectively, and where the system dynamics do not change over time. The state sequence $\{X_k\}_{k\geq 0}$ is a Markov chain and the observed sequence $\{Y_k\}_{k\geq 0}$ follows a HMM.

Due to the finite alphabet and time-invariance assumptions, we have the following:

1. The probability distribution $p(x_0)$ on the initial state X_0 can be represented by a vector, which we denote by $\boldsymbol{\pi}$.
2. The distribution of X_{k+1}, conditioned on X_k, i.e., $p(x_{k+1}|x_k)$, is completely described by the *transition probability matrix* of the Markov chain:

$$\mathsf{A}(x,x') = \mathrm{P}(X_{k+1} = x'|X_k = x\}, \quad x, x' \in \mathcal{X}. \tag{15.55}$$

3. The distribution of Y_k, conditioned on X_k, i.e., $p(y_k|x_k)$, is completely described by the *emission probability matrix*:

$$\mathsf{B}(x,y) = \mathrm{P}(Y_k = y|X_k = x\}, \quad x \in \mathcal{X},\ y \in \mathcal{Y}. \tag{15.56}$$

A graphical representation of the finite alphabet time-invariant HMM is given in Figure 15.3.

For this HMM, the NLF update equations (15.50)–(15.52) can be computed in closed form as follows. Starting with $p(x_0|y_0^{-1}) = \boldsymbol{\pi}(x_0)$, the measurement update is given by

$$p\left(x_k|y_0^k\right) = \frac{\mathsf{B}(x_k,y_k)\, p\left(x_k|y_0^{k-1}\right)}{p\left(y_k|y_0^{k-1}\right)} \tag{15.57}$$

with

$$p\left(y_k|y_0^{k-1}\right) = \sum_{x_k\in\mathcal{X}} \mathsf{B}(x_k,y_k)\, p\left(x_k|y_0^{k-1}\right). \tag{15.58}$$

The time update is given by

$$p\left(x_{k+1}|y_0^k\right) = \sum_{x_k\in\mathcal{X}} \mathsf{A}(x_k,x_{k+1})\, p\left(x_k|y_0^k\right). \tag{15.59}$$

Using (15.57)–(15.59), we may compute MMSE filter and predictor estimates:

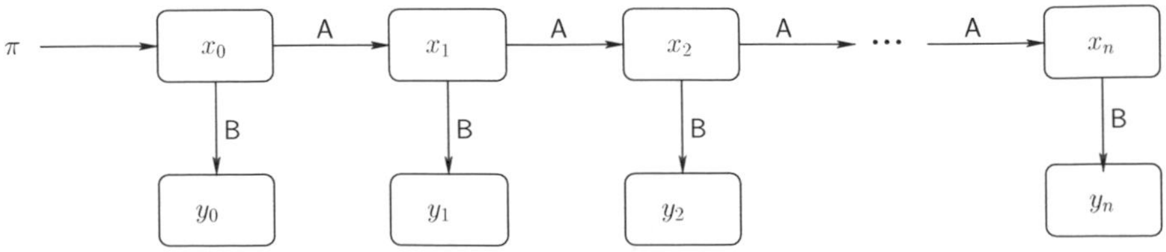

Figure 15.3 Finite alphabet time-invariant HMM

$$\hat{x}_{k|k}^{\mathrm{MMSE}} = \mathbb{E}\left[X_k | Y_0^k = y_0^k\right] = \sum_{x_k \in \mathcal{X}} x_k \; p\left(x_k | y_0^k\right),$$
$$\hat{x}_{k+1|k}^{\mathrm{MMSE}} = \mathbb{E}\left[X_{k+1} | Y_0^k = y_0^k\right] = \sum_{x_{k+1} \in \mathcal{X}} x_{k+1} \; p\left(x_{k+1} | y_0^k\right). \tag{15.60}$$

We may also compute MAP filter and predictor estimates:

$$\hat{x}_{k|k}^{\mathrm{MAP}} = \arg\max_{x_k \in \mathcal{X}} p\left(x_k | y_0^k\right),$$
$$\hat{x}_{k+1|k}^{\mathrm{MAP}} = \arg\max_{x_{k+1} \in \mathcal{X}} p\left(x_{k+1} | y_0^k\right). \tag{15.61}$$

In addition to the filtering and prediction problems, which can be considered online (causal) estimation problems, there are three other offline estimation problems that are of great interest in applications of HMMs, particularly in error control coding and speech signal processing. These are:

1. MAP estimation of the entire state sequence X_0^n from the entire observation sequence Y_0^n, which is solved by the *Viterbi algorithm* (see Section 15.5.1).
2. Determining the conditional distribution of X_k, for $k \in \{0, 1, \ldots, n\}$, given the entire observation sequence Y_0^n, which is solved by the *forward-backward algorithm* (see Section 15.5.2).
3. Estimation of the HMM parameters $\boldsymbol{\pi}$, A, and B from the entire observation sequence Y_0^n, which is solved by the Baum–Welch algorithm (see Section 15.5.3).

15.5.1 Viterbi Algorithm

The MAP estimation of X_0^n from Y_0^n arises in a variety of applications, and Viterbi derived a remarkable practical algorithm for solving it exactly [4]. The probability of the state sequence $x_0^n \in \mathcal{X}^{n+1}$ is given by

$$p\left(x_0^n\right) = \boldsymbol{\pi}(x_0) \prod_{k=0}^{n-1} \mathsf{A}(x_k, x_{k+1}), \tag{15.62}$$

and the conditional probability of the observed sequence y_0^n given the state sequence x_0^n is $p(y_0^n | x_0^n) = \prod_{k=0}^{n} \mathsf{B}(x_k, y_k)$. Hence the joint probability of x_0^n and y_0^n is

$$p(x_0^n, y_0^n) = p(x_0^n) \; p(y_0^n | x_0^n) = \boldsymbol{\pi}(x_0) \prod_{k=0}^{n-1} \mathsf{A}(x_k, x_{k+1}) \prod_{k=0}^{n} \mathsf{B}(x_k, y_k). \tag{15.63}$$

The MAP estimate of interest can be computed as

$$\begin{aligned}\hat{x}_0^{n\,\mathrm{MAP}} &= \arg\max_{x_0^n \in \mathcal{X}^{n+1}} p(x_0^n | y_0^n) \\ &= \arg\max_{x_0^n \in \mathcal{X}^{n+1}} p(x_0^n, y_0^n) \\ &= \arg\max_{x_0^n \in \mathcal{X}^{n+1}} \left[\ln \boldsymbol{\pi}(x_0) + \sum_{k=0}^{n-1} \ln \mathsf{A}(x_k, x_{k+1}) + \sum_{k=0}^{n} \ln \mathsf{B}(x_k, y_k)\right].\end{aligned} \tag{15.64}$$

This maximization apparently requires a search of exponential complexity. Note that the problem does *not* reduce to $(n+1)$ independent problems of computing $\arg\max_{x_k \in \mathcal{X}} p(x_k|y_0^n)$ for $0 \le k \le n$. However, a global maximum of (15.64) can be found in linear time using the Viterbi algorithm. The algorithm was originally designed for decoding of convolutional codes [4], but was soon recognized to be an instance of dynamic programming [5, 6]. The algorithm was previously known in the field of operations research [6] and applies to a variety of problems with similar mathematical structure [7].

Observe that the maximization problem (15.64) may be written in the form

$$\max_{x_0^n \in \mathcal{X}^{n+1}} \left[\mathcal{E}(x_0^n) \triangleq f(x_0) + \sum_{k=0}^{n-1} g_k(x_k, x_{k+1}) \right] \tag{15.65}$$

with

$$f(x) \triangleq \ln \pi(x) + \ln \mathsf{B}(x, y_0),$$
$$g_k(x, x') \triangleq \ln \mathsf{A}(x, x') + \ln \mathsf{B}(x', y_{k+1}), \quad 0 \le k < n-1.$$

The sequence $x_0^n = (x_0, x_1, \dots, x_n)$, which is a realization of states of a Markov chain, is sometimes called a *path*, and the diagram representing all possible paths is called the *state trellis*. Figure 15.4(a) shows an example, with three possible states ($\mathcal{X} = \{0, 1, 2\}$), five time instants ($n = 4$), initial value vector $f = (1, 1, 0)$, and six possible transitions with weights $g(0,0) = 1$, $g(0,2) = 2$, $g(1,1) = 0$, $g(1,2) = 1$, $g(2,0) = 3$, and $g(2,1) = 1$. We set $g(0,1) = g(1,0) = g(2,2) = -\infty$ to disallow these three transitions. In this example, g_k does not depend on k.

Denote the *value function* for state x at time $k = 0$ as $V(0, x) = f(x)$ and at times $k = 1, 2, \dots, n$ by

$$V(k, x) \triangleq \max_{x_0^n : x_k = x} \left[\mathcal{E}(x_0^k) \triangleq f(x_0) + \sum_{u=0}^{k-1} g_u(x_u, x_{u+1}) \right]. \tag{15.66}$$

Since

$$\mathcal{E}(x_0^k) = \mathcal{E}(x_0^{k-1}) + g_{k-1}(x_{k-1}, x_k),$$

we have the forward recursion

$$V(k, x) = \max_{x' \in \mathcal{X}} [V(k-1, x') + g_{k-1}(x', x)], \quad k \ge 1. \tag{15.67}$$

For $k = n$ we have

$$\max_{x_0^n \in \mathcal{X}^{n+1}} \mathcal{E}(x_0^n) = \max_{x \in \mathcal{X}} V(n, x).$$

Denote by $\hat{x}_{k-1}(x)$ a value of x' that achieves the maximum in (15.67). Observe that if a state x at time k is part of an optimal path $\hat{x}_0^n$, then $\hat{x}_k = x$, and $\hat{x}_0^k$ must achieve $V(k, x) = \max_{x_0^k} \mathcal{E}(x_0^k)$. Otherwise there would exist another path x_0^n such that $\mathcal{E}(x_0^k) > \mathcal{E}(\hat{x}_0^k)$ and $x_k^n = \hat{x}_k^n$, and hence $\mathcal{E}(x_0^n) - \mathcal{E}(\hat{x}_0^n) > 0$, which would contradict the optimality of $\hat{x}_0^n$. By keeping track of the optimal predecessor state $\hat{x}_{k-1}(x)$ and tracing back from time k to time 0, we obtain a path that we call the *survivor path*.

Once the forward recursive procedure is completed, the final state achieving $\max_x V(n, x)$ is identified, and backtracking yields the optimal path $\hat{x}_0^n$. The complexity of the algorithm is $O(n|\mathcal{X}|)$ storage (keeping track of the predecessors $\hat{x}_{k-1}(x)$ for all k, x and the most recent value function) and $O(n|\mathcal{X}|^2)$ computation (evaluating all possible state transitions at all times).

Figure 15.4(b)–(e) displays the value function $V(k, x)$, $x \in \mathcal{X}$, at successive times $k = 0, 1, 2, 3, 4$, together with the corresponding surviving paths.

15.5.2 Forward-Backward Algorithm

We now consider the problem of determining the conditional distribution of X_k, for $k \in \{0, 1, \ldots, n\}$, given Y_0^n, i.e., that of determining the posterior probabilities $\mathrm{P}(X_k = x|Y_0^n = y_0^n)$, for $k = 0, 1, \ldots, n$ and $x \in \mathcal{X}$. The solution to the problem is instrumental in solving the problem of estimating the HMM parameters from Y_0^n, as we shall see in Section 15.5.3, for which we will also need to determine $\mathrm{P}(X_k = x, X_{k+1} = x'|Y_0^n = y_0^n)$ for each $k = 0, 1, \ldots, n$ and $x, x' \in \mathcal{X}$.

Define the shorthand notation:

$$\gamma_k(x) \triangleq \mathrm{P}(X_k = x|Y_0^n = y_0^n), \tag{15.68}$$

$$\xi_k(x, x') \triangleq \mathrm{P}(X_k = x, X_{k+1} = x'|Y_0^n = y_0^n), \quad x, x' \in \mathcal{X}. \tag{15.69}$$

Note that γ_k is the first marginal of ξ_k. The forward-backward algorithm allows efficient computation of these probabilities. We begin with

$$\gamma_k(x) = \mathrm{P}(X_k = x|Y_0^n = y_0^n) = \frac{\mathrm{P}(X_k = x, Y_0^n = y_0^n)}{\sum_{s\in\mathcal{X}} \mathrm{P}(X_k = s, Y_0^n = y_0^n)}, \quad 0 \le k \le n. \tag{15.70}$$

Write the numerator as a product of two conditional distributions,

$$\mathrm{P}(Y_0^n = y_0^n, X_k = x) \overset{(a)}{=} \underbrace{\mathrm{P}(Y_0^k = y_0^k, X_k = x)}_{\mu_k(x)} \underbrace{\mathrm{P}(Y_{k+1}^n = y_{k+1}^n|X_k = x)}_{\nu_k(x)}$$

$$= \mu_k(x)\nu_k(x), \quad 0 \le k < n, \tag{15.71}$$

where (a) follows from the Markov chain $Y_0^k \to X_k \to Y_{k+1}^n$. For $k = n$, we let $\nu_n(x) \equiv 1$. Combining (15.70) and (15.71) we have

$$\gamma_k(x) = \frac{\mu_k(x)\nu_k(x)}{\sum_{s\in\mathcal{X}} \mu_k(s)\nu_k(s)}. \tag{15.72}$$

The first factor in the product of (15.71) is

$$\mu_k(x) = \mathrm{P}(Y_0^k = y_0^k, X_k = x), \quad x \in \mathcal{X}, 0 \le k \le n,$$

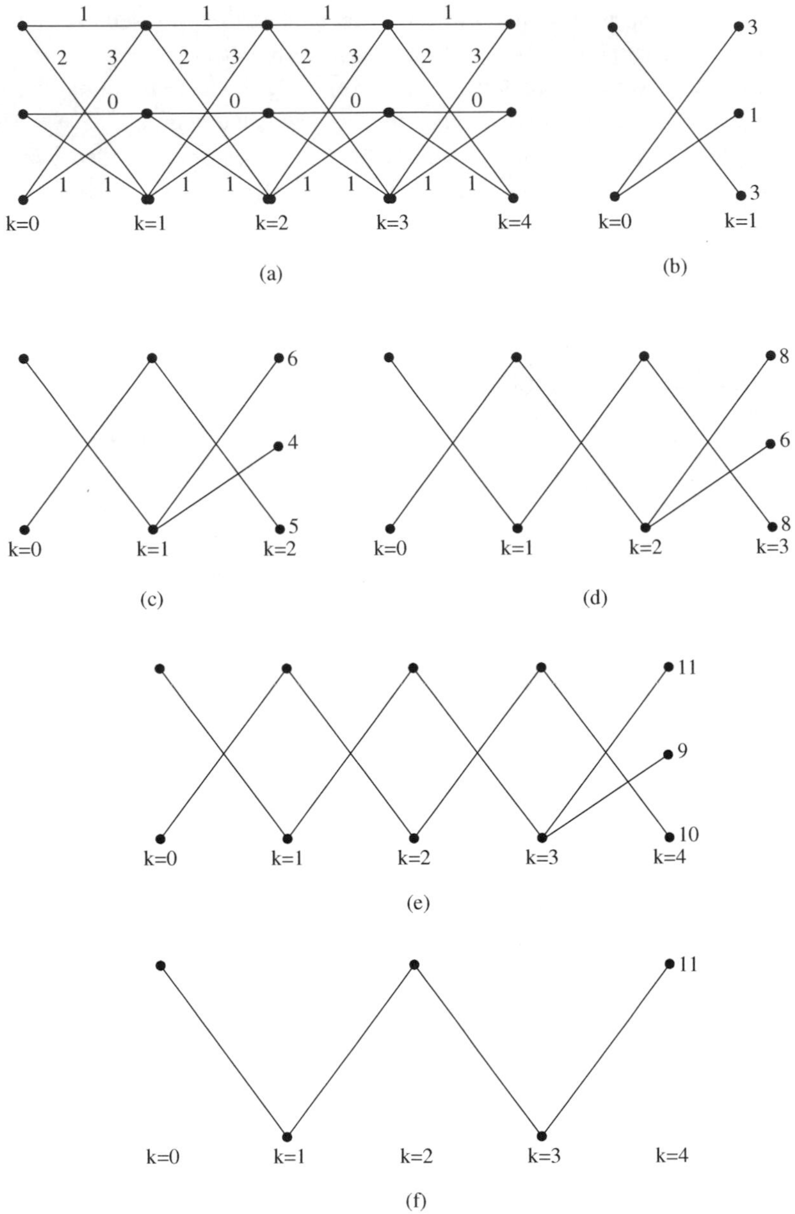

Figure 15.4 Viterbi algorithm: (a) Trellis diagram; (b)–(e) evolution of the Viterbi algorithm, showing surviving paths and values $V(k, x)$ at times $k = 0, 1, 2, 3$; (f) optimal path $\hat{x}_0^4 = (0, 2, 0, 2, 0)$ and its value $\mathcal{E}(\hat{x}_0^4) = 11$

for which we derive a *forward recursion*.[1] The recursion is initialized with

$$\mu_0(x) = \mathrm{P}(Y_0 = y_0, X_0 = x) = \boldsymbol{\pi}(x)B(x, y_0).$$

[1] Also note that $\mu_n(x) = \mathrm{P}(Y_0^n = y_0^n, X_n = x)$, hence $p(y_0^n) = \sum_{x \in \mathcal{X}} \mu_n(x)$.

For $k \geq 0$ we express μ_{k+1} in terms of μ_k as follows:

$$\begin{aligned}
\mu_{k+1}(x) &= \mathsf{P}(Y_0^{k+1} = y_0^{k+1}, X_{k+1} = x) \\
&\overset{(a)}{=} \mathsf{P}(Y_0^k = y_0^k, X_{k+1} = x)\,\mathsf{P}(Y_{k+1} = y_{k+1}|X_{k+1} = x) \\
&= B(x, y_{k+1}) \sum_{x' \in \mathcal{X}} \mathsf{P}(Y_0^k = y_0^k, X_{k+1} = x, X_k = x') \\
&\overset{(b)}{=} B(x, y_{k+1}) \sum_{x' \in \mathcal{X}} \mathsf{P}(Y_0^k = y_0^k, X_k = x')\mathsf{P}(X_{k+1} = x|X_k = x') \\
&= B(x, y_{k+1}) \sum_{x' \in \mathcal{X}} \mu_k(x')\mathsf{A}(x', x), \quad k = 0, 1, \ldots, n-1,
\end{aligned} \tag{15.73}$$

where (a) holds because $Y_0^k \to X_{k+1} \to Y_{k+1}$ forms a Markov chain, and (b) because $Y_0^k \to X_k \to X_{k+1}$ forms a Markov chain.

The second factor in the product of (15.71) is

$$\nu_k(x) = \mathsf{P}(Y_{k+1}^n = y_{k+1}^n | X_k = x), \quad x \in \mathcal{X},\ 0 \leq k < n. \tag{15.74}$$

Starting from $\nu_n(x) \equiv 1$, we have the following *backward recursion*, expressing ν_{k-1} in terms of ν_k for $1 \leq k \leq n$:

$$\begin{aligned}
\nu_{k-1}(x) &= \mathsf{P}(Y_k^n = y_k^n | X_{k-1} = x) \\
&= \sum_{x' \in \mathcal{X}} \mathsf{P}(Y_k^n = y_k^n, X_k = x' | X_{k-1} = x) \\
&\overset{(a)}{=} \sum_{x' \in \mathcal{X}} \mathsf{P}(Y_k^n = y_k^n | X_k = x')\,\mathsf{P}(X_k = x' | X_{k-1} = x) \\
&\overset{(b)}{=} \sum_{x' \in \mathcal{X}} \mathsf{P}(Y_{k+1}^n = y_{k+1}^n | X_k = x')\,\mathsf{P}(Y_k = y_k | X_k = x')\,\mathsf{P}(X_k = x' | X_{k-1} = x) \\
&= \sum_{x' \in \mathcal{X}} \nu_k(x')\mathsf{B}(x', y_k)\mathsf{A}(x, x'), \quad t = n, n-1, \ldots, 1,
\end{aligned} \tag{15.75}$$

where (a) holds because $X_{k-1} \to X_k \to Y_k^n$ forms a Markov chain, and (b) because $Y_{k+1}^n \to X_k \to Y_k$ forms a Markov chain.

Next we derive an expression for

$$\begin{aligned}
\xi_k(x, x') &= \mathsf{P}(X_k = x, X_{k+1} = x' | Y_0^n = y_0^n) \\
&= \frac{\mathsf{P}(Y_0^n = y_0^n, X_k = x, X_{k+1} = x')}{\sum_{s, s' \in \mathcal{X}} \mathsf{P}(Y_0^n = y_0^n, X_k = s, X_{k+1} = s')}.
\end{aligned}$$

We have

$$\begin{aligned}
&\mathsf{P}(Y_0^n = y_0^n, X_k = x, X_{k+1} = x') \\
&\overset{(a)}{=} \mathsf{P}(Y_0^{k+1} = y_0^{k+1}, X_k = x, X_{k+1} = x')\,\mathsf{P}(Y_{k+2}^n = y_{k+2}^n | X_{k+1} = x') \\
&\overset{(b)}{=} \mathsf{P}(Y_0^k = y_0^k, X_k = x)\,\mathsf{P}(X_{k+1} = x' | X_k = x)\,\mathsf{P}(Y_{k+1} = y_{k+1} | X_{k+1} = x')\,\nu_{k+1}(x') \\
&= \mu_k(x)\mathsf{A}(x, x')\mathsf{B}(x', y_{k+1})\nu_{k+1}(x'),
\end{aligned}$$

where (a) holds because $(Y_0^{k+1}, X_k) \to X_{k+1} \to Y_{k+2}^n$ forms a Markov chain, and (b) because $Y_0^k \to X_k \to X_{k+1} \to Y_{k+1}$ forms a Markov chain. Hence, for $0 \le k \le n$ and $x, x' \in \mathcal{X}$,

$$\xi_k(x, x') = \frac{\mu_k(x)\mathsf{A}(x, x')\mathsf{B}(x', y_{k+1})\nu_{k+1}(x')}{\sum_{s,s' \in \mathcal{X}} \mu_k(s)\mathsf{A}(s, s')\mathsf{B}(s', y_{k+1})\nu_{k+1}(s')}. \tag{15.76}$$

Scaling Factors Unfortunately the recursions in (15.75) and (15.76) are numerically unstable for large n because the probabilities $\mu_k(x)$ and $\nu_k(x)$ vanish exponentially with n and are sums of many small terms of different sizes. The following approach is more stable. Define

$$\alpha_k(x) = \mathrm{P}(X_k = x | Y_0^k = y_0^k), \tag{15.77}$$

$$\beta_k(x) = \frac{\mathrm{P}(Y_{k+1}^n = y_{k+1}^n | X_k = x)}{\mathrm{P}(Y_{k+1}^n = y_{k+1}^n | Y_0^k = y_0^k)}, \tag{15.78}$$

$$c_k = \mathrm{P}(Y_k = y_k | Y_0^{k-1} = y_0^{k-1}). \tag{15.79}$$

Then

$$\gamma_k(x) = \alpha_k(x)\, \beta_k(x),$$
$$\xi_k(x, x') = c_k \alpha_k(x)\mathsf{B}(x, y_k)\mathsf{A}(x, x')\beta_k(x').$$

A forward recursion can be derived for α_k and c_k, and a backward recursion for β_k (see Exercise 15.6).

The time and storage complexity of the algorithm is $O(n|\mathcal{X}|^2)$.

15.5.3 Baum–Welch Algorithm for HMM Learning

Finally, we address the learning problem of estimating the HMM parameters $\boldsymbol{\theta} \triangleq (\boldsymbol{\pi}, \mathsf{A}, \mathsf{B})$. The solution to this problem was developed by Baum and Welch in the 1960s and was later recognized to be an instance of the EM algorithm [9].

We derive the Baum–Welch algorithm using the EM framework described in Section 14.11. We take (X_0^n, Y_0^n) to be the complete data; hence the state sequence X_0^n is viewed as the latent (missing) data.

The marginal probability of y_0^n is obtained from (15.63) as

$$p_{\boldsymbol{\theta}}(y_0^n) = \sum_{x_0^n \in \mathcal{X}^n} \boldsymbol{\pi}(x_0) \prod_{k=0}^{n-1} \mathsf{A}(x_k, x_{k+1}) \prod_{k=0}^{n} \mathsf{B}(x_k, y_k). \tag{15.80}$$

Indicating the dependence on $\boldsymbol{\theta}$ explicitly, we obtain the incomplete-data and complete-data log-likelihoods respectively as

$$\ell_{\mathrm{id}}(\boldsymbol{\theta}) = \ln p_{\boldsymbol{\theta}}(y_0^n) = \ln \sum_{x_0^n \in \mathcal{X}^{n+1}} \boldsymbol{\pi}(x_0) \prod_{k=0}^{n-1} \mathsf{A}(x_k, x_{k+1}) \prod_{k=0}^{n} \mathsf{B}(x_k, y_k)$$

and

$$\ell_{\text{cd}}(\boldsymbol{\theta}) = \ln p_{\boldsymbol{\theta}}(x_0^n, y_0^n) = \ln \pi(x_0) + \sum_{k=0}^{n-1} \ln \mathsf{A}(x_k, x_{k+1}) + \sum_{k=0}^{n} \ln \mathsf{B}(x_k, y_k).$$

The function $\ell_{\text{id}}(\boldsymbol{\theta})$ is nonconcave and typically has many local extrema.

The E-step in the mth stage of the EM algorithm is given by

$$\begin{aligned} Q(\boldsymbol{\theta}|\hat{\boldsymbol{\theta}}^{(m)}) &= \mathbb{E}_{\hat{\boldsymbol{\theta}}^{(m)}}[\ell_{\text{cd}}(\boldsymbol{\theta})|Y_0^n = y_0^n] \\ &= \mathbb{E}_{\hat{\boldsymbol{\theta}}^{(m)}}[\ln \pi(X_0)\,|Y_0^n = y_0^n] + \mathbb{E}_{\hat{\boldsymbol{\theta}}^{(m)}}\left[\sum_{k=0}^{n-1} \ln \mathsf{A}(X_k, X_{k+1})\,|Y_0^n = y_0^n\right] \\ &\quad + \mathbb{E}_{\hat{\boldsymbol{\theta}}^{(m)}}\left[\sum_{k=0}^{n} \ln \mathsf{B}(X_k, Y_k)\,|Y_0^n = y_0^n\right]. \end{aligned} \tag{15.81}$$

To perform the M-step, we note that the cost function in (15.81) is the sum of three terms involving the variables $\boldsymbol{\pi}$, A, and B separately. Hence the problem can be split into three independent problems. To solve each of these problems, we use properties of the cross-entropy and entropy of pmfs. Recalling the definition (6.6), the entropy of a pmf $p(z)$, $z \in \mathcal{Z}$, where $\mathcal{Z}$ is a finite set, is defined as

$$H(p) \triangleq -\sum_{z} p(z) \ln p(z). \tag{15.82}$$

The cross-entropy of a pmf $p(z)$, $z \in \mathcal{Z}$ relative to another pmf $q(z)$, $z \in \mathcal{Z}$ is defined as

$$H(p, q) \triangleq -\sum_{z} p(z) \ln q(z). \tag{15.83}$$

The cross-entropy satisfies the following extremal property. For any two pmfs p and q, we have

$$\sum_{z} p(z) \ln q(z) = \sum_{x} p(z) \ln p(z) - D(p\|q) \le \sum_{z} p(z) \ln p(z),$$

where $D(p\|q)$ is the Kullback–Leibler divergence between the pmfs p and q (see Section 6.1). The upper bound is achieved by $q = p$. The inequality can be rewritten as

$$-H(p, q) \le H(p) \tag{15.84}$$

with equality when $q = p$, i.e., $-H(p, q)$ is maximized over q, when $q = p$.

We employ (15.84) as follows. The first term of the cost function,

$$\mathbb{E}_{\hat{\boldsymbol{\theta}}^{(m)}}[\ln \pi(X_0)\,|Y_0^n = y_0^n] = \sum_{x \in \mathcal{X}} P_{\hat{\boldsymbol{\theta}}^{(m)}}(X_0 = x|Y_0^n = y_0^n) \ln \pi(x),$$

is a negative cross-entropy and is maximized by

$$\hat{\pi}^{(m+1)}(x) = P_{\hat{\boldsymbol{\theta}}^{(m)}}(X_0 = x|Y_0^n = y_0^n), \quad \forall x \in \mathcal{X},$$

which is the posterior distribution of X_0, given $Y_0^n = y_0^n$ and the current estimate of $\boldsymbol{\theta}$.

The second term in (15.81) is

$$\begin{aligned}
&\mathbb{E}_{\hat{\boldsymbol{\theta}}^{(m)}}\left[\sum_{k=0}^{n-1} \ln \mathsf{A}(X_k, X_{k+1})\,|Y_0^n = y_0^n\right] \\
&= \sum_{k=0}^{n-1} \sum_{x,x' \in \mathcal{X}} P_{\boldsymbol{\theta}^{(m)}}(X_k = x, X_{k+1} = x' | Y_0^n = y_0^n) \ln \mathsf{A}(x, x') \\
&= \sum_{x \in \mathcal{X}} \left[\sum_{x' \in \mathcal{X}} \sum_{k=0}^{n-1} P_{\boldsymbol{\theta}^{(m)}}(X_k = x, X_{k+1} = x' | Y_0^n = y_0^n) \ln \mathsf{A}(x, x')\right].
\end{aligned}$$

Since both $\frac{1}{n}\sum_{k=0}^{n-1} P_{\boldsymbol{\theta}^{(m)}}(X_k = x, \cdot | Y_0^n = y_0^n)$ and $A(x, \cdot)$ are probability distributions for each $x \in \mathcal{X}$, the second term is maximized by applying the extremal property of cross-entropy for each $x \in \mathcal{X}$. The solution is

$$\begin{aligned}
\hat{\mathsf{A}}^{(m+1)}(x, x') &= \frac{1}{Z(x)} \sum_{k=0}^{n-1} P_{\boldsymbol{\theta}^{(m)}}(X_k = x, X_{k+1} = x' | Y_0^n = y_0^n) \\
&= \frac{\sum_{k=0}^{n-1} P_{\boldsymbol{\theta}^{(m)}}(X_k = x, X_{k+1} = x' | Y_0^n = y_0^n)}{\sum_{k=0}^{n-1} P_{\boldsymbol{\theta}^{(m)}}(X_k = x | Y_0^n = y_0^n)}, \quad \forall x, x' \in \mathcal{X},
\end{aligned}$$

where $Z(x)$ is the normalization factor that ensures $\hat{\mathsf{A}}^{(m+1)}(x, \cdot)$ is a valid probability distribution for each $x \in \mathcal{X}$.

The third term of (15.81),

$$\begin{aligned}
&\mathbb{E}_{\hat{\boldsymbol{\theta}}^{(m)}}\left[\sum_{k=0}^{n} \ln \mathsf{B}(X_k, Y_k)\,|Y_0^n = y_0^n\right] \\
&= \sum_{x \in \mathcal{X}} \left[\sum_{y \in \mathcal{Y}} \sum_{k=0}^{n} P_{\boldsymbol{\theta}^{(m)}}(X_k = x, Y_k = y | Y_0^n = y_0^n) \ln \mathsf{B}(x, y)\right] \\
&= \sum_{x \in \mathcal{X}} \left[\sum_{y \in \mathcal{Y}} \sum_{k=0}^{n} P_{\boldsymbol{\theta}^{(m)}}(X_k = x | Y_0^n = y_0^n)\, \mathbb{1}\{y_k = y\} \ln \mathsf{B}(x, y)\right],
\end{aligned}$$

is similarly maximized by

$$\begin{aligned}
\hat{\mathsf{B}}^{(m+1)}(x, y) &= \frac{1}{Z(x)} \sum_{k=0}^{n} P_{\boldsymbol{\theta}^{(m)}}(X_k = x | Y_0^n = y_0^n)\, \mathbb{1}\{y_k = y\} \\
&= \frac{\sum_{k=0}^{n} P_{\boldsymbol{\theta}^{(m)}}(X_k = x | Y_0^n = y_0^n)\, \mathbb{1}\{y_k = y\}}{\sum_{k=0}^{n} P_{\boldsymbol{\theta}^{(m)}}(X_k = x | Y_0^n = y_0^n)}, \quad \forall x \in \mathcal{X}, y \in \mathcal{Y},
\end{aligned}$$

where $Z(x)$ is the normalization factor that ensures $\hat{\mathsf{B}}^{(m+1)}(x, \cdot)$ is a valid probability distribution for each $x \in \mathcal{X}$.

The M-step can be written more compactly as

$$\hat{\boldsymbol{\pi}}^{(m+1)}(x) = \gamma_0^{(m)}(x), \tag{15.85}$$

$$\hat{\mathsf{A}}^{(m+1)}(x, x') = \frac{\sum_{k=0}^{n-1} \xi_k^{(m)}(x, x')}{\sum_{k=0}^{n-1} \gamma_k^{(m)}(x)}, \tag{15.86}$$

$$\hat{\mathsf{B}}^{(m+1)}(x, y) = \frac{\sum_{k=0}^{n} \gamma_k^{(m)}(x)\, \mathbb{1}\{y_k = y\}}{\sum_{k=0}^{n} \gamma_k^{(m)}(x)}, \tag{15.87}$$

where

$$\gamma_k^{(m)}(x) \triangleq P_{\boldsymbol{\theta}^{(m)}}(X_k = x | Y_0^n = y_0^n), \tag{15.88}$$

$$\xi_k^{(m)}(x, x') \triangleq P_{\boldsymbol{\theta}^{(m)}}(X_k = x, X_{k+1} = x' | Y_0^n = y_0^n), \quad x, x' \in \mathcal{X} \tag{15.89}$$

are the current estimates (at iteration number k) of the posterior probabilities $\gamma_k(x)$ and $\xi_k(x, x')$ of (15.68) and (15.69), respectively. The probabilities (15.88) and (15.89) can be evaluated in a computationally efficient way using the forward-backward algorithm of Section 15.5.2.

Exercises

15.1 *Linear Innovations.* Prove Corollary 15.1

15.2 *Kalman Filtering with Identical State and Observation Noise.* Suppose in the Kalman filter model of Section 15.2, the noise in the state equation and observation equations are identical at each time, i.e.,

$$U_k = V_k, k = 0, 1, \ldots.$$

Rederive the Kalman filter/predictor equations, and the Riccati equation for the predictor covariance update under this model.

15.3 *Kalman Filtering with Control.* Suppose we modify the state equation in the Kalman filtering model as

$$X_{k+1} = F_k X_k + G_k U_k + W_k,$$

where $\{G_k\}_{k=0}^{\infty}$ is a known sequence of matrices. Find appropriate modifications to the Kalman filtering recursions for the case where, for each k, the control U_k is an affine function of the observations Y_0^k.

15.4 *Kalman Smoothing.* For the Kalman filtering model, suppose we wish to estimate X_j from Y_0^k, for $0 \le j \le k$. Consider the estimator defined recursively (in k) by

$$\hat{X}_{j|k} = \hat{X}_{j|k-1} + K_k^{\mathrm{s}}(Y_k - H_k \hat{X}_{k|k-1}),$$

where

$$K_k^{\mathrm{s}} = \Sigma_{k|k-1}^{\mathrm{s}} H_k^\top \left[H_k \Sigma_{k|k-1}^{\mathrm{s}} H_k^\top + R_k \right]^{-1}$$

and

$$\Sigma_{k+1|k}^{\mathrm{s}} = \Sigma_{k|k-1}^{\mathrm{s}} [F_k - K_k H_k]^\top,$$

with

$$\Sigma^{s}_{j|j-1} = \Sigma_{j|j-1},$$

where $\hat{X}_{k|k-1}$, $\Sigma_{k|k-1}$, and K_k are as in the one-step prediction problem.

Show that

$$\Sigma^{s}_{k|k-1} = \mathrm{Cov}((X_j - \hat{X}_{j|k-1}), X_k),$$

and that

$$\hat{X}_{j|k} = \mathbb{E}[X_j | Y_0^k].$$

15.5 *Extended Kalman Filter Simulation.* Redo the simulations example described in Figure 15.1 for the following two cases:

(a) The state and measurement noise variables are Gaussian, with $q = 0.1$, $r = 0.3$, and $\omega_0 = 1$,

(b) The state noise is Gaussian but the measurement noise is non-Gaussian, with mixture pdf:

$$V_k \sim 0.5\,\mathcal{N}\left(\sqrt{\frac{r}{2}}, \frac{r}{2}\right) + 0.5\,\mathcal{N}\left(-\sqrt{\frac{r}{2}}, \frac{r}{2}\right).$$

Note that $\mathrm{Var}(V_k) = r$. Assume that $r = q = 0.1$ and $\omega_0 = 1$.

(c) Compare the tracking performance in parts (a) and (b).

15.6 *Forward-Backward Recursions.* Derive forward recursions for α_k and c_k, and a backward recursion for β_k as defined in (15.77)–(15.79).

15.7 *Baum–Welch.* Show that if the state transition probability matrix $\mathsf{A}(x, x')$ is initialized to 0 for some entry (x, x'), then that entry will remain zero for all iterations of the Baum–Welch algorithm.

15.8 *Baum–Welch with Symmetric Transition Matrix.* In some problems the state transition probability matrix A is known to be symmetric. Derive the appropriate modifications to the Baum–Welch algorithm for estimating symmetric A.

15.9 *Viterbi Algorithm.* Find the sequence $\boldsymbol{x} \subset \{0, 1, 2\}^5$ that maximizes

$$\mathcal{E}(\boldsymbol{u}) = \begin{cases} \sum_t u_t & \text{if } \prod_t u_t \neq 0 \\ -\infty & \text{else,} \end{cases}$$

where $u_t = t - 3 + |x_{t+1} - x_t|$ for $t = 1, 2, 3, 4$. Show the trellis diagram and the surviving paths at times $t = 2, 3, 5$.

15.10 *Two HMMs.* Consider the following extension of HMMs. Let $\boldsymbol{U} = (U_1, U_2, \ldots, U_n)$ and $\boldsymbol{V} = (V_1, V_2, \ldots, V_n)$ be two Markov chains with the same binary alphabet $\mathcal{X} = \{0, 1\}$ and respective transition probability matrices $\mathsf{A} = \begin{pmatrix} 2/3 & 1/3 \\ 1/3 & 2/3 \end{pmatrix}$ and $\mathsf{Q} = \begin{pmatrix} 1/2 & 1/2 \\ 1/2 & 1/2 \end{pmatrix}$. Assume $U_1 = V_1 = 1$. A fair coin is flipped. If the result is "heads" $\boldsymbol{U}$ will be selected as the state sequence $\boldsymbol{X}$ for a HMM with output alphabet $\mathcal{Y} = \{0, 1\}$ and emission probability matrix $\mathsf{B} = \mathsf{A}$. If the result is "tails" $\boldsymbol{V}$ will be selected as the state sequence $\boldsymbol{X}$ for a HMM with output alphabet $\mathcal{Y} = \{0, 1\}$ and emission probability matrix $\mathsf{B} = \mathsf{A}$.

Let $n = 3$ and $\boldsymbol{Y} = (1, 1, 1)$. What is the posterior probability that the coin flip was "heads"?

References

[1] B. D. O. Anderson and J. B. Moore, *Optimal Filtering*, Prentice Hall, 1979.

[2] M. S. Grewal, L. R. Weill, and A. P. Andrews, *Global Positioning Systems, Inertial Navigation, and Integration*, John Wiley & Sons, 2007.

[3] A. Doucet and A. M. Johansen, "A Tutorial on Particle Filtering and Smoothing: Fifteen Years Later," in *Handbook of Nonlinear Filtering*, pp. 656–704, 2009.

[4] A. J. Viterbi, "Error Bounds for Convolutional Codes and an Asymptotically Optimum Decoding Algorithm," *IEEE Trans. Inf. Theory*, Vol. 13, pp. 260–269, 1967.

[5] J. K. Omura, "On the Viterbi Decoding Algorithm," *IEEE Trans. Inf. Theory*, Vol. 15, No. 1, pp. 177–179, 1969.

[6] M. Pollack and W. Wiebenson, "Solutions of the Shortest-route Problem: A Review," *Oper. Res.*, Vol. 8, pp. 224–230, 1960.

[7] G. D. Forney, "The Viterbi Algorithm," *Proc. IEEE*, Vol. 61, No. 3, pp. 268–278, 1973.

[8] R. Bellman, "The Theory of Dynamic Programming," *Proc. Nat. Acad. Sci.*, Vol. 38, pp. 716–719, 1952.

[9] L. R. Welch, "The Shannon Lecture: Hidden Markov Models and the Baum–Welch Algorithm," *IEEE Inf. Theory Society Newsletter*, Vol. 53, No. 4, pp. 1, 10–13, 2003.

Appendix A Matrix Analysis

This appendix includes a number of basic results from matrix analysis that are used in this book. For a comprehensive study of this subject, the reader is referred to the book by Horn and Johnson [1].

A (deterministic) $m \times n$ matrix A is rectangular array of scalar (real-valued[1]) elements:

$$\mathsf{A} = \begin{bmatrix} \mathsf{A}_{11} & \mathsf{A}_{11} & \dots & \mathsf{A}_{1n} \\ \mathsf{A}_{21} & \mathsf{A}_{21} & & \mathsf{A}_{2n} \\ \vdots & & \ddots & \vdots \\ \mathsf{A}_{m1} & \mathsf{A}_{m2} & \dots & \mathsf{A}_{mn} \end{bmatrix}.$$

Special cases are a *square* matrix, for which $m = n$, *row vector* for which $m = 1$, and *column vector* for which $n = 1$. (Deterministic) vectors are denoted by lower-case boldfaced symbols, e.g., $\boldsymbol{a}$ and $\boldsymbol{b}$.

The sum of $m \times n$ matrices A and B is an $m \times n$ matrix C whose elements are given by

$$\mathsf{C}_{ij} = \mathsf{A}_{ij} + \mathsf{B}_{ij}, \quad i = 1, 2, \dots, m, \quad j = 1, 2, \dots, n.$$

Matrix addition is *commutative*, i.e., $\mathsf{A} + \mathsf{B} = \mathsf{B} + \mathsf{A}$.

The product AB of two matrices A and B is only defined if the number of columns of A is equal to the number of rows of B. In particular, if A is an $m \times n$ matrix and B is an $n \times k$ matrix, then $\mathsf{C} = \mathsf{AB}$ is an $m \times k$ matrix with elements

$$\mathsf{C}_{ij} = \sum_{\ell=1}^{n} \mathsf{A}_{i\ell}\mathsf{B}_{\ell j}, \quad i = 1, 2, \dots, m, \quad j = 1, 2, \dots, k.$$

Matrix multiplication is not necessarily commutative (e.g., BA may not be defined even if AB is), but it is *associative*, i.e., $\mathsf{A}(\mathsf{BC}) = (\mathsf{AB})\mathsf{C}$, and *distributive*, i.e., $\mathsf{A}(\mathsf{B}+\mathsf{C}) = \mathsf{AB}+\mathsf{AC}$.

The *transpose* of an $m \times n$ matrix A is an $n \times m$ matrix denoted by $\mathsf{A}^\top$, whose elements are given by

$$\left[\mathsf{A}^\top\right]_{ij} = \mathsf{A}_{ji}, \quad i = 1, 2, \dots, n, \quad j = 1, 2, \dots, m.$$

A matrix A is said to be *symmetric* if $\mathsf{A}^\top = \mathsf{A}$. The transpose $(\mathsf{AB})^\top$ of the product AB is the product of the transposes in reverse order, i.e., $\mathsf{B}^\top\mathsf{A}^\top$.

[1] We restrict our attention to real-valued matrices here, with the understanding that all of the results in this appendix are easily generalized to complex-valued matrices [1].

An $n \times n$ *diagonal* matrix D, denoted by diag$[\mathsf{D}_{11}, \mathsf{D}_{22}, \ldots, \mathsf{D}_{nn}]$, is a matrix whose off-diagonal entries are equal to zero, i.e.,

$$\mathsf{D} = \begin{bmatrix} \mathsf{D}_{11} & 0 & \ldots & 0 \\ 0 & \mathsf{D}_{22} & \ddots & \vdots \\ \vdots & \ddots & \ddots & 0 \\ 0 & \ldots & 0 & \mathsf{D}_{nn} \end{bmatrix}.$$

A special case is an $n \times n$ *identity* matrix, denoted by I or I_n, whose diagonal elements are all equal to 1, i.e.,

$$\mathsf{I} = \begin{bmatrix} 1 & 0 & \ldots & 0 \\ 0 & 1 & \ddots & \vdots \\ \vdots & \ddots & \ddots & 0 \\ 0 & \ldots & 0 & 1 \end{bmatrix}.$$

A square matrix is said to be *unitary* if $\mathsf{A}^\top \mathsf{A} = \mathsf{I}$.

The *inverse* of a square matrix A (when it exists) is denoted by A^{-1}, and satisfies

$$\mathsf{A}\mathsf{A}^{-1} = \mathsf{A}^{-1}\mathsf{A} = \mathsf{I}.$$

If A, B are invertible matrices

$$(\mathsf{AB})^{-1} = \mathsf{B}^{-1}\mathsf{A}^{-1}.$$

An important result regarding matrix inverses is the following:

LEMMA A.1 Matrix Inversion Lemma [1]. *Asssuming all inverses used exist, we have*

$$\begin{aligned}(\mathsf{A} + \mathsf{BCD})^{-1} &= \mathsf{A}^{-1} - \mathsf{A}^{-1}\mathsf{B}(\mathsf{D}\mathsf{A}^{-1}\mathsf{B} + \mathsf{C}^{-1})^{-1}\mathsf{D}\mathsf{A}^{-1} \\ &= \mathsf{A}^{-1} - \mathsf{A}^{-1}\mathsf{BC}(\mathsf{I} + \mathsf{D}\mathsf{A}^{-1}\mathsf{BC})^{-1}\mathsf{D}\mathsf{A}^{-1} \\ &= \mathsf{A}^{-1} - \mathsf{A}^{-1}\mathsf{B}(\mathsf{I} + \mathsf{CD}\mathsf{A}^{-1}\mathsf{B})^{-1}\mathsf{CD}\mathsf{A}^{-1}.\end{aligned}$$

Note that in Lemma A.1 the matrices B and D need not be invertible, and for the second and third lines to hold the matrix C also need not be invertible.

Special cases of Lemma A.1 include

$$(\mathsf{A} + b\boldsymbol{c}\boldsymbol{d}^\top)^{-1} = \mathsf{A}^{-1}\left(\mathsf{I} - \frac{\boldsymbol{c}\boldsymbol{d}^\top}{\frac{1}{b} + \boldsymbol{d}^\top\mathsf{A}^{-1}\boldsymbol{c}}\mathsf{A}^{-1}\right)$$

and

$$(\mathsf{A} + \mathsf{B})^{-1} = \mathsf{A}^{-1} - \mathsf{A}^{-1}(\mathsf{A}^{-1} + \mathsf{B}^{-1})^{-1}\mathsf{A}^{-1}.$$

The *determinant* of an $n \times n$ matrix A is denoted by det(A) or $|\mathsf{A}|$, and is given by

$$|\mathsf{A}| = \mathsf{A}_{11}\tilde{\mathsf{A}}_{11} + \mathsf{A}_{12}\tilde{\mathsf{A}}_{12} + \cdots + \mathsf{A}_{1n}\tilde{\mathsf{A}}_{1n},$$

where the *cofactor* $\tilde{\mathsf{A}}_{ij}$ of A equals $(-1)^{j+i}$ times the determinant of the $(n-1) \times (n-1)$ matrix formed by deleting the ith row and jth column of A. A necessary and sufficient condition for the existence of the inverse of a matrix is that its determinant is nonzero. The determinant of a square matrix is the same as that of its transpose, and for two $n \times n$ matrices A and B

$$|\mathsf{A}\,\mathsf{B}| = |\mathsf{A}|\ |\mathsf{B}|\,.$$

The *trace* of an $n \times n$ matrix A is denoted by Tr(A), and equals the sum of its diagonal elements:

$$\mathrm{Tr}(\mathsf{A}) = \sum_{i=1}^{n} \mathsf{A}_{ii}.$$

The trace is a linear operation on matrices. In particular

$$\mathrm{Tr}(\mathsf{A} + \mathsf{B}) = \mathrm{Tr}(\mathsf{A}) + \mathrm{Tr}(\mathsf{B}).$$

An interesting and useful property of the trace is the following. Given two matrices A and B whose product AB is a square matrix (which implies that BA is well-defined and is also square),

$$\mathrm{Tr}(\mathsf{AB}) = \mathrm{Tr}(\mathsf{BA}).$$

A symmetric matrix A is said to be *positive definite* (denoted by $\mathsf{A} \succ 0$) if

$$\boldsymbol{x}^\top \mathsf{A}\boldsymbol{x} > 0, \qquad \text{for all nonzero } \boldsymbol{x},$$

and is said to be *positive semi-definite* or *nonnegative definite* (denoted by $\mathsf{A} \succeq 0$) if

$$\boldsymbol{x}^\top \mathsf{A}\boldsymbol{x} \geq 0, \qquad \text{for all nonzero } \boldsymbol{x}.$$

For two symmetric matrices A and B of the same size, we use $\mathsf{A} \succ \mathsf{B}$ to denote that $\mathsf{A}-\mathsf{B} \succ 0$ (similarly, $\mathsf{A} \succeq \mathsf{B}$). A necessary and sufficient condition for a symmetric matrix A to be positive definite is if leading principal minors are all positive. The kth leading principle minor of a matrix M is the determinant of its upper-left $k \times k$ submatrix.

A *Toeplitz* matrix has constant values along all of its diagonals, e.g.,

$$\mathsf{A} = \begin{bmatrix} a & u_1 & \dots & u_{n-2} & u_{n-1} \\ \ell_1 & a & u_1 & \dots & u_{n-2} \\ \ell_2 & \ell_1 & a & u_1 & \vdots \\ \vdots & \dots & \ddots & \ddots & \vdots \\ \ell_{n-1} & \ell_{n-2} & \dots & \dots & a \end{bmatrix}.$$

A special case of a Toeplitz matrix is a *circulant* matrix, for which each row is a circular rotation by one of the previous row.

Eigen Decomposition Let A be a $n \times n$ matrix. A nonzero vector $\boldsymbol{u}$ is an *eigenvector* of A with *eigenvalue* λ if

$$\mathsf{A}\boldsymbol{u} = \lambda \boldsymbol{u}.$$

When a matrix is applied to an eigenvector, it simply scales it. Rearranging this, we see

$$(\mathsf{A} - \lambda \mathsf{I})\boldsymbol{u} = \boldsymbol{0}. \tag{A.1}$$

This implies that if A has eigenvalue λ, then the matrix $\mathsf{A}-\lambda I$ is not invertible (otherwise, we would have $\boldsymbol{u} = 0$, which would be a contradiction), i.e.,

$$\det(\mathsf{A} - \lambda \mathsf{I}) = 0. \tag{A.2}$$

The expression $\det(\mathsf{A} - \lambda I)$ is known as the *characteristic polynomial* of A, and has degree n. By solving the *characteristic equation* (A.2) we can find all of the eigenvalues of A. In general, there will be at most n distinct eigenvalues of A (some may be complex valued). An eigenvalue can have multiplicity greater than one, in which case we will have fewer than n distinct eigenvalues. Once we have an eigenvalue, we can find the eigenvectors corresponding to that eigenvalue using (A.1).

Some properties of eigenvalues and eigenvectors of *symmetric* matrices are as follows [1]:

1. The sum of eigenvectors corresponding to the same eigenvalue is also an eigenvector for that eigenvalue, and a scaled version of that eigenvector is still an eigenvector for that eigenvalue. (This holds even if A is not symmetric.)
2. The eigenvalues and eigenvectors are real.
3. If $\boldsymbol{u}_1$ and $\boldsymbol{u}_2$ are eigenvectors corresponding to two different eigenvalues λ_1 and λ_2, then the eigenvectors are *orthogonal*, i.e., $\boldsymbol{u}_1^\top \boldsymbol{u}_2 = 0$.
4. If an eigenvalue λ has multiplicity k, then we can find k mutually orthogonal eigenvectors for that eigenvalue.

The eigenvalues of A are typically indexed in decreasing order:

$$\lambda_1 \geq \lambda_2 \geq \cdots \geq \lambda_n.$$

By the above properties, for symmetric A, we can find corresponding eigenvectors $\boldsymbol{u}_1, \ldots, \boldsymbol{u}_n$ that are *orthonormal*, i.e., they are orthogonal and unit norm. One of the most important results in matrix analysis is the following.

LEMMA A.2 Eigen Decomposition. *An $n \times n$ symmetric matrix A can be written in terms of its eigenvectors and eigenvalues as*

$$\mathsf{A} = \sum_{i=1}^{n} \lambda_i \boldsymbol{u}_i \boldsymbol{u}_i^\top.$$

The form of the eigen decomposition given in Lemma A.2 is often called the outer-product from the eigen decomposition. In many cases, we work with a matrix representation of the eigen decomposition. Let

$$\mathsf{U} = [\boldsymbol{u}_1\, \boldsymbol{u}_2 \ldots \boldsymbol{u}_n]$$

and D be the $n \times n$ diagonal matrix with diagonal entries $\lambda_1, \ldots, \lambda_n$, i.e.,

$$\mathsf{D} = \begin{bmatrix} \lambda_1 & 0 & \ldots & 0 \\ 0 & \lambda_2 & \ddots & \vdots \\ \vdots & \ddots & \ddots & 0 \\ 0 & \ldots & 0 & \lambda_n \end{bmatrix}.$$

Then, the eigen decomposition of A can be written as

$$\mathsf{A} = \mathsf{UDU}^\top. \tag{A.3}$$

Note that

$$\mathsf{U}^\top = \begin{bmatrix} \boldsymbol{u}_1^\top \\ \boldsymbol{u}_2^\top \\ \vdots \\ \boldsymbol{u}_n^\top \end{bmatrix}$$

and that $\mathsf{U}^\top\mathsf{U} = \mathsf{I}$, i.e., U is a *unitary* matrix. Note that $\mathsf{U}^{-1} = \mathsf{U}^\top$. We can invert the relationship given in (A.3) and write

$$\mathsf{D} = \mathsf{U}^\top\mathsf{A}\mathsf{U}.$$

Thus eigen decomposition can also be regarded as a way to diagonalize the matrix A.

Finally, we note that the trace and determinant of an $n \times n$ matrix A can be conveniently written in terms of it eigenvalues $(\lambda_1, \lambda_2, \ldots, \lambda_n)$ as

$$\mathrm{Tr}(\mathsf{A}) = \sum_{i=1}^{n} \lambda_i, \quad |\mathsf{A}| = \prod_{i=1}^{n} \lambda_i.$$

Also, a necessary and sufficient condition for positive definiteness of a symmetric matrix is that all of its eigenvalues are positive.

Matrix Calculus The *gradient* with respect to a matrix A of a scalar-valued function $f(\mathsf{A})$ is a matrix of the partial derivatives of f with respect to the elements of A:

$$\left[\nabla_{\mathsf{A}} f(\mathsf{A})\right]_{ij} = \frac{\partial f}{\partial \mathsf{A}_{ij}}.$$

This definition obviously applies even in the case where A is a row or column vector. Some examples of gradient calculations that follow from this definition are:

$$\begin{aligned}
&\nabla_{\mathsf{A}} \mathrm{Tr}(\mathsf{A}) = \mathsf{I}, \\
&\nabla_{\mathsf{A}} \mathrm{Tr}(\mathsf{BA}) = \mathsf{B}^\top, \\
&\nabla_{\mathsf{A}} \mathrm{Tr}(\mathsf{AB}) = \mathsf{B}, \\
&\nabla_{\mathsf{A}} \mathrm{Tr}(\mathsf{BA}^{-1}) = -\left[\mathsf{A}^{-1}\mathsf{BA}^{-1}\right]^\top, \\
&\nabla_{\mathsf{A}} \exp\left(\boldsymbol{x}^\top \mathsf{A}\boldsymbol{x}\right) = \boldsymbol{x}\boldsymbol{x}^\top \exp\left(\boldsymbol{x}^\top \mathsf{A}\boldsymbol{x}\right), \\
&\nabla_{\mathsf{A}} |\mathsf{A}| = |\mathsf{A}| \left(\mathsf{A}^{-1}\right)^\top, \\
&\nabla_{\mathsf{A}} \ln |\mathsf{A}| = \left(\mathsf{A}^{-1}\right)^\top, \\
&\nabla_{\boldsymbol{x}} \boldsymbol{x}^\top \boldsymbol{y} = \nabla_{\boldsymbol{x}} \boldsymbol{y}^\top \boldsymbol{x} = \boldsymbol{y}, \\
&\nabla_{\boldsymbol{x}} \boldsymbol{x}^\top \mathsf{A}\boldsymbol{x} = 2\mathsf{A}\boldsymbol{x},
\end{aligned}$$

$$\nabla_{\boldsymbol{x}} \operatorname{Tr}\left(\boldsymbol{x}\boldsymbol{y}^\top\right) = \nabla_{\boldsymbol{x}} \operatorname{Tr}\left(\boldsymbol{y}\boldsymbol{x}^\top\right) = \boldsymbol{y}.$$

In these equations we have assumed that matrix inverses exist wherever they are used.

References

[1] R. A. Horn and C. R. Johnson, *Matrix Analysis*, Cambridge University Press, 1990.

Appendix B Random Vectors and Covariance Matrices

We use the following notation. The mean of a random vector $\boldsymbol{Y} \in \mathbb{R}^n$ is denoted by

$$\mu_{\boldsymbol{Y}} = \mathbb{E}(\boldsymbol{Y}) \quad \in \mathbb{R}^n,$$

and its covariance matrix by

$$\mathsf{C}_{\boldsymbol{Y}} = \mathbb{E}[(\boldsymbol{Y} - \mu_{\boldsymbol{Y}})(\boldsymbol{Y} - \mu_{\boldsymbol{Y}})^\top] = \mathrm{Cov}(\boldsymbol{Y}, \boldsymbol{Y}) \quad \in \mathbb{R}^{n\times n}.$$

The cross-covariance matrix of two random vectors $\boldsymbol{X} \in \mathbb{R}^m$ and $\boldsymbol{Y} \in \mathbb{R}^n$ is denoted by

$$\mathsf{C}_{\boldsymbol{XY}} = \mathbb{E}[(\boldsymbol{X} - \mu_{\boldsymbol{X}})(\boldsymbol{Y} - \mu_{\boldsymbol{Y}})^\top] = \mathrm{Cov}(\boldsymbol{X}, \boldsymbol{Y}) \quad \in \mathbb{R}^{m\times n}.$$

The following properties hold:

- Any covariance matrix is nonnegative definite (see Appendix A for definition): $\mathsf{C}_{\boldsymbol{Y}} \succeq 0$. If $\mathsf{C}_{\boldsymbol{Y}}$ has full rank, then it is positive definite, i.e., $\mathsf{C}_{\boldsymbol{Y}} \succ 0$.
- Any covariance matrix is symmetric: $\mathsf{C}_{\boldsymbol{Y}}(i, j) = \mathsf{C}_{\boldsymbol{Y}}(j, i)$ for all $i, j = 1, \ldots, n$.
- If $\mathsf{C}_{\boldsymbol{Y}}$ has full rank, the inverse covariance matrix $\mathsf{C}_{\boldsymbol{Y}}^{-1}$ exists and is also positive definite and symmetric.
- Given two vectors $\boldsymbol{X} \in \mathbb{R}^m$ and $\boldsymbol{Y} \in \mathbb{R}^n$ and two matrices $\mathsf{A} \in \mathbb{R}^{l\times m}$ and $\mathsf{B} \in \mathbb{R}^{n\times p}$, we have the linearity property $\mathrm{Cov}(\mathsf{A}\boldsymbol{X}, \mathsf{B}\boldsymbol{Y}) = \mathsf{A}\,\mathrm{Cov}(\boldsymbol{X}, \boldsymbol{Y})\,\mathsf{B}^\top$.
- The distribution of a Gaussian random vector $\boldsymbol{Y}$ is determined by its mean and covariance matrix. If the covariance matrix has full rank, then $\boldsymbol{Y}$ has a pdf, which is given by

  $$p_{\boldsymbol{Y}}(\boldsymbol{y}) = (2\pi)^{-n/2}|\mathsf{C}_{\boldsymbol{Y}}|^{-1/2} \exp\left\{-\frac{1}{2}(\boldsymbol{y} - \mu_{\boldsymbol{Y}})^\top \mathsf{C}_{\boldsymbol{Y}}^{-1}(\boldsymbol{y} - \mu_{\boldsymbol{Y}})\right\},$$

  where $|\cdot|$ denotes matrix determinant.

Appendix C Probability Distributions

In this book we encounter several classical probability distributions. Their properties are summarized below.

1. Exponential distribution with scale parameter $\sigma > 0$, denoted by $\mathrm{Exp}(\sigma)$:
$$p_\sigma(x) = \frac{1}{\sigma} e^{-x/\sigma}, \quad x \geq 0. \tag{C.1}$$
The mean and standard deviation of X are both equal to σ.
2. Laplace distribution with scale parameter $\sigma > 0$, denoted by $\mathrm{Lap}(\sigma)$:
$$p_\sigma(x) = \frac{1}{2\sigma} e^{-|x|/\sigma}, \quad x \in \mathbb{R}. \tag{C.2}$$
The mean of X is zero and its variance is equal to $2\sigma^2$.
3. Multivariate Gaussian distribution with mean μ and covariance matrix C, denoted by $\mathcal{N}(\mu, \mathsf{C})$. Its pdf takes the form
$$p(x) = (2\pi)^{-d/2}|\mathsf{C}|^{-1/2} \exp\left\{-\frac{1}{2}(x-\mu)^\top \mathsf{C}^{-1}(x-\mu)\right\}, \quad x \in \mathbb{R}^d. \tag{C.3}$$
For $d = 1$, the complementary cdf (right tail probability) is denoted by $Q(x)$.
4. Chi-squared distribution with d degrees of freedom, denoted by χ_d^2. The random variable $X = \sum_{i=1}^d Z_i^2$ where $\{Z_i\}_{i=1}^d$ are i.i.d. $\mathcal{N}(0, 1)$. Its pdf takes the form
$$p(x) = \frac{1}{2^{d/2}\Gamma(d/2)} x^{d/2-1} e^{-x/2}, \quad x \geq 0 \tag{C.4}$$
where $\Gamma(t) \triangleq \int_0^\infty x^{t-1}e^{-x}\,dx$ is the Gamma function, which satisfies $\Gamma(n) = (n-1)!$ for any integer $n \geq 1$, $\Gamma(1/2) = \sqrt{\pi}$, and the recursion $\Gamma(t+1) = t\Gamma(t)$ for all $t > 0$. The mean and variance of X are equal to d and $2d$, respectively. For $d = 2$, $X \sim \mathrm{Exp}(2)$. Using integration by parts, the tail probability of the χ_n^2 distribution can be expressed as
$$\mathrm{P}\{\chi_d^2 > t\} = \begin{cases} 2Q(\sqrt{t}) & : d = 1 \\ 2Q(\sqrt{t}) + \frac{e^{-t/2}}{\sqrt{\pi}} \sum_{k=1}^{(d-1)/2} \frac{(k-1)!(2t)^{k-1/2}}{(2k-1)!} & : d = 3, 5, 7, \ldots \\ e^{-t/2} \sum_{k=0}^{d/2-1} \frac{(t/2)^k}{k!} & : d = 2, 4, 6, \ldots. \end{cases} \tag{C.5}$$
5. Noncentral chi-squared distribution with d degrees of freedom and noncentrality parameter λ, denoted by $\chi_d^2(\lambda)$. The random variable $X = \sum_{i=1}^d Z_i^2$ where $\{Z_i\}_{i=1}^d$

are i.i.d. $\mathcal{N}(\mu_i, 1)$, and the squares of the means $\{\mu_i\}_{i=1}^d$ sum to λ. The pdf of X takes the form

$$p_\lambda(x) = \frac{1}{2}\left(\frac{x}{\lambda}\right)^{(d-2)/4} e^{-(x+\lambda)/2} I_{d/2-1}(\sqrt{\lambda x}), \quad x \geq 0, \tag{C.6}$$

where $I_n(t)$ is the modified Bessel function of the first kind and order n:

$$I_n(t) \triangleq \frac{(t/2)^n}{\sqrt{\pi}\Gamma(n+1/2)} \int_0^\pi e^{t\cos\phi} \sin^{2n}\phi \, d\phi.$$

In particular, $I_0(t) = \frac{1}{\pi}\int_0^\infty e^{t\cos\phi}\, d\phi$. The mean and variance of X are equal to $d+\lambda$ and $2d+4\lambda$, respectively. As a special case, $\chi_d^2(0)$ is the (central) chi-squared distribution of (C.4).

6. Rice (or Rician) distribution, denoted by Rice(ρ, σ^2). The random variable $X = \sqrt{Z_1^2 + Z_2^2}$ where Z_1 and Z_2 are independent $\mathcal{N}(\mu_1, \sigma^2)$ and $\mathcal{N}(\mu_2, \sigma^2)$ random variables, and $\mu_1^2 + \mu_2^2 = \rho^2$. The pdf of X is

$$p_{\rho,\sigma}(x) = \frac{x}{\sigma^2}\exp\left\{-\frac{x^2+\rho^2}{2\sigma^2}\right\} I_0\left(\frac{\rho x}{\sigma^2}\right), \quad x \geq 0. \tag{C.7}$$

If $\sigma = 1$, then $X^2 \sim \chi_2^2(\rho^2)$. If $\rho = 0$, then $X \sim$ Rayleigh(σ) and $X^2 \sim$ Exp$(\sqrt{2}\sigma)$.

7. Gamma distribution, denoted by $\Gamma(a, b)$. The random variable X has pdf

$$p_{a,b}(x) = \frac{b^a}{\Gamma(a)} x^{a-1} e^{-bx}, \quad x > 0. \tag{C.8}$$

8. Rayleigh distribution with scale parameter $\sigma > 0$:

$$p_\sigma(y) = \frac{y}{\sigma^2} e^{-\frac{y^2}{2\sigma^2}}, \quad y \geq 0. \tag{C.9}$$

9. Poisson distribution with parameter $\lambda > 0$, denoted by Poi(λ):

$$p_\lambda(x) = \frac{\lambda^x}{x!} e^{-\lambda}, \quad x = 0, 1, 2, \ldots. \tag{C.10}$$

Appendix D Convergence of Random Sequences

This appendix reviews classical notions of convergence for random sequences and some important properties. Let $\boldsymbol{X}_n, n = 1, 2, \ldots$, be a sequence of random variables in $\mathbb{R}^m$.

The sequence $\boldsymbol{X}_n$ converges in probability to $\boldsymbol{X}$ if

$$\forall \epsilon > 0: \lim_{n\to\infty} \mathrm{P}\{\|\boldsymbol{X}_n - \boldsymbol{X}\| < \epsilon\} = 1, \tag{D.1}$$

where the choice of the norm is immaterial. The sequence $\boldsymbol{X}_n$ converges *a.s.* to $\boldsymbol{X}$ if

$$\mathrm{P}\left\{\lim_{n\to\infty} \boldsymbol{X}_n = \boldsymbol{X}\right\} = 1, \tag{D.2}$$

or equivalently,

$$\forall \epsilon > 0: \lim_{n\to\infty} \mathrm{P}\{\forall k \geq n : \|\boldsymbol{X}_k - \boldsymbol{X}\| < \epsilon\} = 1, \tag{D.3}$$

or equivalently,

$$\forall \epsilon > 0: \mathrm{P}\left\{\bigcup_{n=1}^{\infty}\bigcap_{k=n}^{\infty}\{\|\boldsymbol{X}_k - \boldsymbol{X}\| < \epsilon\}\right\} = 1. \tag{D.4}$$

The sequence $\boldsymbol{X}_n$ converges in the mean square to $\boldsymbol{X}$ if

$$\lim_{n\to\infty} \mathbb{E}\|\boldsymbol{X}_n - \boldsymbol{X}\|^2 = 0, \tag{D.5}$$

where $\|\cdot\|$ is the Euclidean norm.

Continuous Mapping Theorem [1, p. 8] Assume the random sequence $\boldsymbol{X}_n \in \mathbb{R}^m$ converges to $\boldsymbol{X}$ in probability, almost surely, or in distribution. Let $\boldsymbol{g}(\cdot)$ be a continuous mapping. Then $\boldsymbol{g}(\boldsymbol{X}_n)$ converges to $\boldsymbol{g}(\boldsymbol{X})$ in probability, almost surely, or in distribution, respectively.

Delta Theorem [2, Ch. 2.6; 3, p. 136] Fix $\boldsymbol{\mu} \in \mathbb{R}^m$ and assume the random sequence $\boldsymbol{X}_n$ satisfies $\sqrt{n}(\boldsymbol{X}_n - \boldsymbol{\mu}) \stackrel{d.}{\to} \mathcal{N}(\mathbf{0}, \mathsf{C})$. Let $\boldsymbol{g} : \mathbb{R}^m \to \mathbb{R}^m$ be a mapping with nonzero Jacobian matrix J. Then $\sqrt{n}(\boldsymbol{g}(\boldsymbol{X}_n) - \boldsymbol{g}(\boldsymbol{\mu})) \stackrel{d.}{\to} \mathcal{N}(\mathbf{0}, \mathsf{JCJ}^\top)$.

Slutsky's Theorem [1, p. 4]

(a) If $X_n \in \mathbb{R}$ converges to X in distribution and $Y_n \in \mathbb{R}$ converges to c in probability, then $X_n Y_n$ converges in distribution to cX.

(b) If $X_n \in \mathbb{R}$ converges to X in distribution and $Y_n \in \mathbb{R}$ converges to $c \neq 0$ in probability, then X_n/Y_n converges in distribution to X/c.
(c) If $X_n \in \mathbb{R}$ converges to X in distribution and $Y_n \in \mathbb{R}$ converges to c in probability, then $X_n + Y_n$ converges in distribution to $X + c$.

References

[1] A. DasGupta, *Asymptotic Theory of Statistics and Probability*, Springer, 2008.
[2] T. S. Ferguson, *A Course in Large Sample Theory*, Chapman and Hall, 1996.
[3] P. K. Sen and J. M. Singer, *Large Sample Methods in Statistics*, Chapman and Hall, 1993.

Index